Phloem

Methods and Protocols

Second Edition

Edited by

Johannes Liesche

Institute of Biology, University of Graz, Graz, Austria

Sanna Sevanto

Earth & Environmental Sciences Division, Los Alamos National Laboratory, Los Alamos, NM, USA

Humana Press

Editors
Johannes Liesche
Institute of Biology
University of Graz
Graz, Austria

Sanna Sevanto
Earth & Environmental Sciences Division
Los Alamos National Laboratory
Los Alamos, NM, USA

ISSN 1064-3745 ISSN 1940-6029 (electronic)
Methods in Molecular Biology
ISBN 978-1-0716-5487-3 ISBN 978-1-0716-5488-0 (eBook)
https://doi.org/10.1007/978-1-0716-5488-0

This Humana imprint is published by the registered company Springer Science+Business Media, LLC, part of Springer Nature.
The registered company address is: 1 New York Plaza, New York, NY 10004, U.S.A.

If disposing of this product, please recycle the paper.

Preface

Within higher plants, the phloem fulfills a vital role of transporting molecules and signals between organs. Its existence has been known for centuries. The structure of the phloem, specifically the sieve elements, was first described in 1837 by the forest botanist Theodor Hartig. The term "phloem" was later introduced by the botanist Carl Nägeli in 1858. With its critical role in supporting and maintaining plant function, the phloem remains at the center of plant biology research. Yet, challenges in studying this intricate, sensitive, and structurally complex tissue have too often left it ignored when developing theories of plant function.

This book builds upon the foundation established by the first edition of *Phloem - Methods and Protocols*, continuing its mission to document and share methods for studying the phloem with the plant biological community. It includes both entirely new chapters introducing innovative techniques and updated protocols that refine and expand upon those presented in the first edition. The methods compiled here cover the whole spectrum of scientific techniques from molecular biology to anatomical studies, imaging and tracer-based analyses to theoretical and modeling approaches. The protocols also address the whole range of functional aspects of the phloem, from molecular-scale processes to its role in whole plant performance and responses to stress.

Some of the methods have been used for decades, while others represent cutting-edge techniques that have been developed and published only recently. This book aims to provide practical and sufficiently detailed presentation of the methods to make them accessible to scientists with different backgrounds. At the same time, it provides enough context of methodological principles to help researchers select the most appropriate methods for their specific questions. We hope that this book inspires researchers to combine techniques from different subfields of phloem studies to enhance our knowledge on this sensitive and complex tissue. Recent discoveries about the interactions between the phloem and the surrounding tissues, including the xylem, as well as the impact of this volumetrically small tissue in controlling and coordinating resource allocation in plants, are truly remarkable. Undoubtedly, there is still much to discover about this fascinating and essential component of plant biology. We hope that the methods presented in this book can contribute to these discoveries.

Graz, Austria — *Johannes Liesche*
Los Alamos, NM, USA — *Sanna Sevanto*

Contents

Contributors

WILLIAM W. ADAMS III • *Department of Ecology and Evolutionary Biology, University of Colorado, Boulder, CO, USA*
BENJAMIN A. BABST • *Arkansas Forest Resources Center, University of Arkansas, Monticello, AR, USA; College of Forestry, Agriculture, and Natural Resources, University of Arkansas, Monticello, AR, USA*
MARILYN BALL • *Research School of Biology, Australian National University, Canberra, Australia*
SAMANTHA BARKER • *Department of Agronomy, Center for Plant Biology, Purdue University, West Lafayette, IN, USA*
KAI BARTUSCH • *Institute of Molecular Plant Biology, Department of Biology, ETH Zürich, Zürich, Switzerland*
SPENCER T. BEHMER • *Department of Entomology, Texas A&M University, College Station, TX, USA*
CHIARA BERNARDINI • *Department of Agricultural, Food, Environmental and Animal Sciences, University of Udine, Udine, Italy*
LJUDMILLA BORISJUK • *Leibniz-Institute of Plant Genetics and Crop Plant Research, Seeland-Gatersleben, Germany*
LOUIS BROUSSARD • *Institut de Recherche en Horticulture et Semences, Université d'Angers, Beaucouzé, France*
JEAN-PIERRE CALIMAN • *Sinar Mas Agro Resources and Technology Research Institute (SMARTRI), Pekanbaru, Indonesia*
THIBAUD CAYLA • *Institute Jean-Pierre Bourgin for Plant Sciences, AgroParisTech-INRAE, Versailles, France; Ecole Doctorale Sciences du Végétal (ED 567), Université Paris-Saclay, Gif/Yvette, France*
LI-QING CHEN • *Department of Plant Biology, University of Illinois at Urbana-Champaign, Urbana, IL, USA*
CHRISTOPHER M. COHU • *Intrinsyx Environmental, Sunnyvale, CA, USA*
MASAKO DANNOURA • *Graduate School of Agriculture, Kyoto University, Kyoto, Japan*
BARBARA DEMMIG-ADAMS • *Department of Ecology and Evolutionary Biology, University of Colorado, Boulder, CO, USA*
CHEN DENG • *Department of Biological Sciences, Sungkyunkwan University, Suwon, Republic of Korea*
DOMINIQUE DEROME • *Department of Civil and Building Engineering, Université de Sherbrooke, Sherbrooke, QC, Canada*
SYLVIE DINANT • *Institute Jean-Pierre Bourgin for Plant Sciences, Université Paris-Saclay, AgroParisTech-INRAE, Versailles, France*
JEAN-CHRISTOPHE DOMEC • *Nicholas School of the Environment, Duke University, Durham, NC, USA; Bordeaux Sciences Agro, INRAE, ISPA, Villenave d'Ornon, France*
DANIEL EPRON • *Graduate School of Agriculture, Kyoto University, Kyoto, Japan*

RICHARD A. FERRIERI • *Missouri Research Reactor Center and Department of Chemistry, University of Missouri, Columbia, MO, USA*
ALEXANDRA C. U. FURCH • *Matthias Schleiden Institute of Genetics, Bioinformatics and Molecular Botany, Friedrich-Schiller-University Jena, Jena, Germany*
SABRINA GENNIS • *Department of Physics, Technical University of Denmark, Kgs. Lyngby, Denmark*
JESS GERSONY • *Department of Biological Sciences, Smith College, Northampton, MA, USA*
ARTHUR GESSLER • *Forest and Soil Ecology Research, Swiss Federal Research Institute WSL, Birmenddorf, Switzerland; Institute of Terrestrial Ecosystems, ETH Zürich, Zürich, Switzerland*
ROBERT J. GREBENOK • *Department of Biology, Canisius University, Buffalo, NY, USA*
ANDRE GÜNDEL • *Leibniz-Institute of Plant Genetics and Crop Plant Research, Seeland-Gatersleben, Germany; Department of Ecology, Environment and Plant Sciences, Stockholm University, Stockholm, Sweden*
MEINA GUO • *Institute of Agricultural Resources and Regional Planning, Chinese Academy of Agricultural Sciences, Beijing, China*
JIAMIN HAO • *College of Life Sciences, Northwest A&F University, Yangling, China*
URI HOCHBERG • *Soil, Water, and Environmental Sciences, Volcani—Agricultural Research Organization, Ramat Yishai, Israel*
TEEMU HÖLTTÄ • *Institute for Atmospheric and Earth System Research, University of Helsinki, Helsinki, Finland; Department of Forest Sciences, University of Helsinki, Helsinki, Finland*
PASCAL HUNZIKER • *Department of Plant and Environmental Sciences, University of Copenhagen, Frederiksberg, Denmark*
KAARE H. JENSEN • *Department of Physics, Technical University of Denmark, Kgs. Lyngby, Denmark*
HONGSHI JIN • *Department of Biological Sciences, Sungkyunkwan University, Suwon, Republic of Korea*
GABRIEL KATUL • *Department of Civil and Environmental Engineering, Duke University, Durham, NC, USA*
JI-YUN KIM • *Department of Biological Sciences, Sungkyunkwan University, Suwon, Republic of Korea*
SOL-BI KIM • *Department of Biological Sciences, Sungkyunkwan University, Suwon, Republic of Korea*
YUKI KONDO • *Department of Biological Sciences, Graduate School of Science, The University of Osaka, Toyonaka-shi, Osaka, Japan*
CHRISTINA KÜHN • *Institute of Biology, Humboldt University Berlin, Berlin, Germany*
MATTHIAS LANGER • *Leibniz-Institute of Plant Genetics and Crop Plant Research, Seeland-Gatersleben, Germany*
ROZENN LE HIR • *Institute Jean-Pierre Bourgin for Plant Sciences, Université Paris-Saclay, AgroParisTech-INRAE, Versailles, France*
HANNAH LEVENGOOD • *Department of Agronomy, Center for Plant Biology, Purdue University, West Lafayette, USA*
AMIT LEVY • *Citrus Research and Education Center, University of Florida, Lake Alfred, FL, USA; Department of Plant Pathology, University of Florida, Gainesville, FL, USA*
JIANQIU LI • *College of Life Sciences, Northwest A&F University, Yangling, China*
JOHANNES LIESCHE • *Institute of Biology, University of Graz, Graz, Austria*

ANNA LINTUNEN • *Institute for Atmospheric and Earth System Research, University of Helsinki, Helsinki, Finland*
GERTRUD LOHAUS • *Molecular Plant Science/Plant Biochemistry, University of Wuppertal, Wuppertal, Germany*
JUAN MANUEL LOSADA • *Institute for Mediterranean and Subtropical Horticulture -La Mayora- (IHSM La Mayora-CSIC-UMAI), Malaga, Spain*
RUDY LUKMAN • *Sinar Mas Agro Resources and Technology Research Institute (SMARTRI), Pekanbaru, Indonesia*
MING-FANG LV • *Institute of Virology and Biotechnology, Zhejiang Academy of Agricultural Science, Hangzhou, China*
PETER MINCHIN • *New Zealand Institute of Plant and Food Research, Motueka, New Zealand*
JULIANA MIRANDA • *Department of Agronomy, Center for Plant Biology, Purdue University, West Lafayette, USA*
ONNO MULLER • *Institue of Bio-and Geosciences, IBG-2, Forschungszentrum Jülich GmbH, Jülich, Germany*
CAROLINE MÜLLER • *Department of Chemical Ecology, Bielefeld University, Bielefeld, Germany*
MAZEN NAKAD • *School of Engineering, Lebanese American University, Byblos, Lebanon*
GUNDULA A. NOLL • *Institute of Plant Biology and Biotechnology, University of Muenster, Muenster, Germany; Fraunhofer Institute for Molecular Biology and Applied Ecology IME, Muenster, Germany*
MICHITAKA NOTAGUCHI • *Bioscience and Biotechnology Center, Nagoya University, Nagoya, Japan; Graduate School of Bioagricultural Sciences, Nagoya University, Nagoya, Japan; Department of Botany, Graduate School of Science, Kyoto University, Kyoto, Japan*
JONAS NÖTZOLD • *Matthias Schleiden Institute of Genetics, Bioinformatics and Molecular Botany, Friedrich-Schiller-University Jena, Jena, Germany*
R. ANJAL P. NUGROHO • *Institut de Recherche en Horticulture et Semences, Université d'Angers, Beaucouzé, France; Sinar Mas Agro Resources and Technology Research Institute (SMARTRI), Pekanbaru, Indonesia*
KENTARO OKADA • *Bioscience and Biotechnology Center, Nagoya University, Nagoya, Japan*
KOJI OKAYASU • *Graduate School of Bioagricultural Sciences, Nagoya University, Nagoya, Japan*
HENK OPPERMANN • *Matthias Schleiden Institute of Genetics, Bioinformatics and Molecular Botany, Friedrich-Schiller-University Jena, Jena, Germany*
MARIA FERNANDA OVALLE • *Department of Agronomy, Center for Plant Biology, Purdue University, West Lafayette, USA*
MARCELO RODRIGO PACE • *Instituto de Biología, Departamento de Botánica, Universidad Nacional Autónoma de México, Coyoacán, Mexico City, Mexico*
TEEMU PALJAKKA • *Institute for Atmospheric and Earth System Research, University of Helsinki, Helsinki, Finland*
TIN YAU PANG • *German Diabetes Center (DDZ), Leibniz Institute for Diabetes Research, Düsseldorf, Germany*
MARIA PARK • *Department of Ecology, Evolution, and Behavior, University of Minnesota, St. Paul, MN, USA; Department of Organismic and Evolutionary Biology, Harvard University, Cambridge, MA, USA*

RAQUEL PARTELLI FELTRIN • *Independent Researcher, Vancouver, BC, Canada*
CAROLINE PLAIN • *UMR Silva, AgroParisTech-INRA, Université de Lorraine, Nancy, France*
STEPHANIE K. POLUTCHKO • *Department of Ecology and Evolutionary Biology, University of Colorado, Boulder, CO, USA*
DUSTIN M. RAY • *Biology Department, University of Minnesota Duluth, Duluth, MN, USA*
JACOBO ROBLEDO • *Citrus Research and Education Center, University of Florida, Lake Alfred, FL, USA; Emerging Pathogens Institute, University of Florida, Gainesville, FL, USA; Global Food Systems Institute, University of Florida, Gainesville, FL, USA; Department of Plant Pathology, University of Florida, Gainesville, FL, USA*
HARDY ROLLETSCHEK • *Leibniz-Institute of Plant Genetics and Crop Plant Research, Seeland-Gatersleben, Germany*
WENYUAN RUAN • *State Key Laboratory of Efficient Production of Forest Resources/National Engineering Research Center of Tree Breeding and Ecological Restoration, College of Biological Sciences and Technology, Beijing Forestry University, Beijing, China*
YANN SALMON • *Institute for Atmospheric and Earth System Research, University of Helsinki, Helsinki, Finland*
JESSICA A. SAVAGE • *Biology Department, University of Minnesota, Duluth, MN, USA*
MICHAEL J. SCHUELLER • *Missouri Research Reactor Center and Department of Chemistry, University of Missouri, Columbia, MO, USA*
ALEXANDER SCHULZ • *Department of Plant and Environmental Sciences, University of Copenhagen, Frederiksberg, Denmark*
RABEA SCHWEIGER • *Department of Chemical Ecology, Bielefeld University, Bielefeld, Germany*
SANNA SEVANTO • *Earth and Environmental Sciences Division, Los Alamos National Laboratory, Los Alamos, NM, USA*
SHUNJI SHIMADZU • *Department of Biological Sciences, Graduate School of Science, The University of Osaka, Toyonaka-shi, Osaka, Japan*
JARED J. STEWART • *Department of Ecology and Evolutionary Biology, University of Colorado, Boulder, CO, USA*
EFTHYMIA SYMEONIDI • *School of Biological Sciences, University of Utah, Salt Lake City, UT, USA*
GUILHERME TAMASSIA • *Institute of Biosciences, University of São Paulo, São Paulo, Brazil*
GUILLAUME TCHERKEZ • *Institut de Recherche en Horticulture et Semences, Université d'Angers, Beaucouzé, France; Research School of Biology, Australian National University, Canberra, Australia*
ELISABETH TRUERNIT • *Institute of Molecular Plant Biology, Department of Biology, ETH Zürich, Zürich, Switzerland*
CHRISTOPHER VINCENT • *Horticultural Sciences Department and Citrus Research and Education Center, IFAS, University of Florida, Gainesville, FL, USA*
STEFFEN WAGNER • *Leibniz-Institute of Plant Genetics and Crop Plant Research, Seeland-Gatersleben, Germany*
STACY WELKER • *Citrus Research and Education Center, University of Florida, Lake Alfred, FL, USA*
TORSTEN WILL • *Institute for Resistance Research and Stress Tolerance, Julius-Kühn-Institute (JKI), Federal Research Centre for Cultivated Plants, Quedlinburg, Germany*

LISA S. WROBEL • *Institute of Plant Biology and Biotechnology, University of Muenster, Muenster, Germany*
LI XIE • *Analysis Center of Agrobiology and Environmental Sciences, Zhejiang University, Hangzhou, China*
KEKE YI • *State Key Laboratory of Efficient Production of Forest Resources/National Engineering Research Center of Tree Breeding and Ecological Restoration, College of Biological Sciences and Technology, Beijing Forestry University, Beijing, China*
LOUIS YOUSSEF • *Department of Plant Science, McGill University, Montreal, QC, Canada*
CANKUI ZHANG • *Department of Agronomy, Center for Plant Biology, Purdue University, West Lafayette, IN, USA*
HENG-MU ZHANG • *Laboratory of Virology, Institute of Virology and Biotechnology, Zhejiang Academy of Agricultural Science, Hangzhou, China*
QIANQIAN ZHANG • *State Key Laboratory of Efficient Production of Forest Resources/National Engineering Research Center of Tree Breeding and Ecological Restoration, College of Biological Sciences and Technology, Beijing Forestry University, Beijing, China*
MATTHIAS R. ZIMMERMANN • *Matthias Schleiden Institute of Genetics, Bioinformatics and Molecular Botany, Friedrich-Schiller-University Jena, Jena, Germany*
MACIEJ A. ZWIENIECKI • *Department of Plant Sciences, University of California, Davis, CA, USA*

Part I

Visualizing and Quantifying Phloem Structure

Chapter 1

Preparation of Tissue Sections for Light-Microscopic Analysis of Phloem Anatomy

Marcelo Rodrigo Pace

Abstract

Accurate analysis and description of plant tissues often rely on the preparation of high-quality anatomical slides, a task that becomes particularly challenging when dealing with heterogeneous tissues such as phloem, which contains both soft and rigid components. This chapter provides a comprehensive protocol outlining key techniques for the optimal preparation of phloem tissue samples for light microscopy. The protocol encompasses essential steps such as fixation, softening, embedding, sectioning, staining, and mounting, and is adaptable for examining phloem and adjacent tissues in both woody and herbaceous stems and roots. Studying phloem anatomy is crucial for understanding nutrient transport, plant development, and responses to environmental stress, offering insights into both fundamental plant biology and practical applications in agriculture and forestry.

Keywords Bark, Phloem, Sieve element, Structure–function relationship, Sample preparation, Light microscopy, Staining, Dyes, Sectioning

1 Introduction

Botanical histology has been in practice since the dawn of plant microscopy, as shown in some of the classic seventeenth-century works of Grew [1], Malpighi [2], and Van Leeuwenhoek (reviewed by Baas [3]). Yet, producing slides suitable for accurate examination of plant tissues is not trivial, especially for heterogeneous tissues, such as the bark of trees, shrubs, and lianas. The bark typically contains a combination of soft, nonlignified cells, such as the phloem sieve elements and parenchyma, and of hard cells with lignified secondary walls, such as the sclerenchyma (for a detailed, illustrated explanation of each cell type, refer to the IAWA Bark Committee [4]). The term bark is nonspecific and encompasses all tissues outside the vascular cambium, namely the secondary phloem (also known as inner bark), commonly the remnants of primary tissues such as the cortex and primary fiber strands, and the periderm (known as outer bark [5]). Sectioning bark tissues without adequate preparation typically

Johannes Liesche and Sanna Sevanto (eds.), *Phloem*, Methods in Molecular Biology, vol. 3066,
https://doi.org/10.1007/978-1-0716-5488-0_1,

results in a collection of fragments, powders, or thick sections where details of the cell types are barely seen and difficult to describe. What follows is the description of some simple preparatory techniques to avoid these outcomes. The focus is on bark, especially the secondary phloem, although the techniques presented here can be easily adopted to any plant tissue or organ.

The methods described here consist of six steps: fixation, softening, embedding, sectioning, staining, and mounting (Fig. 1). The protocol describes the procedure for both large and small samples, with further details on how to proceed with different kinds of samples in the notes section. Fixation is the process of killing cells while preserving the cell structure as close to the living conditions as possible. Since the bark contains numerous nonlignified, fragile cells, which collapse when dried, the best samples are those fixed immediately in the field and conserved later in liquid solutions. Softening is not necessary for all samples. Rather, it makes certain bark samples easier to section. Sample embedding facilitates thin sectioning. After infiltration and hardening, the samples can be fixated and sectioned in a microtome without damage or distortion. The embedding media stabilizes the anatomical structure and cellular organization. In addition to microtome sectioning, hand sectioning can give good results for phloem analysis, especially when fresh material is available. However, for optimal results, the use of either a sliding or a rotary microtome is recommended. Staining is the essential step to make specific structures of the phloem visible. Different from the study of woods, which are typically stained only in safranin or fuchsin, the study of bark yields superior results when sections are double-stained or stained with a metachromatic stain. This is due to the nature of bark, which contains lignified and nonlignified cells. Bleaching is not recommended for barks, as it removes all cell contents, which are important for differentiating conducting and nonconducting phloem [4]. Once stained, the sections are dehydrated, typically in an ethanol series up to absolute ethanol, rinsed in a solvent, and mounted with a natural or synthetic resin.

Anatomical studies of the phloem remain highly relevant, especially for understanding how phloem architecture can limit plant productivity and shape environmental adaptations [6, 7]. The following protocol enables the interested scientist to produce tissue sections for reliable, high-quality quantification of phloem structure.

2 Materials

2.1 Sample Preparation and Fixation

One of four fixatives (FAA, Glutaraldehyde-formaldehyde, paraformaldehyde, CRAF) can be chosen according to sample size and the intended type of analysis (*see* **Note 1**; Fig. 1).

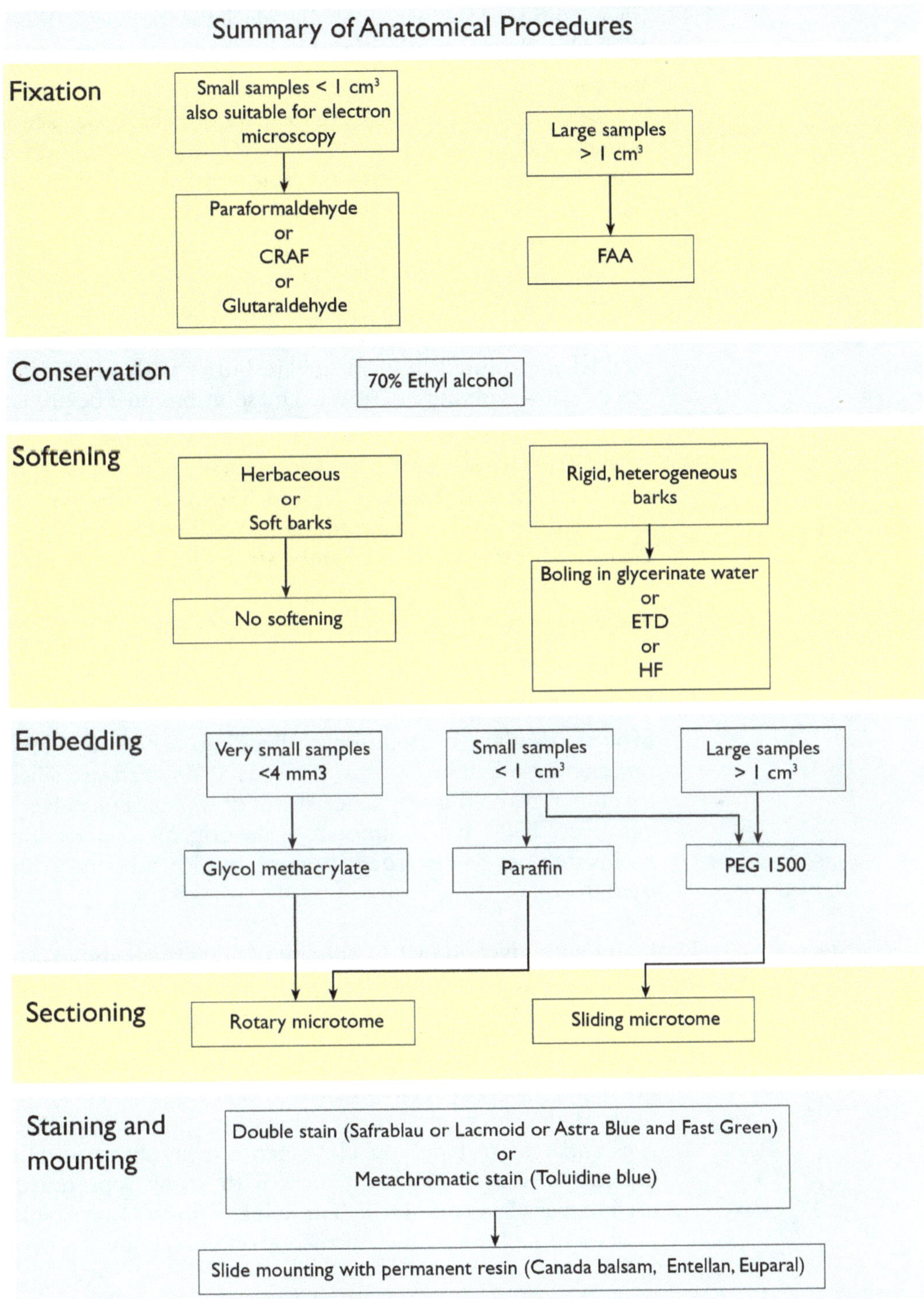

Fig. 1 Flow chart of the sample preparation procedure. *CRAF* chromic acid, acetic acid, formaldehyde, *FAA* formaldehyde, acetic acid, alcohol, *ETD* ethylenediamine, *HF* hydrofluoric acid

1. FAA solution: Mix formaldehyde, glacial acetic acid, 70% ethyl alcohol in the proportion 1:1:18.
2. Paraformaldehyde solution: Place 15 ml of distilled water on a rotary shaker and add 0.8 g of paraformaldehyde, keeping under rotation for 2 h until complete dissolution. Add 1 N NaOH drop by drop until the solution clears. The solution must be prepared fresh every time.
3. Glutaraldehyde-formaldehyde solution: Dissolve 1 g of paraformaldehyde in 50 ml of distilled water at 60–70°C. Add 1 N NaOH drop by drop until the solution clears. Wait until the solution cools down. Mix this solution with 50 ml of 50% glutaraldehyde diluted in a phosphate buffer with pH 7–7.4 to obtain a final volume of 100 ml. The solution must be prepared fresh every time.
4. CRAF III: Mix 30 ml of 10% chromic acid, 20 ml of 10% acetic acid, 100 ml of formaldehyde, and 850 ml of distilled water. The formaldehyde must be added just before usage. The solution without formaldehyde can be stored for long periods.
5. Water with glycerin: Add 1 volume of 87% glycerin to 10 volumes of water. Use for rehydrating dried samples.
6. Sample containers, vials of various volumes with screw-on lids.
7. Rotary or rocker shaker.
8. Molds: Molds are used for embedding. There are several options for obtaining molds. Nowadays, one can use silicone ice cube trays sold in appliance stores or online. For polyethylene glycol 1500, it is common to make origami (paper-folding) molds that can be discarded after use, but ice cube trays can be used, too.

2.2 Softening

1. Water with glycerin (10:1) prepared as described above.
2. Ethylenediamine (ETD) at 4% or 10%, dissolved in distilled water.
3. Hydrofluoric acid (HF) solution: Dissolve one volume of 1 M HF in two volumes of distilled water. Remember to always pour the acid into the water, not the other way around. Extreme care is essential when handling HF, since it is highly toxic. High-quality plastic (PTFE) containers with screw tops must be used; never glass containers. The solution must always be kept under a fume hood. Thick nitrile gloves are recommended.

2.3 Infiltrating and Embedding

1. Polyethylene glycol (PEG) 1500: Place solid PEG pellets in an open jar in an oven at 60°C to liquify within 5–10 min. Solutions with lower concentrations are made by mixing them with distilled water.
2. Paraffin wax.

3. Ethanol at various concentrations. Mix ethanol with distilled water to obtain concentrations of 50%, 70%, and 90%.
4. Historesin: Glycol methacrylate monomer. Can be purchased as a kit containing a hardener (a derivative of barbituric acid).
5. Historesin mold. These molds are either sold by the provider, or alternatively, one can buy small silicone ice cube trays.

2.4 Sectioning

1. Sliding or rotary microtome.
2. Permanent steel microtome knives or a disposable microtome knife.
3. Polystyrene resin: Dissolve packaging Styrofoam in butyl acetate or xylene to make a saturated solution.
4. Hot plate.
5. Haupt's gelatin adhesive: A blended solution of high-quality gelatin, glycerin, and phenol.
6. Coplin staining jar.
7. Xylol or butyl acetate: To remove paraffin from slides.
8. Ethyl alcohol: mix with distilled water to obtain 50%, 70%, and 90% solutions.

2.5 Stains and Stain Procedures

All these stain solutions can be stored in the fridge for many years.

1. Ferric chloride: Dissolve 2% ferric chloride, 1% tannic acid, and 1% sodium bicarbonate in 50% ethanol.
2. 1% sodium bicarbonate solution: Dissolve the sodium bicarbonate in two steps, first dissolve the sodium bicarbonate in water to 2% (w/v) and then dilute with ethanol to obtain a 1% sodium bicarbonate solution in 50% ethanol.
3. Lacmoid solution: Dissolve resorcin blue in 30% ethanol until saturated. Usually, 0.25 g resorcin blue is used for 100 ml of 30% ethanol. Add a few drops of sodium bicarbonate solution to make it alkaline.
4. Safrablau: Dissolve Astra blue and Safranin O in 50% ethanol. Filter each solution before adding one part Safranin O to nine parts Astra blue (1:9). Different concentrations can be used. In some taxa (e.g., *Ipomoea*, Vitaceae), the lignified tissues do not stain as well as those of other taxa. In these cases, it is recommended to use blends with higher Safranine O (2:8, 3:7, and so forth). It can go to the extreme of dissolving a pinch of Astra blue in pure Safranine O solution.
5. Safranin and Fast Green: Dissolve 1 g of Safranin O in 100 ml of distilled water and 0.5 g of Fast Green FCF in 100 ml of 95% ethanol.

6. Aniline Blue: Dissolve Aniline blue in water to obtain a 1% (w/v) aqueous solution.
7. Toluidine Blue: Dissolve 0.05 g Toluidine Blue in 100 ml of phosphate buffer at pH 6.8.

2.6 Permanent Slide Mounting

1. Microscopy slides and coverslips.
2. Mounting medium, e.g., Canada balsam, Entellan, Euparal (*see* **Note 2**).

3 Methods

3.1 Fixation and Sample Preparation

Fixation of Fresh Samples

1. Subdivide the specimens as soon as possible after collection and place them in a container of suitable size, usually a glass vial with a screw-on lid.
2. Add fixative (FAA, glutaraldehyde-formaldehyde, paraformaldehyde, or CRAF) until the sample is completely submerged.
3. Incubate for 12–24 h at 4°C on a rotary or rocker shaker and, if possible, in a vacuum chamber. The minimum incubation time for herbaceous samples is 2 h. The larger and denser a sample, the longer it should be incubated. Usually, 2 days are enough.
4. Place the sample in 70% ethyl alcohol if samples need to be stored for longer periods.

Preparation of Dry Samples

When dry samples, for example, from a xylarium or a herbarium are used, a rehydration step will be needed. Beware that, in most cases, the quality of the sections will be impaired by the natural collapse of thin-walled cells during the drying process, and no quantitative measurement should be performed on such materials.

1. Suspend samples in water with glycerin.
2. Boil for about 2 h. Sinking of a sample indicates that it has been well rehydrated

3.2 Material Softening

Certain barks are more easily sectioned after treatment with a softening agent. The optimal softening procedure has to be determined by testing for each type of sample. Boiling the sample in water with glycerin is the standard procedure for softening stem or root samples. An alternative is ethylenediamine (ETD) treatment, which swells the lignified cell walls of the sclerenchyma, making the overall tissue more homogeneous and less dense, ultimately allowing better sectioning [8, 9].

The swelling produced using ETD is reversed in the subsequent anatomical procedures [9]. This method has been very effective for some families, such as the Bignoniaceae [10, 11], while it has made the bark detach from the wood in a matter of hours in others, such as in the Malpighiaceae [4]. The highly toxic hydrofluoric acid (HF) is the most drastic method for softening and, therefore, is reserved for very hard materials. These include stem samples from certain hardwoods, for example, from the Sapotaceae [8]. Beware that HF will eliminate silica and crystals.

Boiling in Water and Glycerin

1. Boil your sample in water with glycerin for a period of between a few hours and several days.
2. During the softening procedure, periodically test the sample with a razor blade to know when it has reached its desired level of hardness. Samples will vary in timing and treatment.

Softening with Ethylenediamine (ETD)

1. Place the samples in a jar with a screw-on lid and cover with ETD.
2. Incubate samples for 1–3 days in a ventilated oven at 60°C.
3. During the softening procedure, periodically test the sample with a razor blade to know when it has reached its desired level of hardness. Samples will vary in timing and treatment.

Softening with Hydrofluoric Acid (HF)

Beware of the safety procedures described in "Materials"; HF is extremely toxic.

1. Put the sample in an adequate plastic container and submerge it in the HF solution.
2. Incubate for 3–4 days at room temperature under the fume hood.
3. After the treatment, place the samples in a saturated sodium bicarbonate solution to neutralize the acid.
4. Wash thoroughly with water.

3.3 Infiltrating and Embedding of the Samples

The PEG (Polyethylene Glycol) 1500 Method According to Rupp [12]

1. Place the sample in a reasonably large jar (like a tomato sauce jar) and cover the sample with pure dissolved PEG (*see* **Note 3**). Mark the original level of the pure PEG with a water-resistant pen.
2. Fill the jar with water.
3. Place this solution in a ventilated oven (60°C) until all the water evaporates, meaning the amount of liquid in the jar has reached the level marked in Step 1.

4. Store the sample with pure PEG in a vacuum oven for several hours to days.
5. Transfer to a square paper casting mold.
6. Leave at room temperature until solidified. The finished blocks are trimmed and then sectioned using a sliding microtome.

The Paraffin Method According to Johansen [13]

1. Place the sample in a closed container and cover with a solution of one volume of Tertiary butyl alcohol and one volume 50% ethanol.
2. Incubate at room temperature in a rotary or rocking shaker for 2 h.
3. Replace solution and repeat incubation four times with the following solutions:
   Tertiary butyl alcohol and 70% ethanol.
   Tertiary butyl alcohol and 85% ethanol.
   Tertiary butyl alcohol and 95% ethanol.
   Tertiary butyl alcohol and 100% ethanol.
4. After the final incubation, transfer the sample to pure tertiary butyl alcohol and incubate at room temperature in a rotary or rocking shaker for 12 h.
5. Transfer to a new solution of pure tertiary butyl alcohol and keep it within a closed jar.
6. Incubate in a ventilated oven at 60°C for 2 h.
7. Replace alcohol with the following solutions and repeat incubation for each solution:
   i. 3:1 tertiary butyl alcohol-paraffin.
   ii. 1:1 tertiary butyl alcohol-paraffin.
   iii. 1:2 tertiary butyl alcohol-paraffin.
8. After final incubation, replace the solution with pure liquid paraffin and incubate at room temperature on a rotary or rocking shaker for 4 h.
9. Replace once more with pure liquid paraffin, allowing four more hours on the shaker.
10. Transfer the sample to a paper mold.
11. Once hardened, trim the paraffin block, and section on a rotary microtome as described below.

The Glycol Methacrylate Method

Resin embedding allows for obtaining thin sections (3 μm or less) without tissue distortion, which may be necessary for observing the fragile cells of the phloem [14–16]. However, to obtain a perfect historesin embedding, the samples must be kept small, not exceeding $2 \times 2 \times 2$ mm (8 mm^3). Typically, all resins are purchased as a kit with specific, easy-to-follow instructions provided by the supplier.

Here, we describe what is typical for most of the resins available on the market. An adjustment of the standard protocol, the addition of polyethylene glycol 400 to the resin, facilitates obtaining ribbons from sectioning resin-embedded samples [17]. If possible, do all steps within a vacuum chamber at 4°C.

1. Dehydrate your small samples in an ethanol series up to absolute ethanol.
2. Transfer the samples to a solution of absolute ethanol mixed with the embedding solution (1:1) for several hours up to 1 month, depending on the permeability of the material. Thick sclereids may be nearly impermeable to the glycol methacrylate resin and will need a long time to infiltrate.
3. Transfer the samples to the historesin solution and incubate for several hours up to 1 month. Use a vacuum chamber whenever possible at all steps. Note that some resin kits recommend carrying out the procedure at 4°C.
4. Add the hardener and place the sample in a historesin mold.
5. Incubate for 1 day.
6. Place the solid blocks on a holder and cut with a rotary microtome.
7. Place the sections directly on the slide, dropwise and incubate the slide on a hot plate at 55°C until the water has evaporated.
8. Materials sectioned in resin can be stained with Toluidine blue and left without a coverslip. Toluidine blue tends to fade with time, and without a coverslip, the slides can be re-stained.

3.4 Sectioning

Sectioning of Paraffin- and Resin-embedded Samples

1. Check knife sharpness. Regardless of hand sectioning or microtome sectioning, the key to good sections is a very sharp knife. Either disposable knives or sharp permanent steel knives can be used (*see* **Note 4**).
2. Insert the sample in the sliding or rotary microtome. The knife, sample and all the screws in the microtome need to be very tight.
3. In the rotary microtome, adjust cutting speed and angle, make sections, and use tweezers to hold and collect the sections on microscopy slides.
4. If the sample is embedded in resin, place each section on a drop of water on a slide and place the slide on a hot plate (55°C) and incubate until the water has evaporated. This facilitates attachment to the slide.

5. If the sample is embedded in paraffin, place slides upon a hot plate (55°C) and add three drops of Haupt's adhesive and place the paraffin ribbon upon them. Keep for 2 days in a hot paraffin chamber (56°C oven). After that, the slides are treated either with xylol or butyl acetate to remove the paraffin and move to the stains. Because the sections are glued to the slide by the adhesives, they can all be stained simultaneously in a rack in a Coplin staining jar.
6. If using a sliding microtome, start by adjusting the cutting angle. An angle between 15° and 20° gives the best results, even for very dense materials.
7. Adjust the direction in which the knife travels in the sliding microtome towards the block, making it equivalent to between 45° and 90° in relation to the rays. The higher the angle, the lower the impact on the sample; therefore, increase the angle when the bark is harder.
8. Adjust the section thickness to 20 μm on the sliding microtome. However, when the bark or the wood is too hard, make smaller cutting blocks and cut thinner sections (10–15 μm), which will impose less resistance to the knife.

Sectioning of PEG 1500-Embedded Samples

The heterogeneous nature of bark makes it extremely fragile not only while sectioning, but also in the subsequent staining and dehydration procedures. An effective solution to this problem is the use of a polystyrene foam resin placed upon a sample embedded in PEG 1500 [18].

1. In a small vial, place some butyl acetate or xylene and then cut pieces of any packaging Styrofoam and dissolve them in the solvent to make a saturated solution.
2. Install your sample in the microtome and apply some polystyrene foam resin on it with a paintbrush. Let it dry.
3. Perform sectioning. While sectioning, apply a little pressure to the sample with a clean paintbrush to ensure that the sections will not roll up.
4. Transfer sections with a tweezer to a microscope slide or a glass watch with some tap water.
5. Follow the procedures of staining as for other samples.
6. Dehydrate the sections in an ethyl alcohol series (50%, 70%, and 90%, 30 min each step) directly on the slide, so that sample movement is kept minimal.

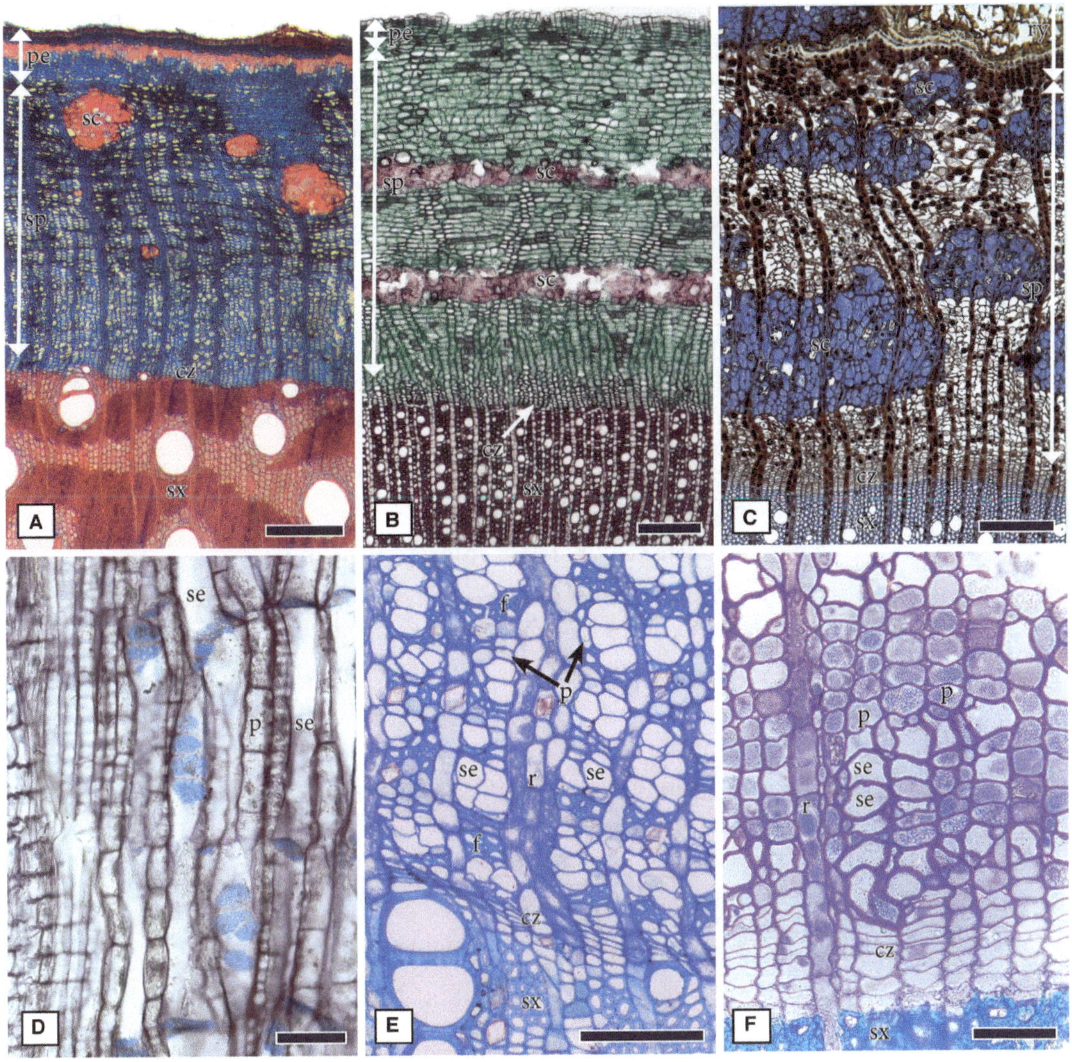

Fig. 2 Photomicrography of barks. All images show transverse sections, except for **D**, which is a longitudinal radial section. (**A**) *Caesalpinia ferrea* (Leguminosae), embedded in PEG 1500, sectioned using a sliding microtome with a polystyrene foam resin, stained with Safrablau, and mounted in Canada Balsam. Lignified tissues stain red and nonlignified tissues blue. (**B**) *Robinsonella discolor* (Malvaceae), sectioned unembedded using a sliding microtome, stained in safranine and fast green (by courtesy of Teresa Terrazas). Lignified tissues stain in red and nonlignified tissues in green. (**C**) *Crataegus intricata* (Rosaceae), stained with ferric chloride and resorcin blue (lacmoid). Lignified tissues stain in blue and nonlignified tissues in brown. (**D**) *Fraxinus americana* (Oleaceae), stained with ferric chloride and resorcin blue (lacmoid), evidencing the pores at the sieve areas, stained in bright blue*. (**E**) *Luehea divaricata* (Malvaceae), embedded in glycol methacrylate, sectioned at the rotary microtome, and stained with toluidine blue. Mounted in Canada Balsam, which reduces the metachromasia. (**F**) *Leucaena leucocephala* (Leguminosae), embedded in glycol methacrylate, sectioned at the rotary microtome, and stained with toluidine blue. Unmounted, with just tap water and a coverslip. Samples shown in **C**, **D** come from the Dr. Ray Evert Collection, which was imaged in collaboration with Carmen R. Marcati and Veronica Angyalossy. Sx = secondary xylem, cz = cambial zone, sp = secondary phloem, pe = periderm, sc = sclereid, ry = rhytidome, se = sieve tube element, p = axial parenchyma, f = fibers, r = rays. Scale bars: **A** = 300 μm, **B**, **C** = 200 μm, **D** = 40 μm, **E** = 100 μm, **F** = 50 μm

7. Remove the Styrofoam resin by carefully washing the sections with butyl acetate or xylene. Note that after the removal of the Styrofoam resin, the sample is very fragile.
8. Carefully add mounting media and the coverslip.

3.5 Stains and Stain Procedures

Ferric Chloride and Resorcin Blue (Lacmoid)

This combination of stains is one of the best for phloem studies. It stains the nonlignified tissues brown, the lignified tissues blue (Fig. 2C), and the callose of the sieve areas bright blue, making it very easy to locate sieve elements and pores of the sieve areas (Fig. 2D).

1. Place the sections in 1% tannic acid for 5–10 min.
2. Wash well in distilled water.
3. Transfer to 2% ferric chloride for 5 min.
4. Wash well in distilled water. At this point, the coloration should be medium to dark gray. Repeat the staining if needed.
5. Transfer the sections to a 1% sodium bicarbonate solution. The coloration should change from gray to brown.
6. Transfer the sections directly to the lacmoid solution.
7. Leave the sections in lacmoid for 12–18 h or more at room temperature.
8. Transfer the section to a sodium bicarbonate solution in 50% ethanol for a few minutes
9. Move to the ethanol dehydrating procedures of mounting (Step 2 in Sect. 3.6).

Safrablau

A solution of Astra blue and Safranin O in 50% ethanol generally yields excellent results for staining bark. This solution is called Safrablau. The lignified tissues stain red and the nonlignified tissues blue (Fig. 2A).

1. Stain sections with Safrablau solution for 5–10 min.
2. Wash well with distilled water and then move to the mounting procedure (see Sect. 3.6).

Safranine and Fast Green FCS

This is one of histologists' preferred staining methods. It stains the lignified tissues in red and the nonlignified tissues in green (Fig. 2B).

1. Stain in 1% Safranin O for 1–12 h.
2. Wash well in distilled water and then dehydrate with an ethanol series up to 95% ethanol. No more than 10 min is needed in each ethanol rinsing step.

3. Counterstain in 0.5% Fast Green FCS for 5–30 s.
4. Wash well in 96% ethanol and move to the permanent mounting procedures (Sect. 3.6).

Aniline Blue

This stain enables the visualization of callose under ultraviolet fluorescent light.

1. Stain sections in 1% aqueous Aniline blue for 1 h.
2. Wash gently with 50% glycerin.
3. Counterstain in 1% aqueous Eosin Y for 15–20 min.
4. Move to the mount procedures (Sect. 3.6). Note that Aniline blue typically fades with time [19].

Toluidine Blue

This metachromatic stain is especially recommended for sections embedded in glycol methacrylate resin, although it also works well on thicker sections obtained from other embedding methods. Typically, the color change is less clear or lost if the material is mounted in permanent resins such as Canada balsam (Fig. 2E). Therefore, it is recommended to have at least some slides unmounted that can be re-stained and observed fresh when needed (Fig. 2F).

1. Place the samples in the Toluidine Blue solution for several minutes.
2. Wash well in distilled water.
3. Either observe the samples freshly stained by placing some drops of 50% glycerin and a coverslip on the slide or move to the permanent mounting procedure (Sect. 3.6), knowing that the color shift will decrease over time.

3.6 Permanent Slide Mounting

1. If not already mounted on a slide, place the sections in the center of a microscope slide using delicate forceps.
2. Dehydrate the samples with an ethanol series and incubation times of 10 min up to absolute ethanol.
3. Wash in absolute ethanol several times to ensure that no water is present. If any water residue is left, it will react with the solvent, creating numerous tiny bubbles that will impair tissue observation.
4. Wash the samples in butyl acetate or xylene under a fume hood. Butyl acetate is less toxic than xylene and toluene and is therefore recommended here.
5. Add a few drops of any permanent mounting media, such as Canada Balsam, Euparal, etc. (*see* **Note 2**), and coverslip. Be careful when adding the coverslip as sections are easily damaged.

4 Notes

1. Bulky samples are preferably fixed in FAA, which has a strong penetrating ability. For small samples (<1 cm^3), basic, noncoagulant fixatives are more adequate and will yield good results for electron microscopy when the objective is the study of organelles such as mitochondria, nucleoplasm and vacuoles.
2. Consider that some mounting media age better than others. Canada balsam is a natural oleoresin obtained from the bark of the balsam fir *Abies balsamea* (Pinaceae) and has been used since at least 1830 [19, 20]. It is one of the best mounting media, since it endures the passage of time without crystallizing. Permount, for instance, should be avoided as it is nonarchival [21].
3. Slower methods can be employed for harder, less permeable specimens, allowing 24 h each in increasing concentrations of PEG, from 20%, 30%, and 40% up to pure PEG [18].
4. To sharpen permanent steel knives, there are two alternatives: either the use of abrasive solutions sold by knife sharpening manufacturers or, as a cheaper alternative, a recently proposed waterproof sandpaper sharpening method [22].

Acknowledgments

I would like to thank Antonio C. F. Barbosa from the Technological Research Institute of São Paulo for instructing me in microtomy, Gisele Costa from the University of São Paulo, Stan Yankowski for important inputs and for proofreading the manuscript, Mark Strong for proofreading the manuscript, Teresa Terrazas for Fig. 1B, the PeterBuck postdoctoral fellowship for funding at the Smithsonian Institution and the Programa de Apoyo a Proyectos de Investigación e Innovación Tecnológica (PAPIIT IA200319—Mexican government).

References

1. Grew N (1682) The anatomy of plants with an idea of a philosophical history of plants. Rawlins, London
2. Malpighi M (1686) Opera omnia. Littlebury, London
3. Baas P (1982) Antoni Van Leeuwenhoek and his observation on the structure of the woody cell wall. IAWA J 3:3–6
4. Angyalossy V, Pace MR, Evert RF, Marcati CR, Oskolski AA, Terrazas T et al (2016) IAWA list of microscopic bark features. IAWA J 37:517–615
5. Evert RF (2006) Esau's plant anatomy: meristems, cells, and tissues of the plant body–their structure, function, and development. John Wiley & Sons, Inc., New Jersey

6. Liesche J, Vincent C, Han X, Zwieniecki M, Schulz A, Gao C, Bravard R, Marker S, Bohr T (2021) The mechanism of sugar export from long conifer needles. New Phytol 230:1911–1924
7. Vincent C, Hussain SB, Losada JM, Liesche J (2025) Source-sink transport as a constraint on photosynthesis and a driver of ecophysiological patterns. Plant Physiol 199:kiaf531
8. Kukachka BF (1977) Sectioning refractory woods for anatomical studies. USDA Forest Service Research Note FPL-0236, Madison
9. Carlquist S (1982) The use of ethylenediamine in softening hard plant structures for paraffin sectioning. Stain Technol 57:311–317
10. Pace MR, Lohmann LG, Angyalossy V (2011) Evolution of disparity between the regular and variant phloem in Bignonieae (Bignoniaceae). Am J Bot 98:602–618
11. Pace MR, Alcantara S, Lohmann LG, Angyalossy V (2015) Secondary phloem diversity and evolution in Bignonieae (Bignoniaceae). Ann Bot 116:333–358
12. Rupp P (1964) Polyglykol als Einbettungsmedium zum Schneiden botanischer Präparate. Mikrokosmos 53:123–128
13. Johansen DA (1940) Plant microtechnique. McGraw-Hill Book Company, Inc., New York and London
14. Bennet HS, Wyrick AD, Lee SW, McNeil JH (1976) Science and art in preparing tissues embedded in plastic for light microscopy, with special reference to glycol methacrylate, glass knives and simple stains. Stain Technol 51:71–97
15. Hamann T, Smets E, Lens F (2011) A comparison of paraffin and resin-based techniques used in bark anatomy. Taxon 60:841–851
16. Yeung EC, Chan CKW (2015) Glycol methacrylate: the art of embedding and serial sectioning. Botany 93:1–8
17. Chamberlain CJ (1935) Methods in plant anatomy. The University of Chicago Press, Chicago.
18. Barbosa ACF, Pace MR, Witovsk L, Angyalossy V (2010) A new method to obtain good anatomical slides of heterogeneous plant parts. IAWA J 31:373–383
19. Neuhaus B, Schmid T, Riedel J (2017) Collection management and study of microscope slides: storage, profiling, deterioration, restoration procedures, and general recommendations. Zootaxa 4322:1–173
20. Loveland RP, Centifanto YM (1986) Mounting media for microscopy. Microscope 34:181–241
21. Ravikumar S, Surekha R, Thavarajah R (2014) Mounting media: an overview. J Dr NTR Univ Health Sci 1(3):S1–8
22. Barbosa ACF, Costa G, Pace MR, Angyalossy V (2018) A simple and inexpensive method to sharpen permanent steel knives for microtomy. IAWA J 39:497–503

Chapter 2

Transmission Electron Microscopy of the Phloem

Pascal Hunziker and Alexander Schulz

Abstract

A universal feature of seed plants is that their phloem consists of a continuous sieve-tube system extending throughout the plant, which is highly pressurized due to its sugar content. This cellular continuity, combined with the pressure-driven, osmotically generated flow in the source leaves, enables assimilates to reach all sink organs. However, these two key features of the phloem, the cellular continuity and high pressure, pose significant challenges when attempting to fix the phloem for transmission electron microscopy. In most cases, the tissue preparation required for fixation triggers rapid wound responses that eventually result in artefacts. This chapter describes the steps necessary to minimize artefact formation during phloem fixation. It includes the preparation of fixatives, a dissection procedure designed to optimize fixative penetration, and their application to both axial and lateral plant organs. Additionally, as an alternative to the conventional fixation of fresh hand sections, we suggest a xylem-assisted perfusion fixation method for herbaceous plants. Following the initial fixation, the subsequent steps of dehydration, embedding, and ultrathin sectioning are carried out using standard procedures, which are briefly discussed. The chapter also addresses sample orientation to obtain transverse and longitudinal sections of the phloem.

Keywords Aldehyde fixatives, Companion cell, Phloem fixation, Perfusion fixation, Pressure release, Sieve element, Transmission electron microscopy, Wound response

1 Introduction

Among all plant tissues, the phloem exhibits the highest degree of specialization, consisting of three or more distinct cell types, each with unique functions [1]. The sieve elements, in particular, contain a nearly molar sugar concentration and a high abundance of plasma membrane aquaporins [2], which together generate a significant osmotic turgor pressure. This high sugar concentration is maintained throughout the phloem network by sugar loading in the leaves and retrieval along the transport pathway. Turgor pressure may partly extend to the companion cells via the plasmodesmata-pore complexes. Phloem parenchyma cells are not part of the high-pressure conducting domain, with the exception of minor veins in symplasmic loaders [3, 4]. This is due to the virtual absence of plasmodesmata between

Johannes Liesche and Sanna Sevanto (eds.), *Phloem*, Methods in Molecular Biology, vol. 3066,
https://doi.org/10.1007/978-1-0716-5488-0_2,

phloem parenchyma and the sieve element-companion cell (SECC) complexes and/or the presence of a pressure valve mechanism in these cells [5].

The dissimilar turgor pressures within the phloem present a significant challenge for its optimal fixation for transmission electron microscopy (TEM). This challenge is further compounded by the fact that the phloem is located deep within plant organs, typically 50–200 μm beneath the epidermis of leaves, stems, and roots. Consequently, preparing phloem samples for TEM requires dissection of the target tissue, which inevitably involves cutting into the functional phloem network. This dissection leads to a sudden release of pressure, followed by the plugging of sieve pores, displacement and rupture of organelles, and the disruption of structural proteins (Fig. 1a). Neither chemical fixation nor rapid freezing can fully resolve these tissue-specific challenges. However, preparatory intrusion can be minimized using plunge freezing followed by freeze substitution of entire *Arabidopsis* seedlings. This approach has achieved unprecedented fixation quality for leaf sieve elements, particularly preserving their plasma membranes and organelles, although phloem parenchyma cells still exhibit freezing artifacts [6]. Plunge freezing, however, is limited by penetration depth, as heat removal follows diffusion laws. High-pressure freezing overcomes this limitation to some extent, achieving penetration depths of up to 300 μm [7]. Even so, the preparation required to immerse plant tissue in freeze protectant and mount it

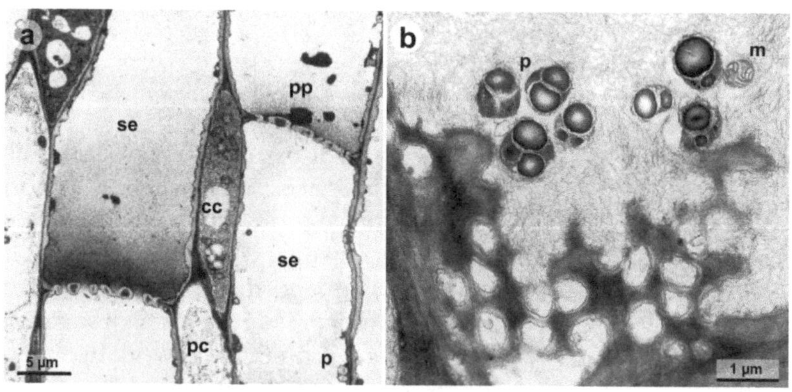

Fig. 1 Phloem is fixed conventionally by submerging fresh hand sections. (**a**) *Cucurbita pepo* stem phloem, taken from the basal end of a 15 mm hand section close to the first incision into the intact plant. Sieve element (se) contents, such as filamentous and persisting P-protein (pp), accumulate on the sieve plates. (cc) companion cell, (pc) parenchyma cell, (p) sieve element plastid (**b**) *Ricinus communis,* oblique section through a sieve plate from the hypocotyl, taken from the middle part of the hand section after discarding the ends. P-protein filaments are evenly distributed in the cytoplasm; the sieve pores appear free, and the organelles are well preserved. (p) sieve element plastids, (m) mitochondrion

in the specimen holder for freezing still involves dissection-induced pressure release within the phloem continuum.

For optimal fixation of the phloem using either chemical or physical methods, the key question is whether it is possible to avoid cutting into the pressurized conducting domain while simultaneously reducing the diffusion distance for chemical fixatives or the removal of heat from the phloem. The answer, unfortunately, is no. Nevertheless, the chemical fixation protocol outlined below achieves a degree of preservation of phloem ultrastructure (see Fig. 1b) that is sufficient to address a range of current questions regarding phloem structure and function.

2 Materials

2.1 Solutions and Buffers

1. Cooking oil to absorb discarded osmium solutions and contaminations from Fixative 2 for chemical waste disposal.
2. Dehydration solvent: 30%, 50%, 70%, 90%, and 100% Acetone.
3. Double-distilled or MilliQ water.
4. Fixation buffer: 0.1, 0.2, and 0.3 M Sodium cacodylate at pH 7.3.
5. Fixative 1 [8]: 4% Paraformaldehyde, 5% glutaraldehyde in 0.1 M cacodylate buffer, pH 7.3, for preparation, *see* Sect. 3.1.
6. Fixative 2: 1% Osmium tetroxide in 0.1 M cacodylate buffer pH 7.3, made from commercially available 4% aqueous solutions (2 or 5 ml ampoules, e.g., from Electron Microscope Sciences), for preparation *see* Sect. 3.1.
7. Post-contrast solution 1: 1%–4% Uranyl acetate in water or, as alternative,aqueous Uranyless.
8. Post-contrast solution 2: Lead citrate, either in airless bottles or to be prepared according to [9].
9. Spurr's resin: 61.3% Nonenylsuccinic anhydride, 23.6% vinylcyclohexene dioxide, 14.2% D.E.R., 0.9% dimethylaminoethanol.

2.2 Consumables

1. Flat embedding molds (e.g., from Agar Scientifics).
2. Gelatine capsules, size 3 (5.5 mm diameter).
3. Glass vials with PE lids (5 or 10 ml).
4. Graded plastic Pasteur pipettes (e.g., from Agar Scientifics).
5. Latex gloves.
6. Microscope slides.
7. Absorbent paper.
8. Parafilm.

9. Petri dishes.
10. Pioloform-film-coated TEM grids (single slot 2 by 1 mm).
11. Polyethylene syringes for freeze storage of the embedding medium.
12. Razor blades.
13. Toothpicks.

2.3 Preparation Tools

1. Diamond tip scriber.
2. Forceps.
3. Preparation needles.
4. Ultramicrotome diamond knife.

2.4 Equipment

1. Dissection microscope.
2. Lab oven attached to a fume hood or exhaust ventilation (range 40–100°C with max ±1°C deviation at 60°C).
3. Shaker.
4. Tissue rotator.
5. Ultramicrotome.
6. Vacuum pump (100 mbar) and desiccator.

3 Methods

Chemical fixation of tissues means a race between the penetration speed of the crosslinking fixative and the development of cellular wound responses. As a rule of thumb, isodiametric cells are best fixed in tissue cubes of less than 1 mm edge length. Thus, the fixative can penetrate from all sides and reach cells in the center so rapidly that their wound response is minimal. However, plant vascular tissue consists of strongly elongated rather than isodiametric cells. In transverse stem sections, all vascular tissues are cut. The cut cells and their neighbors suffer strong wound responses resulting in fixation artifacts. Therefore, fixation of fresh cross sections is not recommended (but *see* **Note 1**). In longitudinal sections, many cells must be penetrated from the side to reach the center. Consequently, the limiting factor for the fixation quality of the central part is the thickness of the longitudinal section. According to the experience in Dietmar Behnke's lab, built upon fixing phloem from several thousand plant species (see, for instance [10]), (i) longitudinal hand sections of phloem tissue should be thinner than half a millimeter. The ends of such a section show strong artifacts in sieve elements and companion cells, such as plasmolysis and displacement of cytoplasmic contents on the next sieve plate, due to the rapid pressure release in the cut sieve element (see Fig. 1a and [11]). This issue is solved (ii) by discarding the ends of longitudinal sections. Finally, (iii) not more than a minute should pass between the first preparatory cut into a stem and immersion of the sample in the fixative. When these three rules are obeyed, images like Fig. 1b, which depicts a sieve plate from *Ricinus* hypocotyl phloem with evenly distributed P-protein filaments and intact plastids and mitochondria, can be achieved. The sieve pores appear free of cytoplasmic constituents.

3.1 Preparation of the Solutions

The fixatives crosslink proteins, compromise membranes, and preserve lipids, which means that they kill plant cells as well as human cells. The fixation buffer contains arsenic and is strongly toxic. Exposure of the skin and respiratory tract to fixatives must be prevented. The embedding resin compounds are carcinogenic before polymerization. Therefore, use gloves, prepare all solutions under a strong fume hood, and follow all safety rules for the disposal of chemical waste.

1. Prepare fixative on the day of fixation, or 1 to 2 days before. Fresh preparation of Fixative 1 is decisive for the fixation quality.
2. Dissolve 4 g paraformaldehyde in H_2O and warm up to 80°C. This results in a milky solution.
3. Add eight droplets 1 M NaOH (solution gets clear) and cool down.
4. Add 20 ml 25% glutaraldehyde.
5. Fill to 100 ml with 0.3 M fixation buffer.
6. Adjust with 1 M NaOH to pH 7.3.
7. Keep refrigerated (4°C) until use.
8. Fixative 2 should be prepared from the ampoule-stored 4% stock solution immediately before the respective fixation step (*see* Sect. 3.2). Dilute 2 ml of the stock solution in 2 ml 0.2 M fixation buffer and 4 ml 0.1 M fixation buffer. For waste disposal of the final 1% osmium solution, (*see* **Note 2**).
9. Prepare the dehydration solutions (30%, 50%, 70%, 90%, and 100%) with water-free acetone when needed. In histological labs with large turnover, keep stocks of the respective solutions.
10. Weigh and combine the components of Spurr's resin in a disposable beaker on a balance, then blend the resin with a magnetic stir bar until completely dissolved.
11. Aliquot the resin into disposable syringes and use it right away. Syringes with unused aliquots should be capped and frozen. Allow frozen syringes to achieve room temperature before opening and applying the resin in order to avoid condensed water precipitation.

3.2 Preparation and Fixation of the Plant Material

Collect all the materials you need for the fixation (plants, Fixative 1, razor blades, forceps, preparation needle, glass vials, and diamond tip scriber) and arrange a working space under the fume hood. When fixing tree phloem outside the lab, organize these items on a portable tray setup.

1. Label or number the vials with the diamond tip scriber (*see* **Note 3**) and fill each vial with 2–5 ml of cold Fixative 1.
2. For fixation of phloem from axial organs, leaves and tree trunks follow the description below at A, B, and C, respectively. As an alternative method to fix stems, petioles and leaves of herbaceous plants, (*see* **Note 1**).
   A. For phloem from axial organs with a diameter larger than 3 mm, such as stems, petioles, and roots, remove them with one transverse cut from the plant (for thinner organs, *see* **Note 4**). While holding the plant with the left hand, supporting the cut end with its index finger, use a razor blade with your right hand to score a line across in 10–15 mm distance from the end, but make sure that you only cut halfway through (left-handed people vice versa). Make longitudinal sections of about 0.5 mm thickness starting from the end. These sections will be 10–15 mm long and slide onto the razor blade because of the line scored into the stem. Discard the first section, mainly containing epidermis and cortex tissue, and immerse the following ones immediately in Fixative 1 using forceps. Any section must be immersed in the fixative within 60 s, counting from the first cut into the plant. Discard the rest of this segment and cut at least 30 mm back before starting a new segment to avoid wound artifacts. Figure 1b is a result of this fixation method.
   B. For major and minor veins of leaves, cut 2 by 2 mm^2, containing the respective veins, on a microscope slide (for the midrib, *see* **Note 5**). Submerge the squares immediately in Fixative 1 using forceps. The vials containing leaf samples should be evacuated in a desiccator as soon as possible for at least 1 h; otherwise, air in the intercellular spaces will hold them at the surface, preventing even penetration by the fixative. Fixation quality should match [12].
   C. For phloem from tree trunks, use a knife to cut a bark piece of 40 by 10 mm and peel it off the wood (secondary xylem) by tearing the cambium. Cut this piece back to 15 mm in length of the mid part. For radial sections, remove the outer bark and make 0.5 mm sections along the radial side with a razor blade. For tangential sections, start at the cambium and make one or two 0.5 mm sections along this side. Discard the rest, since it will only contain obliterated phloem. As in A. and B., the final radial or tangential sections should be immersed in Fixative 1 within 60 s. For example, see [13]).

3. Samples should be fixed for 2–3 h while slowly agitated on a shaker or tissue rotator. If trichomes or enclosed air prevent them from sinking down, evacuate in a desiccator for at least 1 h. If more suitable, the samples can be stored in Fixative 1 at 4–8°C overnight.
4. Wash five times with 0.1 M fixation buffer for 1 h: Remove Fixative 1 from each vial with a Pasteur pipette touching the vial's bottom corner, so that the sections are not aspirated, and add the same amount of fixation buffer immediately. Exchange the fixation buffer four times (*see* **Note 6**). The vials should be slowly agitated throughout all following steps up to embedding (Sect. 3.5).
5. Postfix the samples with Fixative 2 for 1 h at RT, or overnight in the fridge.
6. Wash five times with 0.1 M fixation buffer for 1 h as in Step 5. Dispose of the first three changes in food oil to bind the remaining osmium.

3.3 Dehydration and Infiltration of Samples with Resin

1. Dehydrate the samples through an acetone series (30%, 50%, 70%, and 90%) for 20 min per concentration step after removing the fixation buffer. Close the vials with the lids between the changes from now on. Samples can be stored for several days in 70% acetone in the fridge, if necessary.
2. Dehydrate completely using two changes of water-free acetone for 1 h in total.
3. Remove 1/3 acetone with a Pasteur pipette and add the same amount of Spurr's resin for 30 min.
4. Remove 1/2 of the mixture, add the same amount of resin and incubate for 30 min
5. Remove 2/3 of the mixture, add the same amount of resin and incubate for 30 min
6. Remove all solution, replace with resin (without lid so that the rest of the acetone can evaporate) and incubate for 1 h.
7. Replace with pure resin and incubate for 1 h.
8. Replace with pure resin, close the vials with their lids and incubate overnight.
9. Replace with pure resin and incubate for 2–4 h.

3.4 Oriented Embedding of Samples

Set up a dissection microscope under the fume hood and prepare microscope slides, razor blades, tooth picks, forceps, absorbent paper, and flat embedding molds and/or racks with gelatine capsules (*see* **Note 7**). The intended orientation of the phloem in the TEM decides whether gelatine capsules or flat embedding molds should be used: For cross sections from axial vascular tissue (fixed according to Sect. 3.2.2, approach A and C) and from major and minor leaf veins (fixed according to Sect. 3.2.2, approach B), use flat embedding molds. For longitudinal sections of axial vascular tissue and for paradermal leaf sections, use gelatine capsules.

1. When using gelatine capsules, load the capsule rack with pre-dried gelatine capsules and fill one drop Spurr's resin into each in 5–10 capsules at a time using a plastic Pasteur pipette.
2. When using flat embedding molds, fill the bottom of 5–10 molds at a time with Spurr's resin.
3. Place a longitudinal section from Sect. 3.2, approach **2A** or **C** onto a microscope slide and cut off and discard 3 mm from both ends, which contain phloem damaged by the sudden pressure release during preparation.
4. Divide the middle part into squares of a maximum of 2 by 2 mm using the dissection microscope and a razor blade. The vascular tissues are easily seen as they are stained black by Fixative 2. Ensure each square contains vascular tissue. Fixation and dehydration often make the tissue brittle; a fresh razor blade is recommended.
5. Use toothpicks and/or to slide the squares from the edge of the microscope slide into the molds or capsules. For embedding in molds, place one square each towards the pretrimmed end and gently press it to the bottom. For embedding in gelatine capsules, gently press the square with the toothpick centrally onto the capsule bottom and check its location with the dissection microscope.
6. Fill molds and capsules with resin completely (flat surface).
7. Polymerize the samples in the oven at 60°C for 36 h.
8. Collect all resin-contaminated disposables (microscope slides, Pasteur pipettes, tooth picks, etcetera) and polymerize them as well at 60°C for 36 h. They can then be disposed of as regular waste.

3.5 Trimming and Ultrathin Sectioning

These are routine steps for TEM labs and, therefore, not covered here in detail.

1. After polymerization, let the block cool down.
2. Remove the gelatine with hot water, then trim the blocks to the fixed sample.
3. To get an overview of the sample, make semithin sections, stain them with crystal violet and mount them on microscope slides. This allows for checking the fixation quality and selecting a suitable area for ultrathin sectioning.

4. Continue with ultrathin sectioning (60–80 nm section thickness) and collect the sections on single-slot grids.

4 Notes

1. Even when staying within 1 min for submerging the sections in Fixative 1, the phloem cells will show some signs of artifact formation. We have therefore adapted the method described for petioles in [14] and fixed stems by letting the cut xylem aspire Fixative 1. In our experience, the phloem tissue, including all its cell types, shows an excellent fixation quality in longitudinal and cross sections (see Fig. 2). This is because the distance the fixative has to diffuse from the xylem to the phloem is very short and reaches the sieve-element plasma membrane laterally along its course. Moreover, there is no sudden pressure release with this method, since Fixative 1 has a high osmolality, larger than the osmolality of the phloem sap (see [15]). It is known that aldehyde fixatives stabilize membranes through their protein-crosslinking property, thus preserving a large part of the cell contents (e.g., see [16]).
   a. Remove the stem with one transverse cut from the plant and submerge it immediately in Fixative 1.
   b. Allow the cut plant to transpire the fixative via the xylem for at least 60 min. The plant should stay at room temperature in an illuminated fume hood.
   c. When the top part of the plant starts to wilt, the fixative has been sucked up all the way.
   d. Remove the basal end of the plant by a transverse cut 30 mm above the end, and collect transverse or longitudinal stem sections in a new vial with Fixative 1.
   e. Continue the fixation for 2 h while agitating the vials on a shaker or tissue rotator. The subsequent fixation steps follow Sect. 3.2 from Step 4, except for trimming of the embedded samples (Step 5), which is unnecessary.
2. The final 1% solution should not be stored. Osmium evaporates easily, even from tightly closed bottles, also in the fridge, and will then contaminate the entire cabinet. Osmium from the contaminated vessels and the solutions discarded after the fixation step is best taken up in cheap food oil from the local market.
3. Fixation and dehydration are done in the same glass vials. A diamond-tip scriber should be used since any labeling of vials with acetone-dissolvable markers would disappear.

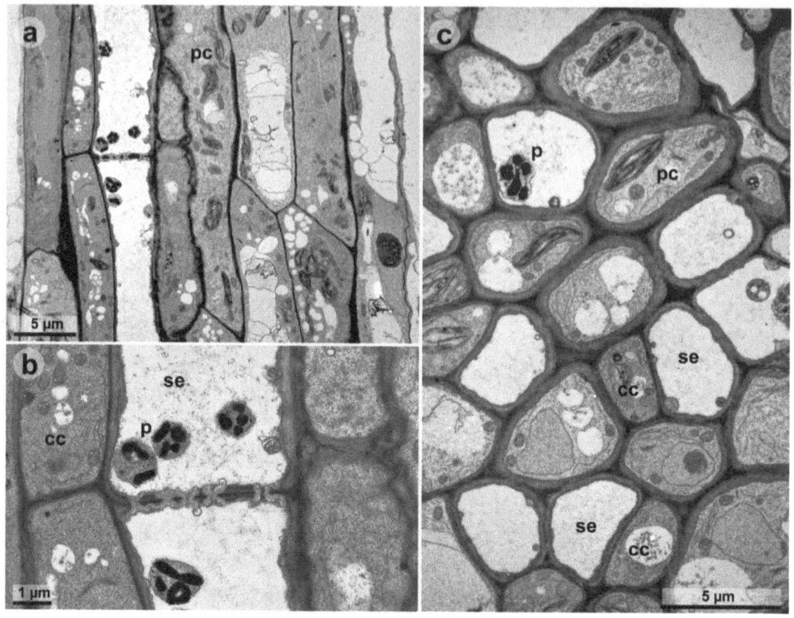

Fig. 2 Phloem fixed by xylem-assisted perfusion (*see* **Note 1**). (**a**) *Arabidopsis thaliana* developing flower stem phloem. The longitudinal overview shows a well-preserved sieve tube and all cell types from the phloem cap (left) to the cambium (right) in the differentiating phloem. (**b**) The close-up from (**a**) shows intact plastids at the sieve plate, the pores of which have just opened. (se) sieve element, (cc) companion cell, and (pc) parenchyma cell. (**c**) *Arabidopsis thaliana* mature flower stem phloem. The cross-section depicts many sieve elements (se), companion cells (cc), and parenchyma cells (pc) without any indication of artifacts

4. Halve axial organs with a diameter between 1 and 3 mm lengthwise and proceed in the same way as in Sect. 3.2, Approach 2A. Cut 10 mm segments from even thinner organs like, e.g., Arabidopsis roots.
5. For midrib phloem, cut the leaf off at the end of the petiole, if any, and isolate the midrib with two parallel cuts from the blade. Proceed as in Sect. 3.2, Approach 2A, starting from the basal end.
6. Thorough washing between Fixative 1 and 2 is necessary to avoid complex formation between the aldehydes and osmium in the tissue.
7. Racks for gelatine capsules should allow light transmission on the dissection microscope to check sample orientation. Disposable ones can be made of cardboard using a hole puncher, the diameter of which fits the no. 3 capsules. For permanent racks, a workshop can easily use 10 mm metal plates and drill rows of cavities. Capsules fit cavities drilled with a 6–6.5 mm borer, about 5 mm deep. Light transmission can be

achieved by continuing to drill with a 3 mm borer. However, racks are easily contaminated with resin that polymerizes in the polymerization oven, which can make it difficult to remove polymerized blocks from permanent racks.

Conflict of Interest

The authors have no conflicts of interest to declare that are relevant to the content of this chapter.

References

1. Jensen KH, Liesche J, Bohr T, Schulz A (2012) Universality of phloem transport in seed plants. Plant Cell Environ 35(6):1065–1076. https://doi.org/10.1111/j.1365-3040.2011.02472.x
2. Stanfield RC, Hacke UG, Laur J (2017) Are phloem sieve tubes leaky conduits supported by numerous aquaporins. Am J Bot 104(5):719–732. https://doi.org/10.3732/ajb.1600422
3. Schulz A (2015) Diffusion or bulk flow: how plasmodesmata facilitate pre-phloem transport of assimilates. J Plant Res 128(1):49–61. https://doi.org/10.1007/s10265-014-0676-5
4. Rockwell FE, Gersony JT, Holbrook NM (2018) Where does Winch flow begin? Sucrose transport in the pre-phloem path. Curr Opin Plant Biol 43:101–107. https://doi.org/10.1016/j.pbi.2018.04.007
5. Ham B-K, Lucas WJ (2014) The angiosperm phloem sieve tube system: a role in mediating traits important to modern agriculture. J Exp Bot 65(7):1799–1816. https://doi.org/10.1093/jxb/ert417
6. Froelich DR, Mullendore DL, Jensen KH, Ross-Elliott TJ, Anstead JA, Thompson GA, Pélissier HC, Knoblauch M (2011) Phloem ultrastructure and pressure flow: sieve-element-occlusion-related agglomerations do not affect translocation. Plant Cell 23(12):4428–4445. https://doi.org/10.1105/tpc.111.093179
7. Karahara I, Kang B-H (2014) High-pressure freezing and low-temperature processing of plant tissue samples for electron microscopy. In: Žárský V, Cvrčková F (eds) Plant cell morphogenesis: methods and protocols. Humana Press, Totowa, NJ, pp 147–157
8. Karnovsky MJ (1965) A formaldehyde-glutaraldehyde fixative of high osmolality for use in electron microscopy. J Cell Biol 27(2):137A–138A
9. Reynolds ES (1963) Use of lead citrate at high pH as an electron-opaque stain in electron microscopy. J Cell Biol 17(1):208. https://doi.org/10.1083/jcb.17.1.208
10. Behnke HD (1995) Sieve-element characters of the Proteaceae and Elaeagnaceae: nuclear crystals, phloem proteins and sieve-element plastids. Bot Acta 108(6):514–524. https://doi.org/10.1111/j.1438-8677.1995.tb00529.x
11. Schulz A (2017) Long-distance trafficking: lost in transit or stopped at the gate. Plant Cell 29(3):426–430. https://doi.org/10.1105/tpc.16.00895
12. Schulz A, Kühn C, Riesmeier JW, Frommer WR (1998) Ultrastructural effects in potato leaves due to antisense-inhibition of the sucrose transporter indicate an apoplasmic mode of phloem loading. Planta 206(4):533–543. https://doi.org/10.1007/s004250050430
13. Schulz A (1992) Living sieve cells of conifers as visualized by confocal, laser-scanning fluorescence microscopy. Protoplasma 166(3–4):153–164. https://doi.org/10.1007/bf01322778
14. Bell K, Mitchell S, Paultre D, Posch M, Oparka K (2013) Correlative imaging of fluorescent proteins in resin-embedded plant material. Plant Physiol 161(4):1595–1603. https://doi.org/10.1104/pp.112.212365
15. Karnovsky MJ (1965) Ultrastructural cytochemistry of cholinesterase activity in cardiac muscle by means of a copper ferrocyanide precipitating method. J Histochem Cytochem 13(1):6
16. Hobro AJ, Smith NI (2017) An evaluation of fixation methods: spatial and compositional cellular changes observed by Raman imaging. Vibrat Spectrosc 91:31–45. https://doi.org/10.1016/j.vibspec.2016.10.012

Chapter 3

Immunolabeling Sieve Element Cell Walls with the LM26 Antibody

Dustin M. Ray and Juan Manuel Losada

Abstract

Sieve elements in the phloem transport carbon and small molecules, such as RNA and phytohormones, throughout the plant body. Understanding the physical dimensions of sieve elements and phloem tissue is thus crucial for predicting how much carbon can be moved at any given time. Quantification of sieve element diameters and areas has previously been performed using transmission electron microscopy, scanning electron microscopy, and light microscopy, but sieve element identification is difficult because the phloem is a heterogeneous tissue. The recently identified LM26 antibody labels a pectin in the sieve element cell wall, allowing the identification of sieve elements and the measurement of their properties, such as diameter, relatively quickly and the quantification of their number in cross sections using image analysis software. Here, we describe methods for immunolabelling sieve elements in fresh or fixed tissue embedded in polyethylene glycol or methacrylate. The protocol is broadly adaptable to various fixation and sectioning methods, provided they do not alter the structure of pectins in the cell wall.

Keywords Histology, Epifluorescence, Phloem, Immunostaining

1 Introduction

The phloem supports plant growth and reproduction by facilitating the movement of carbon and other small molecules, such as RNAs or phytohormones, to areas of low concentration (sinks) via the sieve elements [1]. Prior work has shown that the physical dimensions of sieve elements (length and width) can contribute to resistance to phloem sap movement as they become narrower and shorter with increasing plant height [2, 3]. In nonpermeable tubes with laminar flow, flow resistance specifically scales to the fourth power of sieve element radius [2, 4, 5]. The ability to accurately and precisely measure sieve element dimensions for evaluating the total bulk conductance of sap flow is thus important for understanding the dynamics of carbon movement in the sieve elements. However, quantification of sieve tube number and size is challenging. This is because the phloem is a heterogeneous tissue [6, 7], and identifying and measuring sieve

Johannes Liesche and Sanna Sevanto (eds.), *Phloem*, Methods in Molecular Biology, vol. 3066,
https://doi.org/10.1007/978-1-0716-5488-0_3,

elements can be difficult, time-consuming, and expensive. Historically, sieve element dimensions in woody tissues have been measured from sections stained for light microscopy, with a limited resolution of about 5 μm [8–13], or from preparations for transmission electron microscopy, which allow visualization of finer details ranging from 1 μm to nanometers [14–20]. More recently, the phloem has been investigated using X-ray microtomography [21], or the cell contents of flash-frozen plant material have been dissolved away, revealing sieve plates and allowing for identification and measurement with the scanning electron microscope [16, 22, 23].

The identification of an antibody that selectively binds to pectin primarily located in sieve elements [24] has provided a relatively rapid method for the identification of sieve elements [25]. The antibody works with fresh or fixed tissue, whether or not it has been infiltrated and embedded in polyethylene glycol or methacrylate, allowing for flexibility in sample preparation [25–27]. It appears that the pectin labeled with LM26 is either removed or altered in such a way that complete staining of the sieve element walls (versus only small segments) only occurs when the sieve elements are active [25].

The protocol broadly follows that of most standard immunolabelling protocols in plant sciences [28–30]. Sections are first incubated in a blocking solution to reduce nonspecific staining (Fig. 1). Sieve elements are then labeled with the primary antibody, LM26. The LM26 antibody is not conjugated to a fluorophore and cannot be visualized until labeled with a secondary antibody. This approach allows the user to choose a secondary antibody with a fluorescence spectrum that will suit their needs. Although a fluorescent secondary antibody conjugated to fluorescein isothiocyanate (FITC) is what we use in this protocol (Fig. 2), light microscopy methods are also possible. A secondary antibody conjugated to horseradish peroxidase and visualized with 3-3′diaminobenzine in the presence of hydrogen peroxide would allow visualization of sieve elements under a light microscope under a modified version of the current protocol.

2 Materials

2.1 All Tissue Preparations

Make all solutions fresh on the day the immunolabelling will occur. We used 200 μL of fluid per slide (5 μL per section), but this volume can increase during wash steps, when calculating how much of each buffer to mix. Dehydrate the samples in an ethanol series and mount in DPX mountant, where the fluorescence is retained for approximately 6 months if the slides are stored in the dark. Anti-fade mountants like Citifluor AF-1 can be used to preserve the section fluorescence longer when stored in the dark at 4°C, but are not useful for permanent mounting of sections.

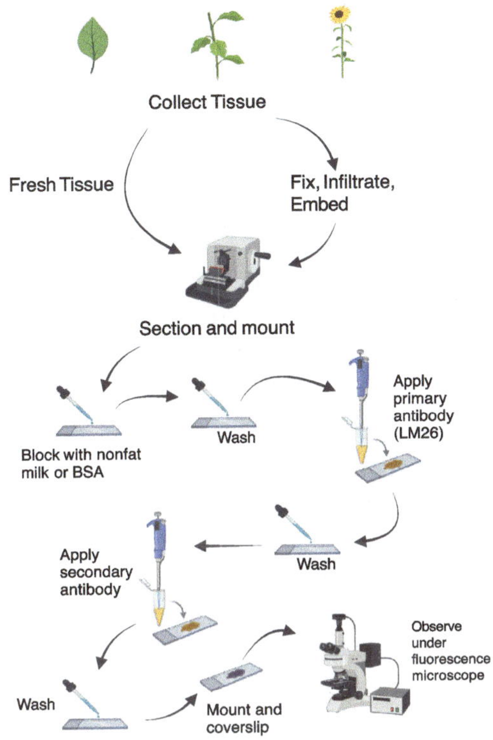

Fig. 1 Diagram of the general process for tissue processing from collection to viewing on an epifluorescence microscope. Created with BioRender

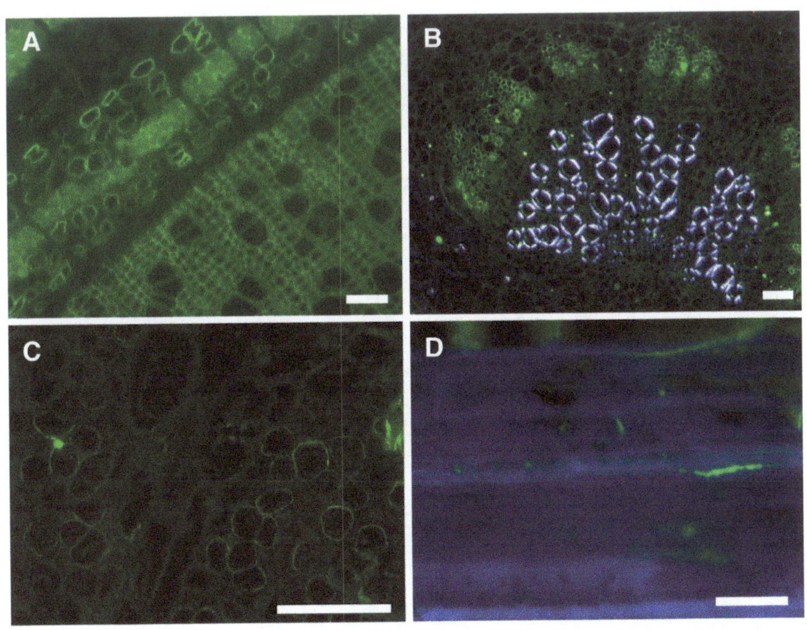

Fig. 2 Examples of tissue stained with LM26. Transverse section of *Populus balsamifera* stem fixed in gluteraldehyde and embedded in PEG (**A**), composite image with LM26 and polarized light of from transverse section of *Atemoya* floral pedicel embedded in Technovit 8100 (**B**), transverse section of *Persea* fruit pedicel embedded in Technovit 8100 (**C**), and longitudinal section of *Arabidopsis thaliana* floral stalk from fresh tissue (**D**). All scale bars are 50 μm

Materials for Making Solutions

1. Stirrers (2–4 per sample).
2. Beakers.
3. Pipettes.
4. Forceps.
5. Glass slides.
6. Micropipettes: 10, 20, 200, and 1000 μL.
7. Paraffin Pen or Tipp-Ex pen.
8. 1× Phosphate-buffered saline (PBS).
9. Ethanol series of 10%, 30%, 50%, 70%, and 95% ethanol in distilled water.
10. Primary Antibody/blocking buffer: LM26 Antibody mixed 1:5 with blocking buffer. Can be stored on the bench while preparing tissue. Invert or pipette to mix.
11. Secondary Antibody/ blocking buffer: Goat anti-rat antibody conjugated to FITC or the desired fluorophore diluted 1:100 with blocking buffer. Invert the tube or pipette to mix. Store on ice until ready to use.
12. Haupt's adhesive.
13. (Optional) Aspirator made from vacuum tubing attached to a Büchner flask (Fig. 3, *see* **Note 1**).
14. Glass coverslips.
15. Ethanol series of 10%, 30%, 50%, 70%, and 95% ethanol in distilled water.
16. Histoclear or similar clearing agent.
17. Citifluor AF-1 mountant or DPX mountant.
18. Optional Toluidine Blue O: 0.1% (w/v) pH 5.5 in 0.1 M sodium phosphate buffer.

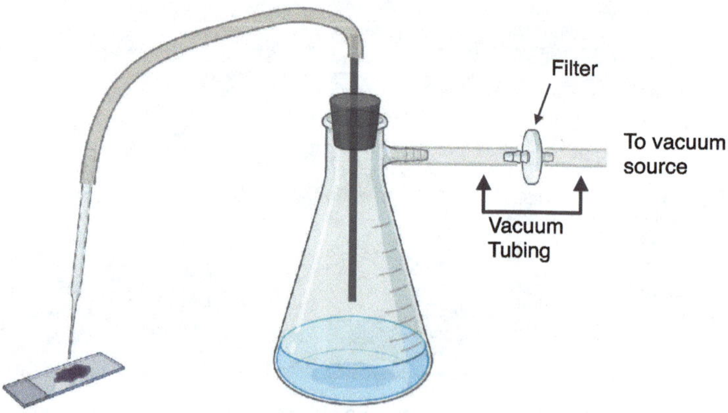

Fig. 3 Schematic of an aspirator made from a Büchner flask, tubing, and a glass pipette. Created with BioRender

19. Optional Sodium phosphate buffer, pH 5.5: For a 500 mL sodium phosphate buffer, put 400 mL distilled water to a beaker with a stir bar. While stirring, add 5.74 g sodium dihydrogen phosphate followed by 0.3 g dibasic sodium phosphate. Add water to 500 mL and stir until dissolved. Check and adjust pH as needed. The solution is stable for up to a year.
20. Optional Calcufluor White: mixed in water 1 mg/mL (w/v).

2.2 Materials for Immunostaining Fresh Tissue, Fixed Tissue, or Tissue Embedded in PEG

1. Blocking Buffer: 20% (w/v) Nonfat milk powder mixed in PBS. Mix on a stir plate, avoiding the production of bubbles from high agitation.

2.3 Materials for Immunostaining Plastic-embedded Material (in Addition to the Above)

1. Bovine Serum Albumin (BSA): Mix 1% in 20% nonfat milk.
2. Refrigerator.
3. Methacrylate resin compatible with immunolocalization. We use Technovit 8100, which includes the embedding solution and the hardener. It also contains plastic covers to create an anoxic environment during hardening.
4. Hard plastic embedding molds.

3 Methods

3.1 Fresh or Fixed Unembedded Tissue

All sample preparation and staining methods can be carried out at room temperature. All sections should be paired with sections from the same material that omit the secondary antibody as negative controls and for subtraction of background autofluorescence if necessary.

1. Using the PAP pen, draw a square or circular well onto the slides (*see* **Notes 2 and 3**). If using Haupt's adhesive, place a thin layer on the slides inside the PAP pen well.
2. Prepare sections using a sliding, vibrating, or rotary microtome. Sections can be placed directly on the prepared slides in a few drops of PBS (if not using Haupt's adhesive) or directly into a small drop of Haupt's adhesive. The PAP pen wells will hold the liquid on the slides. If using Haupt's adhesive, the slides should be dried on a benchtop or slide warmer for 15 min, to adhere the sections to the slide.
3. Aspirate or remove any PBS from the slides with a pipette, then replace with ~200 μL blocking buffer. The sections should be covered by the buffer. Allow the sections to block in the blocking buffer for 30 min (*see* **Notes 4 and 5**).

4. Aspirate or remove the blocking buffer with a pipette. Wash the tissue by replacing the blocking buffer with PBS three times, allowing the PBS to sit on the slides for 5 min each wash.
5. Remove the final PBS wash and replace it with the primary antibody in blocking buffer. One slide should retain the blocking buffer as a negative control.
6. Incubate all slides on the bench top for 90 min (*see* **Note 6**). This can be done without covering the slides on the bench top or in a makeshift humidity chamber made by placing a wet paper towel in a reusable plastic food container.
7. Remove the primary antibody/blocking buffer and wash the slide three times for 5 min per wash as in Step 4.
8. Replace the final PBS wash with the secondary antibody/blocking buffer. Incubate the slides for 90 min (*see* **Note 7**).
9. *Counterstaining*: Calcofluor White—stain 5 min in the dark, then wash with PBS; Toluidine Blue O; stain for 30–60 s, then wash with PBS.
10. If mounting in Citifluor, mount and coverslip the slides.
11. If mounting in DPX or similar, the sections should be adhered to the slide with dental floss, then moved through the ethanol series of 30%, 50%, 70%, and 95% ethanol and then cleared again with 1 min per ethanol dilution and 30 s in the clearing agent. Slides can then be cover-slipped with DPX mountant after careful removal of the dental floss.

3.2 Plastic-Embedded Tissue

All methods can be carried out at room temperature. All sections should be paired with sections from the same material that omit the secondary antibody as negative controls for subtraction of background autofluorescence if necessary.

1. Mount semi-thin sections (2–4 μm) on slides.
2. Using the PAP or TIPP-EX pen, draw a square or circular well onto the slides (*see* **Note 2**). Then place the slides in a humidified chamber (with a wet paper towel at the bottom) and darkened during incubation with the primary antibody.
3. Wash sections with 1× PBS, then incubate them with 5% bovine serum albumin blocking buffer diluted in 1× PBS. We use approximately 10 μL per section. The sections should be covered by the buffer. Allow the sections to block in the blocking buffer for 5 min (*see* **Notes 3 and 4**).
4. Aspirate or remove the blocking buffer with a pipette. Wash the tissue by replacing the blocking buffer with 1× PBS three times, allowing the PBS to sit on the slides for 5 min each time.
5. Remove the final PBS wash and replace it with the primary antibody in blocking buffer, leaving a negative control with blocking buffer rather than the primary antibody.

6. Incubate the slides on the bench top for 60 min (*see* **Note 6**). This can be done without covering the slides or in a makeshift humidity chamber made by placing a wet paper towel in a reusable plastic food container.
7. Aspirate or remove the blocking buffer with a pipette. Wash the tissue by replacing the blocking buffer with 1× PBS three times, allowing the PBS to sit on the slides for 5 min each time.
8. Remove the final PBS wash and replace it with the secondary antibody in blocking buffer, leaving a negative control with blocking buffer rather than the primary antibody. Incubate the slides in the dark for 90 min (*see* **Note 7**).
9. *Counterstaining*: Calcofluor White—stain 2 min in the dark, then wash with PBS.
10. If mounting in Citifluor, mount the slides and cover them with a coverslip.
11. Observe under the fluorescent microscope at the wavelength prescribed by your chosen secondary antibody.

4 Notes

1. A second Büchner flask can be inserted in place of the filter as a vacuum trap to prevent liquids from being drawn into the vacuum source.
2. The fumes from PAP pens can cause irritation, so this step should be done in a well-ventilated area, such as a hood.
3. Tipp-Ex pens can serve as a lower-cost alternative to PAP pens, but care should be taken to avoid contaminating the material.
4. If nonspecific staining is an issue, this blocking step should be empirically increased to a duration that minimizes it. The tradeoff is that longer blocking could prevent primary antibody labeling of the sieve elements.
5. The aspirator rarely removes sections from the slide; tilt the slide to pool liquid within the PAP well, away from any tissue, to first remove the excess. The remaining liquid can then be aspirated from the slide with less risk of pulling up a section.
6. We typically incubate in primary antibody for 90 min, but for tissues where fluorescence appears faint, we have had success with incubations up to 180 min.
7. This can be done on the bench top, but yields the best results if done in the dark, such as by covering the makeshift humidity chamber with foil.

References

1. Münch E (1930) Die Stoffbewegungen in der Pflanze
2. Jensen KH, Berg-Sørensen K, Bruus H et al (2016) Sap flow and sugar transport in plants. Rev Mod Phys 88:035007–63
3. Savage JA, Beecher SD, Clerx L et al (2017) Maintenance of carbohydrate transport in tall trees. Nat Plants 3:965–972
4. Nobel PS (2009) Physiochemical and environmental plant physiology. Elsevier Academic Press
5. Jensen KH, Rio E, Hansen R et al (2009) Osmotically driven pipe flows and their relation to sugar transport in plants. J Fluid Mech 636:371–396
6. Esau K (1939) Development and structure of the phloem tissue. Bot Rev 5:373–432
7. Esau K (1950) Development and structure of the phloem tissue II. Bot Rev 16:67–114
8. Montwé D, Hacke U, Schreiber SG et al (2019) Seasonal vascular tissue formation in four boreal tree species with a focus on callose deposition in the phloem. Front For Glob Change 2:58
9. Gricar J, Prislan P (2022) Seasonal changes in the width and structure of non-collapsed phloem affect the assessment of its potential conducting efficiency. IAWA J 43:1–15
10. Gričar J, Jevšenak J, Giagli K et al (2024) Temporal and spatial variability of phloem structure in Picea abies and Fagus sylvatica and its link to climate. Plant Cell Environ 47:1285–1299
11. Esau K (1948) Phloem structure in the grapevine, and its seasonal changes. Hilgardia 18:217–296
12. Evert RF (1962) Some aspects of phloem development in Tilia americana. Am J Bot 49:659
13. Jyske T and Hölttä T (2014) Comparison of phloem and xylem hydraulic architecture in Picea abies stems. New Phytol 205:102–115
14. Behnke H-D, Kiritsis U (1983) Ultrastructure and differentiation of sieve elements in primitive angiosperms. I. Winteraceae. Protoplasma 118:148–156
15. Evert RF, Murmanis L (1965) Ultrastructure of the secondary phloem of Tilia Americana. Am J Bot 52:95–106
16. Froelich DR, Mullendore DL, Jensen KH et al (2011) Phloem ultrastructure and pressure flow: sieve-element-occlusion-related agglomerations do not affect translocation. Plant Cell 23:4428–4445
17. Prislan P, Mrak P, Žnidaršič N et al (2019) Intra-annual dynamics of phloem formation and ultrastructural changes in sieve tubes in *Fagus sylvatica*. Tree Physiol 39:262–274
18. Silva SC de M, Machado SR, and Rodrigues TM (2025) Sieve element traits differ in pulvini and petioles of Mimosa (Leguminosae, Caesalpinioideae) with different types of leaf movement. Botany 103:1–8
19. Cronshaw J, Esau K (1967) Tubular and fibriallar components of mature and differentiating sieve elements. J Cell Biol 34:801–815
20. Sjolund RD, Shih CY (1983) Freeze-fracture analysis of phloem structure in plant tissue cultures: II. The sieve element plasma membrane. J Ultrastruct Res 82:189–197
21. Jyske TM, Suuronen J-P, Pranovich AV et al (2015) Seasonal variation in formation, structure, and chemical properties of phloem in Picea abies as studied by novel microtechniques. Planta 242:613–629
22. Mullendore DL, Windt CW, Van As H et al (2010) Sieve tube geometry in relation to phloem flow. Plant Cell 22:579–593
23. Truernit E (2014) Phloem imaging. J Exp Bot 65:1681–1688
24. Torode TA, O'Neill R, Marcus SE et al (2018) Branched pectic galactan in phloem-sieve-element cell walls: implications for cell mechanics. Plant Physiol 176:1547–1558
25. Ray DM, Savage JA (2020) Immunodetection of cell wall pectin galactan opens up new avenues for phloem research. Plant Physiol 183:1435–1437s
26. Miguel B-A, Hormaza JI, Losada JM (2021) Conductivity of the phloem in *Mangifera indica* L. Hortic Res 8:150
27. Willats WGT, Steele-King CG, McCartney L et al. (2000) Making and using antibody probes to study plant cell walls. Plant Physiol Biochem 38:27–36
28. Xue J, Bosch M, Knox JP (2013) Heterogeneity and glycan masking of cell wall microstructures in the stems of *Miscanthus* x *giganteus*, and its parents *M. sinensis* and *M. sacchariflorus*. PLoS ONE 8:e82114
29. Knox JP (2008) Revealing the structural and functional diversity of plant cell walls. Curr Opin Plant Biol 11:308–313
30. Hervé C, Marcus SE, and Knox JP (2011) Monoclonal antibodies, carbohydrate-binding modules, and the detection of polysaccharides in plant cell walls. In: Popper ZA (ed) The plant cell wall: methods and protocols. Humana Press, Totowa, NJ, pp 103–113

Chapter 4

Quantification of Callose Deposition in the Phloem of Woody Stems Using Supervised Machine Learning-Driven Automated Image Analysis with the *IlastiKlean* R Package

Jacobo Robledo, Stacy Welker and Amit Levy

Abstract

Callose deposition in the phloem is an innate part of plant development and a response to biotic and abiotic stress, aiding in stress mitigation but potentially also compromising phloem functionality. Measuring callose using aniline blue staining is widely employed, but accurate quantification is hindered by image qualities such as texture and fluorescent artifacts. Here, we describe a method to quantify callose levels in the phloem of woody plants using aniline blue staining, confocal microscopy, and automated supervised machine learning-driven image analysis supported by the IlastiKlean R package. Bark peel samples from woody plants are collected from shoots, stained, and imaged to assess callose deposition. The microscopy images are preprocessed and analyzed using Fiji, Ilastik, and the IlastiKlean R package, which allows accurate quantification of the number, size, and distribution of callose deposits. This quantitative measure can be used to study, screen, and engineer plants that are better adapted to biotic or abiotic stresses, and it serves as an important tool for basic and foundational studies of callose deposition in the phloem.

Keywords Phloem, Sieve elements, Callose deposition, Confocal microscopy, Fiji, ImageJ, Ilastik, Automated image analysis, Machine learning, R

1 Introduction

Callose, a β1,3 glucan polymer, plays a critical role in maintaining structural integrity and mediating immune responses in plant phloem, particularly within sieve elements [1–3]. During periods of biotic and abiotic stress, callose deposition is dynamically regulated, affecting phloem conductivity and overall plant functionality [4, 5]. Traditional methods for quantifying callose deposition, including staining callose with aniline blue and manual counting, as well as binary thresholding based on pixel intensity, are often affected by noise, artifacts, and subjective user calibration [6]. These limitations

These authors contributed equally: Jacobo Robledo, Stacy Welker

Johannes Liesche and Sanna Sevanto (eds.), *Phloem*, Methods in Molecular Biology, vol. 3066,
https://doi.org/10.1007/978-1-0716-5488-0_4,

underscore the need for a more robust, reproducible, and sensitive approach to quantitatively assess callose levels in plant tissues.

In this protocol, we integrate aniline blue staining with high-resolution confocal microscopy and automated image analysis to precisely quantify callose deposits. The methodology employs image preprocessing with Fiji [7], followed by supervised machine learning-based segmentation using Ilastik [6, 8] to differentiate callose from background noise based on features beyond simple brightness, such as texture and scale. Subsequent data cleaning and analysis from Ilastik-generated spreadsheets is conducted using the IlastiKlean package in R, enabling accurate measurement of deposit number, size, and distribution. This technique provides a robust platform for high-throughput analysis of phloem state and plant stress responses.

2 Materials

2.1 Tissue Collection and Preparation

1. Plant material: Woody plants with some mature flush.
2. Scalpel or knife: For sample collection.
3. Collection tubes: 1.5 mL Eppendorf tubes.
4. Ethanol: 85% Ice-cold ethanol for immediate tissue preservation.
5. Tween-20: 0.01%, for rehydration.

2.2 Staining Reagents

Aniline Blue Staining Solution [9]: Dissolve aniline blue to make a 0.1% (w/v) solution in double-distilled water. Prepare a 1 M glycine solution and adjust its pH to 9.5 (using, e.g., KOH). Mix the two solutions in a 2:3 volume ratio (i.e., two parts of the 0.1% aniline blue solution to three parts of the 1 M glycine solution) (*see* **Note 1**).

2.3 Microscopy, Image processing, and Image analysis

1. Confocal microscope: equipped with 10× objective and 405 nm laser for excitation. Aniline Blue fluorescence can be detected using a 475–525 nm band-pass filter.
2. Fiji: Download at https://fiji.sc/. This video tutorial will guide you through the installation process if additional help is needed: https://www.youtube.com/watch?v=K5JRRzpte-8.
3. Ilastik: Download at https://www.ilastik.org/download. Also see the video tutorial here: https://www.youtube.com/watch?v=fgewERekarQ.
4. R: Download at https://cran.r-project.org/bin/windows/base/. RStudio is a free integrated development environment (IDE) for R, which provides a more user-friendly interface with many convenient features. It can be downloaded at https://posit.co/download/rstudio-desktop/.
5. Ilastik plugin for Fiji: Follow instructions here: https://www.ilastik.org/documentation/fiji_export/plugin. A video

tutorial for installing the plugin can be viewed here: https://youtu.be/fgewERekarQ?si=eLD4FeyE-y8kRhwt&t=52.

3 Methods

3.1 Plant Material Collection

1. Use a scalpel or knife to peel sections (approximately 0.5–1 cm wide by 2–3 cm long) of bark from each shoot (*see* **Note 2**). Figure 1 illustrates the appropriate sample quality and collection technique.
2. Immediately place the bark peels into 1.5 mL tubes containing 85% ice-cold ethanol for tissue fixation and destaining (*see* **Note 3**).

Fig. 1 Sequence of steps to choose, collect, and evaluate the quality of bark peels for use in callose quantification. (**A**) Sample collection illustrated with a greenhouse-grown "Hamlin" orange seedling [*Citrus sinensis* (L.) Osbeck]. The ideal sample should come from mature flush that has begun to harden but not lignify. Collect tissue from the main stem. (**B**) Using a sharp knife or scalpel, cut a section of bark away from the stem, keeping the tool at a shallow angle. Use the thumb to hold the sample against the blade, then peel downward to achieve the desired length. **C** Separate the peel from the remainder of the stem and cut off at the bottom of the sample with the blade. (**D–F**) Evaluation of the sample to ensure that the phloem layer has been obtained. (**D**) Cut was too deep; the pith layer has been collected. Can be salvaged by dissecting off the pith. (**E**) Ideal sample. Arrows point to the phloem layer, which can be stained to quantify callose. (**F**) Cut was too shallow. No phloem callose will be visible after staining

3.2 Tissue Preparation and Staining

1. Incubate the ethanol-fixed samples overnight at room temperature to ensure complete destaining (*see* **Note 4**).
2. Rehydrate the samples in 0.01% Tween-20 solution for 1 h.
3. Immerse the rehydrated samples in the aniline blue staining solution for 1 h.
4. Wash the samples briefly with deionized water to remove excess stain, then maintain them in 0.01% Tween-20 until imaging.

3.3 Confocal Imaging

1. Mount the stained samples on microscope slides. A drop of Tween-20 solution can help keep the sample in place.
2. Set up the confocal microscope with a 10× objective, pinhole diameter of 650 and 850 μm (10–12 airy units), 405 nm laser excitation at about 25% power, and a 475–525 nm band-pass filter (*see* **Note 5**).
3. Capture three to five images per bark peel, with three peels collected per plant, resulting in a total of up to 15 images per plant.
4. Convert all images to TIF format for analysis.
5. Critically, name each image file so that the name contains all relevant experimental details (such as plant genotype, treatment, tree ID, peel number, and image number) using a consistent structure with underscores as separators. This ensures proper organization and facilitates downstream statistical analysis. This format is required to perform the downstream analysis pipeline supported by the IlastiKlean R package (*see* **Note 6**).

- *Example file name:*
  GENE1_cultivar1_treatment1_Tree3_Peel2_Image4.tif

3.4 Image Pre-Processing

During automated image analysis, it is critical to reduce the visibility of artificial and natural artifacts so that they will not be counted as callose formations in the tissue. This is done by applying a filter, which also reduces the complexity of the appearance of the callose formations, simplifying the workload for the automated classifier. Fiji, a free and open-source image analysis program [10], is used in this chapter for this task, following the procedure outlined below:

1. Open Fiji. Navigate to the Process menu, scroll down to Batch, and choose Macro from the side menu.
2. In the Input menu, select the folder where the original images are stored.
3. In the Output menu, select a folder where the filtered images will be stored. The original images should be kept unaltered so that the effectiveness of multiple filters can be tested.

4. In the text box, enter the text for a simple macro that applies the filter type and radius. A larger radius will reduce the image's features more, while a small radius will preserve fine details (*see* **Note 7**). A median filter may also be suitable for some datasets. The macro below applies a 3-pixel mean filter to every pixel in the image, replacing each pixel with the average of the 9 pixels within a 3-pixel radius.

   run("Mean...", "radius=3");

3.5 Automated Image Analysis: Supervised Machine Learning (Ilastik)

Callose formations in large batches of confocal images can be accurately quantified by segmenting them from artifacts and background material using supervised machine learning with Ilastik. The images must be segmented in two steps: pixel classification, followed by object classification. For repeat experiments, the same classifier can be used if the tissue type, stain protocol, and confocal settings are the same. If any of these parameters are different, a new classifier should be trained. Even when working with similar samples, the user should always perform the quality control process described in Steps 11–15 to check the classifier results before using them for analysis.

Pixel Classification

1. Open Ilastik. The project creation interface will appear first. In the menu under "Create New Project," choose the Pixel Classification workflow. When prompted, choose a folder and file name for the pixel classification project. When the new project opens, the list of items in the workflow pane on the left side of the screen will direct the user through a series of steps to train the classifier.
2. First, select the training images. There should be at least three images, representing the entire range of variation in callose and artifact levels (high, medium, and low) that are present in the dataset. Depending on the dataset's size and level of variation, it may be beneficial to include more images in the training set. Click on the box with a blue plus sign in the Raw Data menu and select "Add separate Image(s)..." Use the file explorer to navigate to the desired image and repeat this process until the entire training set has been added. If any of the training images are ever moved to another directory, Ilastik will not be able to find them, so training images should not be moved once they are chosen.
3. Next, open the feature selection menu and click "Select Features..." If the user is unfamiliar with the available features, it is beneficial to select all of them first and allow Ilastik to suggest a reduced feature set later. Click and drag the selection boxes until all are highlighted green with a check mark. Unselected features will be blank.
4. Click on the Training menu (see Fig. 2). In the white box on the left, the user designates the color and name of the labels

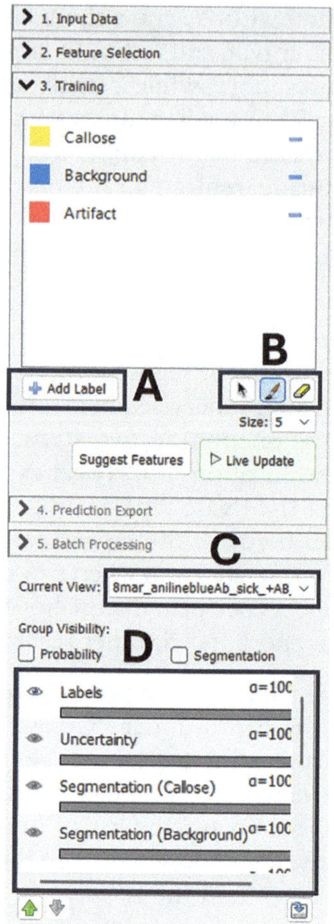

Fig. 2 Training menu. (**A**) Button to add more labels to differentiate types of objects the user wishes to classify. (**B**) Buttons to switch cursor modes. The arrow can be used to navigate around the image, the paintbrush will apply the selected label, and the eraser will remove a label once it has been applied. (**C**) A dropdown menu to allow the user to change which image is active to be labeled in the viewing pane. (**D**) A box which displays layers that can be displayed over the image to reveal more information about user-applied labels, uncertainty, and to check the classification results as new labels are applied

that they will apply to the images to train the classifier. For confocal images of callose formations in plant phloem, click the "Add Label" button three times. Double-click the labels and change them to Callose, Background, and Artifact (*see* **Note 8**). The artifact label is for epidermal or lignin-rich tissue, as shown in the images, and must be differentiated from callose. The box next to the name can also be double-clicked to change the label color, allowing the user to account for differences in color vision and the dye's color.

5. Begin labeling the objects to train the classifier. Select the paintbrush icon. Click the desired label, then click and drag it over the corresponding object in the image. Choose at least

three examples each for callose, artifact, and background. The size of the labeling brush can be adjusted in the box below the paintbrush (*see* **Note 9**). The eraser icon next to the paintbrush can be used to remove erroneous labels. To zoom in or out, press the control key and scroll up or down with the mouse scroll wheel. To navigate to a different area of the image while zoomed in, press the arrow icon, click on the image, and use the arrow keys in the desired direction.

6. When finished labeling the first image, navigate to the next training image by clicking the box next to "Current View." When all training images have received a first round of labels for callose, artifact, and background, the user can evaluate the quality of the classifier by clicking the "Live Update" button (*see* **Note 10**). The classifier uses the user-provided features from earlier in the process and user-labeled pixels to determine the category to which each unlabeled pixel in the image belongs. The classifier also determines this for the other images in the training set as the user chooses them from the "Current View" box.
7. To evaluate the accuracy of the classifier, check the visibility boxes. "Segmentation" will display the final segmentation of the image, where the classifier has placed every pixel in either the "Callose," "Background," or "Artifact" category. This is what will be exported from Ilastik as a usable image file. Below the visibility boxes, there is a list of layers that can be turned on or off with the eye icon, and the transparency can be adjusted with the slider to the right of the eye. The layers include user labels, uncertainty, the probabilities that the classifier uses to segment the image, and the segmentation itself.
8. To refine the classifier, uncheck all layers except for the uncertainty. If the classifier lacked sufficient information to segment pixels into one of the categories, the pixels are brightly highlighted in teal in this layer. Often, this means that the pixels are also inaccurately segmented. Look for highlighted pixels in the uncertainty layer, apply the appropriate label and then recheck the segmentation layer. The uncertainty should decrease, and the segmentation should improve. It is not necessary to remove all uncertainty, just enough to segment the image accurately. Repeat this label-and-check process for all the pictures in the training set until the classifier has accurately segmented callose formations, background, and artifact in every image. Figure 3 illustrates the labeling and checking process.
9. Once the classifier's results are acceptably accurate across all training images, click the "Prediction Export" entry in the

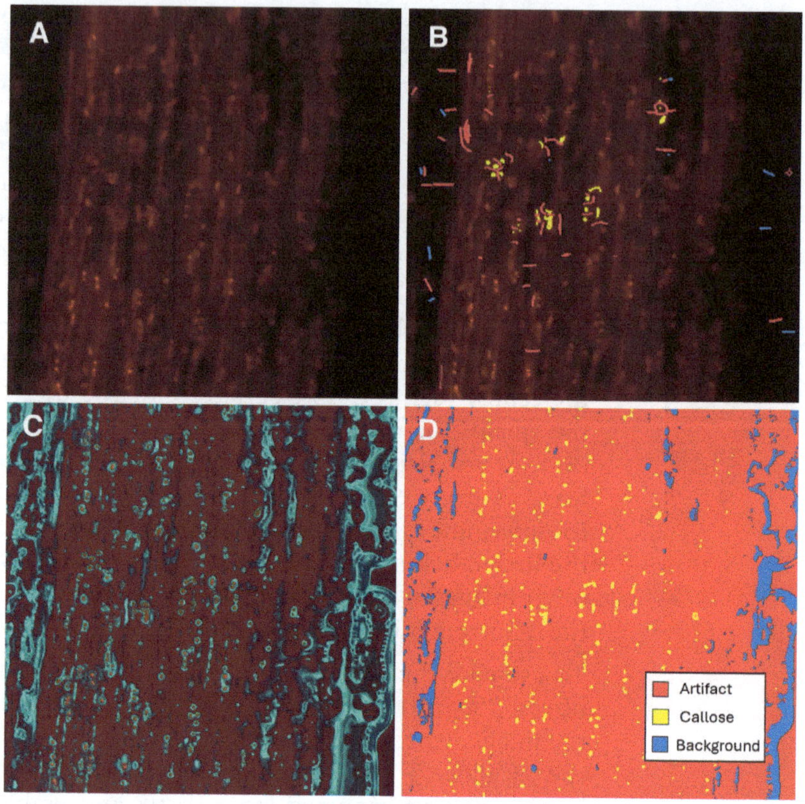

Fig. 3 A sample image of a callose bark peel viewed in each step of the Ilastik classifier training process. (**A**) Image with mean filter applied in Fiji. (**B**) Image with appropriate training labels applied in Ilastik. (**C**) Uncertainty layer made visible to highlight areas in the image where more labels should be considered. (**D**) Final segmented image with label key

workflow pane. In the menu next to "Source," choose Simple Segmentation. The defaults can usually be left in the "Choose Export Image Settings" box. Clicking the "Export All" button will export the Simple Segmentation results for only the training images to the same directory where the training images are located.

10. The "Batch Processing" entry in the workflow pane can be used to run the entire test set through the classifier to obtain segmentation results. Click on the "Batch Processing" button, then click the "Select Raw Data Files" button in the upper left of the window that opens. Select the files from the dataset that were not used for training and click the "Process all Files" button in the workflow pane. It is best to choose a smaller sample of the test set, particularly if it is very large, to perform a quality check before running the classifier on every file.

11. To perform a quality check, run approximately six images from the test data set through the classifier batch process.

Place a copy of the resulting simple segmentation .h5 files in a new, separate folder.

12. Open Fiji and ensure the Ilasik plugin is installed.
13. A macro provided by the Ilastik team will allow the user to quickly open all the .h5 files in the subset and normalize them so that they are human-readable. Navigate to the Plugins menu, scroll down to "New" and select "Macro" from the sidebar. The text below will create a macro that will perform the opening and normalization process for all .h5 files in the directory. Copy and paste this into the macro:

```
#@ File (label = "Input directory", style = "directory") input_dir

processFolder(input_dir);

function processFolder(input_dir) {
        suffix = ".h5";
        list = getFileList(input_dir);
        list = Array.sort(list);
        for (i = 0; i < list.length; i++) {
                if(endsWith(list[i], suffix))
                        processFile(input_dir, list[i]);
                        run("Brightness/Contrast...");
                }
}

function processFile(input, file) {
        inputFilePath = input + File.separator + file;
        print("Processing: " + inputFilePath);
        run("Import HDF5", "select=[" + inputFilePath + "]
        datasetname=/exported_data axisorder=yxc");
}
```

14. In the top menu bar of the macro window, click Run and select the "Run" menu option. The user will be prompted to choose a directory. Navigate to the directory where the subset of .h5 files was placed, and click "Select." The user can save this macro for later use and run it by navigating to the Plugins menu, selecting Macros, and selecting Run from the sidebar. See Fig. 4 for an illustration of the menu navigation in Fiji and an example of an .h5 file that has been adjusted to be human-readable.
15. Carefully examine the segmentation results that have opened in the new windows and compare them to the original images. If the classifier appears to have correctly segmented all the callose formations in the images away from the background and

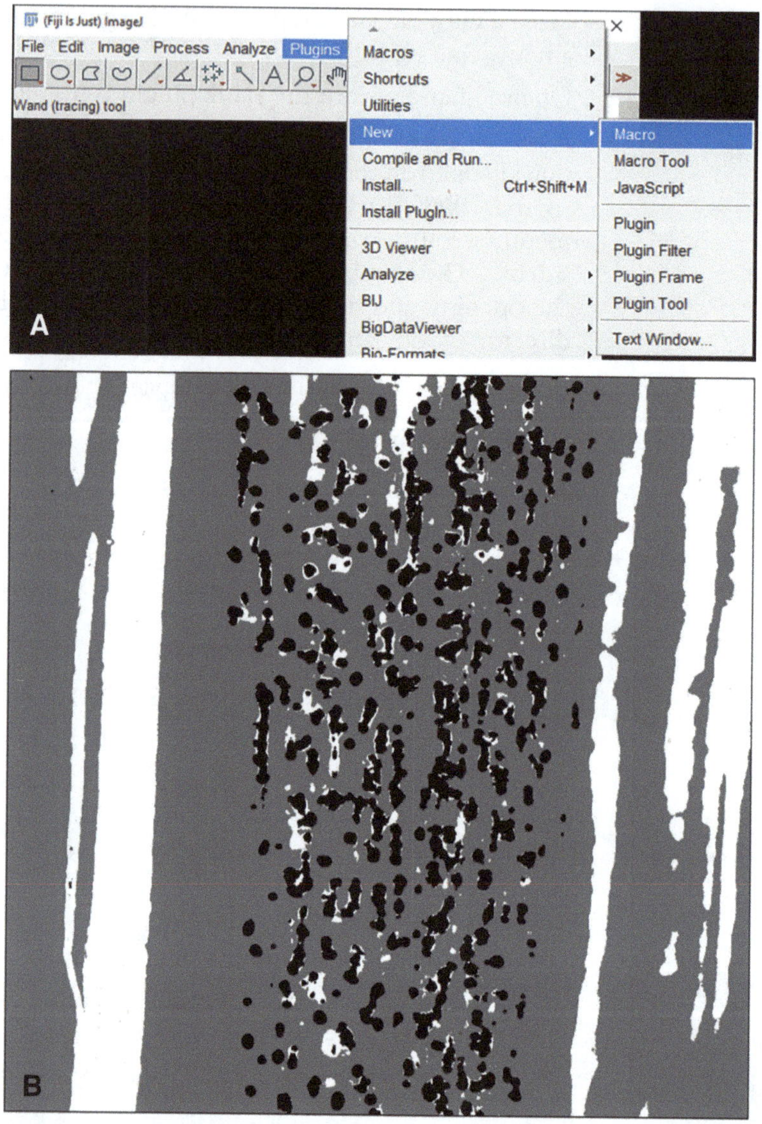

Fig. 4 Using the Ilastik plugin for Fiji to visualize Ilastik output .h5 files. (**A**) Menu navigation to locate the function to create a macro in Fiji. (**B**) An example of segmentation results from Ilastik, which have been normalized with the macro in Fiji. The black objects are callose formations

artifact materials, finish processing all the remaining images in the experimental set and proceed to the object classification workflow in the next section (*see* **Note 11**).

Object Classification

At this point, all the pixels in all the training and test images have been segmented into separate objects. With this step in the workflow, the objects can be classified into either the Callose, Background, or Artifact categories.

1. Open Ilastik and in the main menu, navigate down to the Object Classification workflow options. Click on the option that says Object Classification [Inputs: Raw Data, Segmentation].
2. In the file browser that pops up, navigate to the folder where the training images are stored and save the new object classification file there.
3. Open the Input Data section of the workflow pane on the left. For this workflow, add the training images to the Raw Data tab, and in the Segmentation Image tab, add the simple segmentation .h5 file that was derived from the corresponding training image. The order in which the images and segmentation files are added must match so that Ilastik can associate the correct segmentation with each image.
4. Click on the Object Feature Selection section of the workflow pane. As with pixel classification, it is better to start with the broadest set of features and narrow it down later. Click on the "Select Features" button and click on the "All excl. Location" button in the lower-left corner. The specific location inside the image will likely not be relevant for identifying callose formations.
5. Click on the Object Classification section of the workflow pane. This step will closely resemble the labeling portion of the pixel classification process, except that entire object regions are labeled instead of individual pixels. Designate labels and label the image using the paintbrush, eraser, and arrow tools as described in the previous section. The layer box also works as described above. Use the same label-and-check process using the segmentation and uncertainty layers until the object classifier can accurately segment callose formations, artifacts, and background.
6. Click on the Object Information Export tab in the workflow pane. Here, the user will designate that a .csv file will be exported that contains information about all the objects in every image in the dataset. In the menu next to "Source": choose Object Predictions. The default export image settings in this window are sufficient for most novice users. Click on the "Configure Feature Table Export" button. Ensure that the menu item next to Format is set to CSV. For some purposes, the user may wish to open the Features pane and check some features to be included in the .csv file, such as Mean Intensity. The. csv file will always contain the size of the objects in pixels and the *x*, *y* coordinates for the center of every object. Clicking the "Export All" button will create .h5 and .csv filescontaining information about every object in all the images in the training set.
7. Lastly, click on the "Batch Processing" portion of the workflow. Similar to the pixel classification step, it is prudent to

first run a smaller sample of images from the test set through the classifier. The user can check the quality of the classifier by viewing the .h5 files with the Ilastik plugin in Fiji, once again using the Adjust Contrast feature to normalize the contrast in the imported .h5 file. If the callose formations, background, and artifacts in the sample appear to have been correctly segmented, the rest of the test set can be run through the classifier.

3.6 Statistical Analysis of Ilastik Output Data

The user will now have a secondary dataset consisting of many .csv files, each representing an inventory of all the objects segmented from one image in the dataset. Depending on the size of the primary dataset, this could be hundreds of files with thousands of objects. A programming language such as R is recommended as a systematic and reproducible way to extract meaningful values from this large quantity of information. This step of the workflow will cover the R code needed to quantify the number and size of callose deposits in the dataset, visualize the dataset using graphs, and perform basic, common statistical tests to identify significant differences between treatment groups, if present. The authors have created the IlastiKlean package to assist users with these tasks for the Ilastik outputs.

To ensure complete understanding and success with the script, a group of mock datasets that mimic the output from Ilastik will first be generated (this process will be skipped in the user's pipeline if they have already obtained their own experimental data from previous steps). Created mock datasets will then be exported as .csv files to the user's working directory and reimported as dataframes. The cleaning, visualization, and statistical analysis will be performed on the mock data (for users who successfully obtained their results from Sects 3.4 and 3.6, the cleaning, visualization, and statistical analysis will be performed using that data). For convenience, an explanatory vignette for the IlastiKlean package can be loaded with this code in RStudio (*see* **Note 12**).

Data Preparation and Cleaning

1. First, ensure that the most up-to-date versions of R and RStudio are installed.
2. Open RStudio. A new file should automatically be opened on startup.
3. Specify the working directory, the drive where the .csv files are located and where all outputs from the R script will be saved. The following provides an example drive. The user must replace everything in the quotation marks with their drive location. Note that the backslashes (the default in Windows file paths) must be replaced with forward slashes for use in R.setwd("C:/Users/username/university_working_drive/project_folder/filtered_image_data")
4. Load the necessary packages.

```
# If you don't have devtools installed:
```

```
install.packages("devtools")
# Install the IlastiKlean package:
devtools::install_github("jrobledob/IlastiKlean")
# Load the libraries
library(IlastiKlean)
library(dplyr)
library(stringr)
library(tidyr)
```

5. If you don't have data but want to test out the IlastiKlean package, you can create a mock dataset that mimics typical outputs from Ilastik. If you already have data, this step can be skipped. The function "sim_Ilastik_output" in the package will generate data with the treatment group size and number of images specified by the user. In the wor_dir argument, type the path of the drive where you want the csv files to be created, if it is different from your current working directory:

```
sim_Ilastik_output(samples = 3, pictures = 3)
```

6. For analysis, a single dataframe with all the information about all the objects in every picture in the dataset is needed. The function "read_and_clean_csvs" in the package IlastiKlean will handle this. It reads all .csv files from the working directory, converts all the columns to characters, combines all the common columns into one big data frame, and mutates the data frame so that the file names are attached to each row. Keep in mind that this will load .csv files indiscriminately, so all .csv files that are exported with this script must be moved out of the drive so as not to interfere with the script in the future. In the parentheses, place the directory where the target Ilastik output csv files are located:

```
objects_with_file_names<- read_and_clean_csvs("C:/Users/username/university_working_drive/project_folder/filtered_image_data/")
```

7. File name formats are of paramount importance to the analysis. Each file name should be formatted so that the experimental information is stored in the same order and separated by underscores. The following code provides one way to split a file name into columns so the information can be used as variables in the analysis. Sometimes the microscope imaging software adds unnecessary information at the beginning or end of each file name, which should be removed. The first mutate line in this block will remove text from the beginning, and the second mutate line will remove it from the end. The user must replace the text in the quotation marks with the text in the file

name that they want to remove. If there is no unnecessary text in each file name, this step can be skipped.

```
objects_with_file_names<- objects_with_file_names %>%
mutate(filename = str_replace_all(filename, "dataset.ch00_", "")) %>%
mutate(filename = str_replace_all(filename, ".csv", ""))
```

8. For an experimental dataset, change the variable names and number of names as needed. In the mock dataset, the file name will be broken into three columns. The word "merge" in the too_many portion can be replaced with "debug" to help spot erroneous file names. A too_few portion can also be added to diagnose errors in the same parentheses (which contains the function parameters). For a real dataset, erroneous names can be addressed in the script with a regex expression, but it will be less complicated for the novice user to go back into the folder, fix the file name there, and reload the .csv files.

```
objects_with_file_names<- objects_with_file_names %>%
separate_wider_delim(filename, delim = "_", names = c("plant", "sample", "pic_number"), too_many = "merge")
```

9. Take a dataframe containing object-level classifications and perform key analysis steps using the summarize_callose_data() function of the IlastiKlean package. This function will count the total number of images per group, filter for the target class, calculate the number and average size of detected callose objects per image, and merge these results with a complete reference of all images to ensure images with zero callose are also included. Missing values are filled with zeros to maintain accurate group-level summaries. The complete summary data frame can optionally be saved as a CSV file for further analysis.

```
summary_df <- summarize_callose_data(df = objects_with_file_names, target_class = "Callose", group_vars = c("plant", "sample", "pic_number"), output_csv = "summary_output.csv")
```

Data Visualization and Suggested Statistical Analysis

Run the visualize_and_analyze_callose() function of the IlastiKlean package. This function is a tool for visualizing and statistically analyzing callose count data, typically derived from image-based quantification workflows. Internally, the function begins by generating a clean, customizable boxplot of callose counts, grouped by plant (or any specified grouping variable), to provide a clear visual comparison of their distributions. It optionally saves this plot as a PNG file. It then displays a histogram of the callose count data to give a quick visual impression of the distribution. Next, the function performs a Shapiro–Wilk test to assess whether the callose counts follow a normal distribution. Based on this result, it automatically selects and runs the appropriate statistical test: a Welch two-sample *t*-test if normality is assumed, or a nonparametric Wilcoxon rank-sum test if the data deviate from normality. When there are more than two groups, the function additionally tests for residual normality and variance homogeneity (homoscedasticity); if both assumptions hold, it runs a one-way ANOVA followed by Tukey's HSD post-hoc comparisons, otherwise it applies a Kruskal–Wallis test followed by Dunn's post-hoc comparisons. These are the simplest and most common analyses for this type of data, but more experienced users can carry out further analysis as needed. This straightforward approach is typically sufficient for most users (*see* **Note 12**).

analysis_result <- visualize_and_analyze_callose(df = summary_df, group_col = "plant", count_col = "callose_count", boxplot_filename = "count_box_plot.png")

4 Notes

1. After mixing the aniline blue and glycine solutions, allow the solution to rest for at least 24 h before use. A proper pH adjustment (to 9.5) will cause the solution to shift from blue to a yellowish tint. If this color change does not occur, the pH is likely incorrect and should be verified.
2. When peeling, take care to avoid including xylem tissue, as it may exhibit autofluorescence that could interfere with imaging or analysis. When cutting the sample, use a very sharp blade and ensure your fingers are not in the blade's path to avoid injury.
3. The best samples for the bark peel technique will come from a mature flush. This is tissue that is no longer tender but not yet lignified, ensuring optimal conditions for sampling and reduction of artifacts in images. See Fig. 1 for an example of the proper tissue development stage.
4. To improve fixation efficacy, samples can be vacuum-infiltrated at −80 to −100 kPa for 5–10 min, which helps ethanol penetrate the tissue more effectively. Alternatively, samples may be gently swirled overnight on a rocker or orbital shaker. If samples contain excessive chlorophyll, replacing the ethanol with fresh 85% ice-cold ethanol once or twice during the incubation period may enhance destaining.
5. The greatest challenge in successfully segmenting confocal images with Ilastik is reducing excessive noise that can be erroneously interpreted as object features. Optimize laser intensity and detector gain to stay below saturation, so no signal reaches the maximum intensity. Often, the tissue from the

highest callose treatment group should be used as the control for calibration of the settings.

6. Maintaining this naming convention is essential for batch processing and minimizing confusion during data handling and interpretation.
7. If the automated classifier is struggling to differentiate callose formations, a larger radius can be applied.
8. We have consistently named our target "Callose," and we recommend that users maintain this exact spelling throughout the entire process, especially during the Ilastik labeling step, to minimize syntactic errors in the code.
9. Paintbrush size is an important choice in pixel classification. For callose formations, the user should choose a paintbrush size that covers the entire callose deposit. For the background and artifacts, which are often larger than the largest possible paintbrush size, the user can use the smallest size paintbrush and label just a few pixels at a time. Using the smallest paintbrush size will ensure that pixels are not mistakenly placed into the wrong category. It is often helpful to focus on labeling the boundaries between objects or where different objects occur close together. Paintbrush size is irrelevant in the object classification workflow as long as the manually selected pixels are accurate.
10. When the user presses the "Live Update" button, and new labels are applied to the training images, there can be a long wait while the classifier calculates the new segmentation. A smaller feature set can shorten the calculation time. Ilastik has a "Suggest Features" button located in the middle of the training panel. For novice users, the defaults can be left in the selection boxes. Press the "Suggest Features" button and wait for the calculations to complete. Figure 5 can be used as a guide to navigate the pane.
11. If the results have serious flaws that would affect the counts, the user should add the images with the worst results to the training set. Relabel and rerun the classifier, checking its results again using the macro. This process can be repeated until the segmentation is adequate. If several images must be added to the training set, this can be a sign that the filter applied to the images was not sufficient, and the features in the images are overly complicated. The user should go back to the image pre-processing stage, apply a filter with a larger radius, and re-train a new classifier.
12. Vignettes are long-form guides included with R packages that demonstrate workflows and usage examples, making them a

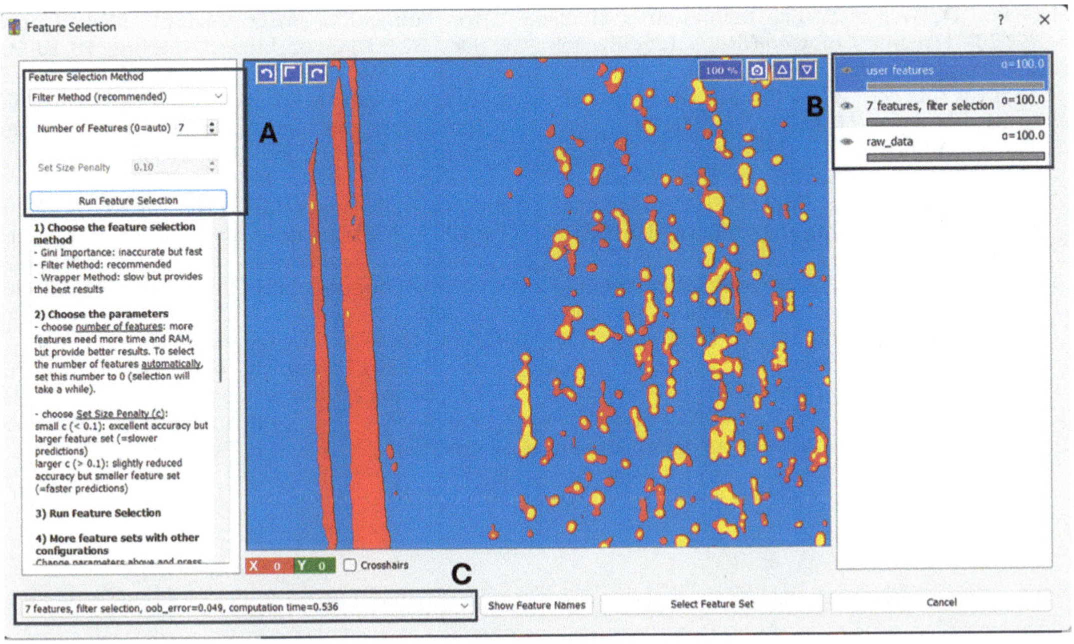

Fig. 5 Ilastik workflow to help users pare down the selected features to only those necessary. (**A**) The user can specify the number of features and the method to be used to select the features. The information in the box below A guides the user through the inputs. (**B**) Comparable layers for the feature set input by the user, sets suggested by Ilastik, and the raw training image. The user can view each layer in the display window at the center by clicking the eye. (**C**) A menu which displays each feature set, its oob ("out-of-bag" or predicted error rate), and its computation time. The user should choose the feature set that has the best classifier results in the display window, the lowest oob, and the shortest computation time, and click the "select feature set" button

key resource for learning how to apply the package effectively. To access the IlastiKlean vignette in R, run:

vignette("Cleaning-and-analysis-of-Ilastik-generated-csv-files-with-IlastiKlean", package = "IlastiKlean")

Acknowledgments

Support was provided by the National Institute of Food and Agriculture (grant 2023-70029-41304).

References

1. Hao P, Liu C, Wang Y et al. (2008) Herbivore-induced callose deposition on the sieve plates of rice: an important mechanism for host resistance. Plant Physiol 146:1810–1820. https://doi.org/10.1104/pp.107.111484
2. Wang Y, Li X, Fan B et al. (2021) Regulation and function of defense-related callose deposition in plants. IJMS 22:2393. https://doi.org/10.3390/ijms22052393
3. Xie B, Wang X, Zhu M et al. (2011) CalS7 encodes a callose synthase responsible for callose deposition in the phloem. Plant J 65:1–14. https://doi.org/10.1111/j.1365-313X.2010.04399.x

4. Achor D, Welker S, Ben-Mahmoud S et al. (2020) Dynamics of *Candidatus* Liberibacter asiaticus movement and sieve-pore plugging in citrus sink cells. Plant Physiol 182:882–891. https://doi.org/10.1104/pp.19.01391
5. Barceló-Anguiano M, Hormaza JI, Losada JM (2021) Conductivity of the phloem in mango (Mangifera indica L.). Horticult Res 8:150. https://doi.org/10.1038/s41438-021-00584-1
6. Welker S, Levy A (2022) Comparing machine learning and binary thresholding methods for quantification of callose deposits in the citrus phloem. Plants 11:624. https://doi.org/10.3390/plants11050624
7. Schindelin J, Rueden CT, Hiner MC, Eliceiri KW (2015) The ImageJ ecosystem: an open platform for biomedical image analysis. Mol Reprod Dev 82:518–529. https://doi.org/10.1002/mrd.22489
8. Berg S, Kutra D, Kroeger T et al. (2019) Ilastik: interactive machine learning for (bio)image analysis. Nat Method 16:1226–1232. https://doi.org/10.1038/s41592-019-0582-9
9. Levy A, Erlanger M, Rosenthal M, Epel BL (2007) A plasmodesmata-associated β-1,3-glucanase in Arabidopsis. Plant J 49:669–682. https://doi.org/10.1111/j.1365-313X.2006.02986.x
10. Schindelin J, Arganda-Carreras I, Frise E et al. (2012) Fiji: an open-source platform for biological-image analysis. Nat Methods 9:676–682. https://doi.org/10.1038/nmeth.2019

Check for updates

Chapter 5

Anatomical Features of Foliar Vasculature as Flux-Capacity Proxies

Jared J. Stewart, Onno Muller, Christopher M. Cohu, Stephanie K. Polutchko, Barbara Demmig-Adams and William W. Adams III

Abstract

We present a procedure for preparing leaf tissue for minor-vein analysis using light and transmission electron microscopy that facilitates quantification of phloem and xylem features that correlate with photosynthetic capacity or transpiration rates, respectively, and thus provide proxies for foliar sugar- and/or water-transport capacity. We suggest here that the numbers and individual sizes of phloem cells of foliar minor veins (sometimes normalized by vein density) and transfer-cell wall ingrowths provide quantitative proxies for the capacity of the leaf to load and export photosynthates. Among apoplastic loaders, for which sugar loading into phloem depends on cell membrane-spanning transport proteins, photosynthetic capacity correlated positively with phloem-cell number and cross-sectional area (within a species or normalized by minor-vein density among different species) and level of cell-wall ingrowths (where present in transfer cells). Among active symplastic loaders, for which sugar loading into the phloem depends on cytosolic enzymes, photosynthetic capacity was positively correlated with individual phloem-cell size and vein density. All these anatomical features and photosynthetic capacity were subject to acclimatory adjustment to environmental conditions, with increased levels of these features associated with higher rates of photosynthesis. The numbers and cross-sectional areas of tracheary elements likewise exhibited acclimation to growth conditions and a strong, positive correlation with transpiration rate, which was greatest when normalized by vein density.

Keywords Phloem, Photosynthesis, Sugar transport, Transpiration, Xylem

Abbreviations

CA	Cross-sectional area of cells per minor vein
CC	Companion cell
CN	Cell number per minor vein
EM	Electron microscopy
PC	Phloem parenchyma cell
SE	Sieve element
TE	Tracheary element
VD	Minor vein density
VP	Vascular parenchyma
XP	Xylem parenchyma

Johannes Liesche and Sanna Sevanto (eds.), *Phloem*, Methods in Molecular Biology, vol. 3066,
https://doi.org/10.1007/978-1-0716-5488-0_5,

1 Introduction

Numerous studies have demonstrated close quantitative relationships between photosynthetic capacity and overall foliar vascular structure and function (e.g., [1–11]). The investigation initially focused on water-importing xylem features and/or parameters, such as vein density (which includes both phloem and xylem tissues), and their relationship to photosynthesis. During the last decade, studies targeted several foliar phloem parameters that correlated closely with photosynthetic capacity [3–5, 10–22]. Since leaf photosynthetic capacity is regulated in response to the rate at which photosynthate is utilized by the plant [23], it is not surprising that the capacity to export photosynthate is closely coordinated with a leaf's maximal photosynthetic capacity [3, 4, 14, 15, 24]. In this chapter, we will (1) present evidence supporting the utility of foliar features as proxies for the flux capacity to export photosynthate (anatomical phloem parameters) or to import and distribute water in the leaf (anatomical xylem parameters), (2) examine the considerable phenotypic plasticity in these features in response to differing environmental conditions, and (3) provide a detailed protocol for measuring and quantifying these vascular parameters. The focus will be on the network of minor veins as the primary location for exchange of molecules between foliar vascular infrastructure and mesophyll tissue [25].

Cell number and cross-sectional area of each cell are two basic anatomical features of phloem and xylem likely to contribute to determining sugar and water flux capacity, respectively, of the leaf vasculature. Examples of the relationship between these vascular features and photosynthetic capacity are illustrated in Fig. 1 using data from three ecotypes of *Arabidopsis thaliana* that developed under different combinations of light intensity and temperature. There was a strong and highly significant correlation between photosynthetic capacity and both the number of sieve elements per minor vein (Fig. 1A) and the cross-sectional area of each of those sieve elements (Fig. 1B). These results suggest that a larger number of wider pipelines can accommodate greater levels of foliar sucrose efflux from a leaf with a higher rate of photosynthate production.

There was likewise a significant positive relationship between transpiration rate and the number of tracheary elements per minor vein (Fig. 1C) as well as the cross-sectional area of each tracheary element (Fig. 1D). Moreover, there is considerable phenotypic plasticity in vascular anatomical and functional traits supported by foliar vasculature in response to growth conditions during leaf development (Fig. 1). For example, leaf development under lower temperature resulted in higher photosynthetic capacities as well as minor veins with expanded phloem infrastructure. On the other hand, leaf development at higher temperatures resulted in higher transpirational water loss and minor veins with expanded xylem.

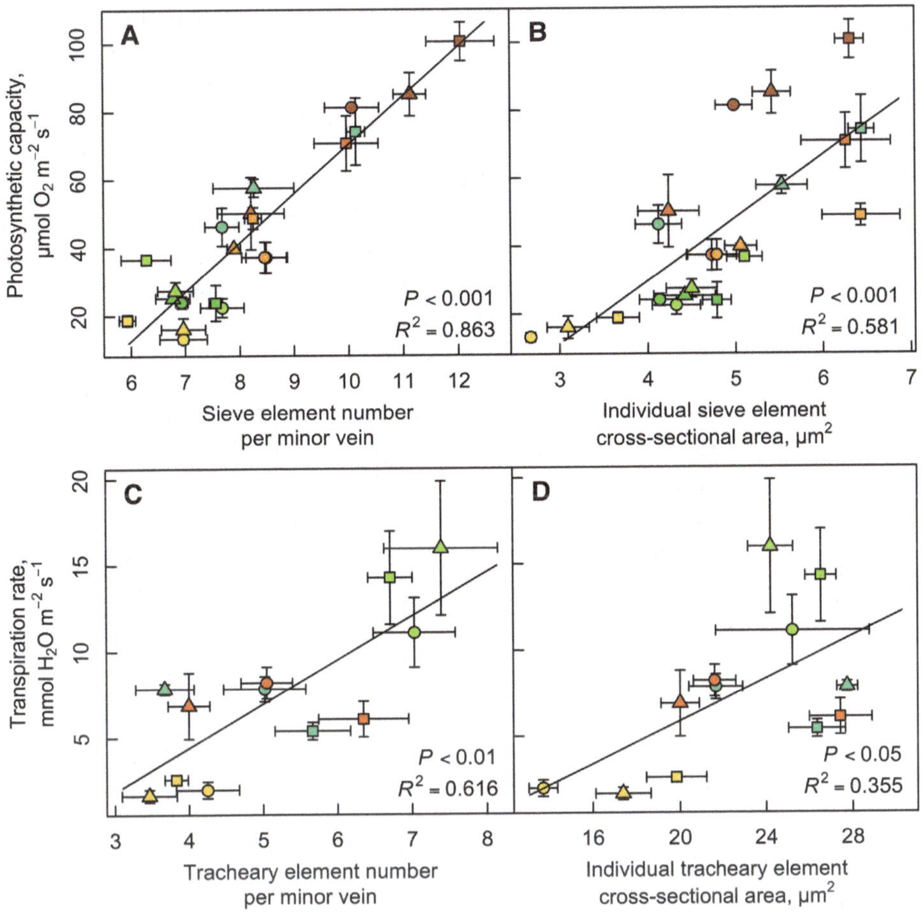

Fig. 1 Relationships between the light- and CO_2- saturated capacity for photosynthetic oxygen evolution on a leaf area basis determined at 25°C and sieve element (**A**) number per foliar minor vein or (**B**) cross-sectional area per cell as well as transpirational water loss and tracheary element (**C**) number per foliar minor vein or (**D**) cross-sectional area per cell in leaves of *Arabidopsis thaliana* ecotypes from Italy (circles), Poland (Columbia-0 wild-type; triangles), and Sweden (squares) that developed under a 9-h photoperiod of seven different photon flux density and temperature regimes: 100 μmol photons $m^{-2} s^{-1}$ at a daytime leaf temperature of 20°C (yellow-orange), 400 μmol photons $m^{-2} s^{-1}$ at a daytime leaf temperature of 14°C (blue-green), 27°C (green), or 36°C (olive-green), and 1000 μmol photons $m^{-2} s^{-1}$ at a daytime leaf temperature of 14°C (brick-red), 20°C (red), or 27°C (orange). Mean values ± standard deviations (photosynthetic capacities and transpiration rates) or ± standard errors (sieve and tracheary element features). Data from Cohu et al. [5, 26], Adams et al. [4, 13], and Stewart et al. [18–20]. Reproduced from Adams et al. [15], in which additional information on the conditions under which gas exchange measurements were conducted can also be found

The levels of significance and correlation coefficients were lower for the relationships between tracheary element features and transpiration than for those between sieve elements and photosynthesis. This difference may arise because water delivered to the leaf not only replaces water lost through transpiration but also supports sucrose export, with the proportion of these latter routes varying with rates of photosynthesis and transpiration. For instance, the Swedish ecotype grown under low temperature (brick-red and blue-green squares) exhibited relatively low transpiration rates (Fig. 1C), while photosynthesis was upregulated (Fig. 1A) as is typical for a winter annual like *A. thaliana*. The data for the Swedish ecotype acclimated to low temperature thus exhibited a considerable departure from the relationships in Fig. 1C, D. With more and wider tracheary elements relative to transpiration rate (Fig. 1C, D), the resulting additional capacity for water transport likely supports greater levels of phloem loading and foliar sucrose export that accompany higher photosynthesis rates of the low-temperature-acclimated Swedish ecotype (Fig. 1A).

The product of cell number per minor vein (Fig. 1A, C) and individual cell cross-sectional area (Fig.1B, D) provides the cross-sectional area of each cell type per minor vein (Fig. 2A, D). As was the case for each of the individual components (Fig. 1) of this composite parameter, the resulting cross-sectional area of the sucrose-exporting pipelines was significantly positively correlated with photosynthetic capacity (Fig. 2A) and cross-sectional area per minor vein of the water pipelines was significantly positively correlated with transpiration rate (Fig. 2D). Moreover, the volume of the network of these pipelines serving a given area of leaf ($mm^3 mm^{-2}$) can be assessed from the product of cross-sectional area of each cell type per minor vein (μm^2) and foliar minor vein density (mm vein length mm^{-2} leaf area). Vein density, while exhibiting no relationship to photosynthetic capacity (Fig. 2B), exhibited a strong and highly significant correlation with transpiration rate (Fig. 2E). This latter relationship presumably resulted because phenotypic plasticity in foliar water import and transpirational loss, as well as of minor-vein xylem features and vein density, varied with precipitation level at the sites of ecotype origin [13, 15]. In contrast, phenotypic plasticity of photosynthesis-related features (leaf thickness, leaf dry mass per area, photosynthetic capacity, and minor vein phloem characteristics) varied with photoperiod and temperature in the sites of ecotype origin (i.e., with habitat latitude [13, 15]).

Notably, the correlation between photosynthetic capacity and the volume-based parameter for sieve elements (sieve element volume per leaf area; Fig. 2C) was slightly lower than for the area-based parameter (Fig. 2A) due to factoring in minor vein density with its nonsignificant relationship with photosynthesis (Fig. 2B). In contrast, the high correlation of minor vein density with transpiration rate (Fig. 2E) caused both significance and correlation coefficient to be greater for the relationship of transpiration rate and the volume-based parameter for tracheary elements (tracheary element volume per leaf area; Fig. 2F) than the area-based parameter (Fig. 2D; see also Fig. 1C, D).

In addition to features of the sieve elements (see above), other phloem cells also exhibit features that correlate with photosynthetic

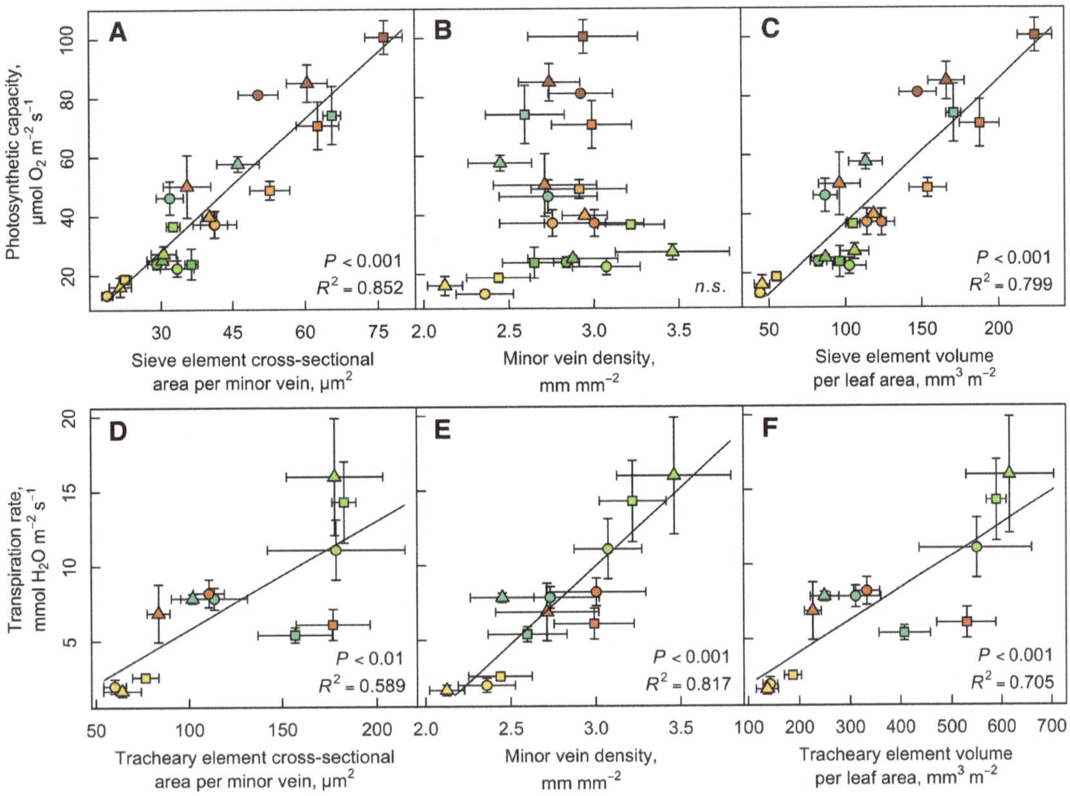

Fig. 2 Relationships between the light- and CO_2- saturated capacity for photosynthetic oxygen evolution on a leaf area basis determined at 25°C and (**A**) sieve element cross-sectional area per minor vein, (**B**) minor vein density, and (**C**) sieve element volume per leaf area as well as transpirational water loss and (**D**) tracheary element cross-sectional area per minor vein, (**E**) minor vein density, and (**F**) tracheary element volume per leaf area in leaves of *Arabidopsis thaliana* ecotypes from Italy (circles), Poland (Columbia-0 wild-type; triangles), and Sweden (squares) that developed under a 9-h photoperiod of seven different photon flux density and temperature regimes. For additional information on growth regimes, measurement conditions, and sources from which the data are drawn, see the legend of Fig. 1. Mean values ± standard deviations (photosynthetic capacities, transpiration rates, and vein densities) or ± standard errors (sieve and tracheary element features). Panels **A** and **D** from Adams et al. [15]. *n.s.* = not significant

capacity and, presumably, sucrose-export capacity. Sieve tubes accumulate sugars and export these from the leaf, thus serving as pipelines for the mass flow of sucrose. Moreover, companion cells and phloem parenchyma cells facilitate the loading of sugar into the sieve elements in species with active phloem loading. In *A. thaliana*, an apoplastic loader, photosynthetic capacity exhibited similarly close correlations with the number (per minor vein) and individual cross-sectional area of companion and phloem parenchyma cells (Fig. 3A, B [5, 15, 27]) as shown for sieve elements in Figs. 1 and 2. Companion and/or phloem parenchyma cells fuel the active loading of sieve elements via cell membrane-spanning transport proteins (sucrose efflux channels and ATPases in the parenchyma cells; ATPases and sucrose-H^+

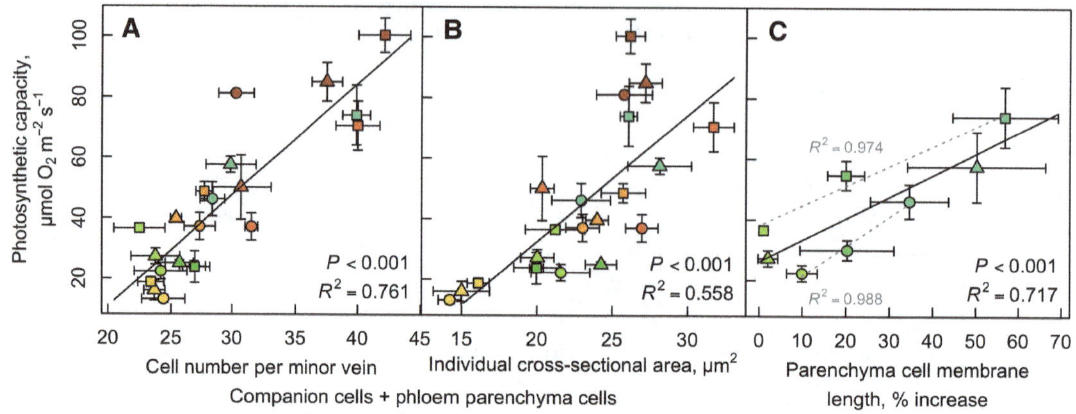

Fig. 3 Relationships between the light- and CO_2- saturated capacity for photosynthetic oxygen evolution on a leaf area basis determined at 25°C and (**A**) the number of companion and phloem parenchyma cells per minor vein, (**B**) the cross-sectional area of companion and phloem parenchyma cells, and (**C**) the estimated percent increase in cell membrane area resulting from minor vein phloem parenchyma cell wall ingrowths in leaves of *Arabidopsis thaliana* ecotypes from Italy (circles), Poland (Columbia-0 wild-type; triangles), and Sweden (squares) that developed under a 9-h photoperiod of seven different photon flux density and temperature regimes. For additional information on growth regimes, measurement conditions, and sources from which the data are drawn, see the legend of Fig. 1. Mean values ± standard deviations (photosynthetic capacities) or ± standard errors (minor vein phloem features). See Adams et al. [4] for the relationship between photosynthetic capacity and phloem parenchyma cell wall ingrowths in the Polish ecotype of *A. thaliana* acclimated to different light intensity growth regimes

symporters in the companion cells; [28–30]). Larger, and especially more numerous, companion and phloem parenchyma cells provide more cell membrane area that can accommodate a greater number of such transport proteins [4], thus presumably providing a greater driving force for phloem loading, photosynthate export from the leaf, and thus photosynthesis itself.

Moreover, an additional feature of phloem parenchyma and/or companion cells could further enhance cell membrane area and the number of transport proteins. In some apoplastically loading species, the walls of companion cells and/or phloem parenchyma cells exhibit inward-facing protrusions or ingrowths that can be quite elaborate [1, 31]; phloem cells with such cell wall invaginations are termed transfer cells. Such ingrowths can result in considerable expansion of membrane area and are particularly evident under growth conditions that stimulate upregulation of photosynthesis [12, 15]. In *A. thaliana*, wall ingrowths only occur in phloem parenchyma cells and are limited to the wall region immediately adjacent to a sieve element, which harbors cell membrane-spanning sucrose-H^+ symporters [32]. Such ingrowths develop to a lesser extent in leaves grown under low light and/or high temperature conditions, under which photosynthetic capacities are low (Fig. 3C). In contrast, the extent of transfer-cell wall ingrowths was enhanced in leaves grown under high light intensities or low temperatures, conditions that resulted in upregulation of photosynthesis (Fig. 3C; [1, 12, 13, 31]). The percent increase in cell membrane (relative to a cell wall with no ingrowths) was positively and linearly correlated with photosynthetic capacity among three ecotypes of *A. thaliana* (Fig. 3C)—with similarly high significance and correlation coefficient as the other phloem features shown in Figs. 1–3 (Figs. 1A, B, 2A, C, and 3A, B). Moreover, a significant correlation between percent increase of cell membrane (between 10% and 220%) and photosynthetic capacity (between 5 and 75 μmol O_2 m^{-2} s^{-1}) was also found among three different species grown under low or high light [12]: *A. thaliana*, pea (ingrowths around the entire companion cell perimeter), and *Senecio vulgaris* (ingrowths around the entire perimeter of both companion and phloem parenchyma cells).

For the winter annual *A. thaliana* grown under a range of different environmental conditions, the number of phloem cells (sieve elements, phloem parenchyma and companion cells) per minor vein exhibited the strongest, most significant linear correlation with photosynthetic capacity (Figs. 1–3; [5, 13, 15]). However, linear relationships between phloem metrics and photosynthetic capacity were stronger when phloem parameters were normalized among different herbaceous annual eudicots by multiplying by vein density (yielding the volume of cells per leaf area or simply the product of cell number per minor vein × vein density; [4, 10, 11, 16, 33]). Increased strength of the relationship was also found for minor vein xylem anatomical metrics versus transpiration rate in *A. thaliana* (Figs. 1 and 2; [15]) and among different species [16]. Moreover, xylem anatomical features and photosynthetic capacity also exhibited significant relationships within and among species in plants grown under moderate temperature [4, 10, 11, 14, 16]. In contrast, there was no correlation between foliar xylem anatomical features and photosynthetic capacity in *A. thaliana* grown over a range of temperatures [4, 13, 27] because this winter annual exhibits opposite trends of temperature acclimation for photosynthesis versus tracheary elements or vein density, with upregulation of photosynthetic capacity at low temperature and upregulation of tracheary element features (cross-sectional area per element and number of elements per minor vein) and vein density at higher temperature (Figs. 1 and 2).

Methods to determine rates of phloem and xylem sap flow through stems have long been employed (for the phloem, see, e.g., Part III of this book). The anatomical metrics characterized in Figs. 1 through 3 of the present chapter complement these functional measurements by providing what can be viewed as proxies for maximal possible fluxes of these tissues based on the correlations between foliar vascular metrics and photosynthesis or transpiration rates presented here, rather than the physics of fluid flow through pipes (see [4, 34, 35]). These relationships between vascular infrastructure and photosynthesis were revealed for the light-saturated photosynthetic capacity of oxygen evolution (determined in the presence of 5% CO_2 that overcomes all resistances [stomatal, cuticular, mesophyll, etc.] to diffusion), a measure of the maximal possible rate of photosynthetic electron transport, carbon fixation, and thus sugar export via foliar phloem loading and transport through sieve elements. Whereas there is no comparable way to determine maximal possible flux for foliar transpirational water loss, the linear relationships between foliar tracheary element metrics and transpiration rate measured under common conditions (Figs. 1C, D and 2D–F) are nonetheless compelling.

To date, just over a dozen annual or biennial herbaceous dicot species with mesophytic leaves have been characterized with respect to the foliar phloem anatomical features shown in this chapter. Possible adjustment of foliar phloem organization has yet to be studied in ferns, monocots, perennial herbaceous species (including seasonally deciduous species), woody plant species, and sclerophytic species, many of which may not employ an active form of phloem loading [36]. Because phloem facilitates export of photosynthate, thereby preventing feedback inhibition of photosynthesis and presumably supporting upregulation of photosynthesis [3, 4], additional studies of foliar phloem anatomical features are warranted. To facilitate application of this approach to other species and growth conditions, a detailed protocol for the quantitative analysis of foliar phloem (and xylem) is outlined below (see Fig. 4). Determination of minor vein density included physical collection, fixation, and clearing of leaf tissue, staining of veins, and quantification of minor vein length per unit leaf area from images obtained using a compound microscope. Quantification of cellular minor vein features included physical collection, fixation, staining, dehydration, and embedding of leaf tissue in resin, followed by sectioning, staining, and image generation using light and/or transmission electron microscopy. A detailed list

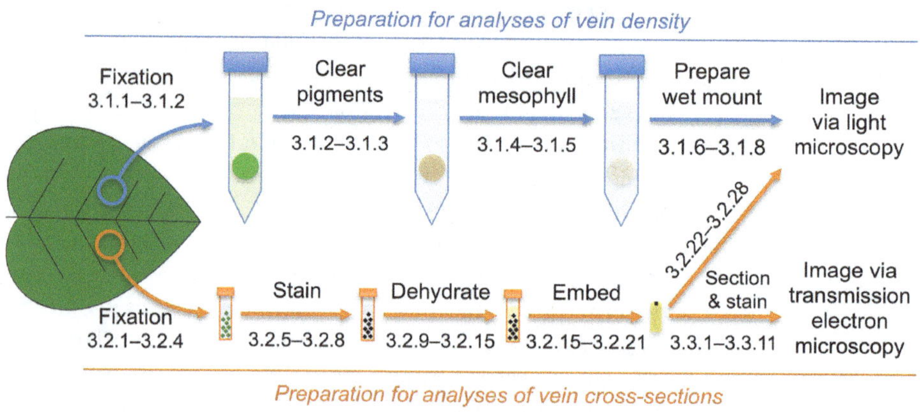

Fig. 4 A schematic illustrating the procedures (with the corresponding numbered steps in the "Methods") for the preparation of tissue for analyses of vein density and vein cross-sections

of relevant metrics and calculations used to derive the parameters is also provided. In addition to the assessment of these parameters, quantification of the proportion of various vascular parameters can be helpful, such as the ratio of tracheary to sieve elements, the product of vein density and the ratio of the latter two (the ratio of tracheary to sieve element volume per unit leaf area), and the ratio of companion cells to sieve elements. Overall, comparison of foliar vascular anatomical features with photosynthesis and transpiration allows exploration of the physiology, ecology, genetics, and evolution of the structure–function relationship between the leaf's primary function (photosynthesis) and the morphological, anatomical, and ultrastructural traits that support it. Such insights may provide tools and directions for crop enhancement through foliar water import and/or sugar export features that support superior photosynthetic performance under variable and extreme climatic conditions [37].

2 Materials

1. Ethanol: 70% (v/v) in water.
2. Sodium hydroxide (NaOH): 5.0% (w/v) aqueous solution (*see* **Note 1**).
3. 15 mL conical tubes with screw caps.
4. Safranin O staining solution: 0.01% (w/v) aqueous solution.
5. Sodium cacodylate buffer: 70 mM aqueous solution; adjust pH to 6.9 using 6 M hydrochloric acid.
6. Glutaraldehyde: 10% (w/v) aqueous solution, EM grade.
7. Paraformaldehyde: 16% (w/v) aqueous solution, EM grade.
8. Fixative solution: Mix 25 mL of sodium cacodylate buffer, 10 mL of 10% glutaraldehyde, 6.2 mL of 16% paraformaldehyde, and 8.8 mL of water.
9. Osmium tetroxide (OsO_4): 4% (w/v) Aqueous solution, EM grade.
10. OsO_4 staining solution: Mix 20 mL of the sodium cacodylate buffer, 10 mL of 4% OsO_4, and 10 mL of water. Prepare fresh.
11. Uranyl acetate: 2.0% (w/v) Aqueous solution.
12. Acetone: 100%, histological grade.
13. Spurr resin [38]: Mix 10 g cycloaliphatic epoxide resin (ERL 4221), 8 g diglycidyl ether of polypropylene glycol (D.E.R. 736), and 25 g nonenyl succinic anhydride (NSA) thoroughly with a magnetic stirrer, then add 0.3 g 2-(dimethylamino) ethanol (DMAE) and mix thoroughly. Prepare fresh.

14. Toluidine blue staining solution: 0.10% (w/v) toluidine blue ($C_{15}H_{16}ClN_3S$) and 1.00% (w/v) sodium tetraborate ($Na_2B_4O_7$) in water.
15. Formvar: 1% (w/v) in ethylene dichloride.
16. Lead citrate: 0.5% (w/v) aqueous solution.
17. 2 mL cryogenic storage tubes.
18. Vacuum desiccator container and vacuum pump.
19. Benchtop sample rocker.
20. Silicone embedding molds with bullet-shaped cavities (approximately 14 × 5 × 4 mm).
21. Drying oven.
22. Glass knives (ultramicrotomy grade) fitted with appropriately sized plastic glass knife boats for water reservoirs for light microscopy.
23. Diamond knives (EM grade) for transmission electron microscopy.
24. Microtome capable of semi-thin (0.7–1.0 μm) sections for light microscopy and ultra-thin (40–80 nm) sections for transmission electron microscopy.
25. 75 mm × 25 mm glass microscope slides with marking area, 22 mm × 22 mm glass microscope cover slips, and a toluene-based mounting medium.
26. Slide warmer.
27. Copper grids: 3.05 mm (diameter) × 0.18 μm (thickness), 75 lines $inch^{-1}$ square mesh.
28. Dissecting microscope.
29. Light microscope with a digital imaging system that has been calibrated with a micrometer glass slide to allow conversion from pixels to meaningful units of measurement (e.g., μm).
30. Transmission electron microscope for quantification of cell wall ingrowths.
31. Desktop computer with ImageJ software [39].

3 Methods

3.1 Minor Vein Density

1. Avoiding major veins and the leaf margin, carefully excise approximately 2 cm^2 of desired leaf tissue that includes only or primarily the two highest-order (minor) veins. For a small-leafed species like *A. thaliana*, the minor veins typically comprise the third and fourth and occasionally fifth order, whereas

for larger-leafed species, minor veins may be sixth and seventh order.

2. Transfer the tissue immediately into a 15 mL screw-cap tube containing 5–10 mL of 70% (v/v) ethanol, which serves as an efficient fixative.
3. Replace ethanol with 5–10 mL of fresh 70% (v/v) ethanol every 12–24 h to remove pigments (*see* **Note2**).
4. After the leaf is cleared, as evidenced by lack of pigments in the solution following 12–24 h, re-hydrate the leaf tissue (e.g., 35% [v/v] ethanol for 5 min, then water for 5 min) and replace water with 5–10 mL of 5% (w/v) NaOH.
5. Replace with 5–10 mL of fresh 5% (w/v) NaOH every 12–24 h until the mesophyll is transparent.
6. Replace 5% (w/v) NaOH with 5–10 mL water.
7. Contrast between vascular and mesophyll tissues can be enhanced by soaking leaf tissue in a dilute (e.g., 0.01% [w/v]) Safranin O aqueous solution for 5–30 min, followed by water for 1–24 h.
8. Mount leaf tissue to a 75 mm × 25 mm glass microscope slide with a small amount of water and a 22 mm × 22 mm glass microscope cover slip, and image approximately 10 mm^2 of the minor-vein network, avoiding major veins, using a compound light microscope fitted with a digital camera system.
9. Import files into ImageJ and convert images from pixels to μm (based on calibration with a micrometer glass slide) via the "Set Scale..." function under the "Analyze" tab.
10. After tracing a minor vein with the "Segmented Line" tool, quantify its length using the "Measure" function under the "Analyze" tab. Each measurement will be logged under the "Length" column in a "Results" window that will appear after the first measurement.
11. Save "Results" as a CSV file or copy and paste these data directly into spreadsheet data files in programs such as Microsoft Excel.
12. Document each measurement using the "Draw" function under the "Edit" tab, and save the worked image as a new file.

3.2 Minor Vein Cross-Sections

1. Cut the excised leaf tissue into segments of approximately 2 mm × 2 mm with a razor blade. To avoid losses in turgor and/or initiate tissue fixation during this process, we recommend cutting leaf tissue in a glass petri dish with a small amount of either water or the fixative solution.

2. Carefully transfer leaf segments into a 2 mL cryogenic tube containing approximately 1 mL of the fixative solution.
3. To facilitate infiltration with the fixative solution, place the cryogenic tubes in a vacuum desiccator and remove the air using a vacuum pump.
4. Re-introduce air into the container slowly such that any air bubbles that have accumulated at the solution surface are not reincorporated into the leaf segments.
5. Replace the fixative solution with cacodylate buffer and repeat/wash samples thrice with cacodylate buffer (*see* **Note 3**).
6. Replace the cacodylate buffer with OsO_4 staining solution and place tubes on a sample rocker in the fume hood for 1 h.
7. Replace the OsO_4 staining solution with distilled water and repeat more times, such that samples are washed thrice with water.
8. Replace water with uranyl acetate solution, let stand for 1 h, and then rinse samples thrice with water.
9. Replace water with 15% (v/v) acetone and let samples stand for 15 min.
10. Replace 15% acetone with 30% (v/v) acetone and let samples stand for 15 min.
11. Replace 30% acetone with 50% (v/v) acetone and let samples stand for 15 min.
12. Replace 50% acetone with 70% (v/v) acetone and let samples stand for 15 min.
13. Replace 70% acetone with 90% (v/v) acetone and let samples stand for 15 min.
14. Replace 90% acetone with 100% acetone and let samples stand for 30 min.
15. Replace with 100% acetone (and let samples stand for 30 min) for a total of three times.
16. Replace 100% acetone with 2:1 acetone:Spurr-resin mixture and leave samples on the rocker overnight.
17. Replace the 2:1 acetone:Spurr-resin mixture with 1:1 acetone:Spurr-resin mixture the following morning and place samples on the rocker for 6–8 h.
18. Replace the 1:1 acetone:Spurr-resin mixture with 100% Spurr resin and leave the samples on the rocker overnight.
19. The following morning, add Spurr resin to each of the bullet-shaped cavities of the silicone embedding molds and carefully transfer individual leaf segments into individual cavities.

20. Arrange each leaf segment to be flush with the tapered end of the respective cavity.
21. Place molds in a drying oven set to 70°C for 8–12 h.
22. Examine the vascular network in embedded leaf samples with a dissecting microscope to identify resin blocks that are likely to yield perpendicular cross-sections of minor veins.
23. The resin surrounding leaf tissue at the end to be sectioned can be removed at this time with a fresh razor blade.
24. Use a microtome with a glass knife to prepare semi-thin (0.7–1.0 μm) cross-sections of the embedded leaf segments.
25. Carefully transfer these cross-sections from the glass-knife water reservoir onto a glass slide, and affix cross-sections to the slide by heating slides at 50–60°C for 30 min.
26. Stain cross-sections with toluidine blue solution for 5–30 min and then rinse with water.
27. After the slides have dried completely, mount a glass cover slip to the slide with a small drop of slide mounting medium.
28. Use a compound light microscope fitted with a digital camera system to image minor veins in the leaf cross-sections with a high-magnification (e.g., 40–100×) objective lens.
29. Open the files with ImageJ software and convert the images from pixels to μm via the "Set Scale..." function under the "Analyze" tab.
30. Identify phloem, xylem, and undifferentiated parenchyma cells within a vein (see Fig. 5 for a typical vein of the winter annual *A. thaliana* and Fig. 6 for typical veins of summer annual apoplastic and active symplastic loaders). For quantification purposes, one can start by identifying all sieve elements as phloem cells that are much smaller than other phloem cells and do not stain as darkly due to their highly simplified protoplasm.
31. Quantify cross-sectional areas of each sieve element using the "Measure" function under the "Analyze" tab after tracing each sieve element perimeter with the "Polygon selections" tool. Each measurement will be logged in the "Area" column of the "Results" window, which appears after the first measurement.
32. Quantify the cross-sectional area of the entire phloem, which consists of companion cells and phloem parenchyma cells (*see* **Note 4**) in addition to sieve elements. Quantify the total cross-sectional areas of tracheary elements, xylem parenchyma, and any undifferentiated vascular parenchyma between the phloem and xylem. Document each measurement using the "Draw" function under the "Edit" tab.

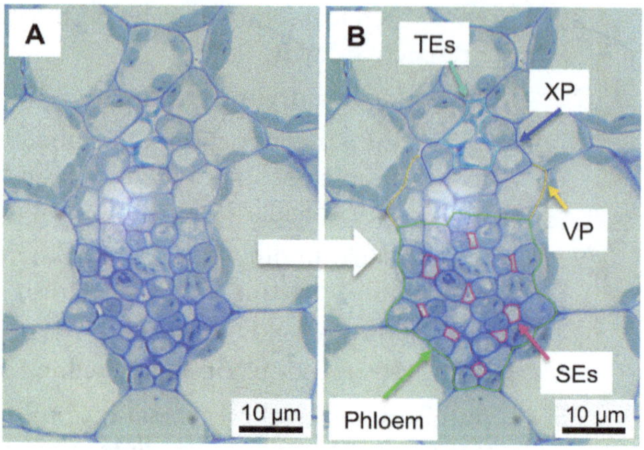

Fig. 5 (**A**, **B**) Light microscopic image of a leaf minor vein cross-section from an *Arabidopsis thaliana* (Columbia-0) plant acclimated to 400 µmol photons $m^{-2} s^{-1}$ (9-h photoperiod) and a leaf temperature of 18°C. The different vascular cell types are outlined in (**B**) and identified as follows: Individual sieve elements (SEs), each of which is smaller and freer of internal structure than the other phloem cells, are outlined in pink. The entire phloem, consisting of companion cells, phloem parenchyma, and sieve elements, is outlined in green. The companion cells are distinguished by a more circular shape and darker (resulting from their denser cytoplasm) color compared to the phloem parenchyma cells. The tracheary elements (TEs), outlined in light blue, are distinguished by their lack of cellular content and thicker cell walls, which stain a slightly lighter hue of blue than the other structures. Xylem parenchyma cells (XP), outlined in dark blue, are adjacent to the tracheary elements. Undifferentiated vascular parenchyma cells (VP; outlined in orange on either side of the vascular bundle) are located between the xylem and phloem tissue. Parameters calculated from this image can be found in Table 1

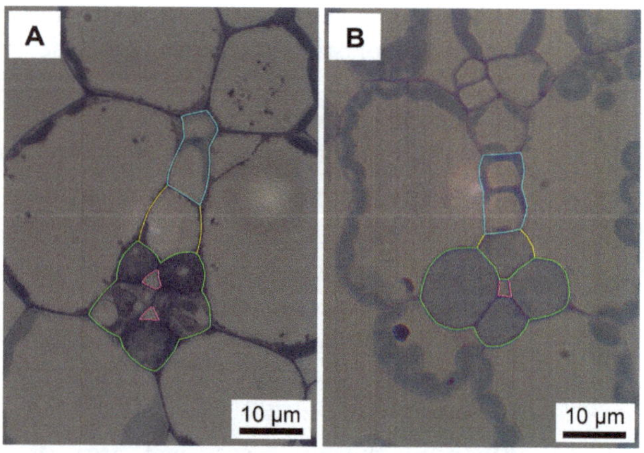

Fig. 6 Light microscopic images of leaf minor vein cross-sections from (**A**) the apoplastic loader sunflower (*Helianthus annuus* L. cv. Soraya) and (**B**) the active symplastic loader watermelon (*Citrullus lanatus* L. cv. Faerie Hybrid) acclimated to 1000 µmol photons $m^{-2} s^{-1}$ (9-h photoperiod) and an air temperature of 25°C (see [16]). The small sieve elements are outlined in pink, the companion (darker cytoplasm) and phloem parenchyma (light cytoplasm) of the apoplastic loader and the intermediary cells of the active symplastic loader are outlined in light green, the tracheary elements are outlined in light blue, and the undifferentiated vascular parenchyma cells are bordered by orange. Parameters calculated from these images can be found in Table 1

33. Also, record the number of each cell type, the process of which can be facilitated with the "Multi-point Tool." Our current protocol [26] involves the quantification of seven to ten minor veins per plant.
34. These same sections can be used to quantify additional leaf anatomical features, such as total leaf thickness, number of palisade cell layers, and mesophyll cell density (for details, see [17]).

3.3 Phloem Transfer Cell Wall Ingrowths

The protocol outlined above yields samples suitable for subsequent analyses of phloem cell ultrastructure via transmission electron microscopy (see also [1, 31]). We have found these techniques particularly beneficial for quantifying cell wall ingrowths of phloem transfer cells (Fig. 7). These same sections are also suitable for the quantification of chloroplast starch content [1] and thylakoid structure [41].

1. Clean copper grids with acetone and allow grids to dry on filter paper with ample room between grids.
2. Apply a drop of formvar to each grid and let the grids dry on filter paper.
3. Fill a 500 mL beaker with water and a 25 mL beaker with 1% formvar.
4. Clean a microscope slide, dip it into the formvar solution to the bottom of the beaker, and let it dry.
5. Using a sharp pin or razor blade, cut a rectangular shape on the microscope slide in the region with dried formvar.
6. Immerse the microscope slide into the beaker with water at an angle such that the formvar film comes off and floats on the water surface.
7. Place the grids carefully on top of the film floating on the water without any overlap. Make sure to place the formvar-treated side of the grid onto the formvar film.
8. Using the resin blocks with minor veins of interest identified via light microscopy, prepare ultra-thin (<100 nm) sections with a microtome and a diamond knife.
9. Rather than using a tool to transfer sections from the knife water reservoir to the copper grid, use the copper grids to directly lift sections from the water reservoir.
10. Stain the sections on the grids with uranyl acetate for 6 min, followed by lead citrate for 4 min. This process can be achieved by placing each copper grid onto a drop of the stain such that sections are in contact with the stain.
11. Rinse the grids gently but thoroughly with water to remove stains and debris.

Table 1

Example data set resulting from the minor veins in Figs. 5A (*Arabidopsis thaliana*), 6A (sunflower), and 6B (watermelon). Cell volumes and cell numbers × vein density were calculated with the mean values of vein density for each group (2.3, 9.0, and 7.4 mm mm^{-2}, respectively)

Metric	*Arabidopsis thaliana*	Sunflower	Watermelon
Cells per minor vein	54	10	7
Cells per minor vein × vein density	126	90	52
Cross-sectional area per minor vein, μm^2	811	427	370
Minor vein volume per leaf area, mm^3 m^{-2}	1887	3840	2722
Phloem cells per minor vein	41	7	4
Phloem cells per minor vein × vein density	95	63	29
Phloem cross-sectional area per minor vein, μm^2	508	270	244
Phloem volume per leaf area, mm^3 m^{-2}	1182	2431	1794
CCs + PCs per minor vein	32	5	3
CCs + PCs per minor vein × vein density	74	45	22
CC + PC cross-sectional area per minor vein, μm^2	475	261	239
CC + PC volume per leaf area, mm^3 m^{-2}	1105	2349	1758
CC & PC cross-sectional area per cell, μm^2	15	52	80
SEs per minor vein	9	2	1
SEs per minor vein × vein density	21	18	7
SE cross-sectional area per minor vein, μm^2	33	9	5
SE volume per leaf area, mm^3 m^{-2}	77	83	36
SE cross-sectional area per cell, μm^2	4	5	5
Xylem cells per minor vein	6	2	2
Xylem cells per minor vein × vein density	14	18	15
Xylem cross-sectional area per minor vein, μm^2	127	74	81
Xylem volume per leaf area, mm^3 m^{-2}	294	668	597
TEs per minor vein	3	2	2
TEs per minor vein × vein density	7	18	15

(Continued)

Table 1 (Continued)

Metric	*Arabidopsis thaliana*	Sunflower	Watermelon
TE cross-sectional area per minor vein, μm^2	56	74	81
TE volume per leaf area, $mm^3\,m^{-2}$	131	668	597
TE cross-sectional area per cell, μm^2	19	37	41

12. Using a conventional transmission electron microscope with a digital camera imaging system, image the transfer cells in each minor vein of each grid. For more information on quantification of ingrowths, see Fig. 7 and Step 6 in Sect. 3.4.

3.4 Metrics and Equations

For examples of the following parameters based on the images in Figs. 5 and 6, see Table 1.

1. *Minor vein density (VD)*: Calculated as total vein length per leaf area from light microscopic images of cleared leaves, vein density, or vein length per unit leaf area, is often reported in $mm\,mm^{-2}$ ([13]; see Table 1).
2. *Cell number per minor vein (CN)*: From each cross-sectional image, count the cell numbers for each of the following cell types: sieve elements (SEs), companion cells + phloem parenchyma cells (CCs + PCs), tracheary elements (TEs), phloem cells (SEs + CCs + PCs), xylem cells (TEs + xylem parenchyma cells [XPs] when present), and all vascular cells (SEs + CCs + PCs + TEs + XPs + undifferentiated vascular parenchyma cells [VPs] when present).
3. *Cell cross-sectional area per minor vein (CA)*: These values represent the total cross-sectional area of a particular cell type per minor vein.:
   i. Sieve elements = CA_{SEs}
   ii. Companion cells + phloem parenchyma cells = $CA_{Phloem} - CA_{SEs}$
   iii. Tracheary elements = CA_{TEs}
   iv. Phloem = CA_{Phloem}
   v. Xylem = $CA_{TEs} + CA_{XPs}$
   vi. All vascular cells = $CA_{Phloem} + CA_{TEs} + CA_{XPs} + CA_{VPs}$
4. *Volume per leaf area*: Multiplying the total cross-sectional area per vein of a particular cell type in $\mu m^2 \times$ vein density (VD) in mm (vein length) $\times\, mm^{-2}$ (leaf area) results in values expressed in mm^3 (cell volume) $\times\, m^{-2}$ (leaf area). This expression is convenient for structure–function investigations using

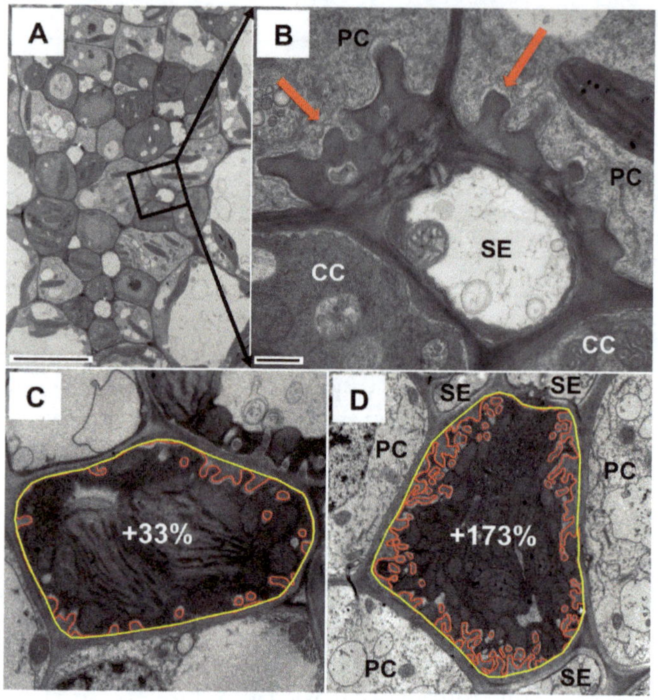

Fig. 7 Transmission electron micrographic images of *Arabidopsis thaliana* (Swedish ecotype) minor vein (**A**) phloem (boxed area delineates the region magnified in **B**) and (**B**) transfer cells (phloem parenchyma cells with wall ingrowths indicated by the red arrows) from a plant that developed under 400 μmol photons $m^{-2} s^{-1}$ (9-h photoperiod) at a leaf temperature of 14°C during the day and 12.5°C during the night, as well as transfer cells (companion cells with wall ingrowths) from foliar minor veins of pea (*Pisum sativum* L. cv. Alaska) plants that developed in low (**C**; 100 μmol photons $m^{-2} s^{-1}$) or high (**D**; 1000 μmol photons $m^{-2} s^{-1}$) light (10 or 14 h photoperiod, respectively, under a 25°C day and 20°C night temperature regime). The length of each yellow line serves as the baseline perimeter of membrane present in the theoretical absence of wall ingrowths and the additional length of the red lines in each represents an estimate of the enhancement of cell membrane area resulting from the cell wall ingrowths (33% in the companion cell from the low light plant and 173% in the companion cell from the high light plant). PC phloem parenchyma cell, SE sieve element, and CC companion cell. Image (**B**) modified from Adams et al. [12] and image (**D**) modified from Amiard et al. [1] Copyright (2005) National Academy of Sciences. Cell wall ingrowths of phloem transfer cells presumably enhance phloem loading by amplifying the surface area available for membrane-bound transporter proteins (e.g., SWEET efflux channels and ATPases in *Arabidopsis thaliana* phloem parenchyma cells; [28, 32]). Although not all species have phloem cells with wall ingrowths, some apoplastic loaders have ingrowths in phloem parenchyma cells localized exclusively to wall areas adjacent to sieve elements (**A**, **B**), other apoplastic loaders have ingrowths around the entire companion cell perimeter (**C**, **D**), and yet other apoplastic loaders exhibit wall ingrowths in both companion cells (around the entire perimeter) and phloem parenchyma cells (portions of walls adjacent to both sieve elements and neighboring companion cells; [31]). Although it is sometimes possible to discern the presence of wall ingrowths in light microscopic images, transmission electron micrographs (for information on electron microscopy, see [40]) are necessary to quantify the extent of ingrowths. Such magnification of cell membrane area by growth under high compared to low light (in three species) or low versus high temperature (in the winter annual *A. thaliana*) was accompanied by greater rates of photosynthesis [1, 12, 15, 31]. Even transfer from one growth condition to another resulted in altered levels of ingrowths in fully expanded, mature leaves, accompanied by a concomitant and proportional change in photosynthetic capacity [1, 12]

photosynthetic and transpiration measurements expressed per m^2 (leaf area) (see Fig. 2C, F).

i. Sieve elements $= CA_{SE} \times VD$

ii. Companion cells + phloem parenchyma cells $= (CA_{Phloem} - CA_{SE}) \times VD$

iii. Tracheary elements $= CA_{TE} \times VD$

iv. Phloem $= CA_{Phloem} \times VD$

v. Xylem $= (CA_{TE} + CA_{XP}) \times VD$

vi. Vascular tissue $= (CA_{Phloem} + CA_{TE} + CA_{XP} + CA_{VP}) \times VD$

5. *Individual cross-sectional area per cell*: For each cell type, these values can be computed by dividing the total cross-sectional area per vein by the total number of this cell type per vein.

   i. Sieve elements $= CA_{SEs} / CN_{SEs}$

   ii. Companion cells and phloem parenchyma cells $= (CA_{Phloem} - CA_{SEs}) / CN_{CC+PC}$

   iii. Tracheary elements $= CA_{TEs} / CN_{TEs}$

6. *Quantification of cell wall ingrowths:* The extent of cell membrane magnification due to wall ingrowths can be quantified using ImageJ (increased length of the perimeter attributable to ingrowths compared to the perimeter in the absence of ingrowths, i.e., red compared to yellow lines in Fig. 7C, D).

4 Notes

1. Use caution and appropriate personal protective equipment throughout sample preparation. Many solutions for the outlined procedure contain chemicals that are acutely toxic, radioactive, carcinogenic, and/or release toxic compounds. Please follow handling, storage, and disposal regulations accordingly.
2. Tissue clearing with ethanol works well for most herbaceous species. For additional information regarding quantification of vein density and/or preparation of leaf tissue for these measurements, please refer to Sack et al. [42] and the protocol cited therein.
3. If sections will only be examined using light microscopy, this staining protocol is not necessary. In this case, skip Steps 3.2.5–3.2.8 (staining with OsO_4 and uranyl acetate) and replace the fixative solution with water thrice.

4. Minor vein cell numbers and sizes vary considerably even across herbaceous eudicots (see [10]). Moreover, the type of phloem cells in minor veins depends on the phloem loading mechanism (Figs. 5 and 6). For example, active symplastic loaders possess specialized companion cells known as intermediary cells in addition to ordinary companion cells and commonly lack phloem parenchyma cells (for a review, see [43]). Classification, quantification, and interpretation of minor-vein phloem anatomy for a given species may thus require prior or concomitant identification of its phloem loading mechanism.

Acknowledgments

We thank Dr. Thomas Giddings, Dr. Véronique Amiard, and Prof. Robert Turgeon for instruction, advice, and valuable discussion. This work was supported by the University of Colorado at Boulder.

Conflict of Interest The authors have no conflicts of interest to declare that are relevant to the content of this chapter.

References

1. Amiard V, Mueh KE, Demmig-Adams B, Ebbert V, Turgeon R, Adams WWIII (2005) Anatomical and photosynthetic acclimation to the light environment in species with differing mechanisms of phloem loading. Proc Natl Acad Sci USA 102:12968–12973
2. Brodribb TJ, Feild TS, Jordan GJ (2007) Leaf maximum photosynthetic rate and venation are linked by hydraulics. Plant Physiol 144:1890–1898
3. Adams WWIII, Cohu CM, Muller O, Demmig-Adams B (2013) Foliar phloem infrastructure in support of photosynthesis. Front Plant Sci 4:194
4. Adams WW III, Stewart JJ, Polutchko SK, Demmig-Adams B (2018) Leaf vasculature and the upper limit of photosynthesis. In: Adams WW III, Terashima I (eds) The leaf: a platform for performing photosynthesis. Advances in photosynthesis and respiration, vol 44. Springer, Dordrecht, pp 27–54
5. Cohu CM, Muller O, Stewart JJ, Demmig-Adams B, Adams WW III (2013) Association between minor loading vein architecture and light- and CO_2-saturated rates of photosynthetic oxygen evolution among *Arabidopsis thaliana* ecotypes from different latitudes. Front Plant Sci 4:264
6. Caringella MA, Bongers FJ, Sack L (2015) Leaf hydraulic conductance varies with vein anatomy across *Arabidopsis thaliana* wild-type and leaf vein mutants. Plant Cell Environ 38:2735–2746
7. Scoffoni C, Chatelet DS, Pasquet-kok J, Rawls M, Donoghue MJ, Edwards EJ, Sack L (2016) Hydraulic basis for the evolution of photosynthetic productivity. Nat Plants 2:16072
8. McAdam SAM, Eléouët MP, Best M et al. (2017) Linking auxin with photosynthetic rate via leaf venation. Plant Physiol 175:351–360
9. Brodribb TJ, Buckley TN (2018) Leaf water transport: a core system in the evolution and physiology of photosynthesis. In: Adams WW III, Terashima I (eds) The leaf: a platform for performing photosynthesis. Advances in photosynthesis and respiration, vol 44. Springer, Dordrecht, pp 81–96
10. Polutchko SK, Stewart JJ, Demmig-Adams B, Adams WWIII (2018) Evaluating the link between photosynthetic capacity and leaf vascular organization with principal component analysis. Photosynthetica 56:392–403
11. Polutchko SK, Stewart JJ, Adams WWIII, Demmig-Adams B (2021) Photosynthesis and foliar vascular adjustments to growth light intensity in summer annual species with

symplastic and apoplastic phloem loading. J Plant Physiol 267:153532
12. Adams WWIII, Cohu CM, Amiard V, Demmig-Adams B (2014) Associations between phloem-cell wall ingrowths in minor veins and maximal photosynthesis rate. Front Plant Sci 5:24
13. Adams WWIII, Stewart JJ, Cohu CM, Muller O, Demmig-Adams B (2016) Habitat temperature and precipitation of *Arabidopsis thaliana* ecotypes determine the response of foliar vasculature, photosynthesis, and transpiration to growth temperature. Front Plant Sci 7:1026
14. Adams WWIII, Stewart JJ, Polutchko SK, Demmig-Adams B (2022) Foliar sieve elements: nexus of the leaf. J Plant Physiol 269: 153601
15. Adams WWIII, Stewart JJ, Polutchko SK, Cohu CM, Muller O, Demmig-Adams B (2023) Foliar phenotypic plasticity reflects adaptation to environmental variability. Plants 12:2041
16. Muller O, Cohu CM, Stewart JJ, Protheroe JA, Demmig-Adams B, Adams WWIII (2014) Association between photosynthesis and contrasting features of minor veins in leaves of summer annuals loading phloem via symplastic versus apoplastic routes. Physiol Plant 152:174–183
17. Muller O, Stewart JJ, Cohu CM, Polutchko SK, Demmig-Adams B, Adams WWIII (2014) Leaf architectural, vascular, and photosynthetic acclimation to temperature in two biennials. Physiol Plant 152:763–772
18. Stewart JJ, Demmig-Adams B, Cohu CM, Wenzl CA, Muller O, Adams WWIII (2016) Growth temperature impact on leaf form and function in *Arabidopsis thaliana* ecotypes from northern and southern Europe. Plant Cell Environ 39:1549–1558
19. Stewart JJ, Polutchko SK, Adams WWIII, Cohu CM, Wenzl CA, Demmig-Adams B (2017) Light, temperature and tocopherol status influence foliar vascular anatomy and leaf function in *Arabidopsis thaliana*. Physiol Plant 160:98–110
20. Stewart JJ, Polutchko SK, Adams WWIII, Demmig-Adams B (2017) Acclimation of Swedish and Italian ecotypes of *Arabidopsis thaliana* to light intensity. Photosynth Res 134:215–229
21. Stewart JJ, Adams WWIII, Cohu CM, Demmig-Adams B (2018) Tocopherols modulate leaf vein arrangement and composition without impacting photosynthesis. Photosynthetica 56:382–391
22. Stewart JJ, Polutchko SK, Demmig-Adams B, Adams WWIII (2018) *Arabidopsis thaliana* Ei-5: altered minor vein architecture compensates for low vein density and supports maintenance of photosynthetic capacity. Front Plant Sci 9:693
23. Adams WW III, Stewart JJ, Demmig-Adams B (2018) Photosynthetic modulation in response to plant activity and environment. In: Adams WW III, Terashima I (eds) The leaf: a platform for performing photosynthesis. Advances in photosynthesis and respiration, vol 44. Springer, Dordrecht, pp 493–563
24. Demmig-Adams B, Stewart JJ, Adams WWIII (2017) Environmental regulation of intrinsic photosynthetic capacity: an integrated view. Curr Opin Plant Biol 37:34–41
25. Cohu CM, Muller O, Demmig-Adams B, Adams WWIII (2013) Minor loading vein acclimation for three *Arabidopsis thaliana* ecotypes in response to growth under different temperature and light regimes. Front Plant Sci 4:240
26. Cohu CM, Muller O, Demmig-Adams B, Adams WWIII (2013) Minor loading vein acclimation for three *Arabidopsis thaliana* ecotypes in response to growth under different temperature and light regimes. Front Plant Sci 4:240
27. Stewart JJ, Muller O, Cohu CM, Demmig-Adams B, Adams WW III (2019) Quantification of foliar phloem infrastructure with microscopy. In: Liesche J (ed) Phloem. Methods and protocols. Methods in molecular biology, vol 2014. Humana Press, New York, NY, pp 55–72
28. Chen L-Q, Qu X-Q, Hou B-H, Sosso D, Osorio S, Fernie AR, Frommer WB (2012) Sucrose efflux mediated by SWEET proteins as a key step for phloem transport. Science 335:207–211
29. Stadler R, Sauer N (1996) The *Arabidopsis thaliana AtSUC2* gene is specifically expressed in companion cells. Bot Acta 109:299–306
30. Ayre BG, Turgeon R (2018) Export of photosynthates from the leaf. In: Adams WW III, Terashima I (eds) The leaf: a platform for performing photosynthesis. Advances in photosynthesis and respiration, vol 44. Springer, Dordrecht, pp 55–79
31. Duan Z, Homma A, Kobayashi M et al. (2014) Photoassimilation, assimilate translocation and plasmodesmal biogenesis in the source leaves of *Arabidopsis thaliana* grown under an increased atmospheric CO_2 concentration. Plant Cell Physiol 55:358–369
32. Amiard V, Demmig-Adams B, Mueh KE, Turgeon R, Combs AF, Adams WWIII (2007) Role of light and jasmonic acid signaling in regulating foliar phloem cell wall ingrowth development. New Phytol 173:722–731
33. Cohu CM, Muller O, Demmig-Adams B, Adams WWIII (2014) Leaf anatomical and photosynthetic acclimation to cool temperature

and high light in two winter versus two summer annuals. Physiol Plant 152:164–173

34. Gleason SM, Blackman CJ, Gleason ST, McCulloh KA, Ocheltree TW, Westoby M (2018) Vessel scaling in evergreen angiosperm leaves conforms with Murray's law and area-filling assumptions: implications for plant size, leaf size and cold tolerance. New Phytol 218:1360–1370
35. Olson ME, Anfodillo T, Gleason SM, McCulloh KA (2021) Tip-to-base xylem conduit widening as an adaptation: causes, consequences, and empirical priorities. New Phytol 229:1877–1893
36. Fu Q, Chen L, Guo Y, Turgeon R (2011) Phloem loading strategies and water relations in trees and herbaceous plants. Plant Physiol 157:1518–1527
37. Demmig-Adams B, Stewart JJ, Adams WWIII (2018) Optimization of photosynthetic productivity in contrasting environments by regulons controlling plant form and function. Intl J Mol Sci 19:872
38. Spurr AR (1969) A low-viscosity epoxy resin embedding medium for electron microscopy. J Ultrastruct Res 26:31–43
39. Schneider CA, Rasband WS, Eliceiri KW (2012) NIH image to ImageJ: 25 years of image analysis. Nat Methods 9:671–675
40. Winey M, Meehl JB, O'Toole ET, Giddings Jr TH (2014) Conventional transmission electron microscopy. Mol Biol Cell 25:319–323
41. Demmig-Adams B, Muller O, Stewart JJ, Cohu CM, Adams WWIII (2015) Chloroplast thylakoid structure in evergreen leaves employing strong thermal energy dissipation. J Photochem Photobiol B Biol 152:357–366
42. Sack L, Caringella M, Scoffoni C, Mason C, Rawls M, Markesteijn L, Poorter L (2014) Leaf vein length per unit area is not intrinsically dependent on image magnification: avoiding measurement artifacts for accuracy and precision. Plant Physiol 166:829–838
43. Zhang C, Turgeon R (2018) Mechanisms of phloem loading. Curr Opin Plant Biol 43:71–75

Part II

Investigating Phloem Content

Check for updates

Chapter 6

EDTA-Facilitated Collection of Phloem Exudates for Chemical Analyses

Rabea Schweiger and Caroline Müller

Abstract

Phloem exudates can be collected using several methods, including the ethylenediaminetetraacetic acid (EDTA)-facilitated method. Plant parts, such as leaves or inflorescences, are cut, and the cut edges are immediately placed into EDTA solution, which prevents plugging of sieve tubes. After this first inoculation, plant parts are transferred to water, into which phloem exudates are collected for a few hours. Inoculation and collection take place in the dark at high relative humidity. The samples can subsequently be used for different chemical analyses of RNA, proteins and various primary and specialized metabolites. This method can be applied to a wide range of plant species and plant parts as well as in various physiological and ecological research contexts.

Keywords Phloem exudate, EDTA-facilitated exudation, Plant parts, Primary metabolites, Specialized metabolites

1 Introduction

Plant phloem sap contains numerous compounds reflecting the long-distance transport, resource allocation, and signal transmission within individuals as well as responses to the environment. Next to RNA and proteins, it contains primary metabolites such as sugars, sugar alcohols and amino acids, as well as taxon-specific specialized (= secondary) metabolites [1, 2]. The chemical composition of phloem sap is decisive for interactions of plants with phloem sap-feeding herbivores, such as aphids and whiteflies. To assess the chemical composition of phloem sap in physiological or ecological studies, phloem exudates are frequently collected. Facilitation of phloem exudation from leaves by ethylenediaminetetraacetic acid (EDTA) has been established by King and Zeevaart [3]. Basically, leaves are cut at the base of the leaf blades or at the petioles, and the cut edges are incubated in EDTA solution in the dark at high humidity. EDTA suppresses damage-induced plugging of phloem sieve pores by callose

Johannes Liesche and Sanna Sevanto (eds.), *Phloem*, Methods in Molecular Biology, vol. 3066,
https://doi.org/10.1007/978-1-0716-5488-0_6,

and proteins, likely because it chelates Ca^{2+} ions released upon plant injury. The open sieve pores allow phloem sap to run out of the plant, which is subsequently collected as phloem exudates after the cut edges are transferred into water. The phloem exudate collection is performed in the dark at high relative humidity to reduce solution uptake via the transpiration stream. Aliquots of phloem exudate can be prepared for various chemical analyses.

Phloem exudates collected via EDTA-facilitated exudation may slightly deviate from the exact phloem sap chemistry [1, 2]. They may contain contamination from surrounding tissues due to cell damage caused by cutting and EDTA. Moreover, enzymes, as well as collection under darkness and high humidity, may result in artifacts. Furthermore, due to unknown exudation/dilution rates, only relative proportions, not absolute concentrations, of compounds can be determined in exudates. Alternatively, phloem sap can be collected via the mouthparts (stylets) of aphids, using aphid stylectomy ([4, 5], *see* Chap. 7), which allows absolute quantification of compounds when the collected phloem sap amount can be quantified. However, stylectomy-collected phloem sap may also contain contaminants or artifacts, such as aphid saliva, and may be influenced by aphid-induced changes in plant chemistry. In contrast to EDTA-facilitated phloem exudate collection, for aphid stylectomy, advanced technical equipment is required; this approach is highly intricate and time-consuming, and phloem sap collection is restricted to the veins the aphids choose for feeding. Phloem exudate collection using EDTA does not require special equipment and delivers an integrated view of the phloem chemistry of the investigated plant part, such as whole leaves or inflorescences. Both techniques are typically performed in the laboratory, while EDTA-facilitated phloem exudation may also be applicable in the field.

Given the important role of the phloem in plant functioning (e.g., transport processes), in stress responses (*see* Chap. 31) and for interactions with phloem sap-feeding herbivores, studying phloem exudates is a substantial part of plant research. Indeed, phloem exudates collected via EDTA-facilitated phloem exudation are analyzed in various research contexts across numerous plant species, using not only leaves but also other plant parts and for chemical analyses of various compounds on different analytical platforms. To understand within-plant transport processes from source to sink organs across shoot development, leaf phloem exudates of tea (Theaceae: *Camellia sinensis*) have been investigated [6]. To characterize plant responses to drought, various metabolites or proteins have been assessed in leaf phloem exudates of pea (Fabaceae: *Pisum sativum*) and barrel medic (Fabaceae: *Medicago truncatula*) [7, 8] as well as in phloem exudates of leaves and inflorescences (ears) of wheat (Poaceae: *Triticum aestivum*) [9]. Research on the induction of silicon (Si), which can act as physical defense in grasses, has revealed that Si is present in phloem exudates of wheat leaves, with evidence for relocation from undamaged to damaged leaves [10]. Moreover, effectors and metabolic responses or signaling in phloem exudates of leaves, being involved in interactions between tomato (Solanaceae: *Solanum lycopersicum*) and whiteflies [11], *Brassica nigra* (Brassicaceae), root nematodes and aphids [12], as well as *Brassica napus* and the pathogen *Xanthomonas campestris* [13] have been studied. Sterols in leaf phloem exudates as well as in aphids have been investigated to gain insights into potential factors influencing the fitness of predatory beetles [14]. Plant part-specific responses to aphids were found in phloem exudate metabolites of wheat [15, 16] and common tansy (Asteraceae: *Tanacetum vulgare*) [17]. The EDTA-facilitated exudation method was also used to examine turgor pressure alongside phloem exudate composition at both leaf and fruit sites in tomato [18].

Here, we describe the EDTA-facilitated collection of phloem exudates based on our experience with different plant species and plant parts. The technique can likely be readily transferred (in modified form) to other plant species and plant parts. Our examples include phloem exudate collections from the monocot wheat [9, 15, 16] as well as from dicot species (Brassicaceae: *Arabidopsis thaliana* [19], *Arabidopsis halleri* [20], and *Tanacetum vulgare* [17]), taken from leaves and/or inflorescences. The suggested steps for phloem exudate collection (Fig. 1) largely follow the publications listed above, with some adaptations due to further optimizations we developed over time. Taken together, we see a high potential for the EDTA-facilitated phloem exudation technique across numerous research fields.

2 Materials

Use chemicals of analytical grade from commercial suppliers, ultrapure water, as well as clean analytical bottles and tubes. Follow the safety and waste-disposal instructions for the used chemicals and tools.

2.1 Plant Material

1. Plant parts for the collection of phloem exudates, such as leaves or inflorescences, depending on the research question (*see* **Note 1**). To avoid stress, which may affect phloem transport, plants should not be moved before cutting the plant parts and phloem exudate collection. Watering the plants shortly before starting the phloem exudate collection may facilitate the exudation.

2.2 Chemicals

1. EDTA solution: 8 mM in ultrapure water, pH 7.0 (*see* **Note 2**). Plan 100 mL for re-cutting of plant parts as well as 1 mL per sample (e.g., 200 mL for 100 samples). *N*-2-hydroxyethylpiperazine-*N*'-2 ethanesulfonic acid (HEPES, 10 mM) may be added to the solution to keep its pH (nearly) constant (e.g., [8]; *see* **Note 2**).

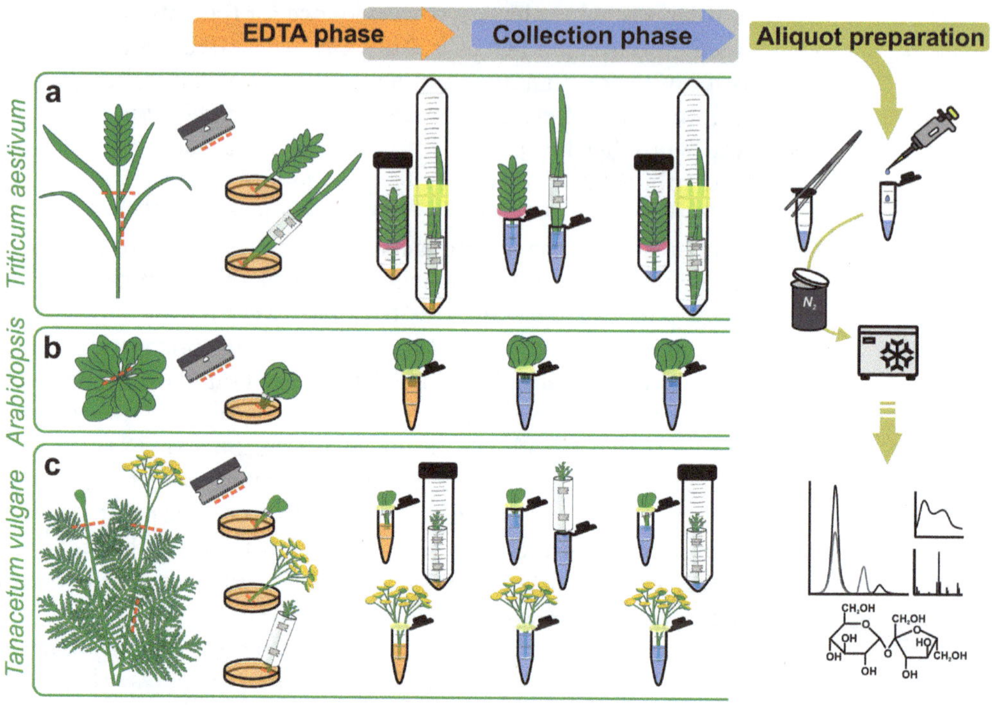

Fig. 1 Steps of the EDTA-facilitated phloem exudate collection technique. The suppression of sieve tube plugging with EDTA ("EDTA phase", left column), the collection of phloem exudates in water ("Collection phase", middle column) and the preparation of aliquots for chemical analyses ("Aliquot preparation", right column) are shown. Exemplarily, potential variations of the set-up are shown for leaves and/or inflorescences of (**a**) *Triticum aestivum* (three leaf blades; one ear), (**b**) *Arabidopsis* (*A. thaliana*, *A. halleri*; 14 rosette leaves), and (**c**) *Tanacetum vulgare* (three leaves; three inflorescences with flower buds; one inflorescence with flowers). The EDTA solution is shown in orange, water in blue, the barrier layer in pink and cover sheets in yellow. The gray rectangle on top indicates steps that have to be performed in the dark

2. Collection solution: ultrapure water (*see* **Note 3**). Plan 100 mL for washing plant parts as well as 1 mL per sample and blank (e.g., 200 mL for 100 samples including blanks). Take this freshly at the day of phloem exudate collection.
3. 70% ethanol.
4. Liquid N_2.

2.3 Tools

1. Gloves.
2. Sharp scalpel or razor blade.
3. Glass dish for re-cutting of plant parts: 5–9 cm diameter.
4. Pipettes and pipette tips: 100 μL to 1 mL.
5. Incubation set-ups (Fig. 1): for each sample, i.e., (bunch of pooled) plant part(s) (*see* **Note 1**), one tube with 1 mL EDTA solution and one tube with 1 mL water as collection solution

(*see* **Note 4**), an appropriate cover for the tubes and (if applicable) a barrier layer (*see* **Note 5**; exemplarily shown in pink for wheat ears in Fig. 1) and a paper strip with tape (*see* **Note 6**); for each blank, one tube with 1 mL water and cover.

6. Washing tube: an additional tube with water for the washing step.
7. Racks for the tubes: one for the tubes with EDTA solution, one for the tubes with water, one for the incubation phase, and a separate rack for the washing tube.
8. Incubation chamber: dark place with circa 20°C and a high relative humidity (ideally ≥90%, but ≥70% may also work) (*see* **Note 7**).
9. Paper bags for the plant part(s): one per sample.
10. Tubes for aliquots for chemical analyses: 1.5 mL, one or more per sample and blank (*see* **Note 8**).
11. Long forceps for the transfer of tubes with aliquots to liquid N_2.
12. Freezer: temperature depending on the stability of the analytes of interest, e.g., −80°C or −20°C.
13. Drying oven.
14. Balance: to weigh dry plant parts.
15. Lyophilizer.
16. Punctured lids that fit on tubes: for lyophilizing aliquots for chemical analyses.

3 Methods

Wear gloves when touching the plants. Clean the scalpel/razor blade between samples with 70% ethanol. Exchange the EDTA solution in the glass dish (for re-cutting) and the water in the washing tube regularly. Arrange a timetable with, for each sample, the onset of incubation in (1) EDTA solution and (2) in water and with the time of freezing aliquots (*see* **Note 9**). For the water collection phase, schedule some blanks in between. Incubation times have to be equal for all samples and blanks.

Some method adjustments may be necessary, especially for new study systems. The suitability of a certain method should be tested with regard to high efficiency of phloem exudation, as well as no/low levels of transpiration-mediated uptake of the solution, and no/few contaminations from surrounding tissues and further artifacts. To evaluate this, visual determination of sieve tube plugging, tracking of (labeled) compounds, and examination of the osmotic pressure and chemical composition of the collection solution may be used

[1, 2, 18]. For example, an EDTA-based method may be compared to a method that includes only water/HEPES, and methods with different EDTA concentrations or incubation durations in water may be contrasted. Comparisons with aphid stylectomy-collected samples or results reported in the literature may also help to validate whether compounds originate from phloem sap [2].

3.1 Suppression of Sieve Tube Plugging using EDTA

1. Cut the plant part(s) with the scalpel/razor blade and immediately submerge the cut edge(s) in the glass dish with EDTA solution (*see* **Notes 1** and **10**).
2. When plant parts are pooled (*see* **Note 1**), align their cut edges in the EDTA solution and tie them up with a paper strip (*see* **Note 11**). Likewise, if needed, wrap plant parts (e.g., broad leaves) and tie them up to fit into the tube (*see* **Notes 6** and **11**). Re-cut the (bunch of) plant part(s), while the cut edges are still submerged in the EDTA solution (*see* **Note 12**). If applicable, prepare the barrier layer (*see* **Note 5**).
3. Rapidly transfer the (bunch of) plant part(s), with the barrier layer (if applicable), into the tube with 1 mL EDTA solution, ensuring complete submergence of the cut edge(s). Cover the tube (*see* **Note 5**; see yellow sheets, tube lids and tubes in upside-down orientation in Fig. 1).
4. Incubate the tubes with the EDTA solution in an upright position in racks in the incubation chamber (*see* **Note** 7) for 2 h (*see* **Note 13**). Note the positions of the tubes so you can find them in the dark later.

3.2 Collection of Phloem Exudates in Water

1. Take the (bunch of) plant part(s) out of the tube and immediately submerge the cut edge(s) in the washing tube with water for 1 min (*see* **Note 13**).
2. Rapidly transfer the (bunch of) plant part(s), with the barrier layer (if applicable), to the tube with 1 mL collection solution (water), ensuring complete submergence of the cut edge(s), and again cover the tubes (*see* **Note 5**).
3. Further incubate the tubes in the incubation chamber for at least 2 h (*see* **Note 14**).
4. In between test samples, process blanks, i.e., tubes with water, which are set up and incubated in the same manner as the sample tubes, but without plant material (*see* **Note 15**).

3.3 Preparation of Aliquots for Chemical Analyses

1. Take the (bunch of) plant part(s) out of the tube and place the plant material (without the paper strip, barrier layer and cover) in a paper bag.
2. Pipette (an) aliquot(s) of the collection solution into (an) 1.5 mL tube(s) (*see* **Notes 8** and **16**) and immediately freeze it/them in liquid N_2 (*see* **Note 17**).

3. When all samples have been collected, transfer the tubes with aliquots to a freezer (e.g., −80°C or −20°C, depending on the stability of the analytes of interest).
4. Optionally: Dry the sampled plant part(s) in the paper bag in a drying oven until complete dryness and weigh it/them (*see* **Note 18**).
5. Perform chemical analysis/analyses with the samples. Often, samples are lyophilized (*see* **Note 19**) and re-dissolved in a solvent. Different analytical techniques may be used (*see* **Note 20**).

4 Notes

1. Leaves are usually used when fully developed, i.e., after the sink-to-source transition. Depending on the size and the planned chemical analyses, it may be necessary to pool several leaves (or other plant parts/organs) from different or (preferably) from the same individual (Fig. 1).
2. Na_2-EDTA or K_2-EDTA are commonly used, with EDTA concentrations of 5–20 mM; we suggest to use/test 8 mM, as the concentration should be high enough for efficient sieve tube plugging, but not too high to prevent (further) damage of plant tissues. The pH value affects the efficiency of exudation and is usually adjusted with NaOH or KOH to 7.0 or 7.5; keeping it constant (e.g., with HEPES) may be important, especially if phloem exudation occurs at high rates and may thus change the pH. The solution may be osmotically adjusted to the osmotic potential of the plant cells to avoid cell bursts (e.g., [21]).
3. If protein and RNA analyses should be done, one may add protease inhibitor and RNase inhibitor (e.g., [11, 22]), respectively, to avoid degradation of the compounds.
4. The size of the tubes needs to be scaled depending on the size and number of plant parts. For small plant parts, 1.5 mL tubes may be used, while larger plant parts require bigger (e.g., 15 or 50 mL or even larger) tubes. The volumes of EDTA solution and water have to be sufficient to ensure submergence of the cut edges of the plant parts.
5. Use a cover to prevent evaporation and contamination of the solutions during incubation. One may wrap an inert, stretchable sheet around the plant part(s) to hold them together and cover them, use a flexible rubber cap to hold them together and cover them, or use a tube lid or cover the system with another tube placed upside down and fixed with a sheet. If contaminations from plant parts below the cover could occur

(e.g., pollen from flowers), a sheet may be additionally used as a barrier layer.

6. Use a paper strip, fixed with tape, to hold pooled plant parts together and wrap plant parts (e.g., broad leaves) to ensure they fit into the tube.
7. To reduce the solution's uptake into the xylem via the transpiration stream [18], incubation is usually done in the dark at high humidity, i.e., when stomata are closed. Use a climate chamber/cabinet or a lightproof (plastic) box that is lined with wet towels. If the plant parts are within closed systems (*see* **Note 5**), the surrounding humidity may be less important. Phloem exudate collection in light under CO_2- and H_2O-saturated conditions has also been described [21].
8. Use tubes that are suitable for freezing in liquid N_2. For each type of chemical analysis, use one tube, or use the same tube for different analyses if the same sample preparation procedure can be applied.
9. To account for the diurnal variation of plant metabolism, use the middle of the light period, i.e., start some hours after the onset of daylight and finish some hours before the night period starts. Samples should be processed as fast as possible, one after the other, but leave enough time between samples to ensure that the later steps (washing, transfer to water, freezing aliquots) are manageable within the standardized incubation time frame.
10. Cut leaves at the base of the leaf blades close to the stem (Poaceae) or at the base of the petiole; cut flowers/inflorescences at their stalks, with sufficient distance to the flower(s). Ensure that plant parts are not squeezed and that no air enters by using a sharp cutting tool and working fast.
11. Gently tie up the plant parts at the upper half with a paper strip secured with tape, ensuring they are not damaged and that the strip is far enough from the cut edge(s) so it does not contact the incubation solutions in later steps.
12. Re-cutting is important to prevent sieve-tube embolism, which would hinder phloem exudation. Depending on the length of the petiole/stem, cut 1 mm to 1 cm.
13. Incubation in EDTA solution to prevent sieve tube plugging is required for 1–2 h, while phloem exudation afterward continues in water (no further plugging occurs at that time) [3]. This allows the use of water as a collection solution, avoiding further plant damage by EDTA and interference of EDTA with analytical methods. The washing step prevents transfer of EDTA to the collection solution. The fill level of the washing tube (water) should be high enough to ensure that

the plant part(s) submerge more than they did in the EDTA solution, which is necessary to completely wash off the EDTA solution.

14. The incubation duration may have to be adjusted to the study system, the concentration of the analytes of interest and the sensitivity of the analytical method(s). It has to be long enough to collect sufficient amounts of compounds, but longer incubation times may exacerbate artifacts. To determine an appropriate incubation duration, use test samples, start with 2 h incubation, and, if needed, increase the incubation duration until the analytes of interest are consistently found.
15. Use blanks to assess potential contaminations from the chemicals and tools. As the transfer of plant parts from the EDTA solution to the collection solution (i.e., water) is difficult to simulate, prepare blanks only for the phase of phloem exudate collection in water.
16. Less than 1 mL is available, as some of the collection solution sticks to the plant. The volume(s) of aliquot(s) depend on the subsequent analytical method(s). We usually use 300 or 400 μL per aliquot.
17. If samples will be lyophilized, use long forceps to place the tube(s) with the bottom part first in the liquid N_2 to ensure that the solution is frozen at the bottom.
18. The dry biomass of the plant parts cannot be used for absolute quantification of compounds due to unknown phloem exudation rates, which may differ in dependence on plant part, developmental stage and prior plant treatment. However, the dry biomass may be used to determine the total aboveground plant dry biomass.
19. For lyophilization, close the tubes with punctured lids. After lyophilization, "fluffy" solid may be visible at the bottom of the tubes due to high concentrations of compounds (e.g., sugars).
20. We exemplarily refer to studies in which phloem exudates were used for various chemical analyses here. Elemental composition may be assessed via ICP-AES [20], while Si can be specifically targeted with the molybdenum method [10]. For RNA, microarray or RNA-seq analysis may be performed [21, 22]. LC-MS is suitable for measurements of proteins [8, 11], phytohormones [13] and various specialized metabolites [15, 17]. Profiling of amino acids may be done using LC-FLD with pre-column derivatization, while GC-MS after sequential methoximation and silylation is suitable for analysis of further (semi-)polar primary metabolites such as sugars and other organic acids [9, 15, 17]. GC-MS may also be used

for the analysis of lipids (e.g., sterols after silylation [14]). For all analytical method(s), a careful analysis and interpretation of the data is required, given the unknown phloem exudation rates and potential artifacts associated with the EDTA-facilitated exudation technique. Comparisons with other methods listed in this book may be useful.

Acknowledgments

We thank Ruth Jakobs and Sandra Regina Lang for their valuable feedback on the manuscript.

Conflict of Interest Statements The authors have no conflicts of interest to declare that are relevant to the content of this chapter.

References

1. Broussard L, Abadie C, Lalande J, Limami AM, Lothier J, Tcherkez G (2023) Phloem sap composition: what have we learnt from metabolomics?. Int J Mol Sci 24:6917
2. Lohaus G (2022) Review primary and secondary metabolites in phloem sap collected with aphid stylectomy. J Plant Physiol 271:153645
3. King RW, Zeevaart JAD (1974) Enhancement of phloem exudation from cut petioles by chelating agents. Plant Physiol 53:96–103
4. Fisher DB, Frame JM (1984) A guide to the use of the exuding-stylet technique in phloem physiology. Planta 161:385–393
5. Kennedy JS, Mittler TE (1953) A method of obtaining phloem sap via the mouth-parts of aphids. Nature 171:528
6. Zhang J, Sun K, Wang Y, Qian W, Sun L, Shen J et al (2024) Integrated metabolomic and transcriptomic analyses reveal the molecular mechanism of amino acid transport between source and sink during tea shoot development. Plant Cell Rep 43:28
7. Blicharz S, Beemster GTS, Ragni L, De Diego N, Spíchal L, Hernándiz AE et al (2021) Phloem exudate metabolic content reflects the response to water-deficit stress in pea plants (*Pisum sativum* L.). Plant J 106:1338–1355
8. Castañeda V, González EM, Wienkoop S (2021) Phloem sap proteins are part of a core stress responsive proteome involved in drought stress adjustment. Front Plant Sci 12:625224
9. Stallmann J, Pons CAA, Schweiger R, Müller C (2022) Time point- and plant part-specific changes in phloem exudate metabolites of leaves and ears of wheat in response to drought and effects on aphids. PLoS ONE 17:e0262671
10. Thorne SJ, Maathuis FJM, Hartley SE (2023) Induction of silicon defences in wheat landraces is local, not systemic, and driven by mobilization of soluble silicon to damaged leaves. J Exp Bot 74:5363–5373
11. Naalden D, Dermauw W, Ilias A, Baggerman G, Mastop M, Silven JJM et al (2024) Interaction of whitefly effector G4 with tomato proteins impacts whitefly performance. Mol Plant Microbe Interact 37:98–111
12. Pajar JA, Otto P, Leonar AL, Döll S, van Dam NM (2024) Dual nematode infection in *Brassica nigra* affects shoot metabolome and aphid survival in distinct contrast to single-species infection. J Exp Bot 75:7317–7336
13. Mamun MA, Lee B-R, Park S-H, Muchlas M, Bae D-W, Kim T-H (2024) Interactive regulation of immune-related resistance genes with salicylic acid and jasmonic acid signaling in systemic acquired resistance in the *Xanthomonas-Brassica* pathosystem. J Plant Physiol 302:154323
14. Ugine TA, Gill HK, Hernandez N, Grebenok RJ, Behmer ST, Losey JE (2021) Predator performance and fitness is dictated by herbivore prey type plus indirect effects of their host plant. J Chem Ecol 47:877–888
15. Bühler A, Schweiger R (2024) Previous infestation by conspecifics leads to a transient increase of the performance of *Sitobion avenae* aphids on wheat leaves. Ecol Entomol 49:476–488
16. Bühler A, Schweiger R (2024) Niche construction and niche choice by aphids infesting wheat ears. Oecologia 206:47–59
17. Jakobs R, Schweiger R, Müller C (2019) Aphid infestation leads to plant part-specific changes

in phloem sap chemistry, which may indicate niche construction. New Phytol 221:503–514

18. Najla S, Vercambre G, Génard M (2010) Improvement of the enhanced phloem exudation technique to estimate phloem concentration and turgor pressure in tomato. Plant Sci 179:316–324
19. Kutyniok M, Persicke M, Müller C (2014) Effects of root herbivory by nematodes on the performance and preference of a leaf-infesting generalist aphid depend on nitrate fertilization. J Chem Ecol 40:118–127
20. Stolpe C, Giehren F, Krämer U, Müller C (2017) Both heavy metal-amendment of soil and aphid-infestation increase Cd and Zn concentrations in phloem exudates of a metal-hyperaccumulating plant. Phytochemistry 139:109–117
21. Deeken R, Ache P, Kajahn I, Klinkenberg J, Bringmann G, Hedrich R (2008) Identification of *Arabidopsis thaliana* phloem RNAs provides a search criterion for phloem-based transcripts hidden in complex datasets of microarray experiments. Plant J 55:746–759
22. Bakirbas A, Castro-Rodriguez R, Walker EL (2023) The small RNA component of *Arabidopsis thaliana* phloem sap and its response to iron deficiency. Plants 12:2782

Check for updates

Chapter 7

Using Aphids to Sample Phloem Sap

Gertrud Lohaus

Abstract

Phloem transport is crucial for plant growth and development because it mediates the transport of metabolites, mineral nutrients, proteins, and secondary compounds between plant tissues. The type of phloem loading with such compounds depends on their concentration in the phloem and the surrounding cells. To investigate these processes, it is very important to collect phloem sap from intact plants and determine the concentrations of the various compounds. This chapter describes aphid stylectomy, in which severed aphid stylets are used to collect phloem sap. Compared to other methods for collecting phloem exudate (e.g., direct stem incisions, EDTA-facilitated exudation), aphid stylectomy is probably the method to collect the most authentic phloem sap. With some modifications, this method can be used for various plant species, including monocots, herbaceous dicots, and tree seedlings. The probability that sufficient phloem sap will exude from the severed aphid stylet is often unpredictable and should therefore be tested for the respective plant species and the respective experimental conditions. It should also be noted that aphid stylectomy is time-consuming and requires specialized equipment and extensive experience.

Keywords Aphid stylectomy, Phloem sap collection, Phloem composition, Phloem metabolite concentration

1 Introduction

Sugars and further assimilates are transported in the phloem from the source to the sink tissues by bulk flow. To investigate phloem transport processes and the composition of phloem sap, the collection of pure phloem sap from intact plants is required. However, sampling phloem sap is difficult because sieve elements (SEs), the transporting cells in the phloem, are hidden within the plant tissue and surrounded by many other cell types. In addition, sieve elements have a very small diameter and are under high hydrostatic pressure. Sudden pressure release can affect many physiological processes in the phloem, such as wound responses and the redistribution of substances between different cell types, leading to a distorted composition of phloem sap [1].

Johannes Liesche and Sanna Sevanto (eds.), *Phloem*, Methods in Molecular Biology, vol. 3066,
https://doi.org/10.1007/978-1-0716-5488-0_7,

Nevertheless, various methods for collecting phloem exudate or sieve tube sap have been developed to investigate the chemical composition of phloem sap. These methods include direct incisions into the stem [2], ethylenediaminetetraacetic acid (EDTA)-facilitated exudation after cutting of petioles [3, 4] or aphid stylectomy [5–7]. Each method has several advantages and disadvantages [8, 9].

Direct incision into the stem to collect exudate has been used in *Ricinus communis* L., cucurbits, and oilseed rape [2, 10], but in many other plant species, spontaneous exudation does not occur. The disadvantages of the incision method include the potential contamination of exudate with substances from other cells and the potential effects of pressure release on many physiological processes [11].

Another method is the cutting of leaves and EDTA-facilitated exudation from the cut petioles [3]. This method is applicable to various plant species. EDTA chelates Ca^{2+} and other divalent cations. It is therefore used to prevent the closure of sieve pores by callose or P-proteins [3]. However, EDTA also impairs membrane integrity and leads to the leakage of metabolites and ions from cells [12]. As a result, EDTA can lead to several artefacts in the composition of the exudate. In addition, enzymatic activity at the cutting edge can change the phloem sap composition. For example, in addition to sucrose, large amounts of hexoses, which are not actually transported in the phloem, have been found [9, 13]. A further disadvantage of the EDTA-facilitated method is that pressure release can affect many physiological processes, potentially leading to changes in metabolite concentrations. Furthermore, it should be noted that this method does not allow for measuring the actual concentrations of components in the phloem sap, but only the composition of solutes in diluted petiole exudate [1, 9].

The aphid stylectomy method utilizes the ability of aphids to puncture a single sieve element [14], and the severed stylet is used like a precisely placed cannula to collect phloem sap. Due to the high hydrostatic pressure in the sieve elements, the phloem sap can exude out into the stylet. The first experiments were carried out with relatively large aphids settled on trees, and their stylets were severed by razor blades [5]. A few years later, experiments with herbaceous plant species with softer tissue were also possible, and radiofrequency or laser was used for microcautery of the smaller aphid stylets [6, 15]. The exuding phloem sap is collected by placing a micro-capillary over the cut end of the stylet [15]. It is important to prevent evaporation during collection, as phloem sap will otherwise become overly concentrated during sampling [15].

Among the methods, aphid stylectomy is probably the least invasive for obtaining phloem sap from intact plants. The reasons for this are that pressure in the sieve tubes changes little after stylet excision [16] and that phloem sap samples collected from different aphid species on the same plant had very similar sugar concentrations [17].

However, the aphid-stylectomy also has several disadvantages: it is generally time-consuming and requires specialized equipment. Also, often only small amounts of phloem sap (10–500 nL) can be obtained and not all plant species exudate sap from cut stylets. When aphids puncture sieve elements, they inject saliva to inhibit wound responses in plant tissue, and phloem sap may be somewhat influenced by substances or enzymes from the aphids' saliva [18]. Therefore, the exudate of an aphid stylet may differ slightly from the phloem sap of an aphid-free plant [5]. To minimize the potential impact of aphids, a small number of aphids should be used per plant/experiment, and the stylets should be cut shortly after they start feeding [8, 15]. Due to possible contamination of the very first exudated phloem sap with proteases or other enzymes, which occurs immediately after cutting the stylet [19], this sap should be discarded if possible. Nevertheless, considering all the advantages and disadvantages, the fluid obtained by aphid stylectomy is probably the most authentic source of phloem sap [13, 16].

The protocol below explains the steps (Fig. 1) for sampling of phloem sap with the laser-aphid-stylet technique. The method can be used for various monocotyledonous and dicotyledonous plants, such as maize [20], barley [21], sugar beet [22], rapeseed [23], or young trees such as oak [24]. Its application to *Arabidopsis* is more difficult because the plants are usually so small that stable placement of a leaf in front of the laser beam and the microscope is often not possible.

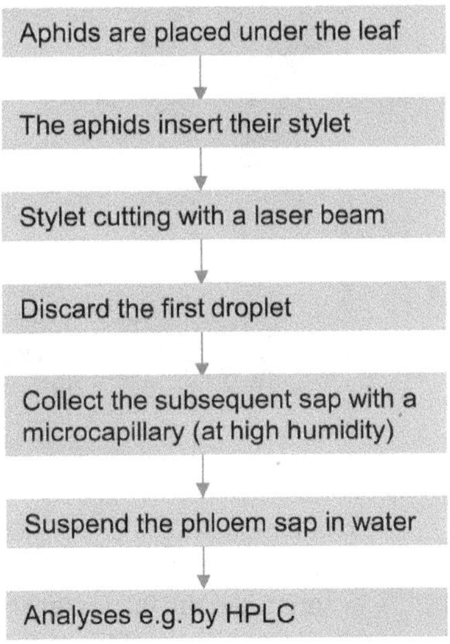

Fig. 1 Flowchart of phloem sap collection by aphid stylectomy

2 Materials

2.1 Plant Material and Aphids

1. Plants in pots are used for the experiments. The plants must be transportable to the workplace where the microscope, laser, and other equipment are located (Fig. 2a).
2. Aphids of the family Aphididae are used for the experiments. It should be noted that certain aphid species feed only on certain plant species. For example, the aphid species bird cherry-oat aphid (*Rhopalosiphum padi*, L.) can be used for maize and barley, and green peach aphid (*Myzus persicae*, Sulzer 1776) for rapeseed or sugar beet.

2.2 Solutions

1. Double-distilled water.

2.3 Disposables

1. Microcapillaries (0.5 μL).
2. Plastic reaction tubes (1.5 mL).

2.4 Cage for the Aphids

1. Homemade aphid cage to keep the aphids in place on the experimental plants before the stylectomy: The cage can be made from a piece of plastic tube (15 mm internal diameter, 10 mm high). The lower side of the plastic tube will be covered

Fig. 2 The photos show the setup of the microscope with the connected laser and the micromanipulators (**a**), the cage with aphids attached under the leaf with a spring clip (**b**), the materials for the aphid cage (**c**), the position of the leaf with the aphids in a holder in front of the microscope objective through which the laser flash is emitted (**d**)

with fine mesh. The edge of the plastic cage is covered with felt.

2. Small Plexiglas discs (2 cm×2 cm).
3. Spring hair clip.
4. Small brush.

2.5 Instruments

1. Stereomicroscope connected with a laser to cut the stylets of feeding aphids (laser device using a neodymium-glass laser or a YAG laser).
2. Micromanipulators with *x*-, *y*-, and *z*-axis movement.
3. Vibration-free benches.
4. Fiber-optic cold light source.
5. Humidifier.
6. A growth lamp for possible plant lighting and to stimulate photosynthesis.

3 Methods

3.1 Collection of Phloem Sap

1. Brush about ten aphids from a host plant into a small aphid cage (*see* **Note 1**). Place the small cage under the leaf of an experimental plant (intact growing plant) and place a small Plexiglas disc (2 cm×2 cm) on top of the leaf (Fig. 2b). Secure the cage with a spring hair clip without damaging the leaf (Fig. 2c). Use only one cage for each plant (*see* **Note 2**).
2. Wait 3–4 h for the aphids to settle and insert their stylets into the sieve tubes of the experimental plant.
3. Keep the plant's leaves illuminated during collection.
4. Remove the cage without disturbing the aphids and place the leaf with the aphids in a holder in front of the microscope lens through which the laser flash is emitted (Fig. 2d). The holder for the leaf should be movable with a micromanipulator so that the stylet of an aphid can be focused in the crosshairs.
5. Sap-feeding aphids (Fig. 3a) can be recognized by the fact that they are rather quiet and sometimes release drops of a sugary liquid called honeydew.
6. To cut the stylet, align the plant and aphid so that the center of the aphid labium is visible. The stylet, consisting of right and left maxillary and right and left mandibles, is visible as a dark brown line running along the front of the labium (Fig. 3b).

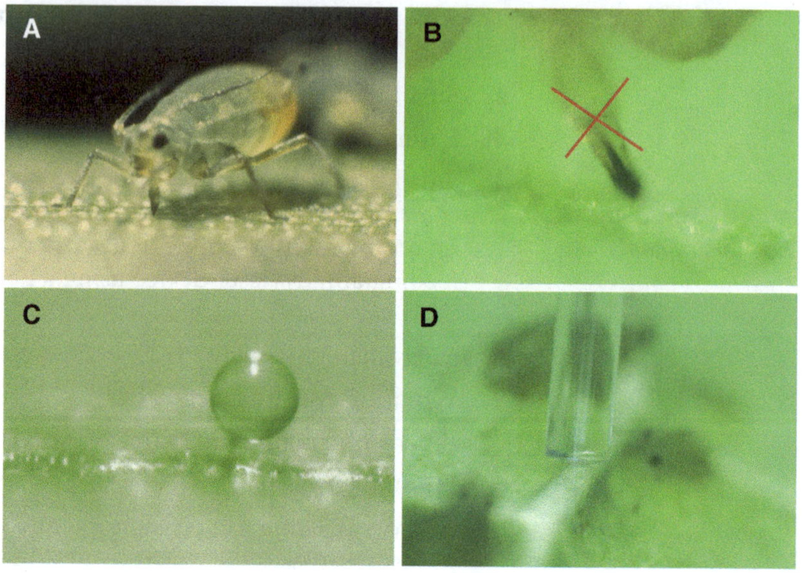

Fig. 3 The photographs show a phloem sap-feeding aphid (**a**), the proboscis of the aphid (the cross marks the spot where a laser beam is used to cut it) (**b**), the cut stylet with an exuding drop of phloem sap (**c**), and the microcapillary for collecting the phloem sap (**d**)

7. Cut the inserted stylets of feeding aphids with a focused laser beam. The laser system is fired through the microscope's objective lens, which is horizontally mounted on a metal rail. Repeat this until a stylet exudes phloem sap (Fig. 3c, *see* **Note 3**).
8. The aphid often moves away from the severed stylet, and the aphid body should be carefully removed from the leaf with a small brush.
9. Using a second micromanipulator, place a microcapillary (total volume 0.5 μL) over the cut end of the stylet to collect the exuded phloem sap (Fig. 3d) over a period of 0.5–4 h, depending on the exudation rate.
10. Evaporation of the phloem sap can be prevented very efficiently by bringing the front edge of the capillary in close contact with the leaf surface and surrounding the back end with a plastic cap [21]. In addition, the humidity around the capillary should be about 80%. Under these conditions, no evaporation from reference capillaries was detectable [20] (*see* **Note 4**).
11. In most cases, the duration of stylet exudation varies from a few minutes to a few hours (*see* **Note 5**), and a successfully severed aphid stylet secretes about 5–500 nL phloem sap. The

exudation rates per stylet range between 20 nL h^{-1} (e.g., sugar beet) and 200 nL h^{-1} (e.g., maize) (*see* **Note 6**).

12. The volume of the phloem sap can be determined by measuring the length it occupies in a calibrated 0.5 μL disposable microcapillary [20].
13. The initial drop of phloem sap after cutting the stylet should be discarded to prevent possible contamination with substances from the aphid salvia [9]. Depending on the phloem sap volume, the samples are ejected from the microcapillary into 50 or 100 μL of ddH_2O and stored at −80°C until analysis (*see* **Note 7**).

3.2 Analysis of Phloem Sap

If sufficient phloem sap can be obtained (at least about 10–20 nL), various analytical methods can be used. Since the concentrations of sugars (e.g., sucrose), amino acids and inorganic ions (e.g., potassium) in the sap are usually very high, these can be analyzed using various HPLC systems with appropriate detection methods [15, 20, 25]. Depending on the plant species, very small amounts of RNAs, proteins, micronutrients, and signal molecules may also be present in phloem sap. Therefore, the analysis of these compounds requires larger amounts of phloem sap in combination with modern and highly sensitive analysis techniques.

4 Notes

1. Stylectomy requires a suitable combination of plant and insect. Aphids of the family Aphididae are the most commonly used [9], but for some plant species, such as rice, planthoppers can also be used [26]. Some aphid species are restricted to plants of a particular plant family or even to just a few plant species.
2. It is advisable to place aphids in a cage on several plants at once, since not all aphids are feeding or the orientation of the stylet or the leaf is not suitable for stylectomy.
3. The probability that phloem sap will actually exude from the severed aphid stylet is often unpredictable and is often about 1 in 10 to 50, depending on the plant species, the aphid species and other conditions. The duration of phloem sap exudation is also usually unpredictable, but it depends at least on the plant species and the growth conditions of the plants. *"Aphid stylectomy can be frustrating. Some days the aphids appear unsettled, and no stylets are cut, and other days stylets are cut but do not exude"* (Quote from [27]).
4. Alternatively, in some applications, the cut stylet is covered with oil to prevent evaporation [16].

5. Only the phloem sap of one cut aphid stylet can be collected per period because the equipment (micromanipulator, microscope and laser) is blocked during sap collection. However, it is possible to collect sap from several cut stylets sequentially.
6. Usually, phloem sap is collected during the light period, but some stylets exudate phloem sap for more than 24 h and therefore phloem sap can be collected from one stylet during both the light period and in the dark period [21]. In the case of barley, for example, a successfully severed aphid stylet exuded phloem sap for such a long time that a series of samples could be taken from a single stylet within up to 72 h. During the illumination period, the phloem sap was exuded at a rate of 50–200 $nL\,h^{-1}$; in the dark, the exudation rate decreased by half, but increased again in the next light period.
7. In the phloem sap of various plant species, the sucrose concentration is between 400 and 1400 mM while the concentration of hexoses (glucose and fructose) is in the lower millimolar range (1–5 mM) [9]. By determining the sucrose or hexose concentration in the phloem sap, it can be checked whether the collected exudate is really phloem sap and has not been altered by enzymatic activities, for example. Glucose and fructose should be detectable only in traces or small amounts in stylectomy samples. If hexoses are found in higher concentrations, the sucrose has probably been hydrolyzed by cleavage enzymes, indicating contamination of the sap.

References

1. Turgeon R, Wolf S (2009) Phloem transport: cellular pathways and molecular trafficking. Annu Rev Plant Biol 60:207–221
2. Hall SM, Baker DA (1972) The chemical composition of *Ricinus* phloem exudate. Planta 106:131–140
3. King RW, Zeevaart JAD (1974) Enhancement of phloem exudation from cut petiols by chelating agents. Plant Physiol 53:96–103
4. Xu Q, Ren Y, Liesche J (2019) Studying phloem loading with EDTA-facilitated phloem exudate collection and analysis. In: Liesche J (ed) Phloem: methods and protocols, methods in molecular biology, vol 2014. Humana Press, New York, pp 125–133
5. Kennedy JS, Mittler JAD (1953) A method of obtaining phloem sap via mouth-parts of aphids. Nature 171:528
6. Barlow CA, McCully ME (1972) The ruby laser as an instrument for cutting the stylets of feeding aphids. Can J Zool 50:1497–1499
7. Fisher DB, Frame JM (1984) A guide to the use of the exuding stylet technique in phloem physiology. Planta 161:385–393
8. Dinant S, Bonnemain JL, Girousse C, Kehr J (2010) Phloem sap intricacy and interplay with aphid feeding. C R Biol 333:504–515
9. Lohaus G (2022) Review primary and secondary metabolites in phloem sap collected with aphid stylectomy. J Plant Physiol 271:153645
10. Walz C, Giavalisco P, Schad M, Juenger M, Klose J, Kehr J (2004) Proteomics of curcubit phloem exudate reveals a network of defence proteins. Phytochemistry 65:1795–1804
11. Zhang C, Yu X, Ayre BG, Turgeon R (2012) The origin and composition of cucurbit "phloem" exudate. Plant Physiol 158:1873–1882
12. Hepler PK (2005) Calcium: a central regulator of plant growth and development. Plant Cell 17:2142–2155
13. Liu DD, Chao WM, Turgeon R (2012) Transport of sucrose, not hexose, in the phloem. J Exp Bot 63:4315–4320

14. Peel AJ (1975) Investigations with aphid stylets into the physiology of sieve tube. In: Zimmermann MH, Milburn JA (eds) Encyclopedia of plant physiology. New series, vol 1. Transport in plants. I. Phloem transport. Springer, New York, pp 171–195
15. Riens B, Lohaus G, Heineke D, Heldt HW (1991) Amino acid and sucrose content determined in the cytosolic, chloroplastic, and vacuolar compartments and in the phloem sap of spinach leaves. Plant Physiol 97:227–233
16. Pritchard J (1996) Aphid stylectomy reveals an osmotic step between sieve tube and cortical cells in barley roots. J Exp Bot 47:1519–1524
17. van Helden M, Tjalingii WF, van Beek TA (1994) Phloem sap collection from lettuce (*Latuca sativa* L.): chemical comparison among collection methods. J Chem Ecol 20:3191–3206
18. Will T, Tjallingii WF, Thonnessen A, van Bel AJE (2007) Molecular sabotage of plant defense by aphid salvia. Proc Natl Acad Sci U S A 104:10536–10541
19. Furch AC, van Bel AJE, Will T (2015) Aphid salivary proteases are capable of degrading sieve-tube proteins. J Exp Bot 66:533–539
20. Lohaus G, Hussmann M, Pennewiss K, Schneider H, Zhu JJ, Sattelmacher B (2000) Solute balance of a maize (*Zea mays* L.) source leaf as affected by salt treatment with special emphasis on phloem re-translocation and ion leaching. J Exp Bot 51:1721–1732
21. Winter H, Lohaus G, Heldt HW (1992) Phloem transport of amino acids in relation to their cytosolic levels in barley leaves. Plant Physiol 99:996–1004
22. Winzer T, Lohaus G, Heldt HW (1996) Influence of phloem transport, N-fertilization and ion accumulation on the sucrose storage in the taproots of fodder beet and sugar beet. J Exp Bot 47:863–870
23. Lohaus G, Moellers C (2000) Phloem transport of amino acids in two *Brassica napus* L. genotypes and one *B. carinata* genotype in relation to their seed protein content. Planta 211:833–840
24. Öner-Sieben S, Lohaus G (2014) Apoplastic and symplastic phloem loading in *Quercus robur* and *Fraxinus excelsior*. J Exp Bot 65:1905–1916
25. Lohaus G, Schwerdtfeger M (2014) Comparision of sugars, iridoid glycosides and amino acids in nectar and phloem sap of *Maurandya barclayana*, *Lophospermum erubescens*, and *Brassica napus*. PLoS One 9:1
26. Fukumorita T, Chino M (1982) Sugar, amino acid and inorganic contents in rice phloem sap. Plant Cell Physiol 23:273–283
27. Gould N, Minchin PEH, Thorpe MR (2019) Using aphids to measure turgor pressure inside sieve elements. In: Liesche J (ed) Phloem: methods and protocols, methods in molecular biology, vol 2014. Humana Press, New York, pp 291–299

Check for updates

Chapter 8

Measurement of Inner Bark and Leaf Osmolality

Teemu Paljakka, Anna Lintunen, Yann Salmon and Teemu Hölttä

Abstract

Sugar transport in the phloem is driven by turgor pressure gradients, which are created by osmotic gradients resulting from sugars loaded into the phloem at the source tissue and unloaded at the sink tissue. Therefore, osmolality is a key parameter that can be used to evaluate sugar status and to indicate the driving force for phloem transport. Here, we describe how osmotic concentration measurements from inner bark and needles of trees can be made. This protocol presents the procedure used by Lintunen et al. [15] and Paljakka et al. [16], and extends it with practical advice and discussion of potential errors and caveats. We describe how to implement this procedure for gymnosperm and angiosperm trees. This method uses mechanical sap extraction with a centrifuge from inner bark and leaf samples that have undergone deep-freeze treatment and thawing. The osmotic potential of these samples is then analyzed using a freezing-point or vapor-pressure osmometer. The aim of these measurements is to study the spatial and temporal dynamics of phloem function.

Keywords Phloem, Osmolality, Osmotic potential, Water potential, Conifer, Method description

1 Introduction

Osmolality is an important quantity in plant function because it is required to draw in water, which is needed to generate turgor pressure that sustains metabolic activity. Furthermore, solute transport in phloem is driven by turgor pressure gradients, which, in turn, are created by osmotic gradients resulting from sugars loaded into the phloem from the source tissue and unloaded in the sink tissue [1, 2]. The phloem water potential is the sum of turgor pressure and osmotic potential. Osmolality is linearly proportional to osmotic potential and independent of temperature, as it is a measure of mass and not volume [3]. Osmolality of the inner bark or leaves is a rather robust measure of the overall contribution of osmotically active solutes and water in the living cells of these tissues. Interpreting the phloem solute content based on more detailed sugar concentration analyses has challenges in terms of comparing absolute values among different methods and laboratories [4]. Moreover, aiming to measure the osmolality of the

Johannes Liesche and Sanna Sevanto (eds.), *Phloem*, Methods in Molecular Biology, vol. 3066,
https://doi.org/10.1007/978-1-0716-5488-0_8,

conducting phloem only is more challenging than measuring the osmolality of the whole inner bark from the sampling point of view [5], as the conducting phloem is a small fraction of the inner bark, and methods for separating the tissues are challenging under field conditions. The phloem is also highly sensitive to damage caused by sampling, e.g., because of pressure release from the sieve cells [6, 7]. Together, these issues create a high risk of error when interpreting experimental data with regard to the dynamics of phloem transport.

The sampling and sap extraction procedure described in this chapter estimates the in situ osmolality of inner bark [1, 5, 8, 9], including tissues from cambium to cork cambium, and the in situ osmolality of conifer needles [10, 11] and angiosperm leaves [8, 12]. This procedure relies on the assumption that solutes from the symplast are released to the apoplast during deep freezing and thawing, followed by mechanical sap extraction [13]. The sampled sap osmolality is a measure of the quantity of all the dissolved substances, such as sucrose, monosaccharides, alcohols, potassium, and other osmotically active solutes, in a given amount of solution. Water potential and tissue properties change along the stem, and between branches and the stem, with a larger proportion of sieve cells toward the stem base [3, 14]. Therefore, differences in solution mixtures are expected at different tree parts. Note that the procedure described here is not targeted to measure the osmolality of the phloem sap transported in the sieve tubes. Instead, the osmolality of the whole inner bark or leaves is measured. Estimates of sap solute content (*see* **Note 1**) and turgor pressure (*see* **Note 2**) are possible with osmolality measurements when underlying assumptions and limitations are taken into consideration.

2 Materials

Materials in this section are described as in the studies of Lintunen et al. [15] and Paljakka et al. [16], and the measurement details with the vapor pressure osmometer as in Salmon et al. [17].

2.1 Sampling

1. Scalpel or knife for collecting inner bark samples.
2. Freezing-tolerant cryo tubes (e.g., 5 mL cryo tube).
3. Dry shipper, cool box with dry ice, or liquid nitrogen for transporting samples, and for the temporal deep-freezing treatment.
4. Deep freezer set to −80°C or a liquid nitrogen tank for preserving the samples for longer times.

2.2 Sap Extraction and Measurement of Osmolality

1. Microcentrifuge, equipped with a 24-place rotor for 1.5 or 2 mL tubes, capable of a speed of 14,000 × *g*.
2. Freezing point osmometer (Osmomat 030 cryoscopic osmometer, Gonotec, Berlin, DE) or vapor pressure osmometer (Vapro 5520, Wescor Inc., Logan, UT, USA).

3. Distilled water and 0.300 mol kg^{-1} NaCl solution for calibration of the freezing point osmometer. For higher sample concentrations than 0.700 mol kg^{-1}, an additional calibration solution with a higher concentration, e.g., NaCl solution of 0.850 mol kg^{-1} (three-point calibration), should be included.
4. Pipettes with disposable tips for moving small amounts of liquid sample (e.g., 20 μL).
5. Tubes for the centrifuge with a liquid-separating column; for example, cellulose acetate-filtered centrifuge tubes (e.g., Spin-X Centrifuge tube filters, 0.45 μm, Costar Inc., US) or silica-based membrane columns can be used.
6. Measurement vessels for the freezing-point osmometer or paper sample discs for the vapor-pressure osmometer.
7. Scalpel and forceps.
8. Distilled water and ethanol solution for cleaning of tools during sample preparation.

3 Methods

3.1 Sampling in the Field

When planning sample positions and the number of replicates, it should be taken into consideration that osmolality values vary spatially and temporally (*see* **Note 3**). In addition, the amount of damage to the tree caused by sampling should be considered if the tree is monitored for longer periods.

1. Remove cork before collecting the inner bark, e.g., with a knife or scalpel (*see* **Note 4**).
2. Collect the inner bark sample, typically a piece of ca. 2 cm^2 (*see* **Note 5**). A scalpel has been found to be a good tool for cutting the sample borders in the bark. Thereafter, the collection could be made with a blunt knife, which allows better bark extraction without cutting into the underlying xylem. Sample collection from branches can be carried out the same way with a similar sample size as in the stem, or with smaller branches by first cutting the branch and then collecting the bark sample by peeling it off the wood surface.
3. Pick leaf samples by hand if the canopy is within reach (*see* **Note 6**). Five pairs of *Pinus sylvestris* needles are often more than enough to extract the required liquid sample (ca. 20–50 μ L or more). Filling the centrifuge tube filter (500 μL) with needles of *Picea abies* branch tips usually results in ca. 20–50 μL of sap sample (Fig. 1A, B). Often, one leaf is enough to obtain a

sufficient amount of sample in angiosperms, but this depends on the species.

3.2 Storage

1. After collection, immediately place the sample into a sealed cryo tube, and place the tube into a dry shipper or dry ice to avoid evaporation and changes in solute composition.
2. Transfer samples to a −80°C freezer or a liquid nitrogen tank if sample extraction is not performed immediately after deep-freeze treatment (*see* **Note 7**).

3.3 Preparation of Samples for Measurements

1. Thaw samples inside the closed tubes at room temperature for ca. 15 min.
2. Before placing bark samples in centrifuge tubes containing the liquid separating column, cut the end of the inner bark sample facing the filter with a scalpel to ease the collection of sap (Fig. 2A). Filtered centrifuge tubes are only used to separate the sap, including the dissolved compounds, from the solid parts of the plant tissue. *P. sylvestris* needles can be cut in half and the cut side set downward against the filter, whereas *P. abies* needles can be cut at needle base when detached from the branch tip and set cut end against the filter (Fig. 1A, B). Angiosperm leaves can be cut perpendicular to the main leaf vein, in, e.g., 5 mm wide slices, and rolled inside the tubes with the cut side downward against the filter. Another option is to grind frozen leaves in a mortar or inside a zipper bag to separate leaf petioles prior to the sample being set inside the tube filter (Fig. 2B). To avoid cross-contamination between samples, sampling equipment, especially scalpel blades, should be cleaned between preparations, e.g., with ethanol solution.
3. Centrifuge at 14,000 × *g* for 10 min for efficient liquid extraction from the sample [9, 15, 16].

Fig. 1 *Picea abies* needles cut from a branch tip (**a**) and placed in a centrifuge tube filter (**b**)

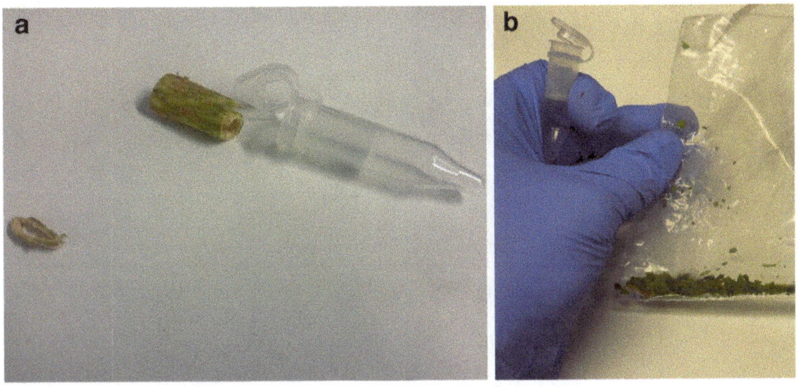

Fig. 2 Inner bark of *Pinus sylvestris* branch, with the filter-facing end toward the tube, to be set inside a filtered centrifuge tube (**a**), and a frozen leaf sample ground inside a zipper bag (**b**)

3.4 Osmolality Measurement with Freezing Point Osmometer

Devices for measuring osmolality differ. One option is to use a freezing point osmometer, where osmolality is defined with the equilibrium freezing point of a solution [15, 16].

1. Calibrate the device following the device manual. In the studies of Lintunen et al. [15] and Paljakka et al. [16], the osmometer was calibrated with a two-point calibration with distilled water and 0.300 mol kg^{-1} solutions of NaCl. Three-point calibration (0, 0.300, and 0.850 mol kg^{-1}) is recommended to better match the measurement range of plant tissue. With this calibration setup, ±1% variance from linearity can be obtained in the calibrated range with ±2% reproducibility in a small sample size (e.g., Osmomat 030, Gonotec, Berlin, DE).
2. Transfer the sap extracted from the tissue to the measurement vessels designed for osmolality measurement using a pipette (*see* **Note 8**). The sap sample should not include air bubbles or visible impurities since they can interfere with the measurement (*see* **Note 9**). A similar liquid sample size is recommended for consistent measurements.
3. Place the vessel in the instrument with the instrument sensor submerged in the sampled sap (Fig. 3A).
4. Measure the osmolality of the sample according to the instrument manufacturer's instructions.
5. Write down the measured osmolality value, which is given in the unit mol kg^{-1}.
6. Rinse the osmometer sensor with Milli-Q or distilled water, and wipe it with lens paper between each measurement.

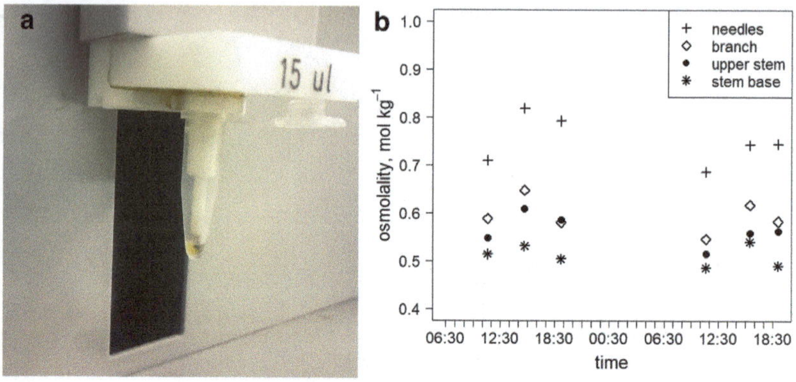

Fig. 3 Extracted sap transferred to the osmolality vessel after centrifugation and is ready for measurement with the freezing point osmometer (**a**). Example set of in situ osmolality values in needles and the inner bark of branches, upper stem and stem base in *Pinus sylvestris* during 2 days in September in Southern Finland (**b**). (Data adapted from Paljakka et al. [15])

3.5 Osmolality Measurement with Vapor Pressure Osmometer

1. Calibrate the device following the device manual.
2. Open the sample chamber and, using forceps, place a single sample disc in the disc holder depression.
3. Carefully, take precisely $10 \pm 1\,\mu L$ from the sample using a pipette with disposable tips (change tips after each sample) and place them on the disc.
4. Close the sample chamber and wait for reading.
5. Write down the measured osmolality value.
6. Remove the disc after the measurements and clean the sample holder with a lint-free tissue.

4 Notes

1. To estimate the amount of osmotically active solutes in the sample [5, 16, 18], one can also measure the relative water content [19], for example, from an adjacent bark piece. Then, in situ osmolality values can be used in the calculation of the osmotically active solutes (osmolality at full saturation) as follows [18].

$$\text{osMOL full saturation} = \text{osMOL in situ} \times \text{RWC} \qquad (1)$$

where osMOL in situ is the in situ osmolality, and RWC is relative water content. Osmotic potential of rehydrated plant tissue is a common measure, e.g., when determining pressure-volume curves [20].

2. Assuming water potential equilibrium between the xylem and inner bark, the inner bark turgor pressure can be estimated by measuring the xylem water potential simultaneously at the same position when collecting the samples for osmolality [3, 8, 21]. From leaves and branches, this can be conducted with a pressure chamber. The method described in this chapter does not separate apoplastic water and is expected to dilute the osmotic potential and turgor pressure estimates of the symplast [22]. Thus, corrections for apoplastic water have been applied for more accurate estimates of osmotic potential in the symplast [23].
3. With the procedure described in this chapter, osmolality in situ has been observed to range from ca. 0.4 mol kg^{-1} in the stem base up to ca. 0.7 mol kg^{-1} in the inner bark of branches, and up to around 0.8 mol kg^{-1} in the needles in *P. sylvestris* and *P. abies* [15, 16] in boreal conditions. These values are similar to previously reported osmolality values in the needles and upper stem of *P. sylvestris* and *P. abies* [5, 11]. Higher values are also expected, especially when trees experience water stress. The daily variation of osmolality can be more than 0.1 mol kg^{-1} depending on the sample location in a tree (Fig. 3B). The mean inner bark osmolality of branches has been reported to be ca. 0.5–0.7 mol kg^{-1} in *Populus tremula* and *Betula pendula* in moist and dry sites around Europe [15]. Signs of seasonal differences in osmolality values have been found at least in conifer needles [16]. Additionally, environmental variables, such as VPD, light, soil water content, and temperature, can be expected to affect osmolality through their effects on source and sink strength and xylem water potential.
4. Removing the cork entirely from a detached bark piece is sometimes challenging, especially in the lower parts of the conifer stem. Additionally, the inner bark of, for example, birch branches has been found difficult to separate from the cork as the branches toward the tips are thin. However, small amounts of cork, with little water or osmotically active solutes, can be expected to affect the osmolality of a sampled sap only marginally.
5. The sample size depends ultimately on the minimum sample size the osmometer requires for measuring, e.g., 15 μL of liquid sample [15, 16]. Salmon et al. [17] used a vapor pressure osmometer for which precisely 10 μL samples are needed. In theory, sample size should not matter as we assume a mixture of solutes per mass. False readings (i.e., instrument returning values that are too high) have been encountered when the amount of extracted sap is too small (less than the minimum sample amount set by the manufacturer). Furthermore, small samples are more prone to evaporation and thus to changes in

solute concentration. The amount of sap required determines the size of the bark sample to be collected. This will vary across locations, tree species, and environmental conditions. For example, the minimum bark sample size of *P. sylvestris* at our site in Southern Finland is ca. 2 cm^2, depending on the sample position in the tree. A piece of this size yields ca. 20–50 μL of sap. The inner bark in the stem of *P. sylvestris* is thicker in the younger stem than at the stem base. Therefore, a smaller piece (e.g., 0.5 cm width × 1.5 cm vertical length) may be enough at higher positions in the stem to acquire the required amount of sample. However, the sufficient sample size might be larger in dry conditions and on dry sites, and this should be tested before the collection. For very dry samples (water potential < −2.5 MPa for *P. sylvestris*), it might be impossible to extract enough sap with centrifugation.

6. Excess water should be removed from the samples to avoid dilution of the extracted sap if the sampling has been conducted in rainy conditions.
7. Freezing at very low temperatures (e.g., −80°C) and thawing are expected to rupture most of the membranes in living cells, releasing solutes into the apoplast (e.g., [13]). Deep-freeze treatment eases mechanical sap extraction but yields a mixture of apoplast and symplast sap from the sampled tissue.
8. After the liquid is extracted by centrifugation, great care has to be taken to prevent evaporation from the small sample volume. The sample number at each set of analyses is recommended to be kept sufficiently small to avoid evaporation between sap extraction and analysis. Additionally, evaporation of the samples should be avoided by closing tubes with lids and by carrying out the measurement procedure from thawing to osmolality measurement as quickly as possible.
9. The freezing point osmometer may sometimes report "spontaneous crystallization" due to impurities in the solution, which affect its freezing point. Because of this, more than one sample should be obtained from each location.

References

1. Münch E (1930) Die Stoffbewegungen in der Pflanze. Gustav Fischer Jena, Jena, pp 124–146
2. Liesche J, Schulz A (2018) Phloem transport in gymnosperms: a question of pressure and resistance. Curr Opin Plant Biol 43:36–42
3. Nobel PS (2005) Physiochemical and environmental plant physiology, 3rd edn. WH Freeman and Company, New York
4. Quentin AG, Pinkard EA, Ryan MG, Tissue DT, Baggett LS, Adams HD et al (2015) Non-structural carbohydrates in woody plants compared among laboratories. Tree Physiol 35(11):1146–1165
5. Rosner S, Baier P, Kikuta SB (2001) Osmotic potential of Norway spruce [*Picea abies* (L.) Karst.] secondary phloem in relation to anatomy. Trees 15(8):472–482
6. Sovonick-Dunford S, Lee DR, Zimmermann MH (1981) Direct and indirect measurements of phloem turgor pressure in white ash. Plant Physiol 68(1):121–126

7. van Bel AJ (2003) The phloem, a miracle of ingenuity. Plant Cell Environ 26(1):125–149
8. Kaufmann MR, Kramer PJ (1967) Phloem water relations and translocation. Plant Physiol 42(2):191–194
9. Devaux M, Ghashghaie J, Bert D, Lambrot C, Gessler A, Bathellier C, Loustau D (2009) Carbon stable isotope ratio of phloem sugars in mature pine trees throughout the growing season: comparison of two extraction methods. Rapid Commun Mass Spectromet 23(16):2511–2518
10. Huttunen S, Kärenlampi L, Kolari K (1981) Changes in osmotic potential and some related physiological variables in needles of polluted Norway spruce (*Picea abies*). Ann Botanici Fennici 18(1):63–71
11. Irvine J, Perks MP, Magnani F, Grace J (1998) The response of *Pinus sylvestris* to drought: stomatal control of transpiration and hydraulic conductance. Tree Physiol 18(6):393–402
12. Scholander PF, Hammel HT, Hemmingsen EA, Bradstreet ED (1964) Hydrostatic pressure and osmotic potential in leaves of mangroves and some other plants. Proc Natl Acad Sci U S A 52(1):119–125
13. Mason TG, Phillis E (1936) The concentration of solutes in sap and tissue, and the estimation of bound water. Ann Bot 50(199):437–454
14. Jyske T, Hölttä T (2015) Comparison of phloem and xylem hydraulic architecture in Picea abies stems. New Phytol 205(1):102–115
15. Lintunen A, Paljakka T, Jyske T, Peltoniemi M, Sterck F, Von Arx G et al (2016) Osmolality and non-structural carbohydrate composition in the secondary phloem of trees across a latitudinal gradient in Europe. Front Plant Sci 7:726. https://doi.org/10.3389/fpls.2016.00726
16. Paljakka T, Jyske T, Lintunen A, Aaltonen H, Nikinmaa E, Hölttä T (2017) Gradients and dynamics of inner bark and needle osmotic potentials in Scots pine (*Pinus sylvestris* L.) and Norway spruce (*Picea abies* (L.) Karst). Plant Cell Environ 40(10):2160–2173. https://doi.org/10.1111/pce.13017
17. Salmon Y, Torres-Ruiz JM, Poyatos R, Martinez-Vilalta J, Meir P, Cochard H, Mencuccini M (2015) Balancing the risks of hydraulic failure and carbon starvation: a twig scale analysis in declining Scots pine. Plant Cell Environ 38(12):2575–2588
18. Takami S, Turner NC, Rawson HM (1981) Leaf expansion of four sunflower (*Helianthus annuus* L) cultivars in relation to water deficits. I. Patterns during plant development. Plant Cell Environ 4(5):399–407
19. Smart RE, Bingham GE (1974) Rapid estimates of relative water content. Plant Physiol 53(2):258–260
20. Bartlett MK, Scoffoni C, Ardy R, Zhang Y, Sun S, Cao K, Sack L (2012) Rapid determination of comparative drought tolerance traits: using an osmometer to predict turgor loss point. Method Ecol Evol 3(5):880–888
21. Hölttä T, Vesala T, Sevanto S, Perämäki M, Nikinmaa E (2006) Modeling xylem and phloem water flows in trees according to cohesion theory and Münch hypothesis. Trees 20(1):67–78
22. Tyree MT (1976) Negative turgor pressure in plant cells: fact or fallacy?. Can J Bot 54(23):2738–2746
23. Gersony JT, Hochberg U, Rockwell FE, Park M, Gauthier PP, Holbrook NM (2020) Leaf carbon export and nonstructural carbohydrates in relation to diurnal water dynamics in mature oak trees. Plant Physiol 183(4):1612–1621

Chapter 9

Infrared Microimaging of Sucrose Distribution in Plant Vascular Tissues

Matthias Langer, Andre Gündel, Steffen Wagner, Hardy Rolletschek and Ljudmilla Borisjuk

Abstract

Sucrose is the primary transport sugar in plants, serving as an essential energy source and signaling molecule. Detection, visualization, and quantification of sucrose in various plant tissues are essential for understanding the metabolic and physiological processes that sustain plant life. Traditional metabolite-mapping techniques have struggled to visualize the quantitative distribution of sucrose at sufficient resolution to distinguish vascular bundles from surrounding tissues. Here, we present a Fourier-transform infrared (FTIR) imaging approach that can visualize sucrose in plant tissues quantitatively at a microscopic resolution (~12 μm). This IR-based, label-free method can be used with both model plants and agriculturally important crops. The assay has a detection range of 20–1000 mM and can map sucrose distribution within complex organs such as stems, leaves, and seeds. Notably, it enables the precise quantification of sucrose levels in vascular tissues. This is a trait of great interest in many current breeding and plant biotechnology approaches aimed at increasing crop yield.

Keywords FTIR, Sucrose imaging, Vascular tissues, Phloem

1 Introduction

Sucrose is a central metabolite in higher plants, functioning not only as a major product of photosynthesis and a primary transport form of fixed carbon, but also as a key signaling molecule regulating gene expression and development [1]. As the predominant transport form of carbon in most plant species, sucrose is synthesized in mesophyll cells and translocated via the vasculature to various sink tissues. Beyond its role as a carbon source, sucrose serves as a major storage carbohydrate, particularly in sugarcane and sugar beet, where it can account for up to 20% of the fresh weight. Quantitative imaging of sucrose, along with other metabolites, is essential for understanding its central role in source-sink relations and plant physiology in general [2].

Johannes Liesche and Sanna Sevanto (eds.), *Phloem*, Methods in Molecular Biology, vol. 3066,
https://doi.org/10.1007/978-1-0716-5488-0_9,

Due to its significance, a wide range of methods for sucrose detection have been developed since the introduction of Dreywood's approach (anthrone reagent) for carbohydrate quantification. Most conventional methods require tissue extraction followed by enzymatic cleavage of sucrose. In colorimetric and fluorometric assays, for example, sucrose is hydrolyzed by invertase or sucrase into glucose and fructose. These are then quantified using glucose oxidase/peroxidase systems [3, 4]. Alternatively, sucrose can be quantified using pulsed amperometric detection [5]. In this case, the technique measures the current produced when electroactive species (like sucrose and other sugars) are oxidized or reduced at an electrode surface. Amperometry is normally combined with either flow-injection or liquid chromatography (LM) applications, where the individual sugars in the tissue extract are separated on the column [6]. For gas chromatography (GC), derivatization is required to enable sucrose quantification [7]. Modern metabolomics approaches based on mass spectrometry (MS), such as LC-MS and GC-MS, enable quantification of sucrose as part of broader metabolite profiling following chromatographic separation [8]. While these approaches are precise and highly informative, they do not preserve spatial information unless they are combined with tissue-specific dissection/extraction methods, such as laser microdissection and pressure catapulting (LMPC), to reconstruct the topographical arrangement of sampling.

The innovative hybrid technologies of matrix-assisted laser desorption/ionization mass spectrometric imaging (MALDI-MSI) and secondary ion mass spectrometric imaging (SIMS-MSI) evolved to provide high-resolution molecular imaging. These techniques are used to study the tissue distribution of isotopically labeled and nonlabeled compounds in ex vivo and in situ plant samples. Unfortunately, the resolution and specificity of these methods are compromised when the detection target, such as sucrose, has a low molecular weight (Mr) due to matrix effects [9].

Visualization of sucrose, sufficient for determining its abundance in different tissues and vascular bundles, remained closely tied to isotopic labeling of sucrose molecules. First successfully applied in the 1960s, it has enabled sucrose localization, measurement, and visualization in plants on a large scale, from tissues to organisms [10]. Currently, positron emission tomography (PET) is used to trace photoassimilate allocation, but so far it lacks the necessary level of chemical and spatial resolution when targeting sucrose [11]. Magnetic resonance imaging (MRI) can measure 13C-labeled sucrose and can also operate label-free in vivo; however, it can be challenged when sucrose is present at low concentrations [12]. While promising, MRI can be limited by tissue magnetic heterogeneity, and metabolic MRI requires expensive equipment [13].

In order to quantitatively visualize sucrose in plant tissue by means of an optical microscope, a method was developed based on complex enzymatic cocktails to specifically assess sucrose directly in tissue cryosections via bioluminescence and single-photon counting [14]. Although this bioluminescence method enabled sucrose imaging with a resolution close to the single-cell level, it remained technically challenging and had limited versatility.

An informative, yet virtuous in application, method is fluorescence resonance energy transfer (FRET), which provides a high level of chemical and spatial resolution [15]. Specific FRET sensors for sucrose are available and are used, for example, to determine sucrose transport activity [16]. However, their practicality for crops is questionable, given the need to genetically transform the plant and certain technical limitations. We conclude that there is still a need for a convenient, high-resolution method of quantifying and visualizing sucrose in vascular tissues.

Over the last decade, infrared (IR) spectroscopy, also known as multispectral imaging or microspectroscopy, has found widespread application in various fields of research. Unlike MRI, which relies on the magnetic resonance of nuclei, and mass spectrometry imaging (MSI), which relies on ionized metabolites, vibrational spectroscopy measures the interaction of light with vibrational modes of molecules. Infrared spectroscopy offers substantial analytical potential and, when coupled to a microscope, provides both chemical and spatial resolution at the low-micrometric level. However, its use in quantitative analysis of metabolite distribution in tissue sections remains limited by the complexity of the resulting spectral matrix, which obscures specific metabolite signatures. Therefore, an algorithmic decomposition is required to recover a given compound-specific signal. Additionally, there has been a lack of means to standardize the analytical procedure and reproduce the data quantitatively.

Over time, it has been demonstrated how these challenges could be overcome to achieve quantitative visualization of sucrose distribution within plant tissues. This has led to the development of a new IR-based method for microimaging of sucrose. Gündel et al. introduced the method for characterizing sucrose distribution in barley seeds, leaves and stems [17]. They demonstrated that the new method enables the quantitative visualization of sucrose distribution in individual vascular bundles. The same approach was used to demonstrate that the presence of certain pathogens in the root stimulates sucrose flux towards infected tissue, providing the pathogen with a carbon source [18]. Finally, a study combining sucrose imaging with vascular chemotyping revealed that vascular tissues are compartmentalized [19]. Fourier-transform infrared (FTIR)-based estimates of the mean sucrose concentration in phloem sap corresponded well with levels obtained from the analysis of aphid style content [20]. IR-based sucrose imaging has become a useful screening tool for identifying mutants/transgenics that differ in their capacity to transport sucrose [21].

This chapter describes the latest method of quantitatively visualizing sucrose in plant tissues using FTIR spectroscopy. The protocol outlines each step of the workflow, from sample preparation and sectioning to spectral data acquisition, and provides an insight into the capabilities and applications of this technology. To complement the protocol and its applications, representative FTIR images are included to illustrate the spatial distribution of sucrose in plant vascular tissues. The examples demonstrate how cryosectioned samples, when analyzed with FTIR microspectroscopy, reveal distinct sucrose accumulation patterns across anatomical features such as the phloem and xylem. As expected, sucrose levels are highest in phloem, with only minimal levels detected in xylem and surrounding parenchyma (Fig. 1). Analysis of soybean petioles, however, revealed that sucrose levels in the cambium ring can also be elevated (exemplified here, see below Fig. 6). The functional significance of this finding remains to be investigated, but showcases the analytical power of FTIR-based sucrose imaging, opening novel views on plant physiology and beyond.

The introduction of new technology necessarily raises the question of whether the outcome accords with what has been measured by established methods. It is important to note that FTIR imaging of sucrose in phloem tissue typically yields lower concentrations than aphid/microsampling techniques, which measure sap from individual sieve elements [22]. This discrepancy arises because FTIR integrates signals from the entire tissue—including cells, sap, walls, and intercellular spaces—rather than just fluid sap. This example shows that different methods are often both exact in what they measure within their respective experimental setups. However, a direct comparison is only possible if the general experimental setup yields similar data

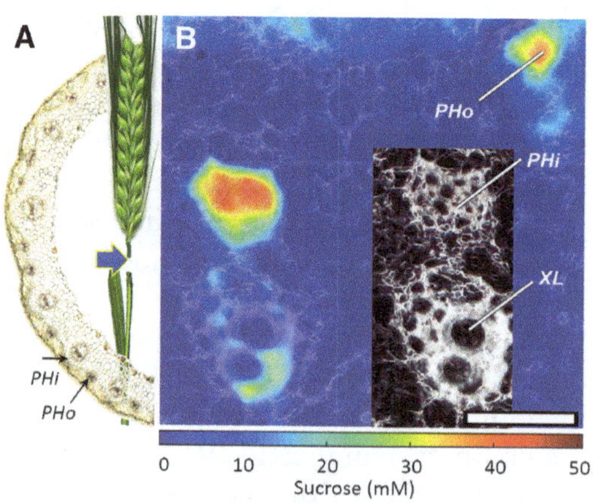

Fig. 1 Quantitative imaging of sucrose in the barley stem. (**A**) The barley spike and a cross-section through the stem. Position of the section is arrowed. (**B**) Color-coded map used to visualize sucrose gradients across the stem. The FTIR image was scaled to match the resolution of the visual image (shown as an inset), revealing the cellular structure of a vascular bundle. Tissue section thickness was 16 µm. Scale bar = 50 µm. Phi Inner vasculature phloem, PHo outer vasculature phloem, XL inner vasculature xylem

under the same context. In the case of sap, the target is sucrose in the context of the liquid fraction extracted from the sample as the average of the sample region of origin over a duration of time, while FTIR sucrose imaging yields sucrose in the context of a total volume of solid and liquid sample fractions at a particular location and time point. So, with sap analysis, we gain an average sucrose concentration from a fraction of a sample region during a time period, while imaging provides a local sucrose concentration over a total sample region at a specific point in time.

The advantages of the presented FTIR-based method in quantification, tissue-specific resolution, and reproducibility are achieved through internal standardization and advanced algorithms for signal normalization. A uniformly distributed internal standard compensates for technical variability, enabling reliable pixel-level quantification. The method is diffraction-limited (~10 μm optical resolution), and artifacts can arise from sample preparation, like recrystallization patterns, especially in cryosections with mixed tissue types or large fluid-filled regions or tissue shrinkage under freeze drying. Good a priori knowledge of major contributors to the infrared signal is essential for selecting an adequate reference spectral library for deconvolution. If major components are underrepresented in the reference model, other components with similar spectral features will become artificially overrepresented.

We are convinced that the FTIR-based, spatially resolved, quantitative imaging method enables deeper insights into mechanisms of carbon allocation, metabolic regulation, and responses to environmental cues, thereby advancing both fundamental research and crop improvement efforts.

2 Materials

1. Adhesive Cryofilm: LMD film 18 (Section-lab Co. Ltd., Hiroshima, Japan) or nonadhesive MMI slides (Molecular Machines & Industries, Eching, Germany).
2. Computer Workstation: High-performance desktop computer for spectral data processing and multivariate analysis.
3. Cryofilm Fitting Tool: From the cryosection preparation kit (Section-lab Co. Ltd., Hiroshima, Japan).
4. Cryomolds: Tissue-Tek Cryomold (Sakura Finetek USA, Inc., Torrance, CA, USA).
5. Cryotome: Equipped with MX35 Premier Microtome Blade (Epredia, Kalamazoo, MI, USA).
6. Data Analysis Software: MATLAB (MathWorks, Natick, MA, USA).

7. Embedding Medium: Super Cryoembedding Medium (SCEM; Section-lab Co. Ltd., Hiroshima, Japan).
8. Heat-conducting cooling block: ThermalTray or CoolRack (Corning, Arizona, USA).
9. Forceps: e.g., Fine-tip, anti-magnetic stainless-steel forceps.
10. Teasing Needle.
11. Freeze-dryer: Suitable for biological tissue dehydration.
12. FTIR Imaging System: e.g., Hyperion 3000 FTIR microscope coupled to an Invenio S FTIR spectrometer (Bruker Optics, Ettlingen, Germany), or comparable instruments from other manufacturers, such as PerkinElmer.

3 Methods

The procedure for preparing frozen sections is based on Kawamoto's film method [23] and adjusted to the prerequisites of IR imaging (tissue thickness) and the handling of plant tissue samples. The quantification of sucrose is done by FTIR spectroscopy [19]. The method was further improved by using adhesive Cryofilm.

3.1 Embedding

1. Cool down a metal-embedding cooling block using liquid nitrogen (Fig. 2A).
2. Fill a Tissue-Tek Cryomold with Super Cryoembedding Medium (SCEM). Fill the mold in one continuous motion to minimize the introduction of air bubbles. Keep the SCEM bottle oriented with the opening facing downward at all times to avoid bubbles forming in the viscous medium from repeated inversion. Ensure the absence of air bubbles, as they may lead to sectioning artifacts or compromise sample integrity (Fig. 2B). If any bubbles appear, remove them by gently guiding them to the surface using fine-tipped forceps or a teasing needle.
3. Carefully position the frozen or freshly collected sample in the embedding medium, verifying the correct orientation for subsequent sectioning (Fig. 2C).
4. Quickly place the Cryomold on the metal embedding cooling block to freeze the embedding medium together with the sample (Fig. 2D).
5. Once completely frozen, take the frozen block out of the Cryomold (Fig. 2E) and fix it to the sample holder of the cryotome for immediate sectioning (Fig. 2F) or keep it in the freezer until all your samples are embedded before continuing with the sectioning.

3.2 Sectioning

1. Prepare the adhesive Cryofilm by cutting stripes corresponding to your sample size (Fig. 3). Please note that wider stripes are more difficult to apply smoothly to the sample and are more prone to wrinkling.
2. The ideal sectioning temperature varies between different sample types, but is usually kept at −15°C to −20°C. To facilitate smooth sectioning, the temperature of the holder and the blade are usually set at a 1°C difference (e.g., −20°C and −19°C, respectively).

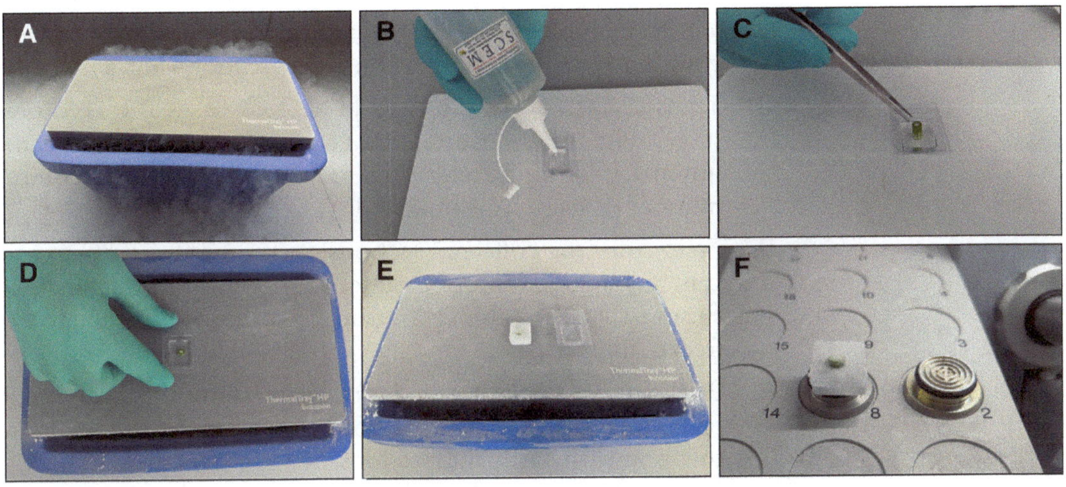

Fig. 2 Stepwise procedure of the cryoembedding workflow: a precooled metal block (**A**) is used to freeze tissue samples placed in SCEM-filled Cryomolds (**B–D**); the frozen block is then extracted (**E**) and mounted for sectioning (**F**)

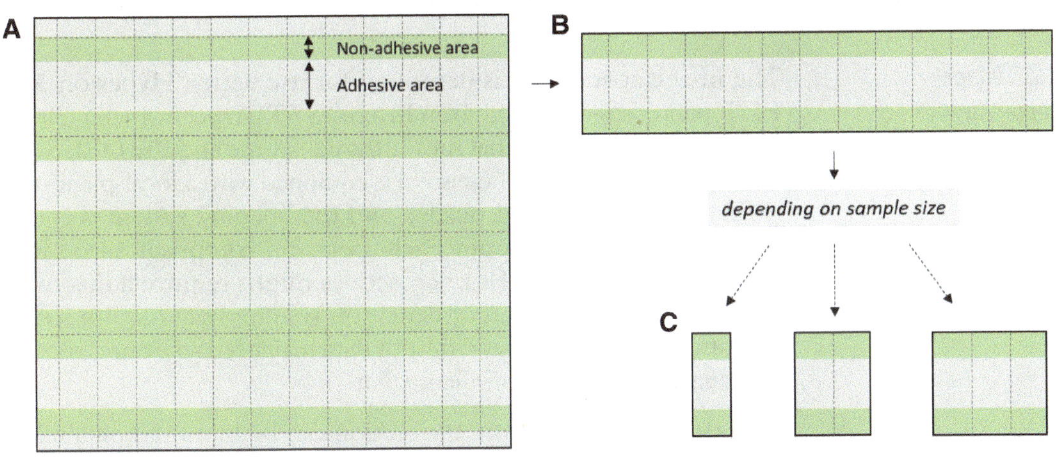

Fig. 3 Cryofilm sheet (**A**) and how to cut it into individual pieces (**B**, **C**), depending on the sample size. (Figure is modified from Kawamoto and Kawamoto [23])

3. If not done at the end of the embedding, now fix the embedded sample to the sample holder of the cryotome (Fig. 4A).
4. Carefully trim the sample until the region of interest is visible on the cutting surface (Fig. 4B).
5. Select an appropriate section thickness (*see* **Note 1**) based on your sample type, and discard the first 1–2 sections to ensure consistent implementation of the selected thickness.
6. Apply the adhesive Cryofilm stripe to the surface of the sample by holding the stripe with your forceps on the nonadhesive side and using the fitting tool to gently press and smooth the film for even adherence (Fig. 4C, D).
7. Cut the sample, including your stripe on top, slowly and at a constant speed to obtain a consistent section (Fig. 4E).
8. Transfer each section to a cold surface (e.g., the cryotome stage, Fig. 4F), ensuring that the adhesive side is facing upward. Continue repeating Steps 5–7 until the required number of sections is obtained.
9. Freeze-dry the collected sections overnight at −20°C to prevent any degradation of targeted compounds during analysis. Ensure that the samples remain cold during transfer to the freeze dryer.
10. Place the sections in a desiccator and allow them to equilibrate to room temperature to prevent condensation. From this point onward, the sections may be kept in darkness at room temperature. Experience has shown that image quality remains consistent for nonreducible or macromolecular major components (sucrose, starch, etc.) for approximately 6–9 months.
11. Screen the collected sections under a microscope for structural integrity and select the best-preserved ones for analysis.

3.3 Image Acquisition

The image acquisition is described for use with a Hyperion 3000 FTIR microscope coupled to an Invenio S FTIR spectrometer (Bruker Optics), using the internal mid-infrared source and the OPUS software (Bruker Optics). The system is equipped with a focal-plane-array (FPA) detector (64 × 64 pixels), and the imaging system is continuously purged with dry air. Each spectrum comprises 64 coadded scans. Using other FTIR microscopes might require adjustments to the protocol, as different benchtop systems may require specific operational procedures, and output data may need to be preprocessed differently depending on file format.

1. For image acquisition, the section can be mounted on a microscope XY stage (or other suitable platform) using mechanical clamps or adhesive tape on its nonadhesive regions. Gentle

tension should be applied to ensure a flat, wrinkle-free, planar surface.

2. Cool down the focal plane array detector using liquid nitrogen.
3. Acquire a visible light (VIS) image of your section or region of interest.
4. Make sure that the signal strength is not oversaturated.
5. Define and measure a region outside of your section (empty light path) in the spectral range of 3847–710 cm^{-1} for the background measurement. It will be subtracted automatically (by OPUS software) from your sample spectrum to improve accuracy.
6. Select your measuring region and start your measurement in transmission mode to obtain the absorbance spectrum in the relevant absorbance region (1800–710 cm^{-1}).

3.4 Internal Standardization, Reference Libraries, and Data Preparation

The hyperspectral FTIR imaging data undergoes a multi-step processing workflow designed to extract chemically meaningful spectral features. This process relies on three preparatory steps to ensure quantitative extraction of chemical features.

Data Pre-Processing

Pre-processing is done to exclude artifact-prone regions, like sample-free areas, and reduce the overall data size to regions of interest.

1. Import OPUS files into MATLAB as ENVI-format hyperspectral images using the multibandread function or the iRootLab toolbox [24]. During this step, an initial pixel-label map (L)

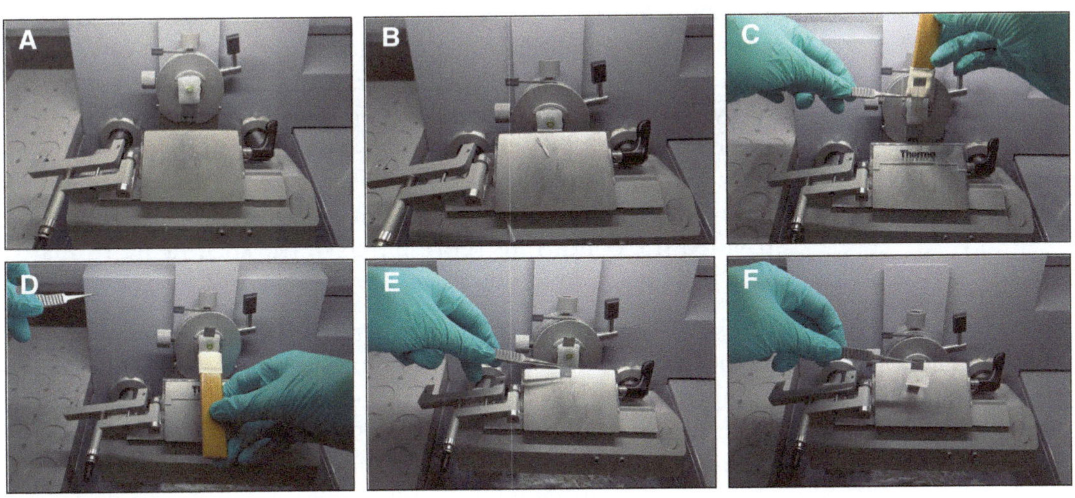

Fig. 4 Overview of key steps in the cryosectioning workflow: the embedded sample is mounted to the cryotome holder (**A**), trimmed to expose the region of interest (**B**), Cryofilm is applied and smoothed (**C, D**), a section is cut (**E**), and transferred to a cold surface with the adhesive side facing up (**F**)

is created either through clustering (e.g., *k*-means) or image segmentation functions (e.g., superpixels in MATLAB).

2. In imaging experiments using Kawamoto's cryofilm, exclude two regions within the fingerprint range because of signal oversaturation. These occur between 1735–1714 and 1278–1242 cm^{-1}.
3. Convert the *n* selected pixel spectra to a 2D matrix M of size *m*x*n* (Eq. 1). Each pixel spectrum is a column vector of *m* absorbance intensities (Abs) for each wavelength λ in cm^{-1} (wave numbers).

$$M = \begin{bmatrix} Abs_{1,1} & \cdots & Abs_{1,n} \\ \vdots & \ddots & \vdots \\ Abs_{m,1} & \cdots & Abs_{m,n} \end{bmatrix}, \; m \in \lambda, \; n \in pixel \quad (1)$$

Internal Standardization

The introduction of a ubiquitous and spatially homogenous internal reference standard is vital prior to modeling. This ensures that any alterations in the imaging setup (e.g., light intensity, filter usage, and temperature shifts) or algorithmic computation will affect the spectral features of the internal standard in the same way the sample features would be affected. Thus, they are essential in the precise quantification of spectral data.

Materials such as MMI membrane slides (Molecular Machines & Industries, Eching, Germany) or the cryofilm used in Kawamoto's film method [23] can serve as internal reference standards (*see* **Note 2**).

Reference Standard Library Preparation

Prior knowledge of the sample's chemical composition is advantageous, as varying embedding methods (e.g., paraffin or resin) or sample types may require specific strategies to assemble a reference library. Typically, the fingerprint region below 1800 cm^{-1} yields the most useful spectral information for sample differentiation. An alternative method for constructing a spectral library involves sourcing reference spectra from digital repositories, such as NIST or INRA, provided the conditions represented in the spectra are similar to those anticipated in the samples.

1. Establish pure-component reference standards by creating a spectral library from pure samples dissolved in water, applying them to a suitable substrate (MMI slide), freeze-drying, and measuring to ensure consistent sample preparation conditions. Liquid samples are recrystallized onto the substrate via lyophilization, while solid samples should be ground into a fine powder and compressed into a translucent/glass-like film before application (*see* **Note 3**). Baseline correction is necessary

to distinguish the IR signature of the substrate (diamond, KBr, PET, PEN, etc.) from the sample spectrum.

2. Post freeze-drying, measure reference components multiple times under storage conditions to accurately assess their hydration behavior in ambient humidity. This is particularly relevant for hygroscopic substances like sugars, which can significantly alter their spectral characteristics due to moisture reabsorption. Smaller molecules are affected more than larger polymers.
3. Collect data until the library comprehensively represents the principal constituents of the imaged sample, including cell wall components, protein proxies, amino acids, lipids, and both soluble and storage compounds in a highly purified form. A compact library enables the use of individual components as proxies for spectrally similar groups, aiding spectral deconvolution. Yet, increasing the library size by adding more pure components raises model complexity and computational demand, necessitating higher spectral resolution or broader spectral ranges to resolve overlapping features effectively.
4. Organize the reference library S similarly to the measurement matrix M, see Eqs. (1) and (2), including the following parameters: baseline and scatter related components (S_{Sc}), the reference spectra of the internal standard (S_{IS}), and a selection of appropriate chemical components (S_{chem}) (*see* **Note 4**).

$$S = [S_{Sc}, S_{IS}, S_{Chem}] \tag{2}$$

5. Calculate component sums. A component is defined as an optically distinct phenomenon (scatter), a chemically distinct structure or group identifiable by a unique spectral fingerprint. For example, if sucrose is represented by several distinct spectral signatures related to its hydration state in the model, then its total contribution is calculated as the sum ($\mathrm{Suc} = \mathrm{Suc}_1 + \mathrm{Suc}_2 + \mathrm{Suc}_n$).
6. If the number of model components in the library is reduced, make sure that prominent features are still reflected by it (*see* **Note 5**). Components with similar signatures should be treated as group parameters to avoid artifacts from modeling them as separate entities.
7. Normalize all spectra within S either to directly reflect a specific total Absorbance per gram compound, a specific maximum Absorbance or an overall equal total Absorbance in the spectral imaging range. Only the first normalization may allow for absolute quantification; all other normalization techniques may only reflect semi-quantitative approaches. Semi-quantitative normalization can be used to quantify absolute difference within one compound or group, but remains semi-quantitative

between different analytes due to the different absolute absorption characteristics of the analytes.

4 Spectral Deconvolution by Modeling

4.1 Deconvolution by Constrained and Unconstrained Spectral Least Squares Regression

The hyperspectral FTIR imaging data undergoes a multi-step processing workflow designed to extract chemically meaningful spectral features. The process is designed to balance efficiency with modeling complexity. As demonstrated in Sect. 3.4, we use up to 140 spectral model vectors in a linear-constrained model. This represents a challenge as images always range between 150,000 and 600,000 pixels, and complex models require longer to compute per individual pixel. Parallel computing can process multiple pixels at once in functions like partial least squares regression (PLS, e.g., plsregress(), MATLAB), while constrained models have to be utilized with other functions (e.g., lsqlin(), MATLAB) that can only process one pixel per operation. Lambert-Beer law driven linear relationships between light absorbance and concentration follow simple linear relationships as described in Eq. (3) for the spectral matrix M.

$$M = CS + E \tag{3}$$

This relationship describes M as the product of the spectral library S (Sect. 3.4) and the coefficient matrix C of size number of spectra in S by n pixels in M, plus an offset or measurement Error E of size $m \times n$. C, as well as S, can be split into its submatrices C_{Sc}, C_{IS}, and C_{Chem}.

Additional constraints required in this model limit C_{IS} and C_{chem} to be nonnegative. This is due to the fact that these represent a quantitative coefficient describing a component's concentration as a percent of the normalized Absorbance per area in S (Sect. 3.4). Since absorbance is a representation of concentration and ultimately mass, real components such as the internal standard (IS) and chemical components (Chem), like sucrose, cannot have a negative mass. PLS regression models do not allow for this limitation and need to be designed to limit negativity to be within the margin of error E. This becomes increasingly difficult with the number of introduced variables, e.g., spectral components. The pipeline designed in Fig. 6 introduces a workaround to this dilemma by accurately deconvoluting a subset of spectra using individual pixel-wise computation via a constrained linear least squares model (lsqlin(), MATLAB). This subset is then used to calibrate a PLS model to extract the coefficient matrix C directly from the raw spectra of the given image.

1. Compress data. Data compression is vital to increase computation speed. Experience has shown that a general compression of images to about 33% of its original pixel size through super-pixel generation (Sect. 3.4) still retains 95%–99% of the original information. By using the generated label matrix *L* of

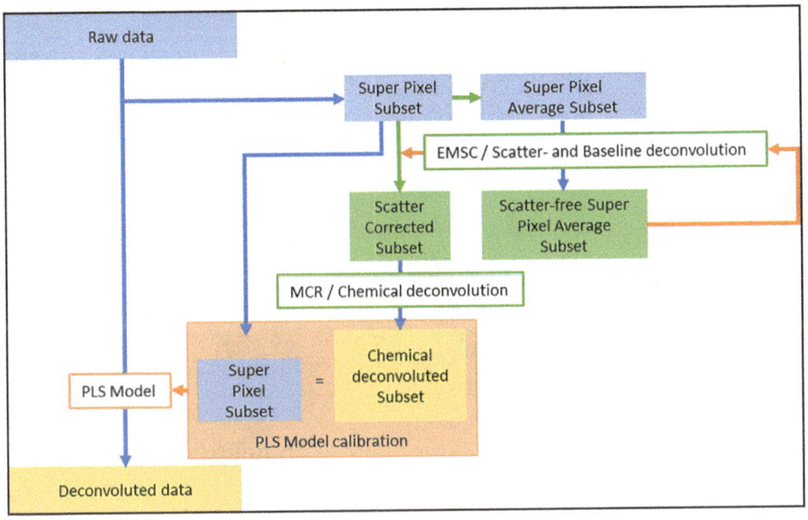

Fig. 5 Schematic overview of the general deconvolution process. After creating a subset derived from compressive labeling of similar pixels into larger super pixels, a 2-step baseline correction is applied using linear constrained remodeling of the measured pixel spectra. Baselines are corrected using EMSC principles in the applied model. The constrained modeling of chemical features is finalized after baseline correction in an MCR-based chemical deconvolution. A simpler PLS model is created based on the subset-derived chemical information and trained on the raw spectra of the subset. The established model is finally applied to the entire image matrix to generate the full image in chemical resolution

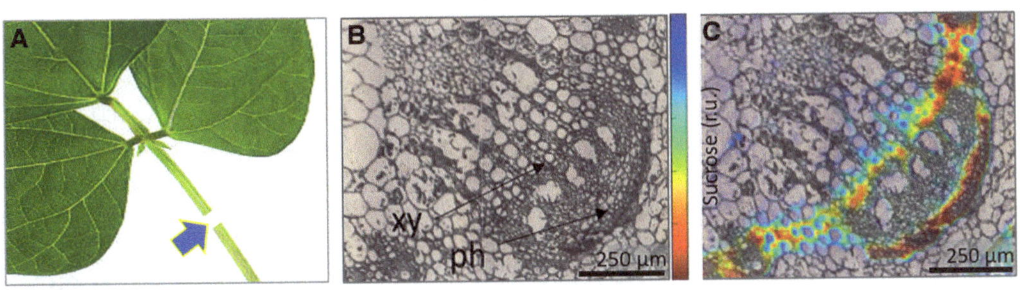

Fig. 6 Spatial distribution of sucrose in soybean petiole tissue. (**A**) The soybean leaves and a dissected petiole. (**B**) Cryosection of a soybean petiole illustrating the anatomical structure of tissues as prepared for FTIR (Fourier-transform infrared) spectroscopic imaging. The section was cut under cryogenic conditions to preserve in situ metabolite localization. (**C**) The sucrose distribution across the same cryosection of the petiole. Sucrose concentration is color-coded. High levels of sucrose (red) colocalized with the vascular bundle and were particularly highest in the phloem and cambium. Scale bar: 250 µm. *ph* phloem, *xy* xylem

the preprocessing done in Sect. 3.4 a first subset creation of $M(L)$ is achieved.

2. $M(L)$ is a sparse pixel matrix of M, with only one spectrum per label in L. This can be the median or average spectra of all spectra with the same label. If the processed $M(L)$ still has more than 10,000 pixels, randomly select a subset of these average pixels.
3. Process the subset of $M(L)$ one spectrum at a time according to the constraints of nonnegativity, using the full matrix S to create the coefficient matrix C iteratively.
4. Use the results of this initial deconvolution to reconfigure upper and lower bounds in the linlsq function (MATLAB) for the computation of C_{Sc} and to generate the scatter-free average subset M_{chem} according to Eq. (4).

$$\overline{M_{chem}(L)} = \overline{M(L)} - \sum_{1}^{i} C_{i,L} S_i = \overline{M(L)} - C_{Sc,L} S_{Sc} - C_{IS,L} S_{IS} \quad (4)$$

5. In order to reduce the complexity of the calculation for the remainder of $M(L)$, substitute the portion of S_{Chem} by a subset of M_{Chem} (*see* **Note 6**).
6. Utilize the individual S_{chem} models to recompute C_{Sc} for the entire subset of $M(L)$.
7. Use the resulting partial coefficient matrix C_{Sc} of C to extract baseline and scatter features from $M(L)$ according to Eq. (5). $\text{Mb}(L)$ is subtracted from $M(L)$ to get the baseline-free chemical spectra $M_{\text{chem}}(L)$, which also includes the chemical features of the internal standard.

$$M_b(L) = \sum_{i+1}^{i+j} C_i S_i = C_{Sc,L} S_{Sc} \quad (5)$$

8. After obtaining the scatter-free sparse pixel-matrix $M_{\text{chem}}(L)$, use a third MCR approach to recalculate the coefficients for C_{IS} and C_{Chem}. Only during this iteration, we can exclude the baseline related variables (Sc) both from S and C, resulting in a complexity reduction of having 17 less variables than the original reference model S. The model matrix S now only contains the subsets S_{IS} and S_{Chem}. The linear constraint of nonnegativity now applies to the full set of reference spectra in S.
9. After computation, sum the individual spectral components corresponding to each chemical species to obtain their absolute absorbance (Eq. 6). For example, if sucrose is represented by seven distinct spectral signatures in the model, then

calculate its total contribution as the sum ($Suc = Suc_1 + Suc_2 + \ldots + Suc_7$).

$$C_{cmpd(j)} = \sum_{j \epsilon cmpd} \left(C(j,:) \sum S(:,j) \right)$$
$$C_{cmpd} = [C_{cmpd,1} \cdots C_{cmpd}] \quad (6)$$

This step can also be applied to any of the previously generated versions of C. The resulting reduced chemical absorbance matrix C_{cmpd} can now be used to calibrate a PLS regression model by determining the PLS coefficient matrix B according to Eq. (6)

$$M(L)B_M + B_0 = C_{cmpd}(L)$$
$$TLAB\ output\ of\ [\sim, \sim, \sim, \sim, B] = \quad plsregress(M, C):$$
$$B = \begin{bmatrix} b_{0,1} & \cdots & b_{0,j} \\ b_{M,1,1} & \cdots & b_{M,1,j} \\ \cdots & \ddots & \cdots \\ b_{M,m,1} & \cdots & b_{M,m,j} \end{bmatrix} = \begin{bmatrix} B_0 \\ B_M \end{bmatrix} \quad (7)$$

B is a coefficient matrix of size $m+1$ x j, the number of components in C. B_0 is represented by the first row in B.

10. B can be applied to the entire pixel matrix M or any of its subsets $M(L)$ to generate the absorbance coefficient matrix C in one operation instead of one pixel at a time.
11. In order to evaluate computational inaccuracies within any step of the deconvolution modeling, statistically compare variance in the coefficients for the internal standard or any of the extracted chemical features. As a representative subset of the spectra was remodeled using both the complex MCR model and the PLS calibration, the two approaches are comparable by Pearson correlation to evaluate their performance metrics, including the correlation coefficient (R), root mean square error (RMSE), and the ratio of performance to deviation (RPD) as a signal-to-noise indicator within an internal validation (see Eqs.8, 9, and 10).

$$R_{cmpd} = corr(C_{cmpd(j,PLS)}, C_{cmpd(j,MCR)});\ pixel \in MCR \bigcup PLS \quad (8)$$

$$RMSE_{cmpd} = \sqrt{mean\left((C_{cmpd(j,PLS)} - C_{cmpd(j,MCR)})^2 \right)};\ pixel \in MCR \bigcup PLS \quad (9)$$

$$RPD_{cmpd} = \frac{median\left(C_{cmpd(j,MCR)}\right)}{RMSE_{cmpd}} \tag{10}$$

This internal validation ensures that the calibration model is robust and reliable when applying it to the entire dataset. It helps confirm that the simplified PLS approach is valid for practical use and identifies any limitations or modeling errors, enabling the user to either adjust and recalculate the model or exclude areas with insufficient performance parameters.

The internal reference standard (IS), namely Kawamoto's cryo-film or the MMI slide, enables quantification of component concentrations across different sets of images and experimental setups by normalizing concentration values of individual compounds relative to the IS. Absolute quantification can be achieved by multiplying the normalized absorbance concentration $C_{\text{cmpd}}(j)$ with a compound-specific absorbance coefficient ε previously generated from optical calibration curves.

$$c_A = \frac{c_j}{c_{IS}} \cdot \varepsilon_{j,n,pixel} \tag{11}$$

It is recommended to have an areal concentration as output for the calculated correlation coefficient in pg μm^{-2}. In order to transfer these values to more conventional molar concentrations, areal concentrations can be converted to a volumetric scale (tissue fresh volume) by division through cryosection thickness (Δz) and the molar mass of the compound in question (M_j; Sucrose, 342.2965 g mol^{-1}). Therefore, the formula is:

$$c_M = \frac{c_A}{\Delta z \cdot M_j} \tag{12}$$

To give an example, a value of 0.244 pg sucrose μm^{-2} corresponds to 44.6 pmol sucrose μm^{-3} tissue (given a section thickness of 16 μm). Note that is a total volume concentration which includes both insoluble, soluble, and air fractions of a given voxel. Thus, it is an under representation of a liquid fraction concentration unless the voxel space is purely occupied by the liquid fraction. In general, it is recommended to use concentrations based on mass/area or volume, since units such as mol/L are difficult to apply to polymers such as starch, cellulose or protein fractions.

Spectral Clustering and Classification

1. Normalize and scale absorbance values (e.g., *z*-score or min–max normalization) across the image to standardize metabolite comparison.
2. Perform super-pixel generation (available in MATLAB 2023b and later) followed by unsupervised clustering (e.g., *k*-means

or hierarchical clustering) on the deconvoluted absorbance values of key metabolites to identify metabolic domains across tissue sections.

3. Assign cluster labels to pixel locations to generate spatial cluster maps corresponding to metabolic domains.

Anatomical Correlation and Masking

1. Overlay clustering results with visible anatomical features from brightfield images or reference absorbance maps (e.g., sucrose for localization in vascular bundle).
2. Apply segmentation masks to define anatomical compartments (e.g., phloem, xylem, and cortex), then extract mean absorbance profiles for each.

4.2 Absolute Quantification and Validation

The following steps can be performed for all targeted components to improve validation accuracy; however, this approach is time-consuming.

1. Test the linear correlation of signal and concentration by freeze-drying frozen drops of sucrose in water with a defined volume across a known area. Dissolve a total of 100 mg of sucrose in 2 mL of water.
2. Prepare aliquots of different concentrations as a dilution series (0.001–20 g L^{-1}).
3. Image the recrystallized sucrose after freeze-drying over several weeks to account for rehydration to ambient levels.
4. Isolate pure spectral components from high signal-to-noise ratio regions to enable the computation of an absorption coefficient.
5. Model the spectral data with sucrose components (raw image), the cryofilm (raw image), and water as parameters.
6. Average the normalized sucrose signal across the imaged area and plot against the average areal concentration from the tested frozen drop, resulting in a linear fit.
7. In order to validate sucrose images after FTIR imaging, laser-dissect contrasting tissue areas from the cryosections, analyze them by, for example, chromatography and quantitatively compare the results with the FTIR results.

4.3 Statistical Image Analysis

Comparisons between digitally dissected tissue regions can be done by nonparametric analysis of the signal distributions within these areas. Extracted features from these regions can be compared using quantile absolute distance (QAD), and signal distributions across the compared areas are deemed significantly different by nonparametric statistical analysis ($p < 0.05$, Mann–Whitney test) and a divergence effect size greater than 0.2 (*see* **Note 7**).

5 Notes

1. Appropriate section thickness depends on both the sample type and the analytical method. For optical methods such as FTIR, the optical density of the tissue influences the maximum feasible section thickness, which typically ranges from 3 to 16 μm when combined with adhesive cryofilm. Therefore, the optimal thickness should be determined in advance to avoid sections that are too thick—leading to excessive optical density—or too thin, which may fall below the detection limit.
2. An effective internal reference standard must be chemically inert, reproducible, and consistent in volume and density across all measurements. It should not exhibit total absorption within any spectral region of interest. Kawamoto's cryofilm does not meet this criterion, requiring the exclusion of affected spectral regions (1735–1714 and 1278–1242 cm^{-1}) from analysis. The thickness of the reference standard should be constant and thin enough for the combined thickness of standard and sample to remain entirely in focus and below the optical depth of field. This is achievable with low-magnification objectives like Bruker's 3.5× and 15× IR objectives. Furthermore, its spectral features should be unique across the majority of the relevant infrared regions, with minimal overlap with the sample's absorption bands. At least one absorption band should be composed entirely—or at least 75%—of the reference signal and exhibit a signal-to-noise ratio greater than 10 to ensure accurate quantification.
3. It is important that the thickness of the pure components aligns with the requirements for transmission spectroscopy. For solid, insoluble substances such as cellulose and starch, we recommend using a diamond compression cell (DC-3, Specac, UK), which allows the analyte to be applied in either solid or dissolved states to a defined surface area, making it ideal for quantitative analysis. Compression screws can be used to compress small sample amounts (from a few nanograms to 1 mg) into a sufficiently thin opaque layer for absorbance spectrum acquisition.
4. In our approach, sucrose is accurately modeled within a library of approximately 120 components (S_{chem}), corresponding to 105 unique chemical structures. The spectral model library S_{Sc} consists of the seven major principal components for a spectral Mie scattering simulation [17], nine experimental components to compensate residual signal related to measurement corrections by OPUS (water vapor, CO_2 bands extracted from air and noise related to empty cryofilm reference images).

5. Note that sucrose can serve as a proxy for other soluble sugars under simplified modeling conditions. When other mono- or oligosaccharides are present, they may influence the sucrose signal unless accounted for by other library components. Care should be taken not to include library spectra of components absent from the sample or lacking distinct infrared absorbance, as such inclusions can increase computational time and model inaccuracies.
6. MCR methods tend to use the purest spectra of a sample set, as this reduces the number of deconvolution variables in many cases. Principal component analysis (PCA), *k*-means clustering, or correlation studies can be used to identify these. In our case, the function pdist2() was used to identify the ten best fitting spectra of M_{chem} for each sample spectrum in $M(L)$ as an individual S_{chem} reference library. This reduces the complexity of the constrained model from 140 to 30 variables.
7. QAD compares the difference of quantiles of two distributions over the entire range of probabilities and is calculated from the mean difference of the quantile functions A^{-1} and B^{-1} of two populations.

Formula: $QAD(A, B) = \int_0^1 | A^{-1}(p) - B^{-1}(p) | dp$

QAD values are always positive and will only become 0 if A and B are identical. Divergence effect size (D) is a probability distance between two cumulative distribution functions A and B,regardless of their context, according to the Equation $D(A||B) = 2x\int_0^1 |B\{A^{-1}(p)\} - p| dp$. Since $D(A||B) \neq D(B||A)$, a symmetric measure of this parameter can be achieved by the Equation $D(A, B) = \frac{1}{2}D(A\|B) + \frac{1}{2}D(B\|A)$. This effect size is a bounded parameter that can take values between 0 (identical) and 1 (no commonalities).

References

1. Rolland F, Baena-Gonzalez E, Sheen J (2006) Sugar sensing and signaling in plants: conserved and novel mechanisms. Plant Cell 18(4):199–225. https://doi.org/10.1105/tpc.105.035063
2. Sonnewald U, Fernie AR (2018) Next-generation strategies for understanding and influencing source-sink relations in crop plants. Curr Opin Plant Biol 43:63–70
3. Stitt M, Quick WP (1989) Photosynthate partitioning in sugar beet leaves in response to light and darkness. Plant Physiol Biochem 27(5):611–620
4. Jones MGK, Outlaw WH, Lowry OH (1977) Enzymic assay of 10^{-7} to 10^{-14} moles of sucrose in plant tissues. Plant Physiol 60(3):379–383
5. Neuburger GG, Johnson DC (1987) Pulsed amperometric detection of carbohydrates at gold electrodes with a two-step potential waveform. Anal Chem 59(1):150–154. https://doi.org/10.1021/ac00128a031
6. Liu H, Singh SK (1992) Determination of sugars in plant tissues by HPLC with pulsed amperometric detection. Analyt Chem 64(21): 2340–2343

7. Belford DS, Rogers AT (1984) Analysis of sugars by gas chromatography. J Chromatogr Sci 22(8):379–382
8. Roessner U et al. (2001) Metabolic profiling and biochemical phenotyping of plant systems. Plant Cell 13(1):11–29
9. Boughton BA, Thinagaran D, Sarabia D, Bacic A, Roessner U (2016) Mass spectrometry imaging for plant biology: a review. Phytochem Rev 15:445–488
10. Schmalstig JG, Geiger DR (1985) Phloem unloading in developing leaves of sugar beet: I. Evidence for pathway through the symplast. Plant Physiol 79(1):237–41. https://doi.org/10.1104/pp.79.1.237
11. Mincke J, Courtyn J, Vanhove C, Vandenberghe S, Steppe K (2021) Guide to plant-PET imaging using 11CO2. Front Plant Sci 12:602550. https://doi.org/10.3389/fpls.2021.602550
12. Borisjuk L, Neuberger T (2025) Perspectives: The look insight–magnet resonance imaging (MRI) of the inner life of plants. J Plant Physiol 309(2025):154502
13. Mayer S, Rolletschek H, Radchuk V, Wagner S, Ortleb S, Gündel A et al (2024) Metabolic imaging in living plants: a promising field for chemical exchange saturation transfer (CEST) MRI. Sci Adv 10(38):eadq4424. https://doi.org/10.1126/sciadv.adq4424
14. Borisjuk L, Walenta S, Rolletschek H, Mueller-Klieser W, Wobus U, Weber H (2002) Spatial analysis of plant metabolism: sucrose imaging within Vicia faba cotyledons reveals specific developmental patterns. Plant J 29:521–530
15. Niittylae T, Chaudhuri B, Sauer U, Frommer WB (2009) Comparison of quantitative metabolite imaging tools and carbon-13 techniques for fluxomics. Methods Mol Biol 553:355–372
16. Chen LQ, Qu XQ, Hou BH, Sosso D, Osorio S, Fernie AR, Frommer WB (2012) Sucrose efflux mediated by SWEET proteins as a key step for phloem transport. Science 335:207–211
17. Guendel A, Rolletschek H, Wagner S, Muszynska A, Borisjuk L (2018) Micro imaging displays the sucrose landscape within and along its allocation pathways. Plant Physiol 178(4):1448–1460. https://doi.org/10.1104/pp.18.00947
18. Walerowski P, Gündel A, Yahaya N, Truman W, Sobczak M, Olszak M, Rolfe SA, Borisjuk L, Malinowski R (2018) Clubroot disease stimulates early steps of phloem differentiation and recruits SWEET sucrose transporters within developing galls. Plant Cell 30:3058–3073
19. Gündel A, Hilo A, Rolletschek H, Borisjuk L (2021) Probing the metabolic landscape of plant vascular bundles by infrared fingerprint analysis, imaging and mass spectrometry. Biomolecules 11:1717
20. Lohaus G (2022) Review primary and secondary metabolites in phloem sap collected with aphid stylectomy. J Plant Physiol 271:153645
21. Radchuk V, Tran V, Hilo A, Muszynska A, Gündel A, Wagner S, Fuchs J, Hensel G, Ortleb S, Munz E et al. (2021) Grain filling in barley relies on developmentally controlled programmed cell death. Commun Biol 4:428
22. Jensen KH, Savage, Jessica A, Holbrook N, Michele (2013) Optimal concentration for sugar transport in plants. J R Soc Interface 10(83):20130055. https://doi.org/10.1098/rsif.2013.0055
23. Kawamoto T, Kawamoto K (2021) Preparation of thin frozen sections from nonfixed and undecalcified hard tissues using Kawamoto's film method. In: Hilton MJ (Ed.) Skeletal development and repair. Humana, New York, NY, 259–281
24. Trevisan J, Angelov PP, Scott AD, Carmichael PL, Martin FL (2013) IRootLab: a free and open-source MATLAB toolbox for vibrational biospectroscopy data analysis. Bioinformatics 29(8):1095–1097. https://doi.org/10.1093/bioinformatics/btt084

Check for updates

Chapter 10

Identification of Phloem Cell Transcriptome via Single-Cell RNA-Sequencing

Sol-bi Kim, Efthymia Symeonidi, Tin Yau Pang, Hongshi Jin and Ji-Yun Kim

Abstract

Single-cell RNA sequencing (scRNA-seq) has emerged as a powerful tool for dissecting cellular heterogeneity in plants at a high resolution. In particular, scRNA-seq has been applied to unravel the complex cell identities and regulatory networks of distinct phloem cell types. Because phloem cells are small and under high osmotic pressure, isolating and maintaining viable protoplasts is a challenge for successful phloem scRNA-seq. Here, we present an optimized protocol for isolating vasculature-enriched protoplasts from mature *Arabidopsis thaliana* leaves. Following tissue harvesting, protoplasting is performed using a cocktail of cell wall-degrading enzymes suspended in an osmotically stabilized buffer. The resulting cell suspension is purified by removing debris through subsequent wash steps and validated using phloem-specific markers. Besides providing a reliable and reproducible method for isolating high-purity phloem-enriched protoplasts, we introduce the basic pipeline for downstream analysis.

Keywords Arabidopsis, Phloem transcriptome, Cell wall digestion, Protoplast isolation, Single-cell RNA sequencing

1 Introduction

The vascular system in plants, comprising xylem, phloem, and associated parenchyma cells, plays a central role in resource allocation and signaling [1]. Despite the importance of vascular cell types in plant development, long-distance signaling, stress response and adaptation, its complex cellular composition and its localization deep within the leaf make targeted transcriptomic profiling of vascular cells particularly challenging [2, 3]. Conventional bulk RNA-seq approaches fail to resolve cell-type-specific expression patterns, underscoring the need for cell-resolution methods [4, 5]. Therefore, to overcome these limitations, single-cell RNA sequencing has recently been applied to study vascular tissues, including the phloem, at cellular resolution.

Johannes Liesche and Sanna Sevanto (eds.), *Phloem*, Methods in Molecular Biology, vol. 3066,
https://doi.org/10.1007/978-1-0716-5488-0_10,

Recent advances in single-cell RNA sequencing (scRNA-seq) have revolutionized our understanding of the cellular heterogeneity and gene expression dynamics at unprecedented resolution [6, 7]. While scRNA-seq technology has been widely applied in various animal systems, its application in plants has posed significant challenges, primarily due to the presence of rigid cell walls that prevent the isolation of intact single cells [2]. To overcome this obstacle, enzymatic removal of the cell wall—a process known as protoplasting—is essential for enabling access to the transcriptomic landscape of individual plant cells [8, 9].

Single-cell transcriptomics has been applied in *Arabidopsis thaliana* and other model plants, providing valuable insights into developmental processes, tissue-specific functions, and cell identity determination [7, 10]. Yet, the success of scRNA-seq experiments in plants depends heavily on the efficiency and reproducibility of the protoplast isolation procedure [10, 11]. Protoplasts must be viable, representative of the tissue of interest, and amenable to downstream processing, including encapsulation or sorting (depending on the single-cell sequencing platform used) for successful capture for reverse transcription and library preparation [12]. Therefore, developing a robust and tissue-specific protoplasting protocol is a critical prerequisite for meaningful single-cell transcriptomic analysis [11].

In this chapter, we present a detailed and optimized protocol for isolating viable vascular-enriched protoplasts from rosette leaves of adult *Arabidopsis* plants grown under short-day conditions [13]. The procedure involves strategic removal of unnecessary cell types (e.g., abaxial epidermis and mesophyll), enzymatic digestion using a cocktail of cell wall-digesting enzymes, and careful wash steps that preserve cell viability and integrity. To ensure the enrichment of vascular cells, we emphasize the use of fluorescent reporter fusion lines, such as *pSWEET11:SWEET11-GFP*, which specifically mark phloem parenchyma cells [13].

This protocol enables consistent and reproducible preparation of protoplasts suitable for downstream scRNA-seq applications [11]. By refining each step—from tissue preparation to final suspension—this method minimizes technical artifacts and maximizes cell-type representation, especially of delicate internal tissues like the vasculature. We further provide a comprehensive downstream analysis pipeline and describe a novel method for assessing metabolic pathway activity at the single-cell level.

2 Materials

2.1 Plant Growth

1. *Arabidopsis* Col-0 plants cultivated for 6 weeks under short-day conditions (8 h light/16 h dark) may be used. Depending on the experimental design, knockout or reporter lines may be used instead of wild-type plants.

2.2 Protoplast Isolation

Enzymes and Reagents

1. Enzymes for cell wall digestion: Cellulase R10 and Macerozyme R10.
2. 1 M $CaCl_2$ stock solution.
3. 0.2 M 2-(N-morpholine)-ethanesulfonic acid (MES) stock solution adjusted to pH 5.7 with KOH.
4. 2 M $MgCl_2$ stock solution.
5. 2 M KCl stock solution.
6. 2 M NaCl stock solution.
7. 0.8 M mannitol stock solution.
8. 10% bovine serum albumin (BSA) stock solution.
9. MMG solution: 0.4 M mannitol, 15 mM $MgCl_2$, 4 mM MES (pH 5.7).

Equipment and Consumables

1. Centrifuge with swing-out rotor suitable for 14 mL round-bottom tubes.
2. Platform shaker: set to 40 rpm.
3. Water bath.
4. Time Tape (for leaf stabilization).
5. 3 M Magic tape: for epidermis removal, or similar tape with high adhesive strength.
6. End-cut pipette tips (wide-bore to minimize shear stress).
7. Fine-tip forceps.
8. Small scissors.
9. Cell strainer: pore size 40 μm.
10. 0.22 μm syringe filters.
11. Hemocytometer.
12. Bright-field light microscope: for cell counting.
13. 14 mL round-bottom tubes.
14. 90 mm diameter Petri dish.
15. Aluminum foil.

3 Methods

3.1 Protoplasting

1. Prepare 20 mM MES (pH 5.7) solution containing 1.5% Cellulase R10, 0.4% Macerozyme R10, 0.4 M D-mannitol, and 20 mM KCl.

2. Warm the solution at 55°C for 10 min to enhance enzyme solubility and to inactivate protease activity.
3. Cool it to room temperature and add 10 mM $CaCl_2$ and 0.1% BSA (*see* **Note 1**).
4. Syringe filter (0.22 μm) the solution directly before use.
5. Place 7–10 rosette leaves with the lower (abaxial) side facing upward on a flat surface, such as an inverted rack or clean bench paper.
6. Stabilize each leaf by affixing a strip of time tape onto the upper epidermal surface.
7. Apply a strip of 3 M Magic tape to the lower epidermal surface and gently peel it away from the time tape to remove the lower epidermal cell layer (Fig. 1).
8. Transfer the peeled leaves, still attached to the Time Tape, into a 90 mm diameter Petri dish containing 10 mL of enzyme solution. Ensure that the peeled (lower epidermal) surface is in contact with the enzyme solution.
9. Continue peeling and transferring leaves until the bottom of the dish is covered with a single layer of peeled leaves.
10. Incubate the leaves in the enzyme solution on a platform shaker at 40 rpm in the dark for 20–60 min to release protoplasts. Cover the Petri dish with aluminum foil to maintain darkness during incubation.
11. Add 5 mL of W5 solution to the dish and swirl gently to rinse and collect residual protoplasts.
12. Using a cut pipette tip, filter the protoplast suspension through a 40 μm cell strainer into a new Petri dish (*see* **Note 2**).

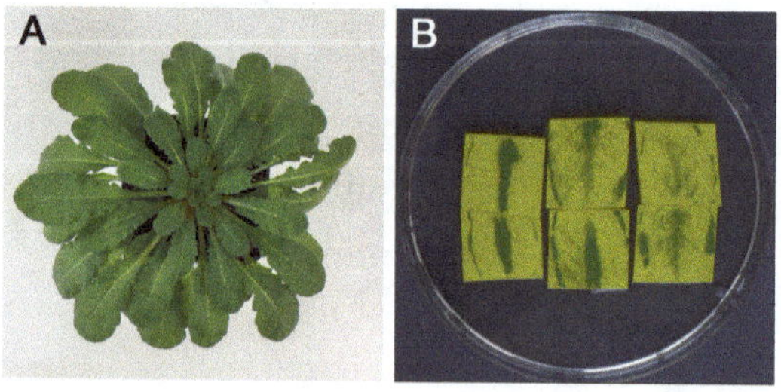

Fig. 1 Plant material preparation and tissue processing for protoplast isolation. (**A**) Morphology of *Arabidopsis thaliana* plants used for protoplasting. Plants were grown under short-day conditions, and source leaves were harvested for vascular cell isolation. (**B**) Abaxial epidermis removal enhances phloem cell accessibility and release efficiency. The peeling process exposes internal vascular tissues and provides visual confirmation of successful tissue preparation during enzymatic digestion

13. Carefully transfer the filtered protoplast suspension into a 14 mL round-bottom tube using an end-cut tip to minimize shear stress.
14. Centrifuge the tube for 10 min at 600 × *g* using a swing-out rotor (*see* **Note 3**).
15. Discard the supernatant and gently resuspend the pelleted protoplasts in 5 mL of W5 buffer.
16. Centrifuge for 5 min at 600 × *g* without brake.
17. Remove the supernatant and resuspend the protoplast pellet in MMG solution using an end-cut pipette tip (*see* **Note 2**).

3.2 Cell Counting

1. Gently mix the protoplast suspension by inverting the tube to ensure even cell distribution.
2. Using a cut pipette tip, take 10 μL of the protoplast suspension and load it into one chamber of a clean hemocytometer.
3. Place the hemocytometer on the stage of a light microscope and focus on the grid.
4. Calculate the average number of cells per square and determine the cell concentration (*see* **Note 4**).
5. Based on the calculated cell density, adjust the protoplast suspension to the desired final concentration (e.g., ~5000 cells/μL) using MMG solution.

3.3 Single-Cell Library Preparation and Sequencing

Following protoplast preparation and counting, single-cell RNA sequencing libraries are prepared using the appropriate platform suited to the experimental design and budget constraints. Several approaches are available, including microfluidics-based platforms (e.g., 10× Genomics Chromium) and high-throughput plate-based methods (e.g., BD Rhapsody) for processing large cell numbers. Traditional plate-based methods (e.g., Smart-seq2) provide deeper per-cell sequencing depth and full-length transcript coverage for smaller cell numbers. These represent just a few examples of available technologies that employ strategies such as droplet encapsulation, combinatorial barcoding, and direct cell isolation. Users should carefully choose the appropriate platform based on the characteristics of the protoplasts, as certain physical properties (e.g., cell size, debris content) may cause compatibility issues with specific technologies (Fig. 2).

3.4 Basic Single-Cell Transcriptome Analysis

Single-cell RNA sequencing data analysis typically follows a standard computational pipeline involving quality control. Here, we provide a brief outline; for comprehensive methodological details, users should refer to specialized bioinformatics resources. Following sequencing, raw reads are processed through quality control steps, including cell filtering (removing low-quality cells and potential doublets), gene

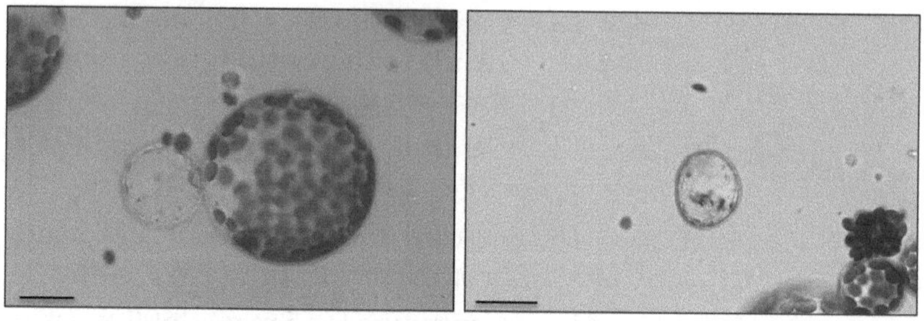

Fig. 2 Assessment of protoplast quality and cellular debris following the isolation procedure. Even after multiple washing steps, small amounts of cellular debris remain in the MMG solution. This residual debris does not interfere with downstream single-cell RNA sequencing applications and cannot be completely eliminated through additional wash cycles. Scale bar represents 10 μm

filtering (excluding low-expressed genes), and mitochondrial gene content assessment. Data normalization and scaling are performed to account for technical variation and differences in library size across cells. Dimensionality reduction techniques (e.g., MAP, t-SNE) are applied to visualize cell populations and identify distinct clusters representing different cell types or states. Differential gene expression analysis between clusters enables the identification of marker genes and the functional annotation of cell populations.

1. **Read alignment:** Align sequencing reads to the *Arabidopsis thaliana* reference genome using tools provided by the provider (e.g., 10× Genomics, Cell Ranger (10× Genomics, California). Output files from multiple replicates can be aggregated into a single gene-cell expression matrix using the mapped-read-depth normalization option.
2. **UMI matrix generation:** The output is a unique molecular identifier (UMI) count matrix, which can be imported into Seurat in R [14, 15] for downstream analysis.
3. **Quality filtering:** During data import with the CreateSeuratObject function, specify thresholds for the minimum number of features (genes) per cell and the minimum number of cells a feature (gene) should be present in to be included in downstream analysis.
4. **Data normalization and scaling:** Normalize the data, and identify scaled and variable features using the SCTransform function. At this stage, confounding sources of variation, such as mitochondrial and/or chloroplastic genes, can be regressed out.
5. **Dimensional reduction:** Perform principal component analysis (PCA) using the RunPCA function in Seurat. The 50 principal components are typically used for downstream analysis and visualization.

6. **Graph construction:** Use the reduced-dimension representation of the data to construct a *K-nearest* neighbor (KNN) graph using the FindNeighbors function in Seurat. This graph captures cell-to-cell similarity based on their principal component scores.
7. **Cell clustering:** Apply the FindClusters function with a resolution parameter defined by the user (resolution: 0.8 in Kim et al. [13]) to partition the KNN graph into discrete clusters, grouping cells that share similar gene expression profiles.
8. **Visualization:** Use uniform Manifold Approximation and Projection (**UMAP**) for two-dimensional visualization of the clustered data, providing a low-dimensional embedding that preserves both local and global transcriptomic relationships. The specific parameters (e.g., the number of principal components, the number of neighbors, and the minimum distance) can be adjusted based on the dataset characteristics and the desired clustering resolution. Kim et al. [13] used the first 10 principal components, 30 nearest neighbors, and a minimum distance of 0.1.

3.5 Pathway Activity Analysis

While differential gene expression analysis reveals information signatures of cell clusters, understanding the broader biological significance of differences between clusters (cell types or cell identities) often requires a higher-level functional interpretation. Here, pathway-based approaches can serve as an invaluable source of information, aggregating gene-level data into biologically meaningful units that represent coordinated cellular processes, signaling cascades, and metabolic functions.

The pathway activity score (PAS) is a metric that quantifies the activation level of biological pathways within cell clusters in scRNA-seq data. By transforming gene-level data into interpretable gene sets representing specific pathways [16], PAS facilitates the analysis of cellular heterogeneity and the elucidation of underlying biological processes. While pathway activity can be represented by RNA read counts of constituent genes, proper normalization of these counts is essential before PAS calculation and inter-cluster comparison. The following sections outline the design principles and computational steps underlying this methodology:

To calculate the PAS, two sources are required: the scRNA-seq dataset and pathway gene annotations. For Arabidopsis, pathway data from *Arabidopsis thaliana* Col-0 ecotype provided in the BIOCYC database [17] can be applied.

1. Normalize gene read counts within each cell to enable cross-cell comparison. Let $g_{i,k}$ denote the mRNA count of gene i in cell k. Here, $g_{i,k}$ is already pre-processed (e.g., by SCTransform in Seurat) so that low-quality cells are removed, and technical

noise from sources, such as cell cycle effects or sequencing platform variations, is regressed out.

2. Normalize the mRNA count of each gene, $g_{i,k}$, by the average gene count in the cell, and this normalized gene count as $g'_{i,k}$.
3. With normalized gene count, $g'_{i,k}$, make cross-cell comparisons of the same gene. We define the mean transcript level of gene, $E_{i,j}$ as

$$E_{i,j} = \frac{1}{n_j} \sum_{k=1}^{n_j} g'_{i,k},$$

where j is the index of the cluster, n_j is the number of cells in cluster j, and k is the index of cells.

4. Calculate relative transcript levels. While $E_{i,j}$, enables comparisons of the RNA expression level between a gene of different clusters, a pathway comprises multiple genes, and the $E_{i,j}$ of different genes are not comparable. Therefore, the relative transcript level $r_{i,j}$ of a gene is calculated as:

$$E_{i,j}:\ r_{i,j} = \frac{E_{i,j}}{\left(\sum_{a=1}^{N} E_{i,a}\right)/N},$$

where N is the number of cell clusters. This normalization makes the average $r_{i,j}$ of gene i across all clusters to be 1. In this way, the expression level between different genes becomes comparable and enables direct summation of $r_{i,j}$.

5. Calculate the PAS of pathway t in cluster j, denoted as $p_{t,j}$, which is a weighted average of the normalized transcript of the pathway's genes:

$$p_{t,j} = \frac{\sum_{i=1}^{m_t} w_i r_{i,j}}{\sum_{i=1}^{m_t} w_i},$$

where m_t is the number of genes in pathway t, while w_i is the weight defined as the reciprocal of the number of pathways that include gene i and ensures a stronger weight for pathway-specific genes. Similar to $r_{i,j}$, $p_{t,j}$ also has an average 1 across all cluster. This weighting assigns the pathway-specific genes—genes that serve fewer pathways—from the nonspecific one, so that the contribution of pathway-specific genes is higher than that of the others in the calculation of PAS.

4 Notes

1. Following incubation of the enzyme mixture at 55°C to dissolve and to inactivate undesired protease activity, the enzyme mixture must be cooled to room temperature before adding BSA and $CaCl_2$. This cooling step prevents protein additive denaturation and ensures optimal membrane stabilization and enzyme function.
2. Wide-bore tips minimize shear-induced stress during pipetting, thereby reducing mechanical damage and preserving protoplast integrity.
3. Swing-out rotors should be used for centrifugation to ensure gentle and uniform pelleting of protoplasts.
4. Calculation of cell concentration should be adjusted based on the specific hemocytometer type the user chooses to use (e.g., Neubauer, Improved Neubauer, Fuchs-Rosenthal, or Bürker chambers).
5. Bulk RNA-seq using leaf tissues and protoplast samples should be performed to identify protoplasting-induced genes [10].

Acknowledgments

We thank Johannes Liesche for reading our manuscript and providing constructive suggestions. This work was supported by the Basic Science Research Program through the National Research Foundation of Korea (NRF), funded by the Ministry of Education (No. RS-2019-NR040081) and by the National Research Foundation of Korea (NRF) grant funded by the Korea government (MSIT) (No. RS-2025-00560593). Hongshi Jin was supported by the National Research Foundation of Korea (NRF) grant funded by the Korean government (MSIT) (No. RRS-2025-24683195).

References

1. van Bel AJ, Helariutta Y, Thompson GA et al (2013) Phloem: the integrative avenue for resource distribution, signaling, and defense. Front Plant Sci 4:471
2. Chen C, Ge Y, Lu L (2023) Opportunities and challenges in the application of single-cell and spatial transcriptomics in plants. Front Plant Sci 14:1185377
3. Liang S, Li Y, Chen Y et al (2023) Application and prospects of single-cell and spatial omics technologies in woody plants. For Res 3:27
4. Bawa G, Liu Z, Yu X et al (2022) Single-cell RNA sequencing for plant research: insights and possible benefits. Int J Mol Sci 23(9):4497
5. Zhu M, Taylor IW, Benfey PN (2022) Single-cell genomics revolutionizes plant

development studies across scales. Development 149(6):dev200179
6. Jovic D, Liang X, Zeng H et al (2022) Single-cell RNA sequencing technologies and applications: a brief overview. Clin Transl Med 12(3):e694
7. Lim J, Park C, Kim M et al (2024) Advances in single-cell omics and multiomics for high-resolution molecular profiling. Exp Mol Med 56(3):515–526
8. Grones C, Eekhout T, Shi D et al (2024) Best practices for the execution, analysis, and data storage of plant single-cell/nucleus transcriptomics. Plant Cell 36(4):812–828
9. Long Y, Liu Z, Jia J et al (2021) FlsnRNA-seq: protoplasting-free full-length single-nucleus RNA profiling in plants. Genome Biol 22(1):66
10. Denyer T, Timmermans MCP (2022) Crafting a blueprint for single-cell RNA sequencing. Trends Plant Sci 27(1):92. https://doi.org/10.1016/j.tplants.2021.08.016
11. Islam MT, Liu Y, Hassan MM et al (2024) Advances in the application of single-cell transcriptomics in plant systems and synthetic biology. Biodes Res 6:0029
12. Zhang K, Liu S, Fu Y et al (2023) Establishment of an efficient cotton root protoplast isolation protocol suitable for single-cell RNA sequencing and transient gene expression analysis. Plant Methods 19(1):5
13. Kim JY, Symeonidi E, Pang TY et al (2021) Distinct identities of leaf phloem cells revealed by single cell transcriptomics. Plant Cell 33(3):511–530
14. Butler A, Hoffman P, Smibert P et al (2018) Integrating single-cell transcriptomic data across different conditions, technologies, and species. Nat Biotechnol 36(5):411–420
15. Satija R, Farrell JA, Gennert D et al (2015) Spatial reconstruction of single-cell gene expression data. Nat Biotechnol 33(5):495–502
16. Xiao Z, Dai Z, Locasale JW (2019) Metabolic landscape of the tumor microenvironment at single cell resolution. Nat Commun 10(1):3763
17. Caspi R, Billington R, Ferrer L et al (2016) The MetaCyc database of metabolic pathways and enzymes and the BioCyc collection of pathway/genome databases. Nucleic Acids Res 4(44):471–480

Chapter 11

Live-Cell Imaging of Fluorescently-Tagged Phloem Proteins with Confocal Microscopy

Thibaud Cayla, Rozenn Le Hir and Sylvie Dinant

Abstract

Confocal laser scanning microscopy can enable observation of phloem cells in living tissues. Here, we describe live imaging of phloem cells in the leaves and roots of *Arabidopsis thaliana* using fluorescently-tagged proteins, either expressed in the vasculature using phloem-specific promoters or constitutively expressed in reference marker lines. We provide instructions for identifying the majority of phloem cell types and a step-by-step protocol for the precise cellular and subcellular localization of phloem proteins.

Keywords Fluorescent markers, Mobile symplasmic probes, Green fluorescent proteins, Live-cell imaging, Microscopy

Abbreviations

CC	Companion cell
CFP	Cyan fluorescent protein
GFP	Green fluorescent protein
PPC	Phloem parenchyma cell
SE	Sieve element
YFP	Yellow fluorescent protein

1 Introduction

The vascular tissues of higher plants provide fast pathways for long-distance transport of water and nutrients between distant organs. Therefore, the phloem and xylem tissues contribute to many important functions in plant growth and development. Noticeably, they encompass critical steps of the primary and secondary metabolism, hormone, peptide and macromolecule signaling, production and delivery of defense molecules, and storage of nutrients or minerals [1, 2]. The vascular tissues are also involved in long-distance signaling. In

Johannes Liesche and Sanna Sevanto (eds.), *Phloem*, Methods in Molecular Biology, vol. 3066,
https://doi.org/10.1007/978-1-0716-5488-0_11,

the phloem, the systemic translocation of proteins, mRNAs, and small RNAs in the phloem is emerging as an important feature of inter-organ communication [3, 4].

Interestingly, many factors acting in vascular tissues accumulate specifically in certain vascular cells, a feature that can be relevant to their in situ activity. For example, the sites of expression of genes encoding myrosinases, enzymes involved in the hydrolysis of glucosinolates into toxic defense compounds, are different from the glucosinolate-rich S cells [5, 6]. Many proteins have been shown to be cell-type specific, even within the phloem tissue, and the discovery of their location is often an important step for understanding their function. For example, the localization of the SWEET11 and SWEET12 sugar transporters in the phloem parenchyma cells in minor and main veins of Arabidopsis leaves provided evidence for a role in the efflux of sugars into the apoplasm [7]. SWEET11 and SWEET12 became the first phloem parenchyma cell markers to be characterized, although they have also been shown to be expressed in the xylem parenchyma cells since [8]. In the leaves of *Arabidopsis*, the cytosolic isoforms of glutamine synthetase, an important enzyme controlling nitrogen remobilization, are present in different phloem cell types and vein orders, supporting complementary functions during leaf development and senescence [9]. Another interesting case is the biosynthesis pathway of monoterpene indole alkaloids in Madagascar periwinkle, which are mainly associated with the internal phloem parenchyma cells [10, 11]. Especially valuable is the visualization of proteins in living cells. For example, live-cell imaging of the SIEVE-ELEMENT-OCCLUSION-RELATED1 (SEOR1) fused to yellow fluorescent protein could be used to study how different agglomeration states affect phloem flow [12].

These observations indicate that intercellular compartmentation is critical for the activity of many proteins acting in the phloem [13]. However, identifying cell types in the phloem can be difficult, firstly because they are located deep within organs beneath several overlapping tissues, and secondly because phloem cells are micrometric in size [14, 15]. Because of the need to unambiguously identify specific phloem cell types, a range of phloem fluorescent markers has been developed [12, 16–20], in combination with a range of phloem mobile tracers [21–24]. In this chapter, we present some of these markers as well as the methods used for in vivo imaging in *Arabidopsis thaliana* [20], although the methods can be adapted to other plants. The following provides an overview of gene constructs and fluorescent dyes that can be used for live-cell imaging of the phloem. The protocol details the procedure for making samples and imaging them on a confocal microscope. Confocal imaging is at a distinct advantage over super-resolution microscopy or immunolocalization, because it affords live-cell imaging and deep imaging, even in 3D, in a convenient way.

2 Materials

Use Arabidopsis transgenic lines expressing fluorescently-tagged proteins either in all organs, under the control of a constitutive promoter such as the CaMV 35 S promoter, or under the control of a phloem specific promoter, such as the *SUC2* or *PP2-A1* promoters (see Chap. 29), for expression in the companion cells, the *SEOR1* or *SEOR2* promoters for expression in the sieve elements, the *SWEET11* and *SWEET12* promoters for expression in the phloem parenchyma cells (Table 1).

2.1 Plant Material

1. Plants grown in vitro on Murashige and Skoog medium (MS/2) or Arabidopsis medium to observe the roots and plants grown either in vitro or in soil to observe the leaves (*see* **Note 1**).

2.2 Microscopy

1. Confocal microscope equipped with excitation lasers of the following wavelengths: 405 and 488 nm, and ideally also 514 and 544 nm. A confocal microscope equipped with transmitted light mode can be used to obtain high-quality bright-field images of the plant samples.
2. Filter or prism spectrometer for multispectral imaging.
3. Objectives with 20X, 40X, and 63X magnification, with water or oil immersion.
4. Regular glass slides and glass coverslip with a thickness of 0.14–0.18 mm.
5. Scalpel and/or razor blades.

2.3 Counterstaining with Vital Dyes

Keep stock solutions at −20°C. See Table 2 for additional information on the dyes.

1. Aniline blue: 0.05% (w/v), dissolve in Sodium phosphate buffer 0.07 M, pH 9; incubate for 10–15 min before mounting. Aniline blue is pH sensitive and effective under alkaline conditions.
2. Sirofluor (Biosupplies): dissolve in 100 mM K_3PO_4 buffer to a final concentration of 25 μg mL^{-1}. Sirofluor is also pH sensitive and more effective under alkaline conditions.
3. FM4-64 (Invitrogen): to prepare the stock solution, dissolve in water to a concentration of 100 μg mL^{-1}. To prepare the working solution, dilute the stock solution in water/buffer to a concentration of 5 μg mL^{-1}. For staining leaf or roots, incubate for 5 min before mounting.
4. DiOC6 (Molecular probes): stock solution: 0.1 mg mL^{-1} in DMSO, For staining, incubate for 1 min with 1 μg mL^{-1} before mounting.

Table 1
Arabidopsis fluorescently-tagged marker lines used for the imaging of phloem in vivo

Arabidopsis transgenic lines	Description	Observations	Reference	Example in phloem research
Lines with non-cell type-specific expression of fluorescently-tagged protein markers				
p35S::ER:YFP	Signal peptide of AtWAK2 and ER retention signal HDEL	ER in SE, CC, and PPC	[25]	[20]
p35S::RbcS:YFP	First 79 aa of the tobacco small unit of Rubisco	Plastids in CC and PPC	[25]	[20]
p35S::GFP-MBD	Microtubule binding domain	Cytoskeleton, microtubules in CC and PPC	[26]	[20]
p35S::FABD2:GFP	FABD2 actin-binding domain	Cytoskeleton, microfilaments in CC and PPC	[27]	[20]
p35S::γTIP:YFP	C-terminus of the tonoplast γTIP	Tonoplast in CC and PPC	[25]	[20]
p35S::H2B:RFP	H2B Human histone	Nucleus in CC and PPC	[28]	[20]
p35S::GFP:LTI6b	LTI6b/RCI2b protein (low temperature inducible protein b)	Plasma membrane of all phloem cell types (SE, CC, and PPC)	[29]	[20]
p35S::COX4:YFP	First 29 aa of yeast ScCOX4	Mitochondria in SE, CC, and PPC	[25]	[20]
p35S::PDLP1a:GFP	PDLP1 (plasmodesmata-located protein 1)	Plasmodesmata	[30]	
Lines expressing tissue- or cell-specific fluorescently-tagged protein marker				
pSUC2::mCitrine:RCI2A	RCI2A (low-temperature-inducible protein a)	Endomembranes in CC and SE	[17]	[17, 20]
pSEOR1::SEOR1:GFP	Sieve element occlusion related protein 1 (SEOR1)	P-protein filament in SE	[12, 18]	[12, 20]

(Continued)

Table 1 (Continued)

Arabidopsis transgenic lines	Description	Observations	Reference	Example in phloem research
Lines with non-cell type-specific expression of fluorescently-tagged protein markers				
pSEOR2::SEOR2:GFP	Sieve element occlusion related protein 2 (SEOR2)	P-protein filament in SE	[18]	[18, 20]
pRTM1::GFP:RTM1	Restricted TEV Movement 1 (RTM1)	Small bodies in the SE (an average of 1 μm diameter)	[31]	[20]
pRTM2::RTM2:GFP	Restricted TEV Movement 2 (RTM2)	Margins of the SE	[31, 32]	[20]
pSUC2::PP2-A1:GFP *pSUC2::PP2-A1:YFP* *pSUC2::PP2-A1:CFP* *pSUC2::GFP:PP2-A1*	Phloem lectin PP2-A1	Periphery of SE bodies	[20]	[20]
pSEOR1::PP2-A1:GFP	Phloem lectin PP2-A1	Periphery of SE bodies	[20]	[20]
pSLI1::eYFP:gSLI1	Small heat shock-like Sieve Element-Lining Chaperone 1 (related to RTM2)	Margins of the SE	[33]	[33]
pSWEET11::SWEET11: GFP	SWEET11 sugar transporter	Plasma membrane of the PPC[a] Specific face of the PPC	[7]	[7, 34]
pSWEET12::SWEET12: GFP	SWEET12 sugar transporter	Plasma membrane of the PPC[a]	[7]	[7]
pSWEET11::SWEET11-2A:GFP	SWEET11 sugar transporter	Plasma membrane of the PPC	[35]	[35]
pSWEET13::SWEET13:YFP	SWEET13 sugar transporter	Plasma membrane of the PPC	[35]	[35]
pMtSEO2::GFP5-ER	ER retention signal HDEL	ER in SE	[12]	[12, 19]
pSUC2::ER-GFP	GFP with the HDEL-retention signal	ER-localized in CC	[16]	[16]
pSUC2::AtSTP9-GFP	tmGFP9: GFP fused to 232 N-terminal amino acids of STP9	Plasma membrane anchored in CC	[16]	[16]

Table 1 (Continued)

Arabidopsis transgenic lines	Description	Observations	Reference	Example in phloem research
Lines with non-cell type-specific expression of fluorescently-tagged protein markers				
pSUC2::SUC2:GFP	tmGFP2: GFP fused to the ORF of SUC2	Plasma membrane anchored in CC	[16]	[16]
Lines expressing soluble cytosolic GFP for studies of symplasmic connectivities				
pSUC2::GFP	Soluble GFP	GFP (27 kDa)	[16]	[16, 19]
pSUC2::ubiquitin: GFP	Soluble GFP-fusion with Plantago major ubiquitin 1	Soluble GFP-variant with increasing molecular mass (36 kDa)	[16]	[16, 19]
pSUC2::GFP:sporamin	GFP fused to the ORF of sporamin, lacking the N-terminal 37 amino acids signal responsible for targeting to the ER	Soluble GFP-variant with increasing molecular mass (47 kDa)	[16]	[16]
pSUC2::aequorin:GFP	GFP fused to the ORF of aequorin Soluble	GFP-variant with increasing molecular mass (48 kDa)	[16]	[16, 19]
pSUC2::GFP-patatin	GFP fused to the ORF of patatin, lacking the N-terminal 23 amino acids signal sequence responsible for targeting to the ER	Soluble GFP-variant with increasing molecular mass (67 kDa)	[16]	[16]
Other lines expressing soluble GFP				
pbZIP9::GFP:GUS	Soluble GFP restricted to PPC	Soluble GFP	[35]	[35]

SE sieve elements, *CC* companion cells, *PPC* phloem parenchyma cells, *ER* endoplasmic reticulum [a]An additional localization has been demonstrated in the xylem parenchyma cells by in situ hybridization and histochemical assay with a *pSWEET11::GUS* or *pSWEET12::GUS* transgenic lines [8]

Table 2
Fluorescent dyes used for imaging the phloem cells and phloem loading or unloading

Dye	λ Excitation	λ Emission	Usage	Reference
Phloem mobile tracer				
5,6 CFDA	488 nm	500–550 nm	Phloem mobile tracer	[24]
HPTS	405 nm	460–570 nm	Phloem mobile tracer	[21]
Esculin	405 nm	420–470 nm	Phloem mobile blue fluorescent probe	[19, 36]
CTER	546 nm	576 nm	Phloem mobile red fluorescent probe	[36, 37]
Subcellular vital dye				
FM4-64	488 nm	701–708 nm	Membranes	[38]
$DiOC_6$	488 nm	515–580 nm	Lipophilic fluorescent dye. Staining of mitochondria, ER and membranes	[39]
MitoTracker Red CMXRos	544 nm	684–764 nm	Mitochondria	[40]
Aniline blue	405 nm	414–476 nm	Callose staining	[17]
Sirofluor	405	420–600	Callose staining	[19]

5. MitoTracker Red CMXRos (Invitrogen) stock solution: 50 μg mL^{-1} in DMSO. For staining, incubate for 15 min in 2.5 μg mL^{-1} before mounting.

2.4 Symplasmic Mobile Phloem Tracers

Keep stock solutions at −20°C. See Table 2 for additional information on fluorescent dyes.

1. 5,6 carboxyfluorescein diacetate (5,6 CFDA) (Invitrogen): stock solution: 60 mg mL^{-1} in water. Use 5 μL for loading after abrasion with Carborundum.
2. 8-hydroxypyrene-1,3,6-trisulphonic acid (HPTS) (Life Technologies): stock solution 5 mg mL^{-1}.
3. Carboxytetraethylrhodamine (CTER): stock solution: 0.8 mM in a 20% acetonitrile solution.
4. Esculin: dissolve to 5 mg mL^{-1} in water.
5. Carborundum powder: 0.037 mm.

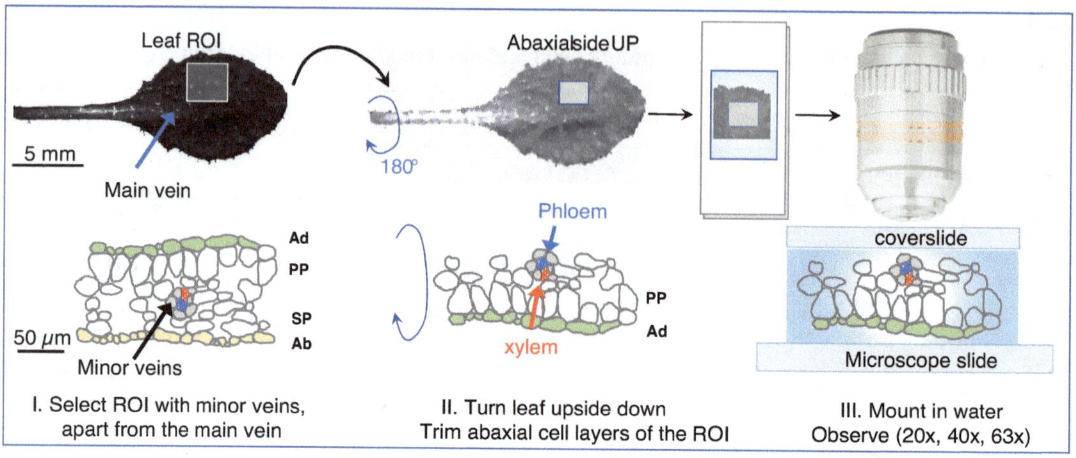

Fig. 1 Preparation of leaf sample. Phloem cells are drawn in blue, xylem cells are drawn in red and bundle sheath cells are filled in gray. ROI region of interest, SP spongy parenchyma, PP palisade parenchyma, Ad adaxial leaf surface (cells filled in green), Ab Abaxial leaf surface (cells filled in yellow)

3 Methods

This protocol combines sample preparation and light microscopy to view the vasculature of the leaves in *Arabidopsis* [20]. It was adapted from the method described for *Vicia faba* [41]. Carry out all the procedures at room temperature.

3.1 Preparation of Leaf Sample (Fig. 1)

1. Harvest the leaves by cutting at the basal part of the leaf petiole. Leaves must be maximum 3 cm long (*see* **Note 2**).
2. Before treatment and microscopic analysis, keep the leaves submerged in water or buffer (*see* **Note 1**).
3. Put the leaf on a glass slide with the abaxial side facing up, and cover with water.
4. Peel the epidermis with a sharp scalpel or a razor blade. Peeling can be done by gently trimming the leaf in the region of interest (ROI). Then, remove the superficial cell layers, including the epidermis, using tweezers. As soon as the epidermis is removed, clean the leaf with water to remove residual epidermal fragments and loose cells. The trimming of the tissues should be done outside of the main vein.
5. Using a scalpel or a razor blade, remove spongy parenchyma by very gently peeling away the cell layers. The leaf lamina should appear almost transparent once done. Clean with water (*see* **Note 3**).
6. Cover the leaf with buffer or the specific dye according to your experiment (*see* **Note 4**). Ensure that the buffer is adjusted to

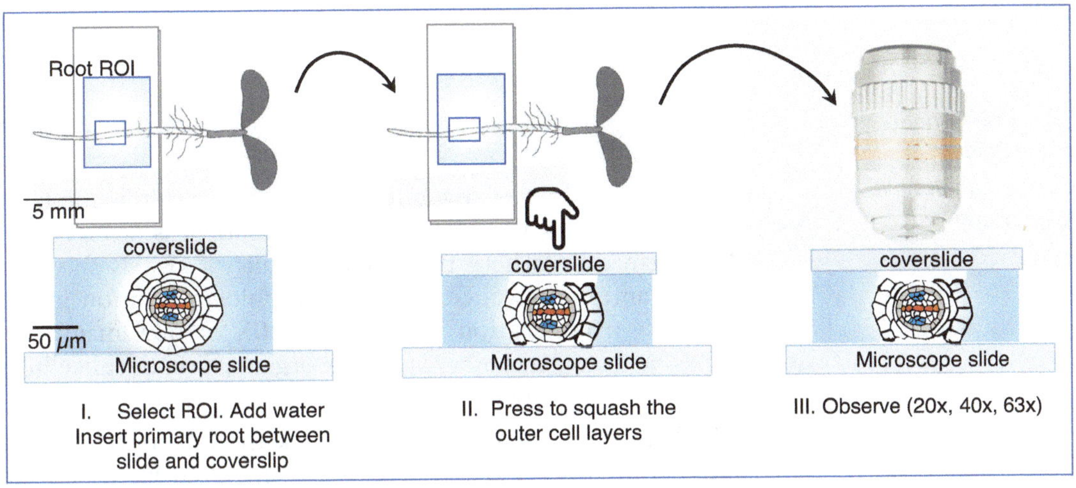

Fig. 2 Preparation of root sample. Phloem cells are drawn in blue, xylem cells in red, and perivascular cells drawn in gray. ROI region of interest

the optimal pH for the respective fluorescent proteins, such as pH 9.5 for GFP-tagged proteins.

7. Add a cover slip and observe immediately (*see* **Note 5**).
8. Observation of the phloem in the leaves can be done on all vein orders of the leaf except for the main vein, which is too thick for good imaging. Veins of order II have generally between 5 and 8 companion cells in a row, interlaced with numerous sieve tubes. Veins of order III are composed of 3 to 4 companion cells in a row, with few sieve elements. In our experience in *Arabidopsis thaliana*, sieve elements are always the thinnest cells in the tissue, with a diameter of about 2–3 µm, making their observation tedious.
9. For checking phloem integrity and functionality, use phloem-mobile fluorescent tracers, such as 5,6 CFDA, HPTS, CTER, or esculin. Add 3 µL of dye solution to the leaf area immediately after preparing it either by peeling off the leaf surface for 5,6 CFDA and HPTS, or by abrasion using powdered Carborundum. The tracers can then be observed in veins downstream of the site of application of the dye.

3.2 Preparation of Root Sample (Fig. 2)

1. Use plantlets grown in vitro for 7 days after sowing.
2. Harvest whole plantlets for observations.
3. Put the primary root, still attached to the plantlet, between a slide and a coverslip in water or buffer mounting (*see* **Note 5**). Use buffer adjusted to the optimal pH for the fluorescent fusion proteins, as GFP fluorescence is pH-sensitive,
4. Observe the main root after gently pressing the tissue between the slide and the coverslip to tear the surface cell layers.

When applying this technique, look out for potential artifacts (*see* **Note 6**).

3.3 Observation of the Phloem In Vivo Using Confocal Microscopy

1. Carry out microscopic analysis of the samples directly after sample preparation (*see* **Note 5**).
2. Using eyepieces, 10X magnification and the transmission light mode, search for the vasculature. The phloem is located near the xylem cells, where the lignified walls of vessels show an intense autofluorescence. To observe phloem location in the leaf, 10X magnification is sufficient. 20X or 63X magnification is necessary to observe phloem organization or subcellular localization of markers.
3. Switch to fluorescence mode to check the quality of samples and identify the regions of interest. Using a red filter, you can check the chloroplast distribution in phloem cells. Chloroplasts are aligned in long strings of 6–10 in companion cells.
4. Switch to confocal mode for the imaging. Optionally, perform multispectral imaging of compatible fluorescently-tagged proteins and dyes (Table 1). Be aware of lignin autofluorescence in xylem and chlorophyll autofluorescence in phloem (*see* **Note 4**).

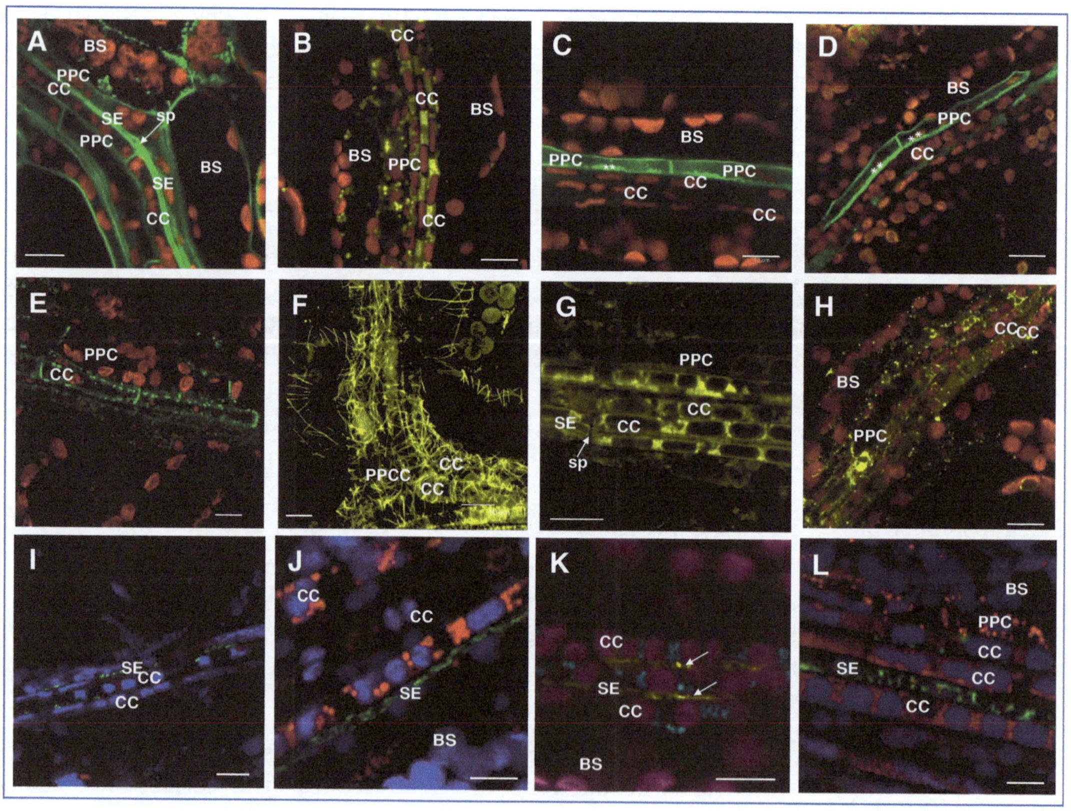

3.4 Identification of Phloem Cell Types

1. Identify sieve elements based on the lack of chlorophyll autofluorescence and their diameter below 5 µm in leaves and roots (Figs. 3 and 4).
2. Identify companion cells in leaves (Fig. 3) by the single string of autofluorescent chloroplasts regularly aligned in the cells, and their close association with sieve elements. The cell diameter is only slightly larger than the chloroplasts' diameters.
3. Identify phloem parenchyma in leaves (Fig. 3) based on their uneven "crinkled" plasma membrane. Also, they often present a number of chloroplasts aligned on one side of the cell. The diameter of these cells is slightly larger than that of the companion cells, but remains generally below 10 µm.
4. Identify phloem perivascular cells in leaves (Fig. 3) based on the location of chloroplasts, easily imaged by their autofluorescence, on the outside border of the phloem tissue. Phloem perivascular cells have a large vacuole, and their diameter is much larger than that of the other phloem cell types, especially of the phloem parenchyma cells with which they can be confused.

Fig. 3 Imaging of GFP-tagged proteins in the phloem of a leaf minor vein. (**a**) Imaging of the plasma membrane in phloem cells in a *p35S::LTi6B:GFP* plant. *LTi6B:GFP* fluorescence can also be used in sieve elements. Phloem parenchyma cells are the largest cells and are located on the edge side of the vasculature. Sieve elements are the thinnest cells and are easily recognizable by their absence of chloroplasts. Companion cells display typical chloroplast alignments. In green: fluorescence of the GFP; In red: autofluorescence of the chloroplasts. (**b**) Imaging of mitochondria in phloem cells in a *p35S::COX4:YFP* plant. In yellow: fluorescence of the YFP; in red: autofluorescence of the chloroplasts. (**c**) Imaging of the plasma membrane in a phloem transfer cell in a *pSWEET11::SWEET11:GFP* plant. In green: fluorescence of the GFP along cell wall ingrowths; in red: autofluorescence of the chloroplasts. (**d**) Imaging of the plasma membrane in a phloem transfer cell in a *pSWEET12::SWEET12:GFP* plant. In green: fluorescence of the GFP along cell wall ingrowths; in red: autofluorescence of the chloroplasts. (**e**) Imaging of plasmodesmata in the phloem cells in a *p35S::PDLP1:GFP* plant. In green: fluorescence of the GFP in plasmodesmata; in red: autofluorescence of the chloroplasts. (**f**) Imaging of cortical microtubules in the phloem parenchyma cells and the companion cells in a *p35S::GFP:MBD* plant. The integrity of the microtubules confirmed that the cells are alive. (**g**) Imaging of the endomembranes in the phloem cells stained with $DiOC_6$. One sieve element is visible in this image. The companion cells show large vacuoles. The mitochondrial membrane is also stained using $DiOC_6$, both in the sieve elements and in the companion cells. (**h**) Imaging of the endoplasmic reticulum in the phloem cells in a *p35S::ER:YFP* plant. (**i**) Imaging of RTM1:labeled sieve element bodies in the sieve elements in a *pRTM1::GFP:RTM1* plant. In blue: chloroplast autofluorescence. In red: mitochondria. In green: RTM1:GFP fluorescence. (**j**) Imaging of the RTM2-labeled parietal layer in a phloem sieve element in a *pRTM2::RTM2:GFP* plant. In blue: chloroplast autofluorescence. In red: mitochondria labeled with Mitotracker Red dye. In green: RTM2:GFP fluorescence. (**k**) Imaging of the P-proteins in a phloem sieve element in a *pSEOR1::SEOR1:GFP* plant. In pink: chloroplast autofluorescence. In light blue: mitochondria labeled with Mitotracker Red dye. In yellow: SEOR1:GFP fluorescence. (**l**) Imaging of the PP2-A1 phloem lectin in a phloem sieve element in a *pSEOR1::PP2-A1:GFP* plant. In blue: chloroplast autofluorescence. In red, mitochondria labeled with Mitotracker Red dye. In green: PP2-A1:GFP fluorescence. In this image, we observe that PP2-A1:GFP fluorescence is located around SE mitochondria. Water-immersion objectives with 20× (**f**), 40× (**c**, **d**, **e**, **h–l**), or 63× (**a**,**b**,**g**) magnification were used. BS bundle sheath, CC companion cell, PPC phloem parenchyma cell, SE sieve element. sp sieve plate. White arrow: P protein filaments, **Cell wall ingrowths. Scale bar = 10 µm

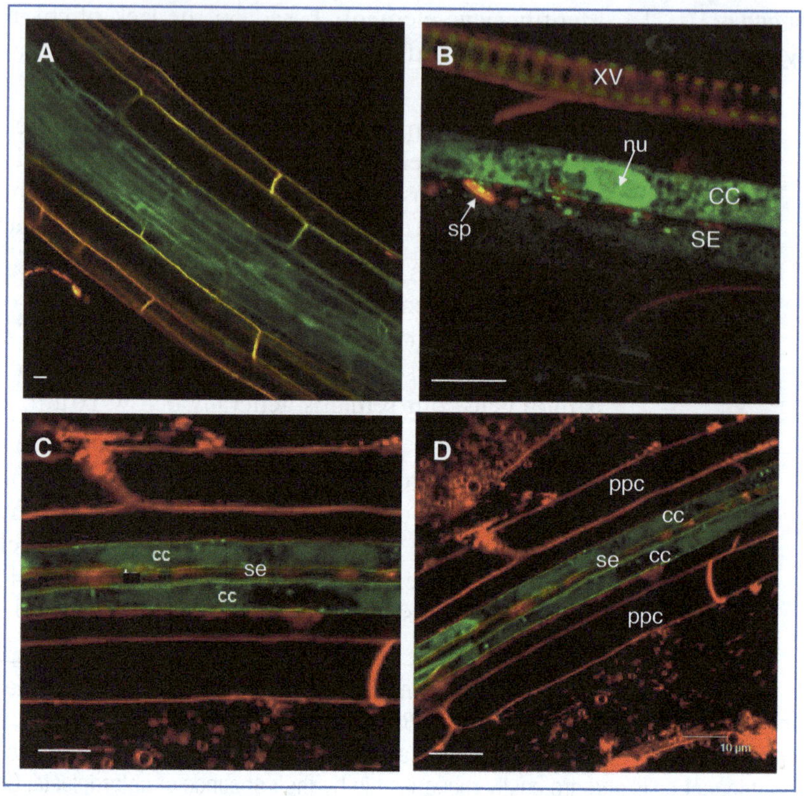

Fig. 4 Imaging of GFP-tagged proteins in the phloem of the primary root. (**a**) Imaging of soluble GFP in the phloem cells in a *pSUC2::GFP* plant. In green: fluorescence of the GFP. In red: fluorescence of FM4-64. (**b**) Imaging of GFP-tagged PP2-A1 in the companion cell and a sieve element in a *pSUC2::PP2-A1:GFP* plant. In the companion cell, GFP-tagged PP2-A1 is mainly located in the cytosol, although some fluorescence was also observed in the nucleus, suggesting nucleo-cytoplasmic shuttling. GFP fluorescence was found both in the companion cells and the sieve elements, indicating in the root the cell-to-cell transport of PP2-A1 from the companion cell to the sieve element. In green: fluorescence of the GFP; In red: fluorescence of callose deposit in a sieve plate stained by aniline blue. (**c**, **d**) Imaging of the plasma membrane in a phloem cell in the root phloem cells of a *pSUC2::RCI2A:YFP* plant. In green: fluorescence of the GFP; in red: fluorescence of FM4-64. Water immersion objectives with 20× (**a**, **c**, **d**) and 63× (**b**) magnification were used. CC companion cell, PPC phloem parenchyma cell, SE sieve element. Scale bar = 10 μm

3.5 Identification of Subcellular Structures in Phloem Cells

1. Observe mitochondria and sieve element reticulum by using the MitoTracker Red CMXRos dye and *Lti6b:GFP* marker, respectively (Fig. 3). Sieve elements contain no chloroplast, no nucleus and no vacuole. Observe the parietal layer, also described as parietal ground matrix, by using *RTM2:GFP* and *YFP:SLI1* markers. No marker is yet available for the observation of the sieve element plastids, but the *GFP:RTM1* marker labels the periphery of sieve element bodies of the same dimension as sieve element plastids and could be a candidate. No marker is yet available to label endocytosis vesicles. Observe P-proteins, which appear as large filamentous bodies, by using *SEOR1:GFP* and *SEOR2:GFP* fluorescent markers.

2. Use standard fluorescent markers to observe the full range of subcellular structures in companion cells, including mitochondria, vacuole, nucleus, plastids, and cytoskeleton.
3. In phloem parenchyma cells that developed into transfer cells, use *SWEET11:GFP* as a marker of plasma membranes associated with cell wall invaginations (Fig. 3). Use standard fluorescent markers to visualize other subcellular structures.
4. Use the chloroplasts' location at the interface with cortical cells to identify phloem perivascular cells, and image them using their autofluorescence. Standard fluorescent markers can also be used in these cells.

4 Notes

1. Plant sampling must be done quickly, and samples should be immediately submerged in water or buffer before treatment and/or staining. The sampling for one set of experiments must always be done at the same time of day. For plants grown in soil, plants should be correctly watered.
2. If possible, use young leaves. Old leaves tend not to be easy to process. Thick leaves are also more difficult to manipulate for imaging phloem tissues. For the observation of leaves in other conditions (senescence, pathogen infection, etc), it might be necessary to optimize the growth conditions (humidity, light quality, photoperiod) to facilitate the leaf preparation.
3. Preparing the sample causes thinning of the leaf lamina. To improve handling, keep the part with peeled epidermis attached to the rest of the leaf or petiole. This enables convenient moving of the sample with tweezers. Before imaging, the sample can be cropped with a razor blade, so that the sample is flat under the microscope.
4. Chlorophyll has an emission spectrum between 650 and 700 nm. Choose dyes, phloem mobile tracers, fluorescently tagged proteins, and imaging parameters to avoid overlap.
5. Once harvested, leaves must be used within an hour, at most. Degradation of the phloem tissue usually appears after 1 h. Once prepared (removal of cell layers), a leaf or a root should be observed within 20 min. Therefore, it is unnecessary to seal the side of the cover slip. When using specific markers, for example, for microtubules, we have observed rapid degradation of phloem vitality after 20 min.

6. Do not use FM4-64 when the roots have been prepared by squeezing. Due to rapid internalization of the dye, all cell components will be labeled.

Acknowledgments

The IJPB benefits from the support of the LabEx Saclay Plant Sciences-SPS (ANR-10-LABX-0040-SPS) managed by the French National Research Agency under an "Investments for the Future" program (ref. ANR-11-IDEX-0003-02). We thank the Imaging and Cytology platform of the Plant Observatory (IJPB Institute, INRAE Versailles-Grignon, France) for light microscopy observations. TC was supported by a PhD fellowship from the Doctoral School ED145 "Sciences du Végétal", Paris—Sud XI University / Université Paris-Saclay .

References

1. Lucas WJ et al. (2013) The plant vascular system: evolution, development and functions. J Integr Plant Biol 55(4):294–388
2. Lee JY, Frank M (2018) Plasmodesmata in phloem: different gateways for different cargoes. Curr Opin Plant Biol 43:119–124
3. Turnbull CGN, Lopez-Cobollo RM (2013) Heavy traffic in the fast lane: long-distance signalling by macromolecules. New Phytol 198(1):33–51
4. Koenig AM, Hoffmann-Benning S (2020) The interplay of phloem-mobile signals in plant development and stress response. Biosci Rep 40(10):1–20
5. Koroleva OA, Davies A, Deeken R, Thorpe MR, Tomos AD, Hedrich R (2000) Identification of a new glucosinolate-rich cell type in Arabidopsis flower stalk. Plant Physiol 124(2):599–608
6. Husebye H, Chadchawan S, Winge P, Thangstad OP, Bones AM (2002) Guard cell- and phloem idioblast-specific expression of thioglucoside glucohydrolase 1 (myrosinase) in Arabidopsis. Plant Physiol 128(4):1180–8
7. Chen L-Q, Qu X-Q, Hou B-H, Sosso D, Osorio S, Fernie AR, Frommer WB (2012) Sucrose efflux mediated by SWEET proteins as a key step for phloem transport. Science 335(6065):207–211
8. Le Hir R et al. (2015) Disruption of the sugar transporters AtSWEET11 and AtSWEET12 affects vascular development and freezing tolerance in Arabidopsis. Mol Plant 8(11):1687–1690
9. Moison M, Marmagne A, Dinant S, Soulay F, Azzopardi M, Lothier J, Citerne S, Morin H, Legay N, Chardon F, Avice J-C, Reisdorf-Cren M, Masclaux-Daubresse C (2018) Three cytosolic glutamine synthetase isoforms localized in different-order veins act together for N remobilization and seed filling in Arabidopsis. J Exp Bot 69(18):4379–4393
10. Burlat V, Oudin A, Courtois M, Rideau M, St-Pierre B (2004) Co-expression of three MEP pathway genes and geraniol 10-hydroxylase in internal phloem parenchyma of Catharanthus roseus implicates multicellular translocation of intermediates during the biosynthesis of monoterpene indole alkaloids and isoprenoid-derive. Plant J 38(1):131–141
11. Simkin AJ, Miettinen K, Claudel P, Burlat V, Guirimand G, Courdavault V, Papon N, Meyer S, Godet S, St-Pierre B, Giglioli-Guivarc'H N, Fischer MJC, Memelink J, Clastre M (2013) Characterization of the plastidial geraniol synthase from Madagascar periwinkle which initiates the monoterpenoid branch of the alkaloid pathway in internal phloem associated parenchyma. Phytochemistry 85:36–43
12. Froelich DR, Mullendore DL, Jensen KH, Ross-Elliott TJ, Anstead JA, Thompson GA, Pélissier HC, Knoblauch M (2011) Phloem ultrastructure and pressure flow: sieve-element-occlusion-related

agglomerations do not affect translocation. Plant Cell 23(12):4428–4445
13. Dinant S, Lucas WJJ (2013) Sieve elements: puzzling activities deciphered through proteomics studies. In: Thompson GA, van Bel AJE (eds) Phloem biochemistry. Wiley-Blackwell, pp 155–185
14. Knoblauch M, Oparka K (2012) The structure of the phloem-still more questions than answers. Plant J 70(1):147–156
15. Truernit E (2014) Phloem imaging. J Exp Bot 65(7):1681–1688
16. Stadler R, Wright KM, Lauterbach C, Amon G, Gahrtz M, Feuerstein A, Oparka KJ, Sauer N (2005) Expression of GFP-fusions in Arabidopsis companion cells reveals non-specific protein trafficking into sieve elements and identifies a novel post-phloem domain in roots. Plant J 41(2):319–331
17. Thompson MV, Wolniak SM (2008) A plasma membrane-anchored fluorescent protein fusion illuminates sieve element plasma membranes in Arabidopsis and tobacco. Plant Physiol 146(4):1599–1610
18. Anstead JA, Froelich DR, Knoblauch M, Thompson GA (2012) Arabidopsis P-protein filament formation requires both AtSEOR1 and AtSEOR2. Plant Cell Physiol 53(6):1033–1042
19. Ross-Elliott TJ et al (2017) Phloem unloading in Arabidopsis roots is convective and regulated by the phloem pole pericycle. Elife 6:1–31
20. Cayla T, Batailler B, LeHir R, Revers F, Anstead JA, Thompson GA, Grandjean O, Dinant S (2015) Live imaging of companion cells and sieve elements in Arabidopsis leaves. PLoS ONE 10(2):0118122
21. Wright KM, Oparka KJ (1996) The fluorescent probe HPTS as a phloem-mobile, symplastic tracer: an evaluation using confocal laser scanning microscopy. 47(296):439–445
22. Knoblauch M, Vendrell M, De Leau E, Paterlini A, Knox K, Ross-Elliot T, Reinders A, Brockman SA, Ward J, Oparka K (2015) Multispectral phloem-mobile probes: properties and applications. Plant Physiol 167(4):1211–1220
23. Knox K, Oparka K (2018) Illuminating the translocation stream. Curr Opin Plant Biol 43:113–118
24. Oparka KJ, Duckett CM, Prior DAM, Fisher DB (1994) Real-time imaging of phloem unloading in the root-tip of Arabidopsis. Plant J 6(5):759–766
25. Nelson BK, Cai X, Nebenführ A (2007) A multicolored set of in vivo organelle markers for co-localization studies in Arabidopsis and other plants. Plant J 51(6):1126–1136
26. Camilleri C, Azimzadeh J, Pastuglia M, Bellini C, Grandjean O, Bouchez D (2002) The Arabidopsis TONNEAU2 gene encodes a putative novel protein phosphatase 2A regulatory subunit essential for the control of the cortical cytoskeleton. Plant Cell 14(4):833–845
27. Voigt B, Timmers ACJ, Šamaj J, Müller J, Baluška F, Menzel D (2005) GFP-FABD2 fusion construct allows in vivo visualization of the dynamic actin cytoskeleton in all cells of Arabidopsis seedlings. Eur J Cell Biol 84(6):595–608
28. Boisnard-Lorig C (2001) Dynamic analyses of the expression of the HISTONE::YFP fusion protein in Arabidopsis show that syncytial endosperm is divided in mitotic domains. Plant Cell 13(3):495–509
29. Cutler SR, Ehrhardt DW, Griffitts JS, Somerville CR (2000) Random GFP::cDNA fusions enable visualization of subcellular structures in cells of Arabidopsis at a high frequency. Proc Natl Acad Sci U S A 97(7):3718–3723
30. Thomas CL, Bayer EM, Ritzenthaler C, Fernandez-Calvino L, Maule AJ (2008) Specific targeting of a plasmodesmal protein affecting cell-to-cell communication. PLoS Biol 6(1):e7
31. Chisholm ST, Parra MA, Anderberg RJ, Carrington JC (2001) Arabidopsis RTM1 and RTM2 genes function in phloem to restrict long-distance movement of Tobacco Etch Virus. Plant Physiol 127:1667–1675
32. Whitham SA (2000) Arabidopsis RTM2 gene is necessary for specific restriction of Tobacco Etch Virus and encodes an unusual small heat shock-like protein. Plant Cell Online 12(4):569–582
33. Kloth KJ, Busscher J, Wiegers G, Kruijer W, Buijs G, Meyer RC, Albrectsen BR, ouwmeester HJ, Dicke M, Jongsma MA (2017) SIEVE ELEMENT-LINING CHAPERONE1 restricts aphid feeding on Arabidopsis during heat stress. Plant Cell 29(10):2450–2464
34. Wei XY, Collings DA, McCurdy DW (2021) Review: More than sweet: new insights into the biology of phloem parenchyma transfer cells in Arabidopsis. Plant Sci 310, 110990.
35. Kim JY, Symeonidi E, Pang TY, Denyer T, Weidauer D, Bezrutczyk M, Miras M, Zöllner N, Hartwig T, Wudick MM, Lercher M, Chen LQ, Timmermans MCP, Frommer WB (2021) Distinct identities of leaf phloem cells revealed by single cell transcriptomics. Plant Cell 33(3):511–530
36. Knoblauch M, Vendrell M, de Leau E, Paterlini A, Knox K, Ross-Elliot T, Reinders A, Brockman SA, Ward J, Oparka K (2015) Multispectral phloem-mobile probes: properties and applications. Plant Physiol 167(4):1211–1220

37. Paultre D (2017) Studies of macromolecular trafficking across Arabidopsis homografts. Dissertation, University of Edinburgh, Edinburgh
38. Vida TA, Emr SD (1995) A new vital stain for visualizing vacuolar membrane dynamics and endocytosis in yeast. J Cell Biol 128(5):779–792
39. Martens HJ, Roberts AG, Oparka KJ, Schulz A (2006) Quantification of plasmodesmatal endoplasmic reticulum coupling between sieve elements and companion cells using fluorescence redistribution after photobleaching. Plant Physiol 142(2):471–480
40. Poot M, Zhang YZ, Krämer JA, Wells KS, Jones LJ, Hanzel DK, Lugade AG, Singer VL, Haugland RP (1996) Analysis of mitochondrial morphology and function with novel fixable fluorescent stains. J Histochem Cytochem 44(12):1363–1372
41. Knoblauch M, van Bel AJE (1998) Sieve tubes in action. Plant Cell 10(1):35–50

Chapter 12

Immunohistochemical Localization of Proteins in the Phloem: Problems and Solutions

Guilherme Tamassia and Christina Kühn

Abstract

The technique of immunolocalization in differentiated phloem cells is a challenging task given their specific functions, anatomy, and cell organellar infrastructure. Incorporation of specific cell wall components in thickened cell walls of phloem cells is often the source of unspecific labeling and erroneous localization data. Furthermore, tissue preservation of cells that are exposed to high osmotic pressure is a very challenging task, encountering many pitfalls. Therefore, special care is required regarding the generation and purification of specific antibodies and in the protocols used for tissue preservation. Here, some useful tips for cautious tissue preparation and antibody purification for immunolocalization are given. In parallel, the main focus should be on the confirmation of immunohistochemical localization data by alternative methods, promoter reporter gene studies, or activity tests.

Keywords Phloem transport, Carbon allocation, Sucrose transporter, Phloem proteins, Immunolocalization, Subcellular localization

1 Introduction

1.1 Phloem Anatomy and How it Constrains Immunohist ochemistry

The specialized anatomy of the phloem cells, which are optimized to fulfill their function of efficient photoassimilate translocation over long distances, poses severe consequences for localization analyses, staining procedures, and measurement of cell coupling. This includes studies of immunohistochemical localization of proteins, which use antibodies raised against a protein-of-interest to visualize its location. Protein localization represents a powerful approach to elucidate structure–function relationships at the tissue or subcellular scale, and to validate proteomic and transcriptomic data obtained from molecular biology analyses. Determining the localization of phloem proteins provides key insights into how the transport of photoassimilates and signaling molecules is spatially organized within phloem cell types, especially in the case of the sieve tube element-companion cell complexes (SECCCs) of flowering plants.

Johannes Liesche and Sanna Sevanto (eds.), *Phloem*, Methods in Molecular Biology, vol. 3066,
https://doi.org/10.1007/978-1-0716-5488-0_12,

Although it is generally assumed that the cell wall of phloem sieve elements (SEs) is typically not lignified, in contrast to xylem tracheary elements, the cell wall can undergo substantial thickening in both mono and dicotelydons [1]. Cell walls of the sieve element are rigid and strengthened to resist the high hydrostatic pressure [2]. Rigid cell walls mean that obtaining thin tissue sections requires sharp knives and great care, especially since strong cell walls tend to become brittle during sample preparation. In addition to sectioning, close attention must be paid to tissue fixation, because the sudden release of hydrostatic pressure inside the SEs can lead to artifacts resulting from defects and alterations of the ultrastructure that are due to wounding and that do not occur under in vivo conditions or in intact tissue. Furthermore, the unique structure and composition of SEs can lead to nonspecific binding of antibodies. This means that, in addition to careful sample preparation, thorough affinity purification of primary antibodies for usage in immunohistochemistry is indispensable.

In the following, the open questions regarding proteins in SEs are briefly discussed before providing detailed information on the best practice to localize proteins with antibodies inside the phloem.

1.2 Specific Protein Composition of Phloem Sieve Elements and the Phloem Sap

The mature sieve tube element consists of non-autonomous cells, which are dependent on the neighboring companion cells (CC). This dependence was metaphorically compared to a hyperactive nurse taking care of a comatose patient by van Bel and Knoblauch [3, 4]. This means that proteins in mature SEs that do not persist from the immature stage are transported from CC. Which proteins localize to mature sieve elements in their native state is still a matter of debate, with only about a dozen proteins having confirmed SE localization. The two main reasons for this are the difficulty of sampling sieve element content [5] and the issue of unspecific protein movement into phloem sap [6].

In the past, a large-scale proteomics approach of pumpkin stem exudates revealed the presence of proteins that function in RNA binding, mRNA translation, ubiquitin-mediated proteolysis, and vesicle trafficking [7]. In this study, the presence of ubiquitin, ribosomal RNA, and proteins was detected in exudates of cucurbits. Thus, translation in mature sieve elements, as well as turnover of proteins and vesicle trafficking, seemed to be feasible [7]. However, it has since been demonstrated that the sampled exudates do mainly stem from the extrafascicular phloem and not from the fascicular phloem that is responsible for sugar translocation [8, 9]. The same applies to a recent study that identified a diverse set of metabolic enzymes in exudate from cucumber and watermelon [10].

In addition to the Cucurbit-specific issue of differentiating between the two types of phloem, it is generally difficult to verify which proteins localize to phloem sieve elements in vivo. The content of mature sieve tubes, as determined by proteomics or immunohistochemical studies has also been proposed to represent mostly "contaminants" from selective degradation of organelles and cytoplasmic structures during sieve element differentiation [11]. Micrograft experiments with Arabidopsis plants have revealed mobility of many GFP-tagged phloem proteins, even those carrying a chloroplast transit peptide. The authors of these studies concluded that there must be an unspecific loss of proteins from companion cells into the sieve elements, after a quite long period of time (10 days). Nevertheless, among the vast number of proteins and mRNAs found in the phloem sap, the authors assumed the presence of a specific set of systemic signals, which were not simply "lost in transit" [12, 13].

Immunohistochemistry remains the best method for proving the presence of a specific protein in SEs. Generation of monoclonal antibodies against several sieve element proteins from *Streptanthus tortuosus* has revealed that the sieve elements contain a specific set of proteins that are absent in the neighboring companion cells [14–16, 2, 17], even at the earliest stages of their differentiation.

1.3 Examples of Protein Traffic Between Companion Cells and Sieve Elements

The follow-up question to "which proteins localize to SEs" is "how do these proteins get into SEs." For solving this question, too, immunohistochemical localization of proteins in CC and SE is a helpful tool.

For example, the potato sucrose transporter StSUT1 protein has been localized to the plasma membrane of the sieve elements by various methods and fixation procedures [18, 19]. The half-life of sucrose transporter proteins is rather short [20, 21], which argues for a continuous de novo transcription, including subsequent plasmodesmal transfer into the SEs. The question of how short-lived sieve element-specific membrane proteins can be targeted to the plasma membrane of sieve elements and be constantly renewed in this non-autonomous cell type remains unanswered. Identification of the interactome of phloem-specific proteins or elucidation of the proteome of plasmodesmata will help to clarify this enigma, as will the validation of protein localizations by immunohistochemistry.

Proteomics can be used to confirm results from immunolocalization. Here, a new proteomic study of phloem sieve elements confirmed the localization of two sucrose transporters from tobacco in phloem sieve elements. The study not only focused on the soluble fraction of phloem proteins without membrane proteins, as often done previously, but introduced a new strategy of sieve element protein sampling [22]. Using the promoter of the sieve element occlusion protein MtSEO2, the enrichment of sieve element samples from stem and leaf tissue of tobacco plants was successful with a quite low degree of contaminants. Interestingly, this technique revealed that two sucrose transporters, NtSUT1 (Nitab4.5_0000531g0040.1) and NtSUT4 (Nitab4.5_0008090g0020.1), were specifically enriched in both the stem-derived as well as the leaf-derived sieve element fraction compared to the surrounding tissues [22]. Thus, the main phloem loader of tobacco, NtSUT1 [23], is targeted to the sieve element plasma membrane. It still remains unknown whether sieve element-specific proteins are translated in neighboring companion cells and transported via plasmodesmata, or alternatively translated early in young and immature sieve elements before differentiation [24].

Pore plasmodesma units (PPUs) are plasmodesmata connecting sieve elements and companion cells. The size exclusion limit (SEL) of PPUs seems to be much larger than for other plasmodesmata. Information about the limits of transport through these plasmodesmata came from the movement of the small FLOWERING LOCUS T (FT) protein, which is called the florigen [25], and transported over long distances via the phloem translocation stream. This protein is a small globular protein of only 20 kDa in size and moves easily from companion cells into the sieve elements when expressed under the companion cell-specific *AtSUC2* promoter. However, when florigen is fused to the GFP or YFP fluorochrome, the mobility and activity of the FT protein are reduced, but it can still move into the conducting elements of the phloem and induce flowering in the shoot apex [26]. Hence, the PPUs between sieve elements and companion cells allow the passage of proteins of at least 47 kDa. However, fusions between FT and two GFPs with an expected size of 74 kDa are not mobile and are not active in flower induction.

The report on florigen translocation demonstrates the importance of mass, or size, in restricting or conferring mobility and bioactivity of companion cell-derived soluble proteins. For membrane-anchored phloem proteins, a very similar size exclusion limit was observed.

When a monomeric yellow fluorescent protein, mCitrine, was N-terminally fused to membrane proteins of different sizes and stably expressed under the control of the companion cell-specific *AtSUC2* promoter in *Arabidopsis thaliana* and *Nicotiana tabacum* [27], the size of these fusion proteins was a limiting factor for the translocation from CCs into the SEs via PPUs. While mCitrin fusion to a small 6 kDa AtRCI2A membrane protein, with only 2 transmembrane spanning domains, yielded stable labeling of the sieve element plasma membrane and luminal vesicles in Arabidopsis and tobacco, fusion to the full-length AtSUC2 sucrose transporter with 12 transmembrane spanning domains did not yield sieve element targeting. A passage via the desmotubule was assumed [27], since it was previously shown to couple the ER of CC and SE [28].

Hence, the SEL of PPUs is the most important selective filter determining phloem mobility of companion-cell-derived membrane-bound (as well as soluble) proteins. However, additional mechanisms to regulate the selectivity of CC-to-SE transfer have been discussed [13].

2 Useful Tips for Plant Immunohistochemistry

2.1 Generation of Antibodies

Purification of the whole proteins for immunization is facilitated by tagging them. In order to avoid the generation of IgG molecules against the tag and not against the protein of choice, it is absolutely necessary to cleave off the tag after purification of the fusion protein, before using it for immunization. For instance, the use of the maltose binding protein (MBP) fused to the membrane proteins efficiently helps purify membrane proteins or accessible parts of them. However, before immunization, it is absolutely necessary to cleave off the MBP via the factor X protease cleavage site, in order to avoid the generation of MBP-specific antibodies.

Besides monoclonal antibodies, the generation of peptide antibodies generally yielded antisera of the highest specificity. Prediction programs can be used to identify amino acid stretches within proteins that are most suitable for peptide synthesis.

2.2 Sample Preparation: Compromise Between Tissue Preservation and Maintenance of the Antigenicity

Besides preserving antigenicity for antigen–antibody recognition, proper analysis under a microscope also requires preserving the integrity of the samples and ensuring the quality of the anatomical sections. Fixation of proteins and protein complexes generally occurs by two different mechanisms: denaturation or cross-link formation. Dehydrants, such as acetone or alcohol, denature proteins and replace free water in cells and tissues. A general distinction can be made between coagulant fixatives and non-coagulant fixatives. Both methods can cause artifacts when applied in plant histological techniques. Cross-linking agents such as formaldehyde are assumed to mask epitopes, whereas the disadvantage of organic acids is the shrinkage of the tissue. Ethanol is considered to be a coagulant that denatures proteins. The major disadvantage of ethanol is the considerable shrinkage of tissue and rather low preservation of ultrastructure [29].

Cross-linking by non-coagulant fixing agents forms both inter and intramolecular cross-links. Therefore, glutaraldehyde and formaldehyde are highly effective in the preservation of the ultrastructure and thus are optimal for electron microscopy. It is suspected that aldehydes mask epitopes. Therefore, the amount of glutaraldehyde should be minimized in the fixative, which is used for immunostaining protocols. The preservation of antigenicity is one of the most important points when choosing the fixative for immunohistochemical experiments. The aim is to find the optimal compromise between tissue preservation and preservation of antigenicity. Direct comparison of both fixation methods was conducted by Milne and coworkers [30], and it was determined that glutaraldehyde–formaldehyde-fixed tissue resulted in more background labeling (Milne et al. [30] supplements). In conclusion, formaldehyde fixation has to be used when protein localization is to be analyzed by electron microscopy, while ethanol–acetic acid causes a lower background when used for light-microscopic analysis at the expense of lower tissue preservation (*see* **Note 1**).

Especially when working with delicate materials such as succulent leaves, whose water-storing cells in the mesophyll are fragile and easily collapsible, the dehydration and embedding steps should be gradual and slow. Good results have been obtained using LR White resin as an embedding medium, which gradually replaces the dehydrating agent. Extreme care is also necessary when placing the samples in gelatin capsules filled with resin for polymerization to ensure they remain in the optimal position for sectioning.

Before proceeding with microtomy, it is important to know whether crystals are present in the plant tissues, as this will dictate the type of knife that should be used in the ultramicrotome, depending on the purpose of the study. For samples intended for immunolocalization by light, fluorescence, or confocal microscopy, 1 μm semi-thin sections are sufficient and can be adequately made with glass knives, even if the material contains crystals. However, for 90 nm ultra-thin sections intended for immunolocalization by transmission electron microscopy, a diamond knife should only be used if the material does not contain crystals. Otherwise, the diamond knife will be severely damaged. A recent study of succulent species containing calcium oxalate crystals [31] demonstrated that ultra-thin sections can be prepared for this type of microscopy using glass knives. However, it should be kept in mind that the knife must be constantly replaced due to the rapid wear of the cutting edge.

Providing a step-by-step protocol of the immunostaining procedures is beyond the scope of this chapter. For general guidelines and recommendations (*see* **Note 2**).

2.3 Antibody Purification

Affinity purification of antisera is generally required before immunolocalization experiments. The easiest way is to drop a certain amount of antigen on nitrocellulose, incubate the crude serum with it, and elute the IgG fraction by a highly acidic pH buffer (pH 2.5) [32]. However, elution of antibodies at highly acidic pH might cause antibody destabilization. Therefore, rapid neutralization of the eluate is required with additional stabilization by BSA or another protein. The disadvantage of this method is that the purified serum is not storable and has a low titer of antibodies. Control experiments revealed that IgG molecules bind efficiently even to an empty nitrocellulose, without any antigen coupled [19]. Therefore, special care should be taken when blocking the nitrocellulose membrane used for antiserum purification or IgG enrichment. This is particularly important when membranes are used several times for the same purpose.

The method used for the antiserum purification seems to be of crucial importance for the final results, because contradictory results can be observed using the same peptide antibody. It was shown that antibody purification via pieces of nitrocellulose (coupled or not with the antigen) often results in labeling of the companion cells, whereas affinity-purification of the same antiserum via Sepharose columns covalently coupled to the antigen, and followed by thorough dialysis of the purified antibody, resulted in labeling of the sieve elements [19]

The binding capacity of a crude antiserum to an empty Sepharose column material was tested in our laboratory [19]. After thorough washing steps, almost no remaining proteins could be eluted from the empty column. Nevertheless, pooled fractions showed weak absorbance at 280 nm. Thus, samples were dialyzed overnight against PBS-T. The processed IgG fraction was tested on a semi-thin section of LR White-embedded plant material containing sieve elements, and no binding could be observed.

In contrast, when the control experiments were performed with empty nitrocellulose (without any peptide or antigen bound to it), the IgG molecules were eluted in significant amounts [19]. This might be due to the high affinity of proteins to the nitrocellulose membrane. The effect of unspecific binding increased after multiple subsequent incubations of the antiserum, or if the blocking of the nitrocellulose membrane was insufficient.

2.4 Signal Enhancement

Weak signals from low-abundant proteins can be enhanced by signal amplification steps, such as biotin–streptavidin: the biotin–avidin interaction is highly efficient and specific, and several small biotin molecules can be attached to a protein or antibody. Thus, the biotinylated protein can bind to more than one avidin molecule, thereby amplifying the signal strength. The immune gold labeling technique can also be a useful method for light-microscopy via particle magnification by silver enhancement (Fig. 1). The length of the silver enhancement reaction is critical for the final particle size. If using gold-coupled secondary antibodies for transmission electron microscopy, special care should be taken in order to avoid any metal material, because precipitation of the metallic silver will be enhanced in the presence of any metals. Usage of gold grids instead of copper grids is recommended if silver enhancement of gold particles should be used.

2.5 What to Do with the High Autofluorescence of Cell Walls and Chlorophyll?

High autofluorescence of the tissue, e.g., from sugar cane leaves (Fig. 2; [33]) or *Convallaria rhizomes*, hampers immunofluorescence labeling. The emission spectra of most of the commercially available fluorochromes interfere with this autofluorescence and are therefore not recommended for immunolocalization studies. Reduction of autofluorescence can be achieved by pre-treatment of the sections with 1.25% Evans Blue for 10 min, or with 0.1% of Toluidine Blue for 5 min. Alternatively, secondary antibodies coupled to a fluorochrome emitting light in the far red, such as Cy5, can be used to evade the autofluorescence signal (I. Medina Borges, unpublished).

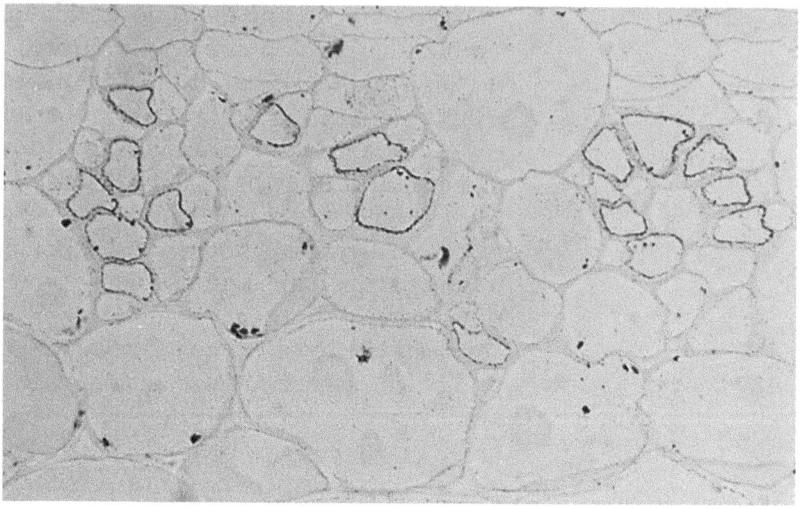

Fig. 1 NtSUT1 in the transport phloem. Strong labeling of the sieve elements of the transport phloem of tobacco leaf petiole cross sections. The relation in size between sieve elements and companion cells is opposite to that in source leaf minor veins. In the transport phloem, the diameter of the sieve elements is bigger than that of neighboring companion cells, and sieve elements are heavily engaged in sucrose retrieval. The diameter of gold particles is enhanced by silver precipitation to increase visibility in the light microscope (400× magnification)

3 Validation and Application

3.1 What is the Best Negative Control for Immunolabeling Reactions?

Control experiments are of an important consideration when performing immunohistochemical experiments, and it is absolutely necessary to ensure the specificity of the antibody by additional tests.

Unfortunately, there is no optimal negative control for immunohistochemical experiments available for plants or animals. Usually, the pre-immune serum from the same immunized animal is used as a negative control. But the absence of the unspecific signals in that case is only a good justification for the choice of the animal used for immunization and does not guarantee the specificity of the immune serum.

Omission of the primary antibody in immunohistochemical experiments can be used to determine unspecific binding of the secondary antibody, but it does not provide information about the specificity of the primary antibody.

Introducing competition to immunoreactions by the supply of an excess of purified antigen can decrease the intensity of a specific signal and provide information about the unspecific reactions. The same principle applies for excessive blocking of the samples with BSA, Tween, or low-fat milk powder, by increasing the concentration of the blocking reagents, which should decrease unspecific binding of

Fig. 2 Sugar cane leaves are highly autofluorescent. The cell wall and chlorophyll autofluorescence of sugarcane leaves show high autofluorescence depending on the excitation wavelength. Due to interference of emission spectra, the usage of most of the commercial secondary antibody conjugates is not recommended. Cross-sections of the sugarcane leaf were excited in a fluorescence microscope with different wavelengths (top: UV, middle: transmission light + UV, bottom: 488 nm)

the antibody. However, this method usually also lowers the signal intensity of the specific signals.

3.2 Reproducibility of the Localization Data

In order to increase the credibility of the data, performing all negative or positive control experiments with the same batch of purified antibodies, or with the same hybridoma of monoclonal antibodies, is recommended. Sodium azide can be added to increase the stability of the purified antibody stocks. Measurements should be taken to avoid contamination of the immunodetection reactions with peroxidase-coupled IgGs, because the peroxidase reaction is efficiently inhibited by NaN_3.

3.3 How to Confirm Localization Experiments? What Can Be Used as a Positive Control?

If the knockout mutant lines or transgenic plants with reduced expression of a given antigen are available, the purified antibody can be tested on SDS-PAGE using transgenic and non-transgenic plant material, followed by immunodetection. In the case of the potato plants with silenced *StSUT1*, the purified antibody recognizes one single band, which is diminished or even disappears in *StSUT1* silenced compared to wild-type plants [34].

Overlapping of protein localization with promoter activity, as revealed by promoter:reporter gene studies, can also be used as a positive control. Immunolocalization of the NtSUT1 protein in the internal, as well as the external phloem of tobacco petiole cross sections (Fig. 3) correlates well with promoter reporter gene studies showing that the SUT1 promoter is active in both phloem types of the bicollateral vascular bundles as well [35].

An even more convincing method to confirm localization data is a functional assay with isolated cells, or RNA and protein detection after laser capture microdissection (LCM) of specific cells. Recently, the presence of sucrose-transporting activity was proven by electrophysiological means on isolated and functionally intact sieve element protoplasts from *Vicia faba* [36], indicating the presence of one or several sucrose-transporting proteins/systems in their plasma membrane. While in *V. faba*, the identity of the sieve element protoplasts could be recognized by the presence of characteristic forisomes [36], this might be more difficult in many other species.

Indirect evidence for phloem/sieve element-specific localization can be obtained from structural and functional studies on mutants or plants with increased or decreased expression of the protein of interest. In the case of the potato sucrose transporter StSUT1, which was immunolocalized to SEs [18] this localization was supported by ultrastructural analysis of *StSUT1* antisense plants [37]. These plants were defective in phloem loading and accumulated soluble sugars and starch in the leaves, as observed by electron microscopy (Fig. 4). The starch granules are visible not only in mesophyll cells, bundle sheath or phloem parenchyma cells, but also in phloem companion cells of minor veins, where they are not detectable in wild-type

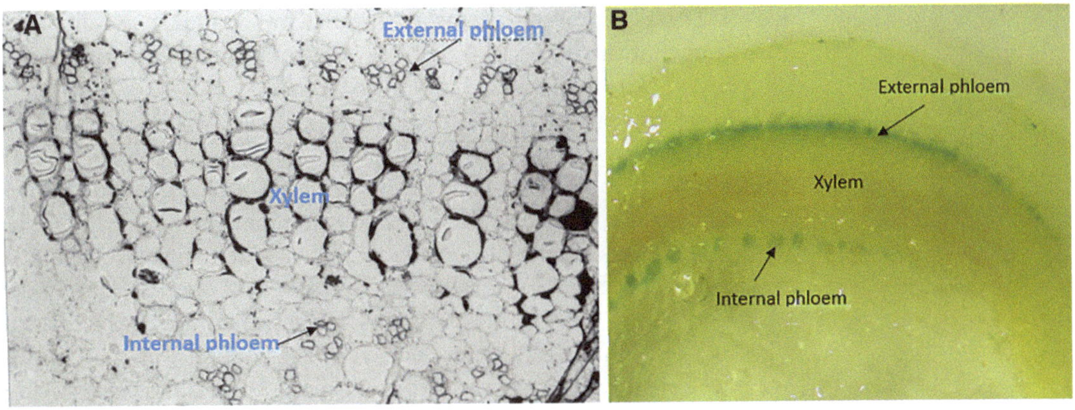

Fig. 3 Correlation of NtSUT1 immunolocalization and promoter::GUS staining. Localization of the sieve elements of the internal and external phloem of petiole cross-sections labeled with SUT1-specific antibodies (**A**) is consistent with SlSUT1 promoter::GUS staining (**B**). Immunodetection was performed with secondary anti-rabbit gold conjugate, followed by the silver enhancement. Thus, the immunolabeling with 10 nm gold particles is detectable even in the bright field mode of a light microscope, and at low magnification (200× magnification)

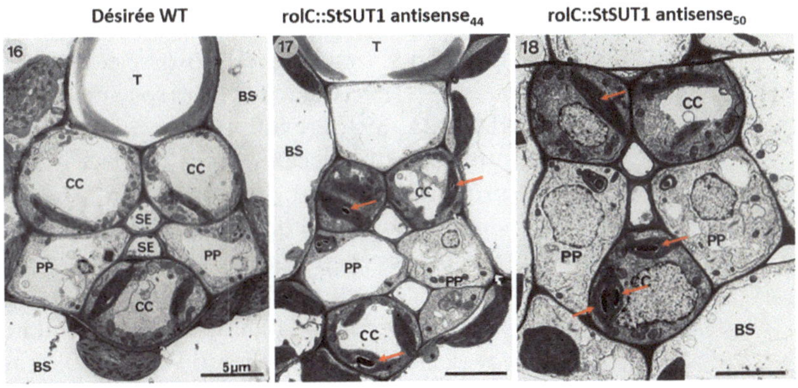

Fig. 4 Functional tests can validate immunolocalization results. Formation of large starch granules due to accumulation of photoassimilates is detectable in source leaves of rolc::*StSUT1*-antisense plants. No such starch granules accumulate in mesophyll cells or in the phloem cells of Désirée wild-type plants. Note that the starch accumulation also occurs in phloem companion cells, where the rolC promoter is active. This indicates that the sucrose transporter StSUT1 exerts its phloem loading function at the boundary of the sieve element plasma membrane. The accumulation of soluble sugars and starch up to the boundary between companion cells and sieve elements is a good confirmation for the sieve-element-specific localization of the transporter. (Reprinted from Schulz et al. [37] with kind permission by Springer Nature (license number 4325910218427))

leaves. Localization of the starch granules in CC in transgenic plants indicates that the block in transport occurs at the SE boundary and, thereby, supports the SE localization of StSUT1.

Another approach to verify the immunolocalization result for a specific protein can be co-localization with reference proteins. In the case of the potato sucrose transporter, its SE localization is supported by the presence of the ATPase, on which sucrose transporter activity depends, in the same cells, which was demonstrated via lead precipitation in the closely related *Nicotiana tabacum* [38]. In *Pisum sativum*, where sucrose transporters are assumed to localize to the plasma membrane of transfer cells, ATPase activity was detected in transfer cells [38], suggesting species-dependent differences in the mechanisms of phloem loading.

A recent case study on phloem loading in the succulent leaves of *Aizoaceae* species (order Caryophyllales) using the immunolocalization technique further illustrates the importance of control experiments [31]. One major difficulty is the lack of specific antibodies against sucrose transporters for these species, so a polyclonal antibody against SUT1 from potato was tested. To ensure the antibody's specificity and reactivity against the homologous proteins of these succulents, SDS-PAGE and Western blot analyses were performed on leaf protein extracts from the target species alongside controls, including potato leaves (positive control) and *Beta vulgaris* leaves, a closely related Caryophyllales species with available SUT1 localization data [21, 39]. Previous observations showed that the anti-StSUT1 antibody only weakly cross-reacts with the homologous BvSUT1 protein and recognizes a very diminished band, making *B. vulgaris* a suitable negative control. Immunolocalization assays using confocal microscopy were conducted, and comparative labeling in both control species reinforced the conclusions drawn from band detection: high background labeling was observed in *B. vulgaris* sections.

Using SDS-PAGE and Western blot analyses to validate immunolocalization results in parallel with well-chosen control species proved essential for distinguishing true signals from possible artifacts. In some cases, apparent specific labeling was detected in sections despite the absence of a detectable band on Western blots of protein extracts from the same species. This underscores the necessity of combining multiple validation strategies, particularly when the cost of generating a new antibody against a species-specific peptide sequence from a non-model species is prohibitive.

4 Applications

The application of the immunolocalization technique in matters of phloem physiology is illustrated in detail by the two examples below.

4.1 Dynamics of Phloem Proteins

One disadvantage of fixation, dehydration, and embedding of the samples is that the result is static, and the dynamics of the localized proteins cannot be visualized. In the case of the plant sucrose transporters, and presumably other phloem proteins, high mobility, endocytosis, and recycling at the plasma membrane have to be taken into consideration [40–42]. Since sieve elements are generally not accessible to live-cell microscopy of fluorescently labeled proteins, results from heterologous expression, for example, in *Nicotiana benthamiana* leaves or *Saccharomyces cerevisiae* cells, have to be validated by immunohistochemistry in embedded, sectioned phloem tissue. That this is possible has been demonstrated for the potato sucrose transporter StSUT1 [43]. Addition of chemicals that stop endocytosis or recycling, such as Brefeldin A [44], prior to fixation, dehydration, and embedding, induced changes in the localization pattern that resemble the effect seen in live-cell microscopy [43]. The same can be observed for the potato sucrose transporter StSUT4, which is also found in mature SEs (Fig. 5). Additionally, pre-treatment with reducing agents such as GSH or DTT affects the subcellular localization of the StSUT4 protein (Fig. 5), potentially indicating a mode of regulation.

4.2 Phloem Unloading in Roots

The technique of immunolocalization can be used to localize proteins in the phloem, whereby phloem function, namely phloem loading and unloading, which is fulfilled by the proteins localized herein can be deduced. Our understanding of how sugars are unloaded from the phloem sieve elements in carbon sink organs, such as roots, was recently extended through studies using esculin, a fluorescent sucrose analog, in Arabidopsis roots that suggested the main unloading activity to be in the protophloem zone. Sugars and other small molecules are suggested to move freely through funnel-type PD [46], suggesting a symplasmic unloading process of sugars in this tissue. Since this result can be assumed to apply to all dicots, apoplasmic unloading of sucrose via StSUT1 in potato would not be necessary, which is supported by the absence of the transporter in this tissue as observed using immunohistochemistry (Fig. 6). Interestingly, the sucrose transporter StSUT1 is present at the plasma membrane of metaphloem sieve elements in roots (Fig. 6), suggesting a role in sucrose reloading along the transport phloem.

In monocotyledons, e.g., *Sorghum*, the sieve elements of the protophloem are, in contrast to Arabidopsis, symplasmically isolated [47]. Here, the PEP2 antibody raised against the first cytoplasmic loop peptide from StSUT1, which is able to cross-react with three sucrose transporters from *Sorghum bicolor*, was used to label the plasma membrane of protophloem sieve elements (PSE). This suggests that the sucrose transporters from *Sorghum* might be involved in apoplasmic phloem unloading of sucrose in the protophloem of roots [30].

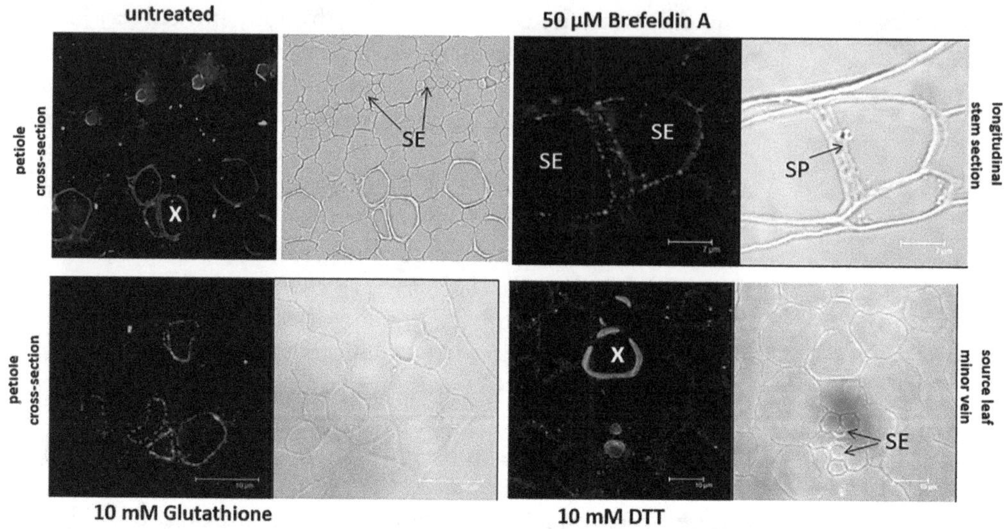

Fig. 5 Subcellular dynamics of the StSUT4 sucrose transporter in mature sieve elements. Pre-treatment of leaf material with either 50 μM brefeldin A (1 h), 10 mM glutathione (2 h), or 10 mM DTT (1 h) before fixation, dehydration, and embedding in LRWhite. Note that the homogenous distribution of StSUT4 at the sieve element plasma membrane is only visible on untreated samples, whereas pre-treatment with reducing agents or BrefA leads to a more patchy redistribution or endocytosis (DTT). Primary antibody, affinity-purified peptide antibody against StSUT4 [45], secondary antibody, anti-rabbit FITC conjugate (Sigma). (Pictures were taken by J. Liesche and C. Kühn)

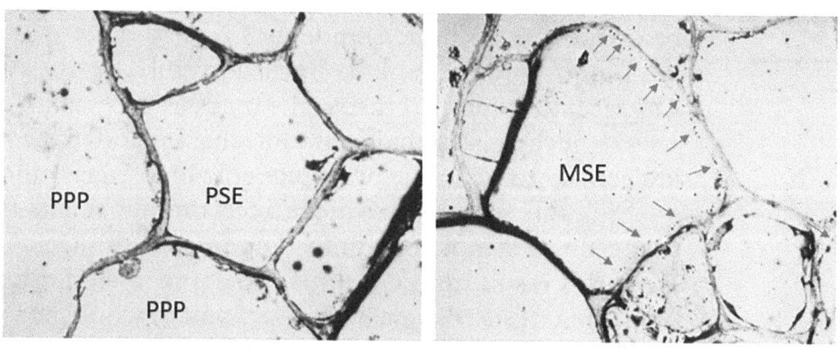

Fig. 6 StSUT1 localizes to SEs of the metaphloem, but it is absent in SEs of the protophloem. Sieve elements of the metaphloem (MSE) of roots are labeled with StSUT1-specific antibodies (arrows), whereas the sieve elements of the protophloem adjacent to pericycle cells (PC) do not immunoreact. Note that the sieve elements of the protophloem (PSE) are extensively connected via plasmodesmata with the surrounding cells; such connections are not visible for MSE. Immunodetection was performed with a secondary anti-rabbit IgG gold conjugate, followed by the silver enhancement, and evaluated by transmission electron microscopy(3800×). Cell wall staining is due to contrasting artifacts. (Pictures were taken by A. Schulz and C. Kühn)

5 Notes

1. As a standardized protocol, fixation in 4% paraformaldehyde in 0.1 M phosphate buffer (pH 7.3), followed by embedding

in LRWhite resin, has yielded reliable results. For samples intended for transmission electron microscopy, add a maximum of 0.1% glutaraldehyde. Post-fixation with osmium tetroxide and post-contrast staining must be omitted.

2. Treat the sections deposited on microscope slides or metal grids with 0.1 M HCl for 10 min. Wash briefly with double-distilled water and block for 30 min with either 2% BSA or 1% powdered milk diluted in PBS buffer (using the latter reduces nonspecific labeling, but also weakens the specific signal). Wash once with PBS and incubate overnight with the primary antibody, whose appropriate dilution must be determined empirically. Wash three times for 5 min each with PBS-T (PBS plus 0.05% Tween 20), with agitation. For the final wash, use PBS without Tween. Proceed to incubate with the secondary antibody for 2 h, determining the dilution empirically. Repeat the washing procedure done after incubation with the primary antibody, but use double-distilled water for the last wash.

Acknowledgments

Critical proofreading by Alexander Schulz, Johannes Liesche, and Pawel Brzezowski was greatly appreciated. We acknowledge helpful discussions with Peer Martin (Humboldt University of Berlin) and Gladys F. A. Melo-de-Pinna (University of São Paulo).

Conflict of Interest The authors have no conflicts of interest to declare.

References

1. Botha CE (2013) A tale of two neglected systems-structure and function of the thin- and thick-walled sieve tubes in monocotyledonous leaves. Front Plant Sci 4:297. https://doi.org/10.3389/fpls.2013.00297
2. Sjølund RD (1997) The phloem sieve element: a river runs through it. Plant Cell 9(7):1137–1146. https://doi.org/10.1105/tpc.9.7.1137
3. van Bel A, Knoblauch M (2000) Sieve element and companion cell: the story of the comatose patient and the hyperactive nurse. Funct Plant Biol 27(6):477–487. https://doi.org/10.1071/PP99172
4. van Bel AJE, Ehlers K, Knoblauch M (2002) Sieve elements caught in the act. Trends Plant Sci 7(3):126–132
5. Turgeon R, Wolf S (2009) Phloem transport: cellular pathways and molecular trafficking. Annu Rev Plant Biol 60:207–221. https://doi.org/10.1146/annurev.arplant.043008.092045
6. Knoblauch AM, Divall MJ, Owuor M, Musunka G, Pascall A, Nduna K, Ng'uni H, Utzinger J, Winkler MS (2018) Selected indicators and determinants of women's health in the vicinity of a copper mine development in northwestern Zambia. BMC Women's Health 18(1):62. https://doi.org/10.1186/s12905-018-0547-7
7. Lin MK, Lee YJ, Lough TJ, Phinney BS, Lucas WJ (2009) Analysis of the pumpkin phloem proteome provides insights into angiosperm sieve tube function. Mol Cell Proteomics 8(2):343–356. https://doi.org/10.1074/mcp.M800420-MCP200
8. Zhang C, Yu X, Ayre BG, Turgeon R (2012) The origin and composition of cucurbit "phloem"

exudate. Plant Physiol 158(4):1873–1882. https://doi.org/10.1104/pp.112.194431
9. Zhang B, Tolstikov V, Turnbull C, Hicks LM, Fiehn O (2010) Divergent metabolome and proteome suggest functional independence of dual phloem transport systems in cucurbits. Proc Natl Acad Sci U S A 107(30):13532–13537. https://doi.org/10.1073/pnas.0910558107
10. Hu C, Ham BK, El-Shabrawi HM, Alexander D, Zhang D, Ryals J, Lucas WJ (2016) Proteomics and metabolomics analyses reveal the cucurbit sieve tube system as a complex metabolic space. Plant J 87(5):442–454. https://doi.org/10.1111/tpj.13209
11. Knoblauch M, Peters WS, Bell K, Ross-Elliott TJ, Oparka KJ (2018) Sieve-element differentiation and phloem sap contamination. Curr Opin Plant Biol 43:43–49. https://doi.org/10.1016/j.pbi.2017.12.008
12. Paultre DS, Gustin MP, Molnar A, Oparka KJ (2016) Lost in transit: long-distance trafficking and phloem unloading of protein signals in Arabidopsis homografts. Plant Cell 28:2016–2025. https://doi.org/10.1105/tpc.16.00249
13. Schulz A (2017) Long-distance trafficking: lost in transit or stopped at the gate?. Plant Cell 29(3):426–430. https://doi.org/10.1105/tpc.16.00895
14. Chou M, Krause KH, Campbell KP, Jensen KG, Sjølund RD (1989) Antibodies against the calcium-binding protein: calsequestrin from *Streptanthus tortuosus* (Brassicaceae). Plant Physiol 91(4):1259–1261
15. Khan JA, Wang Q, Sjølund RD, Schulz A, Thompson GA (2007) An early nodulin-like protein accumulates in the sieve element plasma membrane of Arabidopsis. Plant Physiol 143(4):1576–1589. https://doi.org/10.1104/pp.106.092296
16. Meyer S, Lauterbach C, Niedermeier M, Barth I, Sjolund RD, Sauer N (2004) Wounding enhances expression of AtSUC3, a sucrose transporter from Arabidopsis sieve elements and sink tissues. Plant Physiol 134(2):684–693
17. Wang Q, Monroe J, Sjølund RD (1995) Identification and characterization of a phloem-specific beta-amylase. Plant Physiol 109(3):743–750
18. Kühn C, Franceschi VR, Schulz A, Lemoine R, Frommer WB (1997) Macromolecular trafficking indicated by localization and turnover of sucrose transporters in enucleate sieve elements. Science 275(5304):1298–1300. https://doi.org/10.1126/science.275.5304.1298
19. Kühn C (2011) Sucrose transporters and plant development. In: Geisler M, Venema K (eds) Transporters and pumps in plant signaling. Springer, Heidelberg, pp 225–251
20. He H, Chincinska I, Hackel A, Grimm B, Kühn C (2008) Phloem mobility and stability of sucrose transporter transcripts. Open Plant Sci J 2:1–14. https://doi.org/10.2174/1874294700802010001
21. Vaughn MW, Harrington GN, Bush DR (2002) Sucrose-mediated transcriptional regulation of sucrose symporter activity in the phloem. Proc Natl Acad Sci U S A 99(16):10876–10880. https://doi.org/10.1073/pnas.172198599
22. Liu Y, Vasina VV, Kraner ME, Peters WS, Sonnewald U, Knoblauch M (2022) Proteomics of isolated sieve tubes from Nicotiana tabacum: sieve element-specific proteins reveal differentiation of the endomembrane system. Proc Natl Acad Sci U S A 119 (1). https://doi.org/10.1073/pnas.2112755119
23. Bürkle L, Hibberd JM, Quick WP, Kühn C, Hirner B, Frommer WB (1998) The H+-sucrose cotransporter NtSUT1 is essential for sugar export from tobacco leaves. Plant Physiol 118(1):59–68
24. Sanden NC, Schulz A (2021) Stationary sieve element proteins. J Plant Physiol 266:153511. https://doi.org/10.1016/j.jplph.2021.153511
25. Corbesier L, Vincent C, Jang S, Fornara F, Fan Q, Searle I, Giakountis A, Farrona S, Gissot L, Turnbull C, Coupland G (2007) FT protein movement contributes to long-distance signaling in floral induction of Arabidopsis. Science 316(5827):1030–1033. 1141752 (pii)
26. Mathieu J, Yant LJ, Murdter F, Kuttner F, Schmid M (2009) Repression of flowering by the miR172 target SMZ. PLoS Biol 7(7):e1000148. https://doi.org/10.1371/journal.pbio.1000148
27. Thompson MV, Wolniak SM (2008) A plasma membrane-anchored fluorescent protein fusion illuminates sieve element plasma membranes in Arabidopsis and tobacco. Plant Physiol 146(4):1599–1610. https://doi.org/10.1104/pp.107.113274
28. Martens HJ, Roberts AG, Oparka KJ, Schulz A (2006) Quantification of plasmodesmatal endoplasmic reticulum coupling between sieve elements and companion cells using fluorescence redistribution after photobleaching. Plant Physiol 142(2):471–480. https://doi.org/10.1104/pp.106.085803
29. Cerio R, Belter S, MacDonald DM (1987) The effect of fixation on monoclonal antibody labelling of cell surface antigens in cutaneous tissue. Clin Exp Dermatol 12:79–82
30. Milne RJ, Perroux JM, Rae AL, Reinders A, Ward JM, Offler CE, Patrick JW, Grof CP (2017) Sucrose transporter localization and function in phloem unloading in developing stems. Plant Physiol 173(2):1330–1341. https://doi.org/10.1104/pp.16.01594

31. Tamassia G, Kühn C, Stach T, Melo-de-Pinna GFA (2025) Ecophysiology of phloem loading in succulent leaves with different anatomical patterns of Aizoaceae species (Caryophyllales). Physiol Plant 177(4):e70438. https://doi.org/10.1111/ppl.70438
32. Sauer N, Stadler R (1993) A sink-specific H+/monosaccharide co-transporter from Nicotiana tabacum: cloning and heterologous expression in baker's yeast. Plant J 4(4):601–610
33. Baker RF, Leach KA, Boyer NR, Swyers MJ, Benitez-Alfonso Y, Skopelitis T, Luo A, Sylvester A, Jackson D, Braun DM (2016) Sucrose transporter ZmSut1 expression and localization uncover new insights into sucrose phloem loading. Plant Physiol 172:1876–1898. https://doi.org/10.1104/pp.16.00884
34. Kühn C, Quick WP, Schulz A, Riesmeier JW, Sonnewald U, Frommer WB (1996) Companion cell-specific inhibition of the potato sucrose transporter SUT1. Plant Cell Env 19(10):1115–1123. https://doi.org/10.1111/j.1365-3040.1996.tb00426.x
35. Weise A, Lalonde S, Kühn C, Frommer WB, Ward JM (2008) Introns control expression of sucrose transporter LeSUT1 in trichomes, companion cells and in guard cells. Plant Mol Biol 68(3):251–262. https://doi.org/10.1007/s11103-008-9366-9
36. Hafke JB, Holl SR, Kühn C, van Bel AJ (2013) Electrophysiological approach to determine kinetic parameters of sucrose uptake by single sieve elements or phloem parenchyma cells in intact Vicia faba plants. Front Plant Sci 4:274. https://doi.org/10.3389/fpls.2013.00274
37. Schulz A, Kühn C, Riesmeier JW, Frommer WB (1998) Ultrastructural effects in potato leaves due to antisense-inhibition of the sucrose transporter indicate an apoplasmic mode of phloem loading. Planta 206:533–543
38. Cronshaw J (1980) ATPase in mature and differentiating phloem and xylem. J Histochem Cytochem 28(4):375–377. https://doi.org/10.1177/28.4.6445380
39. Nieberl P, Ehrl C, Pommerrenig B, Graus D, Marten I, Jung B, Ludewig F, Koch W, Harms K, Flugge UI, Neuhaus HE, Hedrich R, Sauer N (2017) Functional characterisation and cell specificity of BvSUT1, the transporter that loads sucrose into the phloem of sugar beet (*Beta vulgaris* L.) source leaves. Plant Biol (Stuttg) 19(3):315–326. https://doi.org/10.1111/plb.12546
40. Krügel U, He HX, Gier K, Reins J, Chincinska I, Grimm B, Schulze WX, Kühn C (2012) The potato sucrose transporter StSUT1 interacts with a DRM-associated protein disulfide isomerase. Mol Plant 5(1):43–62. ssr048 (pii)
41. Krügel U, Veenhoff LM, Langbein J, Wiederhold E, Liesche J, Friedrich T, Grimm B, Martinoia E, Poolman B, Kühn C (2008) Transport and sorting of the *Solanum tuberosum* sucrose transporter SUT1 is affected by posttranslational modification. Plant Cell 20(9):2497–2513. Correction: Plant Cell 2421 (2009) 4059. doi:tpc.108.058271 (pii)
42. Liesche J, Schulz A, Krügel U, Grimm B, Kühn C (2008) Dimerization and endocytosis of the sucrose transporter StSUT1 in mature sieve elements. Plant Signal Behav 3(12):1136–1137
43. Liesche J, He HX, Grimm B, Schulz A, Kühn C (2010) Recycling of Solanum sucrose transporters expressed in yeast, tobacco, and in mature phloem sieve elements. Mol Plant 3(6):1064–1074. ssq059 (pii)
44. Satiat-Jeunemaitre B (1992) Spatial and temporal regulations in helicoidal extracellular matrices: comparison between plant and animal systems. Tissue Cell 24(3):315–334
45. Chincinska IA, Liesche J, Krügel U, Michalska J, Geigenberger P, Grimm B, Kühn C (2008) Sucrose transporter StSUT4 from potato affects flowering, tuberization, and shade avoidance response. Plant Physiol 146(2):515–528. doi: pp.107.112334 (pii) 10.1104/pp.107.112334
46. Ross-Elliott TJ, Jensen KH, Haaning KS, Wager BM, Knoblauch J, Howell AH, Mullendore DL, Monteith AG, Paultre D, Yan D, Otero S, Bourdon M, Sager R, Lee JY, Helariutta Y, Knoblauch M, Oparka KJ (2017) Phloem unloading in Arabidopsis roots is convective and regulated by the phloem-pole pericycle. Elife 6:e24125. https://doi.org/10.7554/eLife.24125
47. Milne RJ, Offler CE, Patrick JW, Grof CPL (2015) Cellular pathways of source leaf phloem loading and phloem unloading in developing stems of Sorghum bicolor in relation to stem sucrose storage. Funct Plant Biol 42(10):957–970. https://doi.org/10.1071/FP15133

Chapter 13

Analysis of Phloem Sterols in Plants using EDTA-Facilitated Exudation and GC–MS

Robert J. Grebenok and Spencer T. Behmer

Abstract

Phytosterols occur in the phloem of vascular plants as free, acylated, and glycosylated derivatives. The phloem is increasingly recognized as an active site of sterol synthesis and transport used to support plant metabolism, while also providing a potential nutritional source for phloem-feeding insects. This chapter describes a method for the collection and quantitative analysis of phloem sterols using an EDTA-facilitated exudation procedure followed by gas chromatography–mass spectrometry (GC–MS). Excised leaves are incubated in an EDTA solution to induce phloem exudation, and sterols from collected exudate are extracted, hydrolyzed to their free forms, derivatized as trimethylsilyl ethers, and analyzed by GC–MS using internal standards for quantification. This method has been successfully applied to several plant species and can be adapted to others with suitable modifications. It provides a reproducible approach for characterizing sterol composition of phloem sap and assessing its biological relevance in plant–insect interactions.

Keywords Phloem, Phytosterol, Phloem sap, EDTA exudation, Sterol quantification, Trimethylsilyl derivatization, Gas chromatography–mass spectroscopy (GC–MS)

1 Introduction

Phytosterols are a structurally diverse class of sterols found in plant tissues. They predominantly occur in three principal forms: [1] free sterols, with an unmodified C3-hydroxyl group; [2] acylated sterols, in which a fatty acid or other aliphatic chain is esterified at C3; and [3] steryl glycosides, which contain a carbohydrate moiety that is linked at C3. The major plant sterols (Fig. 1) include cholesterol, campesterol, ß-sitosterol, stigmasterol, spinasterol, and 22-dihydrospinasterol, among others [1]. Although the functional significance of this structural diversity is not fully understood, these phytosterols are known to stabilize membranes, facilitate lipid raft organizations, act as hormone precursors, and contribute to plant and human health [2]. Vascular plant species differ in both the composition and abundance

Johannes Liesche and Sanna Sevanto (eds.), *Phloem*, Methods in Molecular Biology, vol. 3066,
https://doi.org/10.1007/978-1-0716-5488-0_13,

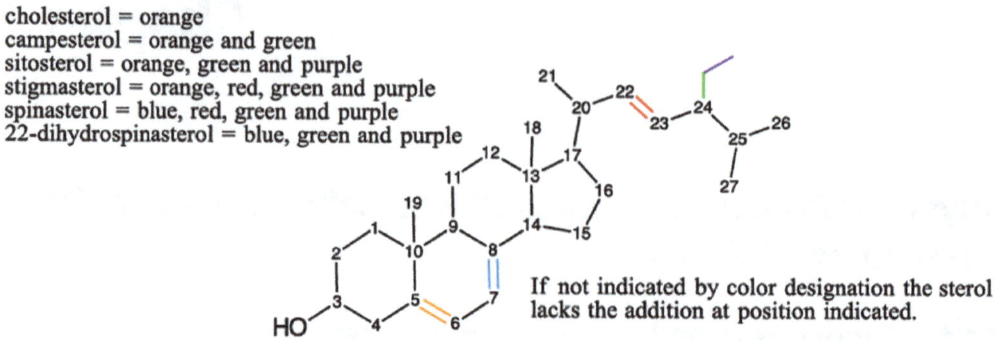

Fig. 1 Representative structures of major plant sterols detected in phloem sap. The example shown illustrates structural differences among cholesterol, campesterol, ß-sitosterol, stigmasterol, spinasterol, and 22-dihydrospinasterol. Variations occur primarily at the C24 side chain and the presence or absence of double bonds within the sterol nucleus. Cholesterol retains the C5–C6 double bond but lacks additional alkylation at C24, whereas other phytosterols exhibit methyl or ethyl substitutions and additional unsaturation. These structural differences account for variation in sterol metabolism and function

of sterols across tissues [2], but sterols have been detected in the phloem of all species examined to date [3–5].

Many animals, including insects, use plants as a primary food source. While most animals can synthesize cholesterol de novo from simple precursors, insects cannot [6, 7]. Instead, they depend entirely on dietary sterols, which they may biochemically modify to meet their physiological needs [7, 8]. Plant vegetative tissues, however, contain only trace amounts of cholesterol. Rather, they maintain comparatively high levels of phytosterols, including the common forms shown in Fig. 1 [1]. Because insects require sterols for normal development, molting, and reproduction [6, 9], they represent an essential nutritional resource. However, herbivorous insect species differ greatly in their capacity to convert plant sterols into cholesterol [7, 8]. For example, grasshoppers can convert ß-sitosterol to cholesterol but are unable to metabolize stigmasterol and other common phytosterols, resulting in reduced growth and survival when fed diets enriched in non-metabolizable sterols [9–13].

Some insects specialize on the phloem (e.g., aphids), and cholesterol has been consistently detected in the phloem of diverse plant species [4]. The presence of cholesterol in the phloem of plants indicates that phloem-feeding insects ingest cholesterol directly when consuming sap, thereby reducing the need for extensive sterol modifications to meet sterol nutritional requirements [14]. Although the physiological basis for cholesterol's occurrence in the phloem remains unclear, its presence is now well established. This finding may partially explain why aphids and other phloem-feeding insects generally possess limited capacity for sterol metabolism [15]. Variation in plant sterol composition and insect sterol metabolism likely constrains insect performance and host range. The low demand for sterol transformation in phloem-feeding species suggests that this feeding strategy evolved to exploit a limited yet readily available sterol source.

In the following, we provide a protocol for the collection and quantitative analysis of phloem sterols using an EDTA-facilitated exudation procedure followed by gas chromatography–mass spectrometry (GC–MS). An overview of the workflow is provided in Fig. 2.

2 Materials

1. Prepare all aqueous solutions with double-distilled, deionized water. Use analytical-grade organic solvents (methanol, hexane, chloroform) and reagents as represented. Pre-equilibrate solvents by mixing prepared solvent combinations with solvent combinations with which they are washed during sample processing. Pre-equilibrated solvents will typically form layers when exposed to the solvents, maintaining great polarity and density differences. Pre-equilibrated solvents and samples

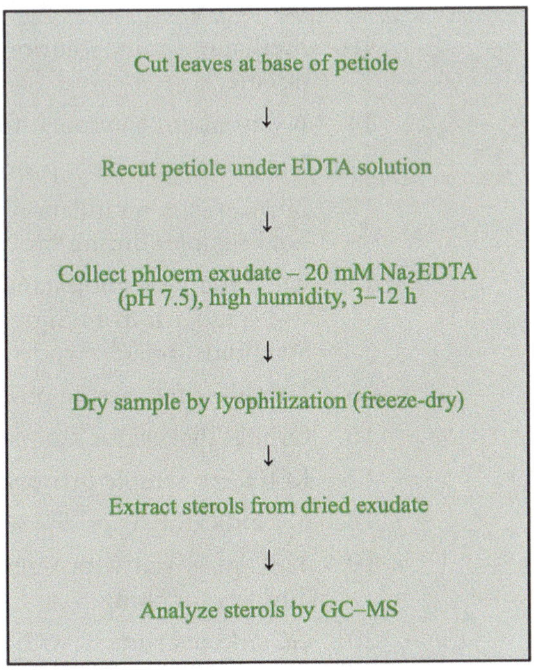

Fig. 2 Workflow for the collection and analysis of phloem sterol. Stepwise summary of the experimental procedure used to obtain and analyze phloem sterols. Leaves are excised and recut under EDTA solution to induce phloem exudation. Exudate is collected in 20 mM Na_2EDTA (pH 7.5) under high humidity for 3–12 h, lyophilized, and extracted for sterols. Sterols are subsequently derivatized and quantified by gas chromatography–mass spectrometry (GC–MS)

are kept at room temperature unless otherwise noted. Plant material: Any vascular plant can be sampled. However, exudates of several leaves need to be pooled in case leaves are small, as noted. For example, when working with *Arabidopsis thaliana*, eight leaves were combined per tube to ensure sufficient exudate for sterol analysis.

2. Razor blade: single- or double-edged.
3. Na^+EDTA solution: for exudate generation.
4. 1.5-mL microcentrifuge tube: for exudate collection.
5. Microcentrifuge tube rack: Standard 80-position tube racks (~20 cm in length) fit easily within a humidity chamber and provide stability during incubation.
6. Humidity chamber: a sealable container lined with water-saturated paper towels to maintain >90% relative humidity.
7. Lyophilizer: for initial processing of collected phloem samples.
8. Fresh BSTFA+TMCS (99:1; Sylon BFT). Store reagents under dry conditions and minimize air exposure to prevent hydrolysis.
9. Hexane: pre-equilibrated with 50% methanol/water
10. Methanol/water-solution: 50% v/v, pre-equilibrated with hexane.
11. A water bath sonicator: for sample processing.
12. Vortexer: for sample processing.
13. 20 mL Glass scintillation vial with screw top lid and cap: to hold samples during processing.
14. Cholestane or cholestanol: to serve as an internal quantitative standard. from Sigma-Aldrich (St. Louis, MO, USA) or Steraloids Inc. (Newport, RI, USA)
15. 10% HCl: for sample processing.
16. Orbital shaker: for sample processing.
17. KOH: for sample processing.
18. pH indicator strips: for sample processing.
19. 2 mL glass vial with screw top lid and cap: to hold samples during processing.
20. GC–MS instrument with mass-specific detector and appropriate column: for sample processing and detection.
21. Fused-silica capillary EC-5 column: 30 m × 0.25 mm i.d., 0.25 μm film thickness (e.g., from Alltech, Nicholasville, KY, USA). For sample separation.

3 Methods

3.1 Phloem Collection

Phloem exudate for sterol analysis is collected from excised leaves using the EDTA-facilitated exudation method [16].

1. Excise leaves from the plant stem using a sharp razor blade (*see* **Notes 1 and 2**).
2. Recut the petiole under 20 mM Na+EDTA solution, pH 7.5, removing approximately 1 mm of petiole tissue from the leaf.
3. Immediately place the cut petioles (leaves) into 1.5-mL microcentrifuge tubes containing 1 mL of 20 mM Na+EDTA solution, pH 7.5 (*see* **Note 3**).
4. Place the tubes in a sealed humid chamber and incubate in the dark for 180 min at 23–25 °C (*see* **Notes 4 and 5**).
5. After incubation, remove the leaves carefully and collect the exudate from each tube.
6. Rinse the cut petioles with distilled water to collect any remaining exudate and EDTA solution. Combine the rinse with the initial exudate for analysis.
7. Proceed immediately to sterol extraction and analysis as described in Sect. 3.2.

3.2 Sterol Extraction

Sterol Extraction—Part 1

Sterols in phloem exudate samples are analyzed using GC–MS after conversion of all sterol forms to their free structural state and derivatization as trimethylsilyl (TMS) ethers. Phloem-derived sterols occur in three major forms: (1) free sterols with an unmodified C3-hydroxyl group, (2) acylated sterols in which a fatty acid or some hydrocarbon is esterified at C3, and (3) glycosylated sterols that possess a carbohydrate conjugated at C3 (Fig. 1). The workflow involves sequential solvent extraction, hydrolysis, derivatization, and internal standardization using cholestane or cholestanol as quantitative markers. Each class of sterol is processed separately during sample preparation, as described in Sects. 3.2.1–3.2.4.

Perform all solvent extractions and derivatizations in a fume hood. Wear gloves and eye protection when handling acids, bases, or trimethylsilyl reagents. Dispose of organic waste according to institutional guidelines. Process samples under subdued light and avoid plastic consumables, when possible, to minimize sterol loss or contamination. Ensure all solvents are freshly equilibrated before use.

1. Collect each phloem sample in a 1.5-mL microcentrifuge tube and dry completely (e.g., by lyophilization).
2. For initial extraction, resuspend the dried sample in 500 µL of 50% methanol/water and add 500 µL of 100% hexane.

3. Facilitate solubilization by sonicating for 3 min in a room temperature water bath sonicator and vortexing for 30 s at high speed.
4. Transfer the 1-mL mixture to a 20-mL glass scintillation vial with a screw-top lid. Add 9 mL of extraction solvent, composed of 4 mL 50% methanol/water and 5 mL 100% hexane.
5. Add 30 μg of cholestane or cholestanol to each sample as an internal quantitative standard.
6. Vortex for 30 s, then incubated 24 h at room temperature in the dark to allow complete sterol partitioning.
7. Transfer the hexane phase (free and acylated sterols) and the methanol/water phase (glycosylated sterols) into separate 20-mL glass vials with screw cap lids.

Glycosylated Sterol Extraction

Process Glycosylated Sterols

1. Evaporate the methanol/water fraction to dryness under nitrogen with gentle heating.
2. Resuspend in 8 mL of 100% methanol containing 10% HCl.
3. Incubate for 2 h at 75 °C with shaking (225 rpm) to hydrolyze glycosidic bonds.

Acylated Sterol Extraction

Process Acylated Sterols

1. Divide the original hexane fraction into two equal parts.
2. Evaporate one aliquot to dryness and then resuspend in 100 μ L of clean hexane.
3. Add 8 mL of 70% methanol/water containing 5% KOH, and incubate for 2 h at 75°C with shaking to saponify (cleave) acyl esters.

Free Sterol Extraction

Process Free Sterols

The remaining half of the hexane fraction (designated the free sterol fraction) requires no hydrolysis and can be processed directly.

Sterol Extraction—Part 2

1. The treated sample is in a volume of conditioned solvent to hydrolyze bonds. This fraction is washed with an equal volume of water-equilibrated hexane, and the hexane wash is transferred to a fresh 20-mL glass vial with a screw cap lid. Free sterols are extracted from all chemically treated fractions using water-equilibrated hexane.
2. Wash the hexane layer with several equal volumes of water until the neutral pH of the hexane layer is achieved (using indicator strips to examine the removed water layer).

3. Evaporate the hexane to dryness under nitrogen and resuspend in 100 μL of clean (non-equilibrated) hexane.
4. Sonicate the sample in a room-temperature water bath sonicator for 30 s to ensure resuspension.
5. Add 200 μL of BSTFA+TMCS (99:1; Sylon BFT, Supelco) to each 100 μL sample. Vortex briefly and then leave to react overnight at room temperature to form trimethylsilyl (TMS) ethers.
6. Following derivatization, add 2 ml volumes of pre-equilibrated 50% Methanol/water and hexane to the sample.
7. Wash the derivatized sterol sample by taking up portions of both solvent layers and expelling them through the mix several times with a pipette.
8. Let the mixed sample sit for several hours (or overnight) at room temperature to allow layers to separate. The hexane layer (top layer) will contain the derivatized sterols, while the 50% methanol/water (lower layer) will contain unused BSTFA reagents.
9. Transfer derivatized sample (hexane layer) to 2 mL glass vial and evaporate to dryness under nitrogen and resuspend in 50 μ L of clean (non-equilibrated) hexane; sonicate for 30 s in a room temperature water bath sonicator to ensure solubility.

3.3 Gas Chromatograph–Mass Spectroscopy (GC–MS)

This section describes the separation and identification of derivatized sterols by GC–MS. Sterols are analyzed as their trimethylsilyl (TMS) ethers using an Agilent 6850 N GC coupled to a 5973 mass-selective detector. Equivalent instrumentation can be used, but might require an adjustment of settings.

1. Adjust the flow of the carrier gas helium to a flow rate of 1.2 mL min^{-1}.
2. Set the following GC temperature program:

- Inlet temperature: 280 °C
- Transfer-line temperature: 290 °C
- Column temperature: 80 °C for 1 min, then ramp 10 °C min^{-1} to 240 °C, followed by 5 °C min^{-1} to 300 °C.
- Hold at 300 °C until all components have eluted.

3. Set the following MS settings:

- Ion-source temperature: 250 °C
- Quadrupole temperature: 180 °C
- Operation in electron-impact (EI) mode with selected-ion monitoring (SIM) for sterol identification (*see* **Note 6**).

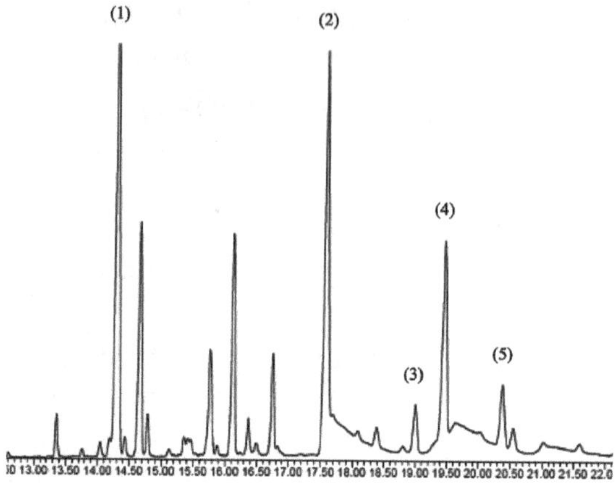

Fig. 3 Typical GC–MS chromatogram and mass spectra of sterol TMS derivatives from phloem exudate. Representative chromatogram showing separation of major sterol trimethylsilyl (TMS) ethers and the internal standard (cholestane). Relative retention times (min from injection) and key ions (m/z, relative intensity %) are indicated for each compound. Cholestane (1) eluted at ~14.3 min (m/z 372 [25%], 357 [50%], 217 [100%]); cholesterol TMS derivative (2) at ~17.6 min (m/z 458 [30%], 368 [80%], 329 [100%], 255 [20%], 129 [100%]); campesterol TMS derivative (3) at ~19.0 min (m/z 472 [30%], 382 [70%], 343 [80%], 271 [25%], 129 [100%]); stigmasterol TMS derivative (4) at ~19.5 min (m/z 484 [30%], 394 [35%], 355 [20%], 255 [40%], 129 [80%]); and β-sitosterol TMS derivative (5) at ~20.4 min (m/z 486 [25 %], 396 [60%], 357 [65%], and 129 [100%]). Quantification was based on peak-area ratios relative to the internal standard

4. Inject 1 μL volume of sample and verify internal standard recovery (*see* **Note 7**) and sterol analysis (see Fig. 3). The detection limit for TMS–sterols is typically in the tens of nanograms range.

3.4 Sterol Identification and Quantification

1. Identify sterols by comparing retention times and mass spectra with authentic standards of cholesterol, campesterol, stigmasterol, β-sitosterol, and others [17].
2. Quantify compounds using the internal standard (cholestane or cholestanol) to correct for recovery and derivatization efficiency.
3. Report relative sterol abundance as the ratio of sterol peak area to internal-standard peak area.

4 Notes

1. Combine exudate from several leaves if necessary to obtain sufficient material for GC–MS analysis. Consider leaf developmental position (source vs. sink) when comparing samples

across treatments or species. Insert an appropriate number of leaves per tube, depending on leaf size. Our experience has shown the following number of leaves per tube to result in reliable measurements: 8–10 mature *Arabidopsis thaliana* leaves; 1 mature tobacco (*Nicotiana tabacum*) leaf; 3–8 mature bean (*Phaseolus vulgaris*) leaves.

2. Sterol contamination: Because leaf cutting disrupts some cells, a small contribution of cellular sterols may contaminate the phloem fraction. As a result, sterol concentrations obtained using this method represent approximate rather than absolute values. Despite this limitation, the method reliably provides sufficient material for comparative and compositional analyses across plant species. However, quantitative values should be considered approximate rather than absolute.
3. Submergence of the cut petiole in the EDTA solution prevents xylem embolism and seal disruption.
4. Incubation duration: Typical exudation times range from 3 to 12 h, depending on plant species and leaf size. Monitor for visible exudate formation and avoid overlong incubations that increase cellular contamination.
5. The dark incubation minimizes photosynthetic activity and wound responses during exudation.
6. Use SIM ions characteristic for each sterol to improve signal-to-noise ratios and detection limits.
7. Include an internal standard (cholestane or cholestanol) in every sample. Typical recovery should be 90%–95%. Values outside this range suggest losses during extraction or derivatization.

Conflict of Interest

The authors have no conflicts of interest to declare.

References

1. Moreau RA, Whitaker BD, Hicks KB (2002) Phytosterols, phytostanols, and their conjugates in foods: structural diversity, quantitative analysis, and health-promoting uses. Prog Lipid Res 41(6):457–500
2. Moreau RA, Nystrom L, Whitaker BD, Winkler-Moser JK, Baer DJ, Gebauer SK, Hicks KB (2018) Phytosterols and their derivatives: structural diversity, distribution, metabolism, analysis, and health-promoting uses. Prog Lipid Res 70:35–61
3. Heyer J, Parker B, Becker D, Ruffino J, Fordyce A, DeWitt M, Grebenok RJ (2004) Steroid profiles of transgenic tobacco expressing an *Actinomyces* 3-hydroxysteroid oxidase gene. Phytochemistry. 65(20):2967–2976
4. Behmer ST, Olszewski N, Sebastiani J, Palka S, Sparacino G, Sciarrino E, Grebenok RJ (2013) Plant phloem sterol content: forms, putative functions, and implications for phloem-feeding insects. Front Plant Sci 4:370

5. Forrest JM, Knights BA (1972) Presence of phytosterols in the food of the aphid *Myzus persicae*. J Insect Physiol 18(4):723–728
6. Grieneisen ML (1992) The biosynthesis of ecdysteroids by prothoracic glands of *Manduca sexta*. PhD dissertation. University of North Carolina, Chapel Hill, NC
7. Behmer ST, Nes WD (2003) Insect sterol nutrition and physiology: a global overview. Adv Insect Physiol 31:1–72
8. Jing X, Behmer ST (2020) Insect sterol nutrition: physiological mechanisms, ecology, and applications. Ann Rev Entomol 65:251–271
9. Jing X, Vogel H, Grebenok RJ, Zhu-Salzman K, Behmer ST (2012) Dietary sterols/steroids and the generalist caterpillar *Helicoverpa zea:* physiology, biochemistry, and midgut gene expression. Insect Biochem Mol Biol 42(11):835–845
10. Behmer ST, Elias DO (1999) The nutritional significance of sterol metabolic constraints in the generalist grasshopper *Schistocerca americana*. J Insect Physiol 45(4):339–348
11. Behmer ST, Elias DO (2000) Sterol metabolic constraints as a factor contributing to the maintenance of diet mixing in grasshoppers (Orthoptera: Acrididae). Physiol Biochem Zool 73(2):219–230
12. Behmer ST, Grebenok RJ, Douglas AE (2011) Plant sterols and host plant suitability for a phloem-feeding insect. Funct Ecol 25(3):484–491
13. Jing X, Grebenok RJ, Behmer ST (2012) Plant sterols and host plant suitability for generalist and specialist caterpillars. J Insect Physiol 58(2):235–244
14. Gilbert LI, Rybczynski R, Warren JT (2002) Control and biochemical nature of the ecdysteroidogenic pathway. Annu Rev Entomol 47:883–916
15. Campbell BC, Nes WD (1983) A reappraisal of sterol biosynthesis and metabolism in aphids. J Insect Physiol 29(2):149–156
16. King RW, Zeevaart JAD (1974) Enhancement of phloem exudation from cut petioles by chelating agents. Plant Physiol 53(1):96–103
17. Rahier A, Benveniste P (1989) Mass spectral identification of phytosterols. In: Nes WD, Parish EJ (Eds.)Eds. Analysis of sterols and biologically relevant steroids. Academic Press, New York, NY, 223–248

Chapter 14

Visual Detection of Inorganic Orthophosphate in Phloem Cells

Wenyuan Ruan, Meina Guo, Qianqian Zhang and Keke Yi

Abstract

The phloem plays a vital role in actively transporting soluble organic compounds—such as sugars, minerals, and proteins—from source to sink tissues, a process driven by turgor pressure and adenosine triphosphate (ATP)-dependent energy. Since ATP levels in phloem cells are closely linked to inorganic orthophosphate (Pi) availability, precise and efficient detection of Pi in these cells is essential for understanding phloem function. To address this, we developed the inorganic orthophosphate staining assay (IOSA), a method for visually detecting Pi in phloem cells. As a demonstration, we have successfully applied IOSA to the quantification of Pi levels in rice stem phloem cells.

Keywords Inorganic orthophosphate (Pi), Phloem, IOSA

1 Introduction

The phloem serves as a critical transport system in vascular plants, facilitating the long-distance movement of soluble organic compounds—including sugars, minerals, and proteins—from source to sink tissues [1]. This process is driven by turgor pressure and adenosine triphosphate (ATP)-dependent energy. Since ATP levels in phloem cells are strongly influenced by inorganic orthophosphate (Pi), the primary form of phosphorus (P) uptake and metabolism in plants [2], Pi availability plays a key regulatory role in phloem function.

Moreover, the phloem acts as a major pathway for nutrient recycling, particularly Pi redistribution from senescing or apoptotic tissues [3]. This mechanism enhances nutrient-use efficiency and helps plants adapt to environmental stresses. Thus, precise and efficient quantification of Pi levels in different phloem cell types is essential for advancing our understanding of phloem physiology.

To address this need, we developed the inorganic orthophosphate staining assay (IOSA), a high-resolution colorimetric method

Johannes Liesche and Sanna Sevanto (eds.), *Phloem*, Methods in Molecular Biology, vol. 3066,
https://doi.org/10.1007/978-1-0716-5488-0_14,

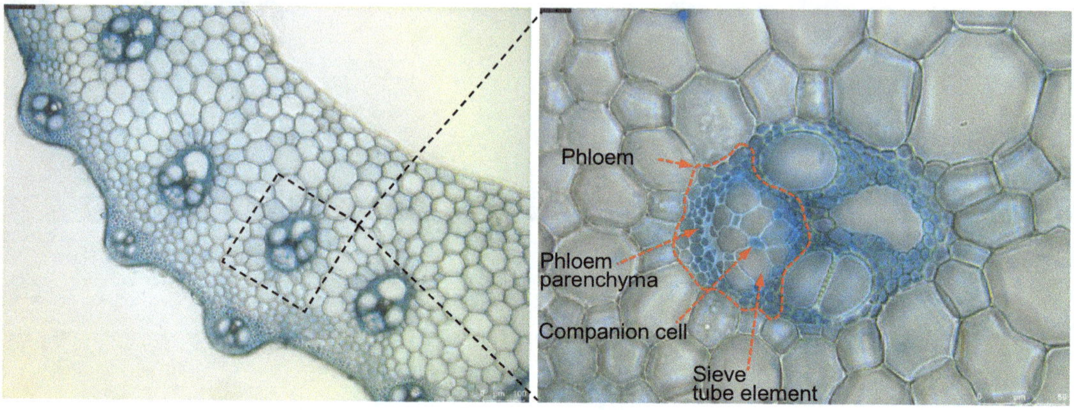

Fig. 1 Pi staining assay of rice stem phloem cells. The blue signal indicates Pi. The heading stage rice stems were used for cross-sectioning and IOSA

for visualizing intracellular Pi in plants (Fig. 1) [4]. In this study, we present a detailed protocol for Pi detection in phloem cells using IOSA, enabling researchers to investigate Pi status with cellular resolution.

2 Materials

1. Wild-type rice (*O. sativa* L. spp. *japonica* Nipponbare) grown under hydroponic conditions, stem at the heading stage (*see* **Note 1**).
2. Agarose solution: 2%–4% (w/w) in water (*see* **Note 2**).
3. Reaction tubes: 1.5 mL volume.
4. Inorganic orthophosphate detecting buffer (IDB): ammonium molybdate (0.81 mM), potassium antimonyl tartrate (0.15 mM); sulfuric acid (0.19 M), L-ascorbate (17 mM) in water. IDB should not be stored for more than two days. Instead, prepare 1 L of a 10× stock solution of all components except L-ascorbate. Before use, dilute the 10× fold reagents to 1× and dissolve the L-ascorbate powder.
5. Clearing buffer: to 160 g Trichloroacetaldehyde, add 20 mL 100% Glycerin and 20 mL H_2O.
6. Vibratome: for sectioning.
7. Light microscope: equipped at least with 10×, 20×, and 40× objectives and a camera.
8. Vacuum pump: must be capable of producing ~500 mbar vacuum.
9. Razor blades, scissors, and tweezers.
10. Pipette.

3 Method

All steps can be performed at room temperature.

3.1 Sample Preparing for IOSA

1. Embed samples in melted agarose solution (~60°C).
2. After cooling, section the embedded samples on a vibratome with a section thickness of 30–80 µm (*see* **Note 3**).
3. Collect several sections in a 1.5 mL reaction tube for the Pi staining assay.

3.2 IOSA Procedures for Sectioned Samples

1. Stain the section with 1 mL IDB for 30 min under vacuum treatment. The reaction time can be varied (*see* **Note 4**). The volume of IDB can also be adjusted according to the number of samples.
2. Completely remove Buffer IDB using a pipette.
3. Add 500 µL of clearing buffer, and incubate for 45 min. If not fully cleared, extend the incubation time.
4. Mount each section on a microscope slide and image it on the light microscope at adequate magnification.

4 Notes

1. Other plant species, including Chinese Pine (*Pinus tabuliformis*), *Arabidopsis thaliana*, liverwort (*Marchantia polymorpha*), and even the algae Catenella, can also be utilized for cellular Pi detection using IOSA. For large tissue samples, specific regions of interest can be selectively sectioned and subjected to IOSA analysis.
2. The agarose concentration should be adjusted according to different sample characteristics. The old and hard tissues should be embedded with a higher agarose concentration, but not more than 5%. For the young and soft tissues, the agarose concentration should be lower, but not less than 2%.
3. The thickness of the advice section should include at least three layers of companion cells for different plant scpecies.
4. The incubation time can be shortened or prolonged according to the Pi abundance in the sample. For example, plants cultivated under Pi-deficient conditions require prolonged incubation time, whereas incubation time for those grown in Pi-sufficient conditions need to be shortened.

Conflict of Interest

The authors have no conflicts of interest to declare.

References

1. Pace MR (2020) Phloem: cell types, structure, and commercial uses. IntechOpen. 10.5772/intechopen.88162
2. Chiou TJ, Lin SI (2011) Signaling network in sensing phosphate availability in plants. Annu Rev Plant Biol 62: 185–206
3. Ding G, Lei GJ, Yamaji N, Yokosho K, Mitani-Ueno N, Huang S, Ma JF (2020) Vascular cambium-localized AtSPDT mediates xylem-to-phloem transfer of phosphorus for its preferential distribution in Arabidopsis. Mol Plant 13(1):99–111
4. Guo M, Ruan W, Li R, Xu L, Hani S, Zhang Q, David P, Ren J, Zheng B, Nussaume L, Yi K (2024) Visualizing plant intracellular inorganic orthophosphate distribution. Nat Plants 10(2):315–326

Check for updates

Chapter 15

Characterization of Viruses in Phloem by Correlative X-Ray Microtomography (μCT)-Volume Electron Microscopy (vEM) Imaging

Li Xie, Ming-fang Lv and Heng-Mu Zhang

Abstract

Studying virus-infected phloem is of significant importance, as it not only enhances our understanding of viral pathogenesis but also leverages viruses as tools to expand knowledge about plant phloem physiology. The uneven distribution pattern of phloem-infecting viruses poses methodological challenges for such studies—requiring both large field of view (FOV) and high-resolution imaging. A comprehensive anatomical analysis of the phloem necessitates global visualization, while resolving viral structures demands local high-resolution observation. This chapter describes a method, the X-ray microtomography (μCT)-volume electron microscopy (vEM) correlative imaging technique, which effectively addresses these methodological requirements, where μCT provides the large FOV for identification of regions of interest, followed by vEM acquisition of high-resolution images. It is a six-step protocol, including: (1) sample preparation, (2) flaw detection, (3) overview imaging by μCT, (4) identifying viral infection regions, (5) high-resolution imaging by vEM, and (6) image processing and analysis. In this workflow, the steps of sample preparation and identification of viral infection regions are critical. This protocol was originally established for investigating Southern rice black-streaked dwarf virus (SRBSDV) infection in rice phloem, with parameters optimized for plant reoviruses. We provide advice on how to adapt the approach for studying other viral infections.

Keywords Phloem, Viral infection, X-ray microtomography, Volume electron microscopy, Correlative imaging

1 Introduction

The phloem serves not only as a conduit for nutrient transport but also as a battleground where host plants defend against pathogen invasion. Investigating the interactions between pathogens (including bacteria, fungi, and viruses) and host plants within the phloem represents one of the key research topics in phloem biology. Viruses exploit the phloem for systemic movement to colonize host plants [1, 2]. As a countermeasure, host plants employ specific mechanisms to confine the infection of certain viruses (such as some geminiviruses

Johannes Liesche and Sanna Sevanto (eds.), *Phloem*, Methods in Molecular Biology, vol. 3066,
https://doi.org/10.1007/978-1-0716-5488-0_15,

and plant reoviruses) within the phloem tissue [3, 4], thereby preventing their spread to other tissues and mitigating the detrimental effects of viral infection. For example, most plant reoviruses other than Rice dwarf virus (RDV) are restricted to phloem tissue upon infection, where they induce phloem cell proliferation leading to tumor formation [5]. The host plant forms a physical barrier that prevents viral exit from the phloem through a restrictive distribution pattern of high-capacity intercellular channels [6]. In the emerging field, some critical questions about viral anchoring mechanisms in the phloem flow for infection establishment, and the regulatory processes governing directional viral movement within the phloem transport network remain to be explored. Elucidating the interaction mechanisms between viruses and host plants in the phloem will not only provide a foundation for developing strategies to control diseases but also offer novel insights into the fundamental processes of phloem-mediated substance transport in plants.

Electron microscopy provides an excellent tool for direct visualization of the cellular composition and ultrastructural changes of phloem under different physiological states [7]. In recent years, the advancement of volume electron microscopy (vEM) has further provided a three-dimensional (3D) perspective. For example, vEM was employed to investigate the enucleation during sieve element (SE) morphogenesis [8] and the mechanism of phloem unloading [9]. Investigation of phloem-infecting viruses faces unique challenges, which impose specific requirements for the selection of appropriate microscopy methods. Viruses are unevenly distributed within the phloem, necessitating a balance between the field of view (FOV) and resolution. A comprehensive anatomical study of the phloem requires a large FOV for tissue- or organ-level observation, while visualization of virus localization, structure, and abundance demands subcellular-resolution imaging of localized areas. The micro-computerized tomography (μCT)-vEM correlative imaging technique is particularly suitable for studying samples requiring both large FOV and high resolution, such as phloem-infecting viruses with heterogeneous distribution patterns [6]. Leveraging the nondestructive nature of μCT imaging (where samples remain intact for subsequent microscopy analyses), the μCT-vEM correlative imaging technique enables us to simultaneously achieve both large FOV and high-resolution imaging on the same specimen, thereby balancing these conflicting requirements. μCT provides wide-field imaging for overview, while vEM delivers high-resolution imaging for minute and critical details. In such a correlation, vEM should take precedence over μ CT to guarantee optimal high-resolution data acquisition, which is particularly vital for virus studies. The resulting 3D reconstruction digitally dissects the specimen through image processing techniques, offering unprecedented insights into the anatomical organization and the spatial relationships between different regions of interest (ROI).

This µCT-vEM correlative imaging technique has been widely applied in multiple research fields, including geology [10] and biology, with biological applications extended to animal tissues [11], plant tissues [6], and even single-celled yeast [12]. It is anticipated that this technique could be further extended to study other pathogens with heterogeneous infection patterns after parameter optimization.

The technical workflow for studying phloem-infecting viruses using µCT-vEM correlative imaging consists of multiple steps. The workflow consists of six sequential steps (Fig. 1): (1) sample preparation, (2) flaw detection, (3) overview imaging by µCT, (4) identifying viral infection regions, (5) high-resolution imaging by vEM, and

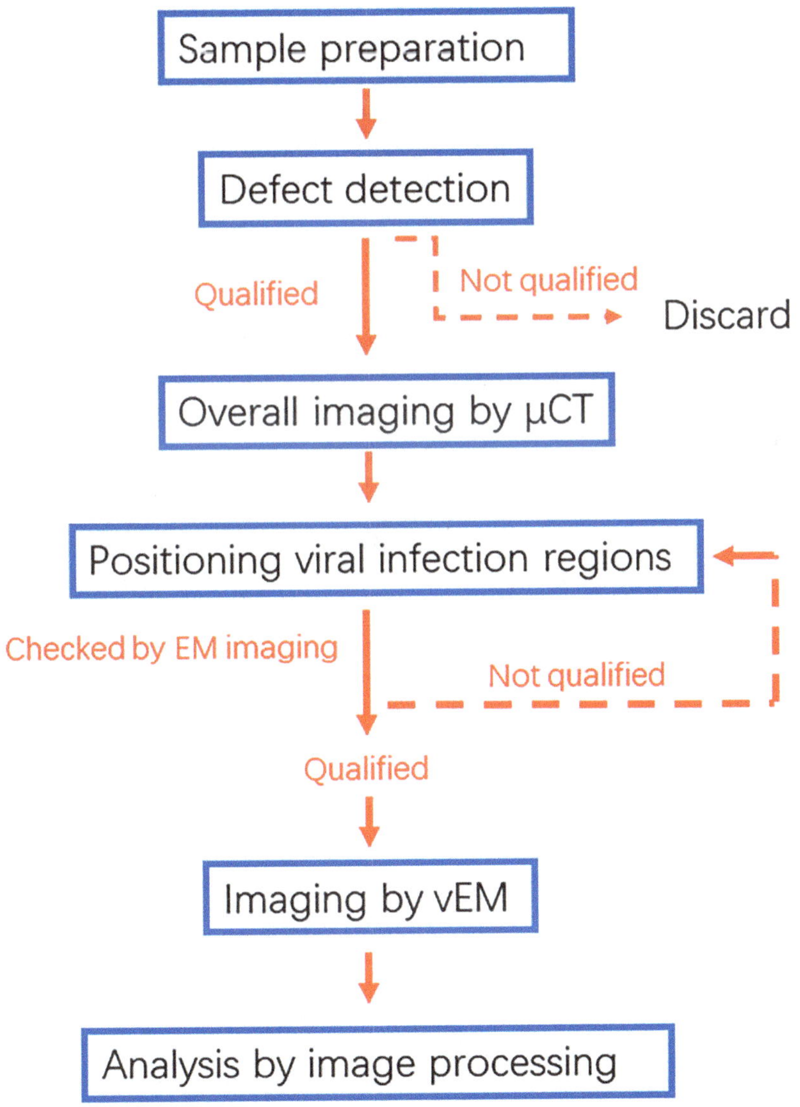

Fig. 1 The workflow of µCT-vEM correlative imaging technique

(6) analysis by image processing. In this workflow, μCT, vEM, and image processing are the three major steps for data collection and the following analysis. Their successful implementation relies on proper sample preparation and identifying the sample for imaging the viral infection regions. Compared to viruses that infect whole organs, in the study of phloem-infecting viruses, the steps of sample preparation and identifying viral infection regions are even more critical for successful μCT-vEM correlative imaging. The step of X-ray flaw detection is employed to examine cracks, bubbles, or other defects within sample blocks to minimize unnecessary vEM work, thus improving efficiency and reducing cost. It is highly recommended to include this step in the workflow when possible. During sample preparation, three major challenges in traditional phloem research must be addressed, including the challenge of sectioning entire SE, the artifacts from phloem pressure release, and the influence of sectioning on plant physiology [7]. The strategies to overcome these challenges have been thoroughly discussed in other chapters of this book and thus will not be reiterated here. In particular, the application of μCT-vEM correlative imaging to study phloem-infecting viruses imposes additional specific requirements on sample preparation. These requirements are designed to mitigate the adverse effects of low signal intensity and high charge accumulation during vEM imaging. The protocol of OTO (Osmium-Thiocarbohydrazide-Osmium) fixation, accompanied by large volume en-bloc staining was employed to meet such requirements by enhancing the specificity of heavy metal staining on membrane systems [13–15]. It not only increases signal intensity by improving contrast but also utilizes the conductive properties of metals to dissipate excess accumulated charges from the imaging area, thereby improving SEM imaging quality. The sample embedding blocks prepared using this protocol can also meet the requirements of μCT imaging, specifically, sufficient differences in X-ray absorption among different parts of the sample, as well as the absence of shaking or deformation due to physiological or dehydration factors. Other sample preparation methods that enhance the heavy metal staining level of membrane systems can also be applied to the μCT-vEM correlative imaging technique, but it is essential to ensure that these methods preserve accurate structural details (*see* **Note 1**). On the other hand, the study of unevenly distributed viruses in phloem using μCT-vEM correlative imaging imposes stringent requirements for pinpointing their precise localization. The strategy employed in this study involves a three-step procedure: (1) identifying regions of coexistence between viral inclusion bodies and phloem cells in μCT 3D images through digital sectioning based on their distinctive characteristics; (2) precisely determining the spatial coordinates of these regions within the sample resin block coordinate system; and (3) selectively removing extraneous material via block trimming to expose the target regions. Notably, this method was established in plant reoviruses, and its efficacy across all virus

species requires further exploration. Therefore, to avoid potential misjudgment of viral localization due to the limited resolution of μ CT imaging (which is insufficient for clear virus visualization), validation through vEM imaging (with resolution adequate for precise virus identification) is essential.

Using μCT-vEM correlative imaging, we mapped the distribution of Southern rice black-streaked dwarf virus (SRBSDV) within the phloem tissues of infected rice plants. The viral infection was found to be confined to tumors formed by proliferating phloem cells. We further classified the infection severity of the viruses across different cell types within these tumors. By analyzing the spatial distribution of intercellular channels with a large size exclusion limit in the rice phloem, we identified a potential physical barrier that may restrict plant reoviruses from exiting phloem cells [6]. This barrier likely explains why most plant reoviruses are confined to phloem-derived tumors [5]. In this study, the large-FOV data obtained by μCT enabled the distinction between the meristematic and neovascularization regions (containing abnormal developed phloem) within tumor tissues (Fig. 2A–D), allowing us to focus our investigation on the neovascularization region (Fig. 2C, D, yellow arrows). Meanwhile, the high-resolution vEM data provided critical insights into the distribution patterns of viral inclusion bodies (Fig. 2E–G) and intercellular channels (Fig. 2H–l), forming the foundation of our conclusions.

2 Materials

2.1 Solutions and Buffers

1. Fixative 1: 2.5% glutaraldehyde in 0.1 M cacodylate buffer, pH 7.3.
2. Fixative 2: 2% Osmium tetroxide in 0.1 M cacodylate buffer, pH 7.3.
3. Fixative 3: 2% Osmium tetroxide in MilliQ water.
4. Cross-linking agent of osmium: 1% Thiocarbohydrazide (TCH) in MilliQ water after filtering (0.22 μm). Store at a constant temperature (65°C) in the dark before use.
5. Washing buffer: 0.1 M cacodylate buffer, pH 7.3.
6. MilliQ water.
7. Dehydration solvent: 30%, 50%, 70%, 80%, 90%, 95% and 100% ethanol, and 100% acetone.
8. Staining solution 1: 1% uranyl acetate in MilliQ water, wrapped in aluminum foil to keep it light-shielded.
9. Staining solution 2: dissolve 0.066 g lead nitrate in 10 mL 0.03 M aspartic acid solution and pH adjusted to 5.0 with 1 N KOH, keeping at 60 °C before use.

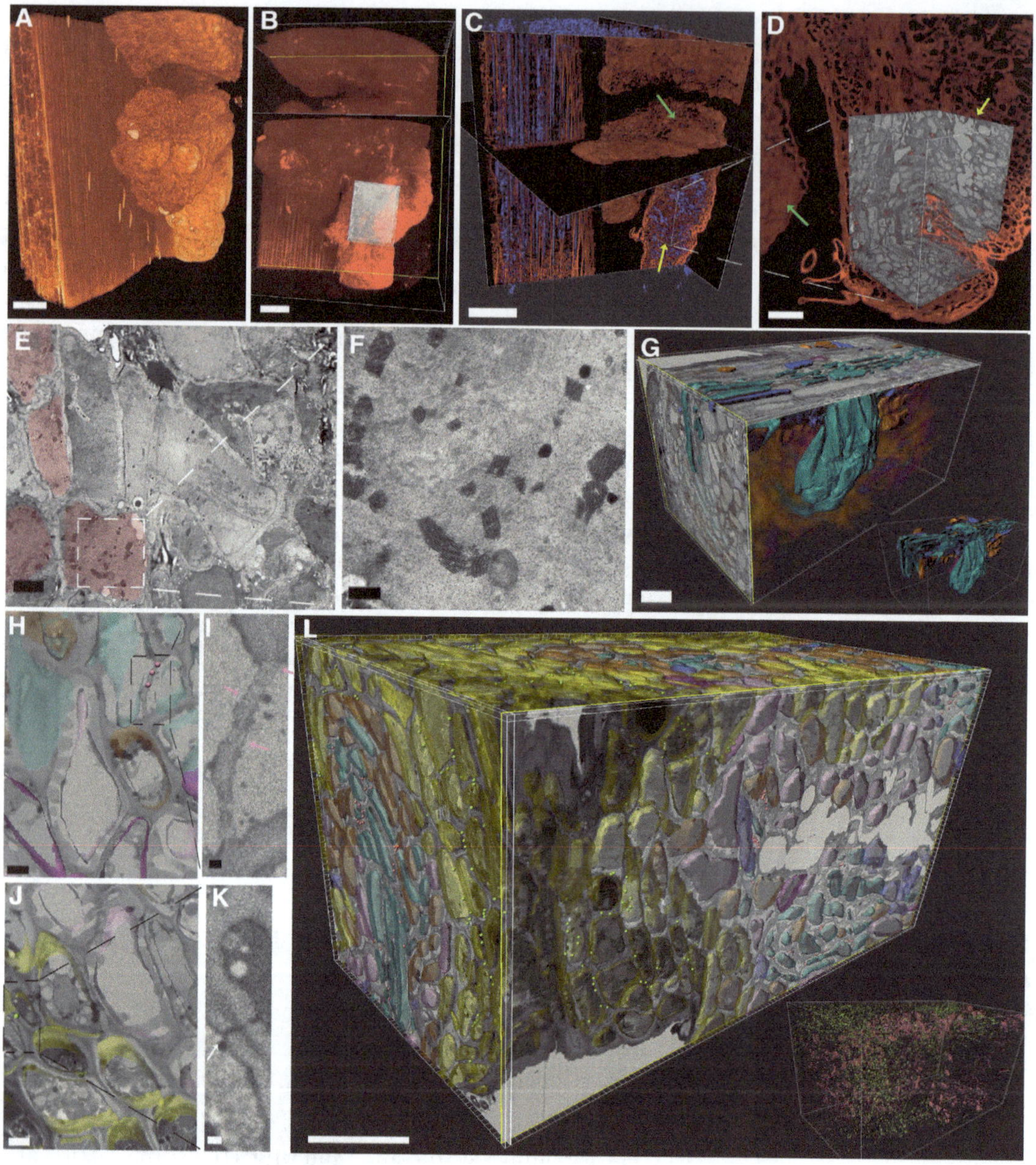

Fig. 2 Application of X-ray microtomography (μCT)–volume electron microscopy (vEM) correlative imaging technique in the study of Southern rice black-streaked dwarf virus (SRBSDV) infection in rice phloem. (**A–D**) Correlative imaging of μCT with vEM showing the region of interest in SRBSDV infection (white box in (**B**, **D**)). μCT can be used to quickly differentiate the meristematic region (green arrows in (**C**, **D**)) from the neovascularization region (yellow arrows in (**C**, **D**)) in the tumor. This correlation enables simultaneous achievement of both large FOV and high-resolution imaging on the same specimen. (**E–G**) vEM can be used in identifying the viral infection region in phloem. High-resolution images showing virus crystals (**F**) which indicate viral infection in the current cell. Integrating the distributions of virus-infected cells (red region in (**E**)), a 3D map of viral infection can be obtained (**G**). (**H–I**) vEM can be used in identifying the distribution of intercellular channels in large capacity. High-resolution images showing the intercellular channels in large capacity ((**H**, **I**) channels indicated by pink arrows in (**I**)) and PD (**J**, **K**, white arrow in (**K**)). Integrating the distributions of different types of intercellular channels, a 3D map of intercellular channels in large capacity (pink and red dots in (**G**)) can be obtained

10. Epon 812 resin: mix 6.3 g Epon 812, 3.3 g DDSA, 3.8 g NMA, and 200 μL DMP-30.

2.2 Disposables

1. 0.22 μm Millipore syringe filter.
2. Flat embedding modules.
3. Double-edged razor blades.
4. Toothpicks.

2.3 Tools

1. Forceps.
2. Abrasive paper.
3. Carbon rod with a diameter of 1–2 mm.
4. Cyanoacrylate superglue.
5. Ultramicrotome diamond knives (SBF-SEM customized, such as Diatome 3VDKN) and diamond trim knives.
6. Carbon tape.
7. Liquid conductive silver gel.
8. Image processing software.

2.4 Equipment

1. Dissection microscope.
2. Microbalance.
3. Lab oven (range 40–100 °C with 1 °C deviation).
4. pH meter.
5. Fume hood.
6. Shaker.
7. Ultramicrotome.
8. High vacuum ion sputter with a platinum target.
9. X-ray tube-based μCT.
10. Scanning electron microscopy with an in-chamber ultramicrotome.

3 Methods

3.1 Sample Preparation

In the use of μCT-vEM correlative imaging, the electron microscopy imaging imposes higher requirements on the samples. Therefore, the sample preparation method needs to accommodate electron microscopy by adopting OTO fixation accompanied by large-volume en bloc staining [13, 14], which ensures sufficient contrast and conductivity for SEM imaging. This sample preparation method yields embedded blocks that have undergone dehydration and contrast enhancement, while also meeting the requirements for X-ray imaging. To prevent artifacts from phloem pressure release due to involuntary dissection, the sampling of virus-infected phloem is performed using a specialized procedure designed to minimize such artifacts, such as the xylem-assisted perfusion method [16, 17]. Additionally, viruses exhibit a preferential tropism for younger plant tissues; therefore, sampling should preferentially target the phloem from tender regions.

1. Sample the phloem tissues and fix them in fixative 1 for over 48 h. Evacuate the phloem samples with a vacuum pump until they sink to the bottom of the tube.
2. Wash samples with cold washing buffer three times (each for 5 min) at room temperature.
3. In the fume hood, immerse samples in fixative 2 for 1 h on ice.
4. In the fume hood, wash samples with MilliQ water three times (each for 5 min) at room temperature.
5. In the fume hood, immerse samples in the 0.22 μm filtered cross-linking agent of osmium solution for 20 min at room temperature. During the treatment, wrap it in aluminum foil to keep it light-shielded. The treatment usually induces bubbles escaping from the sample in solution.
6. In the fume hood, wash samples with MilliQ water three times (each for 5 min) again at room temperature.
7. In the fume hood, immerse samples in fixative 3 for 30 min at room temperature.
8. In the fume hood, wash samples with MilliQ water three times (each for 5 min) again at room temperature.
9. Immerse samples into staining solution 1 at 4 °C overnight, wrapped in aluminum foil to keep them light-shielded.
10. Wash samples with MilliQ water four times (each for 10 min) at room temperature.
11. Immerse samples into staining solution 2 at 60 °C for 40 min (for anti-contamination, *see* **Note 2**).
12. Wash samples with MilliQ water five times (each for 5 min) at room temperature.
13. Dehydrate samples sequentially *d* in 30%, 50%, 70%, 80%, 90%, and 95% ethanol for 15 min each, then in 100% ethanol twice for 20 min each and in 100% acetone twice for 30 min each.
14. Replace 100% acetone with a 3:1 acetone:resin (Epon 812 or Durcpan) mixture for 3 h on a shaker (for the hardness of resin, *see* **Note 3**).

15. Replace 3:1 acetone:resin mixture with 1:3 acetone:resin mixture for 8 h on shaker.
16. Replace 1:3 acetone:resin mixture with pure resin (Epon 812 or Durcpan) overnight on shaker.
17. Carefully transfer individual samples into individual embedding modules by toothpicks. For polymerization, heat them at 60 °C in an oven for over 48 h until the liquid resin becomes solid.

3.2 Selection of Properly Sampled Resin Blocks Qualified for SBF-SEM Imaging by μCT

The differential penetration capacity of liquid resin across various tissue cells occasionally results in void formation within polymerized resin blocks. These voids manifest as artifacts on the cutting surface (Fig. 3A, arrows), compromising SBF-SEM imaging quality. The density contrast between these resin-deficient voids and the surrounding resin matrix enables their detection via X-ray imaging (Fig. 3B, arrows). Therefore, we can employ the non-destructive capability of X-ray imaging to effectively screen sample resin blocks, where only void-free specimens were selected for the following high-quality SBF-SEM imaging. To improve efficiency, multiple resin blocks can undergo inspection scanning simultaneously without polishing (Fig. 3C). The resolution of such batch scanning remains sufficient for detecting void defects (Fig. 3D, arrows).

1. Mount eight intact sample blocks onto the rotational stage of the μCT system with cyanoacrylate superglue. Align the sample in the XY plane relative to the optical axis. Rotate the sample by 90° and make fine adjustments to ensure proper XY-plane alignment.
2. Turn on the X-ray source (usually at 80 kV), perform fine alignment based on the images obtained during sample rotation.

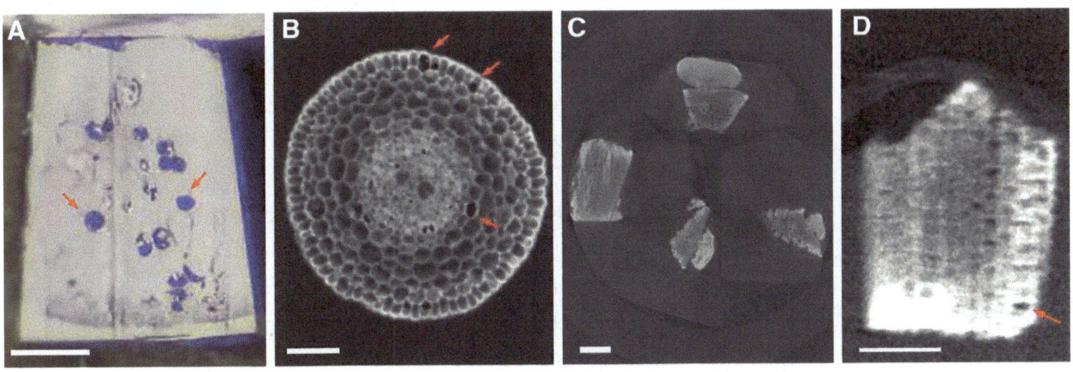

Fig. 3 Void detection of sample resin blocks. (**A**) Void (red arrows) of resin block on the cutting surface. (**B**) Void (red arrows) of resin block detected by μCT imaging. (**C**) Multiple resin blocks detected simultaneously by μCT imaging without polishing. (**D**) Void (red arrows) can also be detected by low-resolution μCT imaging when multiple resin blocks are mounted together for a simultaneously detection

3. Define FOV to fully include all eight sample blocks.
4. Set the imaging parameters as follows: exposure time: 1 s, projection: 801, no filter. Capture projection images both with and without the sample in the beam path and use these to determine the transmission ratio of the current FOV.
5. Initiate data acquisition by scanning the sample without X-ray source warm-up.
6. Perform three-dimensional reconstruction of the data using the filtered-back-projection (FBP) algorithm.
7. Thoroughly examine μCT images for areas with extremely low density, which suggest cracks or bubbles. Exclusively use void-free resin blocks for the following high-definition X-ray imaging and SBF-SEM imaging.

3.3 Overview Imaging of Virus-Infected Phloem by Noninvasive X-Ray μCT

Imaging the pathological changes in phloem cells induced by viruses requires both a large field of view (FOV) and high resolution. To provide a global view with cellular resolution for subsequent volume electron microscopy (vEM) studies, X-ray imaging must achieve a voxel size below 0.35 μm—sufficient for clear visualization of cell walls and reliable discrimination between the vast majority of cells. The methodology described in this chapter was implemented using an X-ray tube-based μCT system, which could alternatively be performed using synchrotron radiation facilities, but cannot be achieved with benchtop CT systems having a voxel size larger than 1 μm. Although nano-CT achieves high resolution, it remains inferior to vEM in performance while incurring significantly higher economic costs. Since vEM will be subsequently employed, the use of nano-CT in this step becomes unnecessary. Moreover, the resin blocks subjected to gradient dehydration are entirely anhydrous, thereby precluding the necessity of critical point drying for dehydration (see [18] for details). The resin blocks stained with heavy metals exhibit sufficient contrast, thereby eliminating the need for phase-contrast imaging to enhance contrast (see [18] for details).

1. Remove the peripheral resin with razor blades and abrasive paper as thoroughly as possible (*see* **Note 4**) and ensure that the sample is exposed.
2. Securely attach the sample block to the end of the carbon rod with cyanoacrylate superglue; wait at least 0.5 h.
3. Mount the sample block-carbon rod assembly on the rotation stage of the μCT system. Position the sample as close as possible to the X-ray source to maximize geometric magnification while maintaining a safe distance to avoid collisions. The optimal working distance is determined by the device specifications, sample dimensions, and the sample-to-carbon rod spatial configuration. Align the sample in the XY plane

relative to the optical axis. Rotate the sample by 90° and make fine adjustments to ensure proper XY-plane alignment.

4. Following the preset conditions specified in the instrument manual, turn on the X-ray source, acquire projection images of the sample, and perform fine alignment based on the projection images obtained during sample rotation.
5. Define the FOV and accordingly set the parameters, including sample-to-detector distance, pixel size, type of objective lens (for optical secondary magnification), and cone/fan angle.
6. Capture projection images both with and without the sample in the beam path, and use these to determine the transmission ratio of the current FOV. Choose the manufacturer-specified filter according to the measured transmission ratio.
7. Using the acquired sample projections, determine the X-ray energy (*see* **Note 5**) and the X-ray photon count, and adjust the exposure time accordingly. Target photon counts should be >7000 while avoiding detector over-saturation.
8. Set the rotation strategy according to the target pixel size, with the number of projections following the manufacturer's specifications to achieve optimal sampling.
9. Initiate data acquisition by scanning the sample. Ensure that the X-ray source warm-up procedure has been completed.
10. Reconstruct the sequential images into three-dimensional data using the standard filtered-back-projection (FBP) algorithm.

3.4 Identifying Viral Infection Regions for Fine Imaging

This subsection describes the procedure for identifying phloem regions suspected of viral infection in μCT images using image processing techniques. Subsequently, the nondestructively X-ray-scanned sample is trimmed to target the candidate region of virus-infected phloem via ultra-microtome, which serves as the ROI for high-resolution vEM imaging. This approach enables correlative μCT and vEM imaging within the same sample. The identification of phloem, xylem, and non-vascular tissues should be based on their distinctive features in μCT images. Compared to non-vascular tissues, vascular cells exhibit a significantly larger length-to-width ratio (typically greater than 10), which can be used to distinguish between the two tissue types. The position of xylem can be determined by the vessels or tracheid elements, which appear in μCT images as directionally elongated hollow cells with unevenly thickened cell walls (Fig. 4, red arrowheads). The position of phloem can be identified by sieve elements (SE). In μCT images, SE appears as directionally elongated hollow cells with slightly swollen connections to adjacent SEs at the short axis, and exhibits strongly thickened cell walls there, namely sieve plates (Fig. 4, high-density lines indicated by red arrows). Viral infection induces the formation of large inclusion bodies, which appear as cloud-like occlusions in hollow SEs in μCT images (Fig. 4, green arrowheads). Therefore, after identifying the phloem, the cloud-like occlusions in hollow SEs can be utilized to localize virus-infected regions. It should be noted that sieve tubes infected by certain viruses fail to maintain their directionally elongated morphology. In μCT images, the diagnostic feature of virus-infected sieve tubes is strongly thickened sieve plates (Fig. 4, red arrows). After identifying the putative virus-infected phloem regions in μCT images, targeted resin trimming is required to precisely localize the ROI. This process involves removing extraneous embedding media and nontarget tissue to expose the ROI for subsequent high-resolution SBF-SEM imaging.

1. Find vascular bundles in μCT images, based on their characteristic high length-to-width ratio of cells. If the low-density cell cavities representing vessels/sieve tubes are not visually distinct, density inversion may be applied to enhance contrast.
2. Identify candidate SE infected by viruses based on two characteristic features in μCT images: strongly thickened sieve plates and cloudy intracellular inclusions. Mark these as an ROI for the following measurements.
3. Set the edge of the sample resin block as the origin (zero point), which serves as the starting position for block trimming.

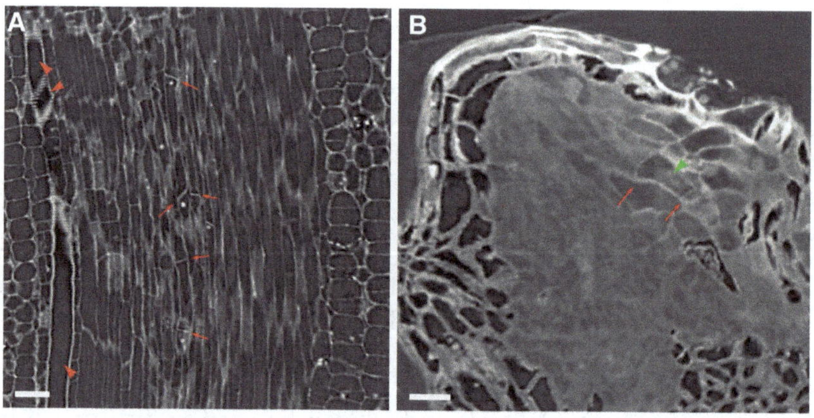

Fig. 4 Identifying xylem, phloem, and virus-infected SE in μCT images. (**A**) Healthy vascular bundle. Xylem is identified by the vessels or tracheid elements, which appear as directionally elongated hollow cells with unevenly thickened cell walls (red arrowheads). Phloem is identified by the sieve elements (SE), which appear as directionally elongated hollow cells with slightly swollen connections at the short axis where they join adjacent SEs, and exhibit strongly thickened sieve plates (high X-ray absorption lines indicated by red arrows). (**B**) Viral infected phloem with abnormal cellular development. SE, failing to develop as a directionally elongated hollow cell, is identified by strongly thickened sieve plates (red arrows). The cloud-like occlusion in SEs (green arrowheads) can be considered as a candidate of viral infection-induced inclusion body, whose characteristics can be used to locate the possible viral infection regions

Measure the distances from the ROI to this origin along the length, width, and height dimensions.

4. To prepare a stable base for ultrathin sectioning, remove the embedded sample block (without surround resin) from the carbon rod and re-embed in an identical liquid embedding resin within a mold of suitable dimensions.
5. Cure the resin thermally to form a properly shaped block, facilitating secure fixation in the in-chamber ultra-microtome during sectioning.
6. Based on the measured distance between the ROI and the trimming starting point obtained above, trim the block to the ROI position, ideally exposing the candidate region of virus-infected phloem at the center of the section.
7. Trim the resin block with a diamond trimming knife (*see* **Note 6**) on an ultra-microtome to produce a sample-containing pillar structure with 1–2 mm height, nearly identical top and bottom dimensions, and the edge length of the cutting surface < 0.5 mm.
8. Polish the cutting surface to a mirror-like finish using a diamond trimming knife.
9. Excise the sample-containing cube carefully from the embedding block using a double-edged razor blade.

3.5 Region of Interest (ROI) Imaging of Virus-Infected Phloem by Serial Block Face Scanning Electron Microscopy (SBF-SEM)

Among various vEM techniques, SBF-SEM currently represents the optimal method for studying viral infection in phloem tissues, as it achieves the best balance between scale, resolution, and ease of image registration [19]. Compared to serial sections transmission electron microscopy (ssTEM) and focus ion beam-scanning electron microscopy (FIB-SEM), while SBF-SEM offers lower resolution (though sufficient for virus identification), it provides a significantly larger FOV, enabling investigation of broader infection patterns. Relative to automatic tape-collecting ultramicrotome-scanning electron microscopy (ATUM-SEM), SBF-SEM demonstrates superior convenience in serial image registration. However, SBF-SEM is more destructive to samples than ATUM-SEM and cannot perform repeated imaging of ROIs under different conditions.

1. Mount the trimmed sample block on an aluminum sample stub using cyanoacrylate superglue. All peripheral gaps were carefully filled with liquid conductive gel to facilitate charge dissipation during SEM imaging.
2. Coat the whole sample with platinum by ion sputtering for 10 min at a power of about 6 W (such as 220 V 30 mA results in a power of 6.6 W) and at a vacuum range of 10^{-1} to 10^{-2} Pa.

3. In order to prevent collisions, make sure the stage of the SEM is lowered, and the diamond knife is located at a safe position. Then mount the sample stub to the in-chamber microtome and check the suitable sample height. Focus on the edge of the diamond knife using the in-chamber optical system under LED illumination. Advance the diamond knife until you can see the knife shadow on the mirror-like finish of the block.
4. Start the advance of the knife until it almost touches the sample surface, while optimizing the sample position.
5. Set the parameters of cutting and imaging, including accelerating voltage, beam current, magnification, spot size, dwelling time of beam, cutting thickness (usually 50 nm for SBF-SEM), and define the FOV (*see* **Note 7**). Backscattered electron (BSE) signals are commonly utilized for imaging. During SBF-SEM imaging, image quality is inversely proportional to the working distance (the sample-to-detector distance). However, this distance must accommodate the diamond knife's clearance between the detector and sample to prevent mechanical interference.
6. Restart cutting to remove the first layer of the cutting surface.
7. Apply a charge-neutralizing gas (typically nitrogen or argon, equipped in SEM) around the freshly cut sample surface to mitigate charge buildup during imaging. The procedure temporarily lowers the vacuum level in the SEM specimen chamber to 10^{-2} Pa.
8. Perform the data acquisition of ROI by imaging on the newly cut block surface rather than the section itself. During imaging, the diamond knife was moved to the side parking position to avoid interference with data acquisition and ensure operational safety. Thus, one complete sectioning-imaging cycle was accomplished.
9. Due to the limited resolution of μCT imaging, identifying viral inclusion bodies carries a risk of misjudgment, which may result in the SBF-SEM imaging area failing to target virus-infected regions. Therefore, before initiating automated SBF-SEM imaging, it is essential to carefully examine several completed images to confirm viral presence. If no viruses are detected, the automated SBF-SEM imaging should be terminated to avoid unnecessary experimental costs.
10. Repeat the automated sectioning-imaging sequence for the next cycle. Since the working distance of the sample remained constant due to the fixed cutting position of the knife, both sectioning and imaging could proceed automatically without refocusing. During sectioning, the sample stage automatically elevates to maintain a constant working distance,

compensating for the height loss due to section removal at the cutting surface. The cycle was repeated iteratively until sufficient *z*-axis depth data were acquired.

11. Registration of the serial EM images by image processing software.
12. Reconstruct the serial images to generate the final volumetric density map.

3.6 Analysis of the Ultrastructural Features of Virus-Infected Phloem by Image Processing

1. Carefully examine the 3D dataset of virus-infected phloem to identify target objects for analysis. Develop an image processing pipeline based on their size, grayscale intensity, texture, and spatial distribution characteristics.
2. Crop the FOV size while maintaining full resolution, or reduce sampling density by deleting images from the stack to reduce memory load and improve computational efficiency.
3. Apply a filter to enhance the signal-to-noise ratio (SNR), facilitating the extraction of critical features during subsequent segmentation steps. For denoising vEM data, nonlocal means filtering [20] is implemented as the method of choice.
4. Extract target objects from images by delineating clear boundaries between regions of interest and background noise or irrelevant areas (*see* **Note 8**). After segmentation, pseudo-colored 3D models are obtained.
5. Perform quantitative measurements and statistical analysis on pseudo-colored models, from which the research conclusions were established.

4 Notes

1. For clear observation of virus structural details under electron microscopy, potassium ferrocyanide is NOT recommended in osmium fixation. An alternative is the rOTO fixation method [14].
2. Staining solution 2, which is also called Walton's lead aspartate, is highly susceptible to reaction with ambient CO_2 (originating from operator respiration), which may lead to sample contamination. During the treatment of lead aspartate, placing a large number of NaOH tablets around the sample centrifuge tubes can effectively mitigate the adverse effects caused by CO_2.
3. To ensure that the section thickness meets the requirements for automated serial sectioning in vEM, the embedding block must possess sufficient hardness. During sample preparation, Spurr's embedding medium should be avoided; instead,

Epon812 or Durcupan resin should be used with the formula "Tough."

4. In X-ray imaging, the more thoroughly the excess embedding resin around the sample is removed, the better the X-ray penetration capability and the higher the image SNR.
5. In imaging of biological samples by X-ray tube-based μCT, a reduction of X-ray energy yields better image SNR while maintaining a qualified transmission ratio and photon counts. This energy reduction can be achieved either by lowering the X-ray tube voltage, or by selecting an appropriate X-ray filter based on transmission ratio considerations, or by both.
6. Glass knives are NOT recommended for trimming sample blocks prior to SBF-SEM imaging, as they may generate microscopic debris on the cutting surface during the trimming process, which could potentially damage the expensive customized diamond knife used for in-chamber sectioning.
7. When setting the imaging ROI (region of interest) of SBF-SEM, it is necessary to balance the FOV and resolution according to the experimental objectives. Attempting to simultaneously achieve both a large FOV and high resolution in a single image would result in excessively large file sizes, compromising the efficiency of subsequent data processing.
8. For automated segmentation of whole cells or organelles, several ready-to-use tools are available, such as Cellpose (specialized in cell) [21] and OrgSegNet (specialized in organelle) [22], which were developed based on deep learning approaches. However, few reported programs currently exist for automated segmentation of virus-related targets. For virus-associated structures with low contrast or complicated architectures, manual segmentation may still be required. When researchers are primarily interested in quantitative and distributional information rather than dimensional or ultrastructural details, segmentation can be omitted. Labeling approaches may suffice to convey these characteristics [6].

Acknowledgments

This work was supported by the National Key R&D Plan in China (2024YFD1400702), the Natural Science Foundation of China (U20A2037, 32001869, and 31972980), the Talent Project of Zhejiang Province (2019R52033), and the Key R&D Program of Zhejiang Province (2019C02018).

Conflict of Interest The authors have no conflicts of interest to declare.

References

1. Folimonova SY, Tilsner J (2018) Hitchhikers, highway tolls and roadworks: the interactions of plant viruses with the phloem. Curr Opin Plant Biol 43:82–88
2. Kappagantu M, Collum TD, Dardick C et al (2020) Viral Hacks of the plant vasculature: the role of phloem alterations in systemic virus infection. Annu Rev Virol 7:10.1–10.20
3. King AM, Adams MJ, Lefkowitz E (2011) Virus taxonomy: ninth report of the International Committee on Taxonomy of Viruses. Elsevier, San Diego, CA, USA
4. Jimenez J, Moreno A, Fereres A (2020) Transmission of phloem-limited viruses to the host plants by their aphid vectors. In: Progress in botany, pp 357–382. Springer, Cham.
5. Lv MF, Xie L, Song XJ et al (2017) Phloem-limited reoviruses universally induce sieve element hyperplasia and more flexible gateways, providing more channels for their movement in plants. Sci Rep 7: 27128
6. Lv MF, Dai YX, Xie L et al (2024) Volume electron microscopy reconstruction uncovers a physical barrier that limits virus to phloem. New Phytol 241:343–362
7. Truernit E (2014) Methods of phloem visualization: a clear future in sight?. Methods Mol Biol 2014:73–79
8. Furuta KM, Yadav SR, Lehesranta S et al (2014) Plant development. Arabidopsis NAC45/86 direct sieve element morphogenesis culminating in enucleation. Science 345:933–937
9. Ross-Elliott TJ, Jensen KH, Haaning KS et al (2017) Phloem unloading in Arabidopsis roots is convective and regulated by the phloem-pole pericycle. Elife 6:e24125
10. Li JH, Li QL, Zhao et al (2022) Rapid screening of Zr-containing particles from Chang'e-5 lunar soil samples for isotope geochronology: technical roadmap for future study. Geosci Front 13:101367
11. Lu Y, Liu J, Li B et al (2024) Spatial patterns of noise-induced inner hair cell ribbon loss in the mouse mid-cochlea. iScience 27:108825
12. Carlson DB, Gelb J, Plashin V, Evans JE (2013) Laboratory-based cryogenic soft X-ray tomography with correlative cryo-light and electron microscopy. Microsc Microanal 19:22–29
13. Seligman AM, Wasserkrug HL, Hanker JS (1966) A new staining method (OTO) for enhancing contrast of lipid-containing membranes and droplets in osmium tetroxide-fixed tissue with osmiophilic thiocarbohydrazide (TCH). JCB 30:424–432
14. Hua YF, Laserstein P, Helmstaedter M (2015) Large-volume en-bloc staining for electron microscopy-based connectomics. Nat Commun 6: 7923
15. Gillies AR, Bushong EA, Deerinck TJ et al (2014) Three-dimensional reconstruction of skeletal muscle extracellular matrix ultrastructure. Microsc Microanal 20:1835 1840
16. Bell K, Mitchell S, Paultre D et al (2013) Correlative imaging of fluorescent proteins in resin-embedded plant material. Plant Physiol 161:1595–1603
17. Hunziker P, Schulz A (2014) Transmission electron microscopy of the phloem with minimal artifacts. Methods Mol Biol 2014: 17–27
18. Suuroen JP, Jyske T (2014) Noninvasive investigation of phloem structure by 3D synchrotron X-ray microtomography. Methods Mol Biol 2014:37–54
19. Denk W, Horstmann H (2004) Serial block-face scanning electron microscopy to reconstruct three-dimensional tissue nanostructure. PLoS Biol 2:1900–1909.
20. Zhang LY, Hu C, Wu SQ et al (2017) An improved non-local means image denoising algorithm. 2017 IEEE International Conference on Information and Automation (ICIA). https://doi.org/10.1109/ICInfA.2017.8079010
21. Stringer C, Wang T, Michaelos M et al (2021) Cellpose: a generalist algorithm for cellular segmentation. Nat Methods 18:100–106
22. Feng XP, Yu ZY, Fang H et al (2023) Plantorganelle Hunter is an effective deep-learning based method for plant organelle phenotyping in electron microscopy. Nat Plants 9:1760–1775

Chapter 16

Go with the Phloem: A Review of the Methods to Collect Plant Phloem Tissue

Hannah Levengood, Juliana Miranda, Samantha Barker, Maria Fernanda Ovalle and Cankui Zhang

Abstract

The study of plant phloem tissue is important for understanding plant physiology, yet the anatomy and small size of the phloem tissue make direct sampling challenging. Due to the absence of a single method that allows for phloem extraction in every species, researchers have developed several strategies to address this drawback. These strategies can be categorized into three groups: classical, advanced, and mechanical extraction methods. Classical methods, including cucurbit drip, EDTA-facilitated exudation, stylectomy, and heterografting, have a long history in phloem research but are generally less precise. Advanced methods, such as TRAP-seq, FACS, scRNA-seq, and LCM, leverage recent technological advancements to provide precise measurements, with the drawback of requiring specialized expertise and costly reagents. Mechanical methods focus on species from which phloem can be directly sampled, *Plantago* and celery. This review discusses these methods, detailing their advantages and disadvantages, and how they can be applied to current phloem studies.

Keywords Carbon allocation, Sugar transport, Phloem exudate, TRAP-seq, Grafting, *Plantago*

1 Introduction

Researchers understand that the vasculature, xylem, and phloem play an essential role in regulating how a plant responds to environmental pressures. While both tissues contribute to water and nutrient allocation, the phloem is particularly important for mediating stress response pathways, including those triggered by pests, diseases, nutrient deficiencies, and other challenges. Investigations into the plant vasculature have revealed key genes and pathways that underlie these responses, offering valuable insights for developing more resilient crop varieties. Therefore, the phloem represents a significant tissue of study for many plant biologists.

The first step of many investigations into the phloem involves the extraction of phloem tissue or the collection of phloem sap. In

Johannes Liesche and Sanna Sevanto (eds.), *Phloem*, Methods in Molecular Biology, vol. 3066,
https://doi.org/10.1007/978-1-0716-5488-0_16,

most species, mechanically isolating pure phloem is challenging due to its complex anatomy. Samples contaminated with cells from the mesophyll or other surrounding tissues are undesirable for molecular, metabolic, or physiological analysis, as they often lead to a misinterpretation of the results. Pure tissue extraction is therefore key to an accurate analysis. Scientists have developed several methods that achieve this purpose.

Many of the known methods for phloem collection have been used reliably in plant vascular research; however, not all methods are appropriate for every plant or study. Therefore, it is important to list the known phloem collection methods and highlight their advantages and disadvantages, so that researchers are better equipped to choose the most appropriate method for their research. The purpose of this chapter is to dissect the different methods of phloem collection that are available. These methods are arranged according to the classifications that are most commonly used in the literature, including classical approaches and advanced techniques. In addition, a few methods for mechanical extraction, specific to celery and species in the *Plantago* genus, are also described. These approaches are detailed in Table 1.

2 Classical Methods

In the early days of plant vascular biology research, researchers implemented several methods to collect phloem tissue. Due to the lack of availability of modern techniques like sequencing or reliable plant transformation, many of these techniques were based on specific biological features endemic to each plant. For example, the cucurbit drip method was based on the observation that some plants, especially those in the *Cucurbita* genus, spontaneously exude phloem sap when cut, allowing the sap to be collected. Not all plants have this feature, as P-protein and callose rapidly seal the incisions. This method has been modified with the addition of EDTA, which chelates the sealing elements and prevents their action, allowing the phloem to be collected. Another example is the aphid stylet collection, which represents a more precise collection method, leveraging a feature of insects that feed directly from the phloem. Over time, however, it became clear that each of these methods contains contaminants that compromise the purity or accuracy. Fortunately, experiments can be designed carefully to optimize their use. Grafting represents a classical technique that has been combined with advanced technologies to facilitate an alternative phloem analysis method.

2.1 Cucurbit Drip

The cucurbit drip method has a long history of being used to study plant vasculature. Several plant species, including cucurbits [1], spontaneously exude phloem sap when a cut is made in their bark

Table 1
An overview of current phloem extraction methods, including their advantages and disadvantages

Method	Description	Advantages	Disadvantages
Cucurbit Drip	Phloem sap can be collected from spontaneously exuding plants with a microcapillary	Simple, inexpensive Allows for the collection of a large amount of material	Not applicable to plants that do not spontaneously exude phloem sap Wounding often contaminates phloem sap
EDTA	EDTA inhibits callose synthesis and P-proteins, allowing for a species-independent method for phloem exudation and collection	Simple, inexpensive Species-independent	Requires an artificial collection environment EDTA can contaminate the purity of the sample with the tissue of surrounding cells
Stylometry	The heads of phloem-feeding insects are severed during feeding, allowing for precise sampling of phloem sap	Precise	Risk of contamination from surrounding cells and insect saliva Technically challenging Requires plant–insect compatibility
Heterografts	The stem of one plant is joined to the rootstock of another, then sequenced to identify exogenous signals	Useful for studying signal mobility. Applicable to monocots and dicots	Contamination can be caused by sampling too close to the graft junction Not all species are graft compatible
TRAP	A technique that requires transgenic plants with the desired cell or tissue-specific tagged ribosomes. Cells are collected by immunoprecipitation and then combined with RNA-sequencing for further analysis	Highly precise capture of vasculature-specific mRNA without contamination	Requires transgenic plants Requires strong promoters Technically demanding, expensive
FACS	A technique of flow cytometry that allows for fluorescently labeled cells to be separated from non-labeled cells of interest	Allows for the collection of any cell type, given that the appropriate promoters are available	Requires a reliable protoplast protocol Requires transgenic plants Enzymatic digestion of cell walls to create protoplasts can misrepresent results

Table 1 (Continued)

Method	Description	Advantages	Disadvantages
scRNA-seq	Sharing similarities with both TRAP-seq in sequencing depth and FACS requirement for protoplasts, this allows for the mRNA of individual cells to be analyzed	The transcriptomics of individual cells can be visualized Does not require transgenic plants	Expensive Technically demanding Requires a reliable protoplast protocol Enzymatic digestion of cell walls to create protoplasts can misrepresent results
LCM	A microscope-based technique that uses an infrared laser to isolate individual cells or tissues. Cells are adhered to a film or membrane and then are isolated from the film for further analysis	Species independent Precise, clean extraction Transgenic plants are not needed	Requires expensive equipment Requires expertise to recognize specific cells Challenging in woody species miRNAs are difficult to recover
Plantago	Pure phloem tissue can be mechanically extracted from the leaves and petiole, due to a layer of endodermal tissue that surrounds the vascular bundle	Simple and efficient Allows for the collection of a large amount of material Plantago is emerging as a new model organism for vasculature studies	Limited to *Plantago* species More difficult to isolate phloem from seedlings Genomic resources exist, but lack functional annotation of other model species
Celery	Phloem can be isolated from petiole tissue due to a structural layer of collenchyma tissue, exposing the xylem and phloem	Cost-effective, simple Xylem and phloem can be individually separated	Celery is unsuitable for gene functional analysis Vasculature cannot be isolated from leaf tissue

Traditional methods include: cucurbit drip, EDTA, stylometry, and Heterografting. Advanced methods include TRAP, FACS, scRNA-seq, and LCM. Methods of mechanical extraction include *Plantago* and Celery

or petiole. Large amounts of this sap can then be collected using a microcapillary [2]. This method can be ideal for researchers studying proteomics and/or metabolomics, or for experiments that require large amounts of starting material to facilitate DNA extraction. For researchers studying plants with this feature, the cucurbit drip method offers a cheap and simple option for phloem collection, which can be advantageous in labs that have limited funding or lack existing infrastructure for the other methods.

The most obvious disadvantage of this method is that it is not applicable to plants that do not spontaneously exude their phloem sap when cut. Since this category includes many of the most used models, like Arabidopsis, this can serve as a major limitation, especially for laboratories that are unable to adapt to new research organisms. Common features of plants that spontaneously exude can include long life cycles, lack of a sequenced genome, or recalcitrance to plant tissue culture or transformation, making them unrealistic models for experimentation. Another concerning drawback of this method is the discovery that the wounding step required for the exudation could contaminate samples with material from surrounding cells, therefore compromising the purity of the sample [1]. While this concern can be mitigated with careful analysis, an alternative mechanism to study phloem may be preferred, especially for studies where sample purity is essential.

While the cucurbit drip method lacks the species independence of other discussed techniques, the low cost and ease of extracting phloem tissue from applicable species make it practical to consider during experimental design. Although this method has been cited as ineffective for capturing immobile signaling components, it is great to obtain mobile phloem signals, making it a valuable technique for long-distance or systemic signaling studies. In cucurbits, sieve elements exist in the vasculature bundles, contributing both internal (vasculature-bundle-associated) and extracellular (entocyclic/ecotocyclic-associated) sap to the sample. Such contamination issues can be mitigated by sampling directly from the smaller droplets that are thought to directly come from vasculature bundles [1]. This can be followed by careful interpretation of experimental results, like adding a step to confirm the presence of compounds specific to the phloem sap, along with the confirmation or absence of markers from surrounding tissues. This will help to verify the origin and purity of the collected samples.

2.2 EDTA

As previously stated, the vasculature of many plant species cannot be directly sampled by using the cucurbit drip method. Callose synthesis proteins work to seal any wound, while P-proteins function to seal off wounded sieve elements of vasculature tissue, which effectively prevents phloem exudation. EDTA-facilitated extraction exists as a solution to this problem. EDTA chelates calcium ions, needed for callose and P-protein action, enabling phloem sap collection from more plant species than the cucurbit drip method. In most protocols, the petioles of leaves are cut and immersed in an EDTA solution and then incubated in the dark for several hours, after which sap can be obtained. Like the cucurbit drip method, EDTA-facilitated extraction is simple, efficient, and well-suited for obtaining large amounts of sap [3]. Because the EDTA solution makes the technique species-independent, it has become especially attractive for many studies [3, 4].

While the EDTA-facilitated extraction method has been popular for many studies that investigate phloem tissue, several concerns have been highlighted that should be considered when using this method [5]. The artificial conditions required for optimal sample collection limit the scope of experiments that can be conducted. They also have the potential to trigger stress responses, which may unintentionally alter the genetic expression and signaling pathways researchers aim to analyze under a given treatment [6]. Additionally, there is evidence that the EDTA solution may impact the purity of the sample, as the chelation of calcium ions may also allow the exudation of surrounding cells, causing contamination [7]. Together, these limitations suggest that EDTA-facilitated extraction may not always provide the most accurate or reliable measurements of phloem composition.

Although researchers are recommended to use caution when interpreting results due to the limitations of the EDTA method, the low cost and convenience, combined with the species-independence of this technique, make it a useful approach to exudate extraction. Studies have shown that meaningful results can still be obtained through altering the procedure so that cellular damage is mitigated by shortening the EDTA incubation times and collecting sap into distilled water [4]. However, even with adjustments, EDTA-facilitated sampling may not be suitable for all applications. For example, experiments that demand rapid sample collection, such as those that deal with RNA, which is easily degraded, or operate on specific time scales, are not suited for this method. Therefore, for these experiments, a different extraction method should be used.

2.3 Stylometry

As previously stated, both the cucurbit drip and the EDTA-facilitated techniques are attractive for their simplicity and low cost. However, each procedure has the major drawback of the possibility of introducing contaminants that compromise sample purity. In contrast, stylometry exists as the most precise and nondestructive classical method of sampling phloem [8]. Researchers discovered that phloem-feeding insects often leverage the internal turgor pressure of a plant to feed on the phloem sap, as opposed to using their own suction. By severing the stylets of these insects during feeding, the sap continues to flow, allowing for the phloem to be directly sampled [9].

While effective in mediating precise sampling, stylometry presents multiple challenges that make it unsuitable for several types of phloem analysis. The primary concern lies in both the technical difficulty of the procedure and, still, the risk of contamination. Severing insect stylets is particularly challenging due to their small size, which makes consistent execution difficult. While phloem sap has traditionally been regarded as compositionally pure, recent proteomic analyses reveal that it is prone to contamination from insect salivary proteins [10] and other artifacts [6]. Because sap collected during this method may also contain plant defense compounds in response to the feeding of the insects, this may be a concern if the research interest is not on plant–insect interactions [2]. Plant–insect compatibility and the restriction to young tissue pose additional limitations, as plant defenses and insect feeding patterns can limit applicability [5]. Furthermore, only a few microliters of sap can be collected, making the method unsuitable for analyses requiring large sample volumes.

Although the challenges presented do contribute to stylometry having limited viability for all types of phloem analysis, it can be an advantageous tool for specific uses. When applied to the study of plant–insect interactions, such as plant responses to insect feeding, this method can still yield valuable insights that inform strategies for developing insect-resistant crops. Another consideration is that it provides the advantage of collecting phloem sap with minimal plant damage, making the method attractive for experiments where destructive sampling is undesirable. However, the potential for sample impurities should always be considered when interpreting results.

2.4 Heterografts

Grafting represents another classical technique that has been repurposed for phloem research, especially for the study of mobile signaling. While the direct extraction of phloem tissue is not possible, meaningful insights into the vasculature have been obtained using this system. Traditionally used for agronomic improvement or the cultivation of fruits in woody crops, heterografting has evolved into a useful tool for the investigation of multiple aspects of plant biology. Heterografts are created by attaching the stem of one type of plant or species to the rootstock of another [11]. Depending on the context of the experiment, samples can be taken from the plant after treatment and then sequenced to identify the presence of exogenous signals. This technique has been especially useful to study mobile signals, including those that are involved in nutrient-limiting [12, 13] or mineral-deficient conditions [14]. Although most studies typically use dicotyledonous species for heterografting purposes, monocot grafting protocols have also been developed [15], suggesting that similar insights could be gained using these plants.

While heterograft systems provide researchers with a robust mechanism for studying the mobile signals that exist between species or genotypes, several points should be considered before they can be effectively adapted to a given study. Additionally, not all combinations of species are compatible, as several species may have difficulty in forming a vascular connection between the rootstock and shoot, which is crucial to graft establishment [11]. Closely related species or accessions are also not ideal, as it can be difficult to identify exogenous signals due to their genetic similarity [16]. Sampling strategy is another important consideration; collecting tissue too close to the graft junction can misrepresent mobile components due to potential graft contamination.

Many of these challenges can be addressed through careful experimental design. For example, the mobility of the identified signals can be verified with RNAseq reads when identifying mobile mRNAs [17], and graft contamination can be minimized by sampling further from the junction. With these precautions in place, heterograft systems remain advantageous for investigating long-distance signaling and associated mobile signals.

3 Advanced Methods

While classical sap collection methods have their advantages, their limitations, especially involving species specificity and sample purity, limit their application across all types of experimental conditions. With modern advances in plant transformation and sequencing, several techniques have been developed to facilitate more precise vasculature tissue analysis. However, the practicality of performing these experiments is limited by the need for specialized training, expensive equipment, and even the need for transgenic plants. The following section summarizes these techniques, termed "advanced" in the literature.

3.1 Translating Ribosome Affinity Purification Sequencing (TRAP-Seq)

The study of plant vasculature can be improved by examining how genes and proteins are regulated in response to experimental conditions. While traditional transcriptomic analyses of the total mRNA population of a cell allow for the characterization of plant responses, post-transcriptional regulations mean that not all transcripts are translated into proteins, which can misrepresent interpretations. Translating ribosome affinity purification (TRAP) can be combined with RNA sequencing (TRAP-seq) to overcome this drawback by isolating only the mRNAs that are actively associated with tagged ribosomes. This provides a direct snapshot of what is actively being translated, called the translatome. By attaching a promoter-driven tag to ribosomes, TRAP enables highly specific isolation of tissues or cell types whenever a suitable promoter is available. Tagged ribosomes are then isolated by immunoprecipitation with an anti-tag, allowing the rescue of tissue-specific mRNAs. Since several phloem-specific promoters have been identified, TRAP is effective for the precise isolation of pure phloem tissue, which has enabled its success when applied to vasculature research [2, 6, 18, 19].

Although TRAP allows for effectively isolating pure vasculature mRNA without damage, it is not appropriate for all experiments. It requires the generation of transgenic plants expressing the tagged ribosomal proteins, making the method unsuitable for species recalcitrant to transformation. The technique is also dependent on the strength and specificity of the promoter driving ribosomal protein expression. Weak or transient promoter activity can reduce yield, sometimes below levels needed for downstream analysis. Additionally, the procedure required to capture the ribosomes is long and requires expensive affinity reagents. As both the expertise and the chemicals may not be accessible to all researchers, a different extraction method is recommended in situations where these issues may arise.

Despite these challenges, TRAP-Seq offers unique advantages that make it a powerful tool for phloem research. By providing a direct assessment of the translational activity in specific cell types without contamination [19–21], it allows researchers to investigate the functional translatome under various conditions. As genomic and genetic resources for target species continue to expand, TRAP-Seq is poised to remain an invaluable technique for studying plant vasculature and gaining insights that might be missed by conventional transcriptomic analyses.

3.2 Fluorescence Activated Cell Sorting (FACS)

Fluorescence-activated cell sorting (FACS) is a type of flow cytometry that allows for fluorescently labeled cells to be separated from non-labeled cells. Experimentally, FACS requires transgenic plant lines containing fluorescently labeled cells or tissues of interest to be grown under the desired experimental conditions. Next, the tissues are enzymatically digested to produce protoplasts and are subsequently sorted using a flow cytometer. Due to the versatility of this method, which can sort virtually any cell given the presence of a cell-specific promoter, it holds strong potential for applications in vascular biology studies. As a result, several studies have successfully used FACS to mediate vasculature analysis [22, 23].

Despite the advantage of having high specificity for isolating vasculature tissues and cells, several concerns have been raised about adapting FACS to different plant systems. Due to its reliance on fluorescence-based sorting, it requires transgenic lines with tissue-specific reporters, which are only readily available for a few model species, such as Arabidopsis [6], while other plants demand custom line development and reliable plant transformation systems [24]. The need to generate protoplasts also yields challenges, as enzymatic digestion of cell walls can stress or damage cells and alter gene expression [25]. Researchers can reduce these effects by using healthy tissues, shortening isolation times, or using more plant material when stress exposure is needed [23]. Additionally, the most common fluorescent markers, including GFP and YFP, can overlap with the autofluorescence of photosynthetic tissues, but this can be addressed by using band-pass emission filters or by applying proper controls.

Due to the issues described above, FACS is most appropriate for widely used plant model systems, such as Arabidopsis, where transgenic GFP lines are readily available. In other, less-used species with established transformation systems, FACS can also be used, although it requires additional effort to establish GFP lines and optimized protoplasting techniques. However, for plant systems that are recalcitrant to transformation, an alternative phloem sampling method is recommended to be used.

3.3 Single-Cell Sequencing (scRNA-seq)

As mentioned above, techniques that allow for the extraction of actively translated mRNA can yield valuable insights, especially for tissues like the phloem, which are traditionally difficult to isolate. However, the requirement of transgenic plants for both TRAP-seq and FACS limits this application to plant species with a reliable transformation system. Similar to both TRAP-seq and FACS, single-cell RNA sequencing (scRNA-seq) is a tool that allows for a deeper resolution of plant tissue on a single-cell level [26]. Plant tissue is enzymatically digested to release protoplasts from cell walls, individually labeled with barcoding beads, and analyzed through RNA sequencing. Plant cells vary in size and shape depending on their type and developmental stage, which highlights the importance of enhanced cellular resolution to capture the nuanced and subtle changes in individual cells [27]. Therefore, applying scRNA-seq to phloem studies could provide valuable insights into complex biological processes, and has been used as a tool for some phloem investigations [28–30].

Even when considering the advantages gained by the high resolution of scRNA-seq, the technique is currently not applicable to all studies. Like FACS, scRNA-seq requires the generation of cell wall–free protoplasts, thus sharing the same limitations of protoplast-based approaches, including restricted applicability to non-model species and the potential for experimental artifacts [26]. Similarly, scRNA-seq shares with TRAP the disadvantages of a lengthy protocol demanding costly reagents in addition to advanced technical expertise. In this respect, it can be even more expensive than TRAP, as individually labeling each cell with a barcode consumes more cells and reagents [27].

Many studies have recognized that the various challenges associated with scRNA-seq mean that it is most suitable for studying specialized cells in model organisms, such as *Arabidopsis* [26, 31]. However, the unique composition of phloem tissue, containing several specialized cell types that each participate in essential biological processes, suggests that scRNA-seq may still be valuable for certain phloem studies. For example, recent work has shown that companion cells play a key role in regulating plant responses to nutrient deficiencies [32]. Achieving single-cell resolution of these cells under varying mineral conditions could enable further insights into their intricate biological functions.

3.4 Laser Capture Micro-Dissection (LCM)

While the other discussed methods represent advanced systems for evaluating relatively pure plant phloem tissue, they both suffer from the constraint of requiring organisms that can be genetically modified, which limits their applicability. Additionally, the requirement of protoplasts for FACS and scRNA-seq also reduces its application in most plants, making it inappropriate for species that are incompatible with that step. For studies that require precision, using plants where transformation and/or protoplast techniques are not available, laser capture microdissection (LCM) exists as an alternative for contamination-free phloem extraction. LCM is a microscope-based technique that involves using an infrared laser to isolate individual cells or tissues. The tissue or cells of interest are identified, cut, and adhered to film or membranes using the laser, and can then be isolated from the film for downstream analysis. LCM is typically used for genomic and transcriptomic analysis at specific tissue or cellular resolution [33] and is broadly applicable across species [34], provided the tissues can be reliably identified by an expert. Experimentally, this technique has been widely applied to study the phloem, including processes related to development [35] and responses to various types of stress, including pathogens [36], obligate biotrophs [37, 38], and symbiotic bacteria [39].

Although broadly applicable, LCM may not be easily implemented in all research programs. The procedure is time-consuming, and the equipment is costly and requires significant expertise to operate and to identify distinct cell populations within plant tissues [40]. While LCM is generally species-independent, it can be particularly challenging in woody species due to their different anatomy, necessitating additional optimization. Most cell types and macromolecules can be effectively isolated with LCM, but high-quality RNAs, including miRNAs, are difficult to recover due to ribonuclease (RNase) activity and require further refinement [41]. Because miRNAs play a crucial role in vascular signaling, studies targeting this area may need alternative isolation methods.

Even when considering its constraints, LCM remains a useful tool for the analysis of plant vasculature. The advantage of cleanly extracting cells or tissue without being limited by the size of the tissue is particularly useful for phloem studies, where the small size of the tissue makes it particularly difficult to extract. Additionally, because LCM does not involve transgenic plants, it exists as a precise option for isolating phloem tissue from plants that are difficult to transform.

4 Mechanical Extraction

As previously stated, many of the described techniques that mediate vasculature extraction were created to overcome the difficulties that exist in direct sampling of the phloem in most species. However, a few species do exist where the mechanical extraction of phloem tissue is possible, without the need for advanced tools. Both celery and species of the *Plantago* genus contain unique anatomical structures that allow pure vasculature tissue to be obtained quickly and easily when compared to other extraction methods. While celery may not be ideal as a model due to the lack of genetic resources, species in the *Plantago* genus have been used in multiple areas of plant vascular biology research [42] and have been suggested as new model organisms for plant vascular biology research [2, 24]. These methods are described below.

4.1 Plantago

Species within the genus *Plantago* have recently been recommended as new plant research organisms [2, 23, 24]. This recommendation is supported by a growing breadth of studies using it as a model. In addition, for some species within the genus, sequencing resources are available [43–45], genetic transformation systems have been established [46, 47], and our group has also shown that the species is compatible with gene editing [48]. Important for plant vasculature research, several species in this genus have vasculature tissue that can be easily mechanically extracted due to the presence of a continuous layer of endodermal tissue that surrounds the vasculature bundles. When the leaf or petiole tissue of these species is broken, the Casparian stripes in the endodermis are ruptured, leaving pure vascular tissue, which can be collected and used in downstream analysis [46]

The mechanical method of *Plantago* vasculature tissue extraction represents a valuable technique. However, it may not be suitable for all vasculature studies. The age of the plant can contribute to the ease of obtaining the vascular tissue. Pulling the petiole tissue from seedlings or young plants can be challenging, as the tender tissues are more delicate, requiring more care to obtain undamaged phloem tissue. Compared to other long-standing models like Arabidopsis, the genomic resources for *Plantago* remain limited. For example, although a chromosome-level genome assembly does exist [43], it has not been functionally annotated, making experiments that rely on greater data availability more difficult.

Even with these limitations, the ease of isolating undamaged vasculature makes *Plantago* a valuable system for phloem research. Although less studied than traditional model species, mounting evidence suggests that findings in this genus can have broad relevance across the plant kingdom [2, 49]. As more researchers recognize the potential of *Plantago*, resources specific to this genus are expected to expand. Taken together, the combined research advantages and the practical advantages of using *Plantago* to study vasculature establish it as a compelling model for future phloem-related studies.

4.2 Celery

Celery (*Apium graveolens*) has traditionally been used in plant cell and structural analyses. It also represents an appropriate system for some phloem studies because of its vascular tissue, which can be mechanically extracted, due to its unique anatomical structure. The petiole contains a structural layer of collenchyma that can be peeled away to expose the vascular bundles, from which intact phloem and xylem can then be separated using a scalpel. The main advantages of this technique are cost-effectiveness and simplicity, making it an attractive alternative for laboratories with limited resources. Although celery has not been used often in phloem research, some studies have taken advantage of this feature for phloem isolation and analysis [50–54].

For basic studies investigating plant phloem tissue, celery may be an appropriate model. However, it presents a challenge when used for more advanced functional and genomic analyses. Resources such as sequenced genomes [55], gene editing [56], and genetic engineering [57] protocols have been reported in the species, but these resources are often unreliable. As celery is open-pollinated, producing highly heterozygous progeny, and contains a large genome (~3.3 Gb), it is not recommended for studies focusing on gene functional analysis. Additionally, although both phloem and xylem can be isolated from petiole tissue, the vasculature cannot be isolated from the leaves, which can limit its applicability. As a result, unlike *Plantago*, celery is not recommended as a model organism for studies investigating the leaf vasculature.

Although celery is not recommended for any in-depth molecular analysis, the ability to quickly and easily extract undamaged vasculature warrants its inclusion in this review. Existing challenges involving its large genome are difficult to address, although transformation challenges could easily be reduced. Several publications involving celery have shown that it is relatively easy to propagate in tissue culture [56, 57]. Micropropagation of genotypes known to be transformation-compatible could yield standardized material for research and overcome existing transformation limitations. For studies that do not focus on functional genomics and require relatively pure tissue, celery may be a good alternative to other methods.

5 Conclusions

The essential role of the phloem tissue, besides nutrient and metabolite transport, is facilitating the mobilization of key defenses and physiological responses to environmental constraints, making its investigation important in the area of plant biology. The difficulty of extracting phloem has prompted researchers to create a few reliable methods for obtaining tissues. As purity is a concern for many experiments, the most appropriate method for each investigation can vary based on the laboratory resources, analyses required, and the plant species used. In this chapter, we have highlighted the advantages and disadvantages of each method in order to better inform researchers in their experimental design.

Many classical methods, based on leveraging distinct biological characteristics of certain plants, are useful because they are cheap and relatively simple. However, despite their convenience, the purity of the samples is often sacrificed, making them not suitable for molecular or metabolic analyses where this feature is critical. Advanced methods solve the purity bottleneck, allowing for high specificity on a tissue or even cellular resolution. However, the high expense and requirement for transgenic plants could be limiting factors. Mechanical methods of extraction serve as ideal methods to gain high sample purity without being cost-prohibitive. However, the limited genetic resources for both *Plantago* and celery can make their methods more labor-intensive, although this is expected to be improved as interest continues to increase for these methods.

Taken together, the methods described offer a variety of approaches and an extensive toolkit useful for any experiment that necessitates phloem extraction. As the phloem tissue is critical for the management of important plant processes, gaining authentic results is vital for generating new discoveries in plant biology. The practical application of these results has the potential to bolster the food security of plants, creating crops that use nutrients more efficiently, are more resistant to diseases, and are more adaptable to a changing environment. Ultimately, these approaches not only deepen our understanding of phloem biology but also pave the way for innovations that strengthen agricultural resilience and sustainability.

Author Contributions

HL, SB, JM, and MO wrote the paper. HL and CZ revised the paper.

Funding This work was supported by the National Science Foundation (EDGE IOS-1923557) and the Agriculture and Food Research Initiative, project award No. 2023-67013-39441, from the U.S. Department of Agriculture's National Institute of Food and Agriculture.

Data Availability No datasets were generated or used for this review paper.

Conflict of Interest The authors declare that they have no competing interests.

References

1. Zhang C, Yu X, Ayre BG, Turgeon R (2012) The origin and composition of cucurbit "Phloem" exudate1[OA]. Plant Physiol 158:1873–1882. https://doi.org/10.1104/pp.112.194431
2. Huang J, Huang Z, Zhou X et al (2019) Tissue-specific transcriptomic profiling of Plantago major provides insights for the involvement of vasculature in phosphate deficiency responses. Mol Genet Genomics 294:159–175. https://doi.org/10.1007/s00438-018-1496-4
3. Xu Q, Ren Y, Liesche J (2019) Studying phloem loading with EDTA-facilitated phloem exudate collection and analysis. In: Liesche J (ed) Phloem: methods and protocols. Springer, New York, NY, pp 125–133
4. Killiny N (2019) Collection of the phloem sap, pros and cons. Plant Signal Behav 14:1618181. https://doi.org/10.1080/15592324.2019.1618181
5. Xia C, Miranda J, Mendoza-Cozatl D et al (2025) Decoding long-distance communication

under mineral stress: advances in vascular signalling and molecular tools for plant resilience. Plant Cell Environ 48:4781. https://doi.org/10.1111/pce.15475

6. Guelette BS, Benning UF, Hoffmann-Benning S (2012) Identification of lipids and lipid-binding proteins in phloem exudates from *Arabidopsis thaliana*. J Exp Bot 63:3603–3616. https://doi.org/10.1093/jxb/ers028
7. Ogden AJ, Bhatt JJ, Brewer HM et al (2020) Phloem exudate protein profiles during drought and recovery reveal abiotic stress responses in tomato vasculature. Int J Mol Sci 21:4461. https://doi.org/10.3390/ijms21124461
8. Dinant S, Bonnemain J-L, Girousse C, Kehr J (2010) Phloem sap intricacy and interplay with aphid feeding. C R Biol 333:504–515. https://doi.org/10.1016/j.crvi.2010.03.008
9. Kennedy JS, Mittler TE (1953) A method of obtaining phloem sap via the mouth-parts of aphids. Nature 171:528–528. https://doi.org/10.1038/171528a0
10. Pan C, He X, Xia L et al (2024) Proteomic analysis of salivary secretions from the tea green leafhopper, *Empoasca flavescens* fabrecius. Insects 15:296. https://doi.org/10.3390/insects15040296
11. Ferreira Torres L, López De Andrade SA, Mazzafera P (2021) Split-root, grafting and girdling as experimental tools to study root-to shoot-to root signaling. Environ Exp Bot 191:104631. https://doi.org/10.1016/j.envexpbot.2021.104631
12. Notaguchi M, Higashiyama T, Suzuki T (2015) Identification of mRNAs that move over long distances using an RNA-seq analysis of Arabidopsis/*Nicotiana benthamiana* heterografts. Plant Cell Physiol 56:311–321. https://doi.org/10.1093/pcp/pcu210
13. Thieme CJ, Rojas-Triana M, Stecyk E et al (2016) Correction: corrigendum: endogenous Arabidopsis messenger RNAs transported to distant tissues. Nat Plants 2:16195. https://doi.org/10.1038/nplants.2016.195
14. Xia C, Huang J, Lan H, Zhang C (2020) Long-distance movement of mineral deficiency-responsive mRNAs in *Nicotiana Benthamiana*/tomato heterografts. Plants 9:876. https://doi.org/10.3390/plants9070876
15. Reeves G, Tripathi A, Singh P et al (2022) Monocotyledonous plants graft at the embryonic root–shoot interface. Nature 602:280–286. https://doi.org/10.1038/s41586-021-04247-y
16. Xia C, Zheng Y, Huang J et al (2018) Elucidation of the mechanisms of long-distance mRNA movement in a *Nicotiana benthamiana* /tomato heterograft system. Plant Physiol 177:745–758. https://doi.org/10.1104/pp.17.01836
17. Paajanen P, Tomkins M, Hoerbst F et al (2025) Re-analysis of mobile mRNA datasets raises questions about the extent of long-distance mRNA communication. Nat Plants 11:977–984. https://doi.org/10.1038/s41477-025-01979-x
18. Mustroph A, Zanetti ME, Jang CJH et al (2009) Profiling translatomes of discrete cell populations resolves altered cellular priorities during hypoxia in Arabidopsis. Proc Natl Acad Sci U S A 106:18843–18848. https://doi.org/10.1073/pnas.0906131106
19. Reynoso MA, Blanco FA, Bailey-Serres J et al (2013) Selective recruitment of mRNA s and miRNA s to polyribosomes in response to rhizobia infection in *Medicago truncatula*. Plant J 73:289–301. https://doi.org/10.1111/tpj.12033
20. Heiman M, Kulicke R, Fenster RJ et al (2014) Cell type–specific mRNA purification by translating ribosome affinity purification (TRAP). Nat Protoc 9:1282–1291. https://doi.org/10.1038/nprot.2014.085
21. Thellmann M, Andersen TG, Vermeer JE (2020) Translating ribosome affinity purification (TRAP) to investigate *Arabidopsis thaliana* root development at a cell type-specific scale. J Vis Exp. https://doi.org/10.3791/60919
22. Galbraith DW, Anderson MT, Herzenberg LA (1998) Chapter 19: flow cytometric analysis and FACS sorting of cells based on GFP accumulation. In: Methods in Cell Biology. Elsevier, pp 315–341
23. Iyer-Pascuzzi AS, Benfey PN (2010) Fluorescence-activated cell sorting in plant developmental biology. In: Hennig L, Köhler C (eds) Plant developmental biology: methods and protocols. Humana Press, Totowa, NJ, pp 313–319
24. Levengood H, Zhou Y, Zhang C (2024) Advancements in plant transformation: from traditional methods to cutting-edge techniques *and* emerging model species. Plant Cell Rep 43:273. https://doi.org/10.1007/s00299-024-03359-9
25. Reyna-Llorens I, Ferro-Costa M, Burgess SJ (2023) Plant protoplasts in the age of synthetic biology. J Exp Bot 74:3821–3832. https://doi.org/10.1093/jxb/erad172
26. Bawa G, Liu Z, Yu X et al (2022) Single-cell RNA sequencing for plant research: insights and possible benefits. Int J Mol Sci 23:4497. https://doi.org/10.3390/ijms23094497
27. Luo M, Cao Y, Hong J (2025) Opportunities and challenges in the application of single-cell transcriptomics in plant tissue research. Physiol

Mol Biol Plants 31:199–209. https://doi.org/10.1007/s12298-025-01558-6
28. Kim J-Y, Symeonidi E, Pang TY et al (2021) Distinct identities of leaf phloem cells revealed by single cell transcriptomics. Plant Cell 33:511–530. https://doi.org/10.1093/plcell/koaa060
29. Liu J, Majeed A, Mukhtar MS (2022) Rooting through single-cell sequencing in phloem pole cells. Commun Biol 5:1194. https://doi.org/10.1038/s42003-022-04180-4
30. Liu X, Roszak P, Helariutta Y (2024) The challenge of defining rare genetic programs by single-cell RNA sequencing: Insights from phloem studies. Mol Plant 17:22–25. https://doi.org/10.1016/j.molp.2023.12.012
31. Islam MT, Liu Y, Hassan MM et al Advances in the application of single-cell transcriptomics in plant systems and synthetic biology. Biodes Res 6:0029. https://doi.org/10.34133/bdr.0029
32. Xia C, Huang J, Zhou X et al (2025) The crucial roles of phloem companion cells in response to phosphorus deficiency. Plant Cell Environ 48:4327–4340. https://doi.org/10.1111/pce.15421
33. Janíček T, Hobza R, Hudzieczek V (2023) Laser capture microdissection: from genomes to chromosomes, from complex tissue to single-cell analysis. In: Heitkam T, Garcia S (eds) Plant cytogenetics and cytogenomics. Springer, US, New York, NY, pp 163–175
34. Blokhina O, Valerio C, Sokołowska K et al (2017) Laser capture microdissection protocol for xylem tissues of woody plants. Front Plant Sci 07:1965. https://doi.org/10.3389/fpls.2016.01965
35. Berkowitz O, Xu Y, Liew LC et al (2021) RNA-seq analysis of laser microdissected *Arabidopsis thaliana* leaf epidermis, mesophyll and vasculature defines tissue-specific transcriptional responses to multiple stress treatments. Plant J 107:938–955. https://doi.org/10.1111/tpj.15314
36. Zhu X, Wang W, Crowley DE et al (2017) The endophytic bacterium Serratia sp. PW7 degrades pyrene in wheat. Environ Sci Pollut Res Int 24:6648. https://doi.org/10.1007/s11356-016-8345-y
37. Hacquard S, Delaruelle C, Legué V et al (2010) Laser capture microdissection of uredinia formed by *Melampsora larici-populina* revealed a transcriptional switch between biotrophy and sporulation. Mol Plant Microbe Interact® 23:1275–1286. https://doi.org/10.1094/MPMI-05-10-0111
38. Tremblay A, Li S, Scheffler BE, Matthews BF (2008) Laser capture microdissection and expressed sequence tag analysis of uredinia formed by Phakopsora pachyrhizi, the causal agent of Asian soybean rust. Physiol Mol Plant Pathol 73:163–174. https://doi.org/10.1016/j.pmpp.2009.06.001
39. Roux B, Rodde N, Jardinaud M et al (2014) An integrated analysis of plant and bacterial gene expression in symbiotic root nodules using laser-capture microdissection coupled to RNA sequencing. Plant J 77:817–837. https://doi.org/10.1111/tpj.12442
40. Mahalingam M (2018) Laser capture microdissection: insights into methods and applications. In: Murray GI (ed). Laser capture microdissection. Springer, New York, New York, NY, 1–17pp
41. Butler AE, Matveyenko AV, Kirakossian D et al (2016) Recovery of high-quality RNA from laser capture microdissected human and rodent pancreas. J Histotechnol 39:59–65. https://doi.org/10.1080/01478885.2015.1106073
42. Levengood H, Smith L, Gillis S et al (2025) Plantago species are emerging model organisms for functional genomics and stress biology. Plant Cell Rep 44:142. https://doi.org/10.1007/s00299-025-03530-w
43. Herliana L, Schwerdt JG, Neumann TR et al (2023) A chromosome-level genome assembly of *Plantago ovata*. Sci Rep 13: 1528. https://doi.org/10.1038/s41598-022-25078-5
44. Lyu S, Mei Q, Liu H et al (2023) Genome assembly of the pioneer species *Plantago major* L. (Plantaginaceae) provides insight into its global distribution and adaptation to metal-contaminated soil. DNA Res 30:dsad013. https://doi.org/10.1093/dnares/dsad013
45. Mehmood F, Li M, Bertolli A et al (2024) Comparative plastomics of plantains (Plantago, Plantaginaceae) as a tool for the development of species-specific DNA barcodes. Plants 13:2691. https://doi.org/10.3390/plants13192691
46. Pommerrenig B, Papini-Terzi FS, Sauer N (2007) Differential regulation of sorbitol and sucrose loading into the phloem of *Plantago major* in response to salt stress. Plant Physiol 144:1029–1038. https://doi.org/10.1104/pp.106.089151
47. Levengood H, Dou Y, Fan J et al (2023) *Agrobacterium tumefaciens*-mediated genetic transformation of narrowleaf plantain. J Vis Exp. https://doi.org/10.3791/64777
48. Levengood H, Smith L, Gillis S et al (2025) Plantago species are emerging model organisms for functional genomics and stress biology. Plant Cell Rep 44:142. https://doi.org/10.1007/s00299-025-03530-w
49. Pommerrenig B, Feussner K, Zierer W et al (2011) Phloem-specific expression of yang cycle genes and identification of novel yang cycle enzymes in *Plantago* and *Arabidopsis*. Plant Cell 23:1904–1919. https://doi.org/10.1105/tpc.110.079657

50. Daie J (1986) Kinetics of sugar-transport in isolated vascular bundles and phloem tissue of celery. J Am Soc Hortic Sci 111:216–220
51. Landouar-Arsivaud L, Juchaux-Cachau M, Jeauffre J et al (2011) The promoters of 3 celery salt-induced phloem-specific genes as new tools for monitoring salt stress responses. Plant Physiol Biochem 49:2–8. https://doi.org/10.1016/j.plaphy.2010.09.008
52. Noiraud N, Maurousset L, Lemoine R (2001) Identification of a mannitol transporter, AgMaT1, in celery phloem. Plant Cell 13:695–705. https://doi.org/10.1105/tpc.13.3.695
53. Vilaine F, Palauqui JC, Amselem J et al (2003) Towards deciphering phloem: a transcriptome analysis of the phloem of *Apium graveolens*. Plant J 36:67–81. https://doi.org/10.1046/j.1365-313X.2003.01855.x
54. Reardon W, Gallagher P, Nolan KM et al (2014) Different outcomes for the MYB floral symmetry genes *DIVARICATA* and *RADIALIS* during the evolution of derived actinomorphy in Plantago. New Phytol 202:716–725. https://doi.org/10.1111/nph.12682
55. Singh M, Nara U, kaur K et al (2022) Genetic, genomic and biochemical insights of celery (*Apium graveolens* L.) in the era of molecular breeding. J Appl Res Med Aromat Plants 31:100420. https://doi.org/10.1016/j.jarmap.2022.100420
56. Liu J-X, Li T, Wang H et al (2022) CRISPR/Cas9-mediated precise targeted mutagenesis of phytoene desaturase in celery. Hortic Res 9:uhac162. https://doi.org/10.1093/hr/uhac162
57. Song G-Q, Loskutov AV, Sink KC (2007) Highly efficient Agrobacterium tumefaciens-mediated transformation of celery (*Apium graveolens* L.) through somatic embryogenesis. Plant Cell Tissue Organ Cult 88:193–200. https://doi.org/10.1007/s11240-006-9190-3

Part III

Measurement of Phloem Transport

Check for updates

Chapter 17

Detecting Rapid Changes in Carbon Transport and Partitioning with Carbon-11 (^{11}C)

Benjamin A. Babst, Richard A. Ferrieri and Michael J. Schueller

Abstract

Noninvasive techniques to measure phloem transport of carbon will be crucial to efforts to engineer improved crop yields, which are highly dependent on carbon partitioning. Phloem, which is buried in the interior of the plant, is highly sensitive to tissue damage. Here we describe nondestructive methods using carbon-11, fed to leaves as $^{11}CO_2$, as a tracer to track export of recently fixed carbon from leaves, transport speed through the phloem, and distribution or partitioning throughout the plant.

Keywords Photosynthate transport, Phloem, Carbon-11, Whole-plant ^{11}C partitioning, Carbon allocation

1 Introduction

Delivery of photosynthates (i.e., carbonaceous biochemicals that are produced largely in photosynthetic leaves) to nonphotosynthetic tissues is essential for plant growth. Since photosynthates are primarily transported through the phloem, engineering phloem transport is an enticing strategy for plant improvement [1–4]. In the post-genomics era, tools to measure phloem physiology, such as photosynthate transport and partitioning to different parts of the plant, are essential to probe the mechanisms and regulation of phloem transport. Since phloem function can be easily disrupted by invasive techniques (Chaps. 1–3), noninvasive techniques are crucial for a better understanding of this dynamic process. Sensitive detection of positron-emitting isotopes, such as carbon-11 (^{11}C), presents opportunities to determine photosynthate transport and partitioning within live plants. When ^{11}C is administered to intact leaves

In memory of our friend and colleague, whose genius made it possible for us to carry out our science, Dr. Michael J. Schueller (1972 - 2025).

Johannes Liesche and Sanna Sevanto (eds.), *Phloem*, Methods in Molecular Biology, vol. 3066,
https://doi.org/10.1007/978-1-0716-5488-0_17,

as $^{11}CO_2$, it is naturally assimilated via the photosynthetic machinery and subsequently incorporated into a variety of biochemicals, including soluble carbohydrates and amino acids, as well as cellulose and phenolics [5–8]. In vascular plants, a large fraction of these soluble photosynthates can be exported rapidly from leaves to sink tissues such as young developing leaves and roots via the phloem. Carbon-11 offers unique qualities, affording researchers the opportunity to monitor phloem transport in real time [9–12]. On the one hand, its mode of decay by positron emission with subsequent annihilation to gamma radiation allows for noninvasive measurement of changes in the isotope's spatial distribution over time, without the destructive sampling required for stable carbon isotopes (e.g., ^{13}C). The nondestructive measurement combined with the short radioactive half-life of ^{11}C ($t_{½}$ = 20.4 min) can allow for repeated studies on the same plant, enabling examination of changes in transport and partitioning as a function of environmental cues or stages of development. Whereas recent improvements in dye-based methods have utility (Chaps. 15 and 22), especially where cellular or subcellular resolution is required [13, 14], they are imperfect tracers since they rely on non-biological chemicals as a proxy for plant carbohydrates. ^{11}C can be considered a true tracer in that it is incorporated into and allows detection of biochemicals biosynthesized by the plant.

While these methods have been available only to a very limited number of investigators in the past, there is potential to expand their availability. The cyclotrons needed to produce ^{11}C are now widely distributed (often associated with universities and hospitals). Recently, a portable $^{11}CO_2$ trap was developed to allow safe and easy transportation of $^{11}CO_2$ over short distances from the site of production for use in plant radiotracer labs [15], providing potential to make these methods more widely available and accelerate progress in understanding and engineering phloem transport.

Here, we describe approaches for using ^{11}C to measure several features that underpin phloem transport, including photosynthetic fixation as $^{11}CO_2$, export of recently fixed ^{11}C from the load zone over time, speed of ^{11}C transport through the phloem, and carbon partitioning to various sink tissues within the plant. We also highlight three case studies that have used dedicated radiation detectors as well as imaging technology, including autoradiography and positron emission tomography (PET), to carry out these measurements.

2 Materials

All equipment for producing and administering ^{11}C to plants must be assembled, checked for leaks, and fitted with appropriate shielding (e.g., lead or similar dense material) to minimize radiation exposure of personnel prior to use, and to comply with local radiological safety rules and laws. Systems can be easily designed such that exposure of personnel to radioactivity is negligible. We include several approaches for imaging and detection of ^{11}C within plants, which could be used separately or in combination. Typical radiotracer experiments can be conducted for up to 2 h, and so it is essential that all equipment is prepared and in place prior to beginning each experiment.

2.1 Equipment for the Production of Carbon-11

1. A charged particle accelerator, such as a cyclotron, capable of producing a >10 MeV proton beam at a current of tens of microamps.
2. Compressed N_2 gas (ultrahigh purity 99.9995%, with trace 0.1% O_2 gas) for loading the irradiation target.
3. Compressed air supply for the load leaf and for carrying $^{11}CO_2$ to the plant.
4. CO_2 trap to hold $^{11}CO_2$ for short periods and then release a sharp pulse of $^{11}CO_2$ to the plant in a stream of air. The trap consists of a 3–5 cm tube or column packed with at least 90–120 g molecular sieve (4Å, 80–100 mesh), with frits to contain the material at both ends, and a heating element capable of rapidly raising the trap temperature to about 250°C to rapidly release the $^{11}CO_2$ from the molecular sieve [15]. Valves should be placed on both ends of the trap such that the trap can be isolated during heating, and the airstream can continue flowing to the plant (Fig. 1; *see* **Notes 1–3**).

2.2 Radiotracer Containment

1. Controlled environment chamber for the plant: Slight negative pressure is required to exhaust any potentially leaked radioisotope, as well as lead shielding, which serves to minimize both user exposure to radioactivity and interference from environmental radiation.
2. Lighted leaf cuvette: standard leaf cuvette like those used for IRGA-based photosynthesis measurements. The cuvette should be lined with a foam gasket to seal against the leaf so that only a single leaf is exposed to $^{11}CO_2$.

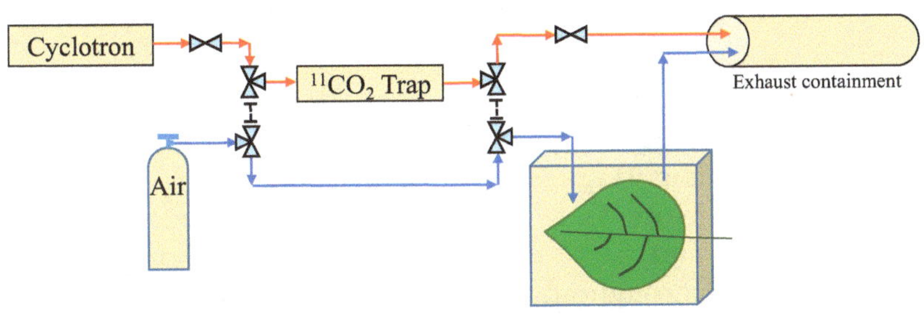

Fig. 1 Schematic drawing showing the typical plumbing setup for trapping $^{11}CO_2$ from the cyclotron target and releasing the radiotracer to a leaf cuvette in a steady stream of air

3. Tubing: Standard copper or polytetrafluoroethylene (PTFE) tubing with air-tight fittings to connect the cyclotron to the $^{11}CO_2$ trap, the $^{11}CO_2$ trap to the leaf cuvette, and the leaf cuvette to an exhaust containment device, e.g., a large balloon (50–100 L) that can expand with little backpressure as it takes in exhaust air (Fig. 1).

2.3 Detectors

1. Fixation detector: The fixation detector is typically a small PIN Diode detector, which is constructed of an undoped intrinsic semiconductor material sandwiched between p-type and n-type semiconductor materials. These detectors are commercially available (e.g., Carrol Ramsey Associates, Berkeley, CA, USA). The detector can be very small in size (typically 10 × 5 × 2 mm), allowing very easy integration into the underside of the leaf cuvette for the detection of radioactivity from the $^{11}CO_2$ administered. While these detectors are not very sensitive to radioactivity, they work well in the 3.7 MBq-to-3.7 GBq sensitivity range and are well suited for measuring the $^{11}CO_2$ pulse released from the trap and the amount of ^{11}C that remains fixed in the leaf tissue minutes after the pulse is swept through the cuvette. Additionally, the decline in the ^{11}C signal from this detector, when corrected for radioactive decay, indicates the amount of radiotracer exported from the leaf over time.
2. Transport detectors: These detectors typically rely on inorganic scintillators such as NaI or CsI salts that are either diode-backed, providing a sensitivity range of 0.037 MBq-to-3.7 MBq, or backed by a photomultiplier (PMT) providing sensitivity below 0.037 MBq. Solid-state detectors, such as cadmium telluride detectors, can also be used instead of scintillation detectors. Typically, all these detectors require some manner of shielding with lead or similar high-density material to collimate the field-of-view (FOV) these detectors see (Fig. 2A). Due to the weight of the shielding, physical support must be provided to hold the detectors in place on the plant stems or leaves, such as clamps and rings stands, or other support structure (Fig. 2B). Often, several detectors are placed along the stem and aligned with regions of sink tissues providing time-activity data on transport.
3. Positron emission tomography (PET): PET can be used together with, or separately from, collimated radiation detectors (Fig. 2C, D) for dynamic imaging of ^{11}C transport (*see* **Note 4**), to determine transport speeds within the PET FOV, and for 3-dimensional imaging of the distribution of ^{11}C throughout the plant at any time during the experiment. While PET is dynamic, the downside is that spatial resolution is limited to a few millimeters (Fig. 3). This is because the energetic

Fig. 2 Panels (**A**) and (**B**) show how lead collimated CsI diode detectors can be strategically positioned to monitor the movement of radioactivity from the load leaf encased in the cuvette (**B**). Panel (**C**) shows the same leaf cuvette attached to a maize leaf with a dedicated plant PET imager that can be positioned to vertically translate along the stem and/or root of the plant while acquiring data. Panel (**D**) shows what a typical PET image might look like for a large grass imaged in a Siemens HR+ human scanner. The 3mm spatial resolution of this scanner was adequate to show ^{11}C occupancy of the stem nodes, and in dynamic mode, one can observe filling and emptying of these stem features as ^{11}C-photosynthates transport through the phloem en route to sink tissues

positron from ^{11}C can travel a few millimeters before it annihilates with an electron, giving off two coincident gamma rays at 511 keV energy (equivalent to the rest mass energy of an electron). Because PET measures gamma radiation, which penetrates all tissue, the technique is noninvasive in nature. Software for 3-D image reconstruction and analysis is available and can be purchased with commercial PET systems (e.g., ECAT). Freeware and open-source software are also available (e.g., http://amide.sourceforge.net/; http://www.turkupetcentre.net/petanalysis/sw_pet.html; https://www.radiology.pitt.edu/pet-software-resources.html).

4. Autoradiography: Unlike PET, this approach to imaging is static. It can provide information on the spatial distribution of radioactivity within plant tissues and relies on using flexible acetate plates coated with a photostimulable phosphor material, which, when laid across the tissue, can capture the positron energy before it annihilates. The energy capture process results in electronic excitation of the phosphor material (Fig. 4A). The plate is then removed from the tissue and placed in the imaging instrument, where it is scanned with a neon laser at 580 nm wavelength resulting in light emission at 400 nm wavelength due to photoluminescence (see inset Fig. 4A) that is recorded in 2-D coordinates by a CMOS or CCD image sensor. Commercial imaging instruments are available today (e.g., Amersham Typhoon, GE Healthcare, Piscataway, NJ, USA; or Sapphire™, Azure

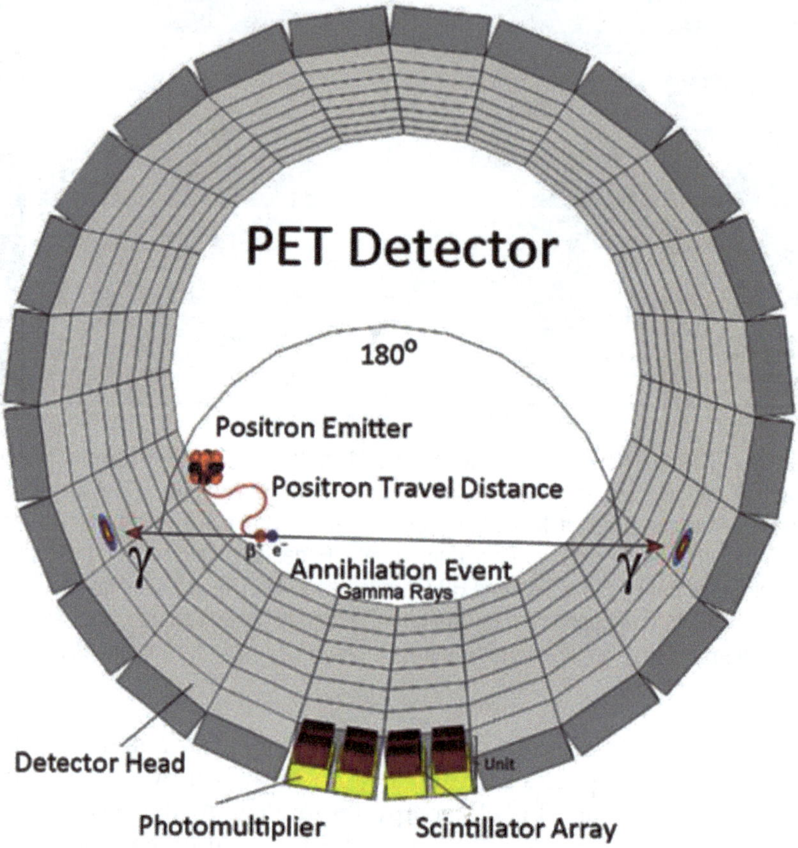

Fig. 3 Schematic drawing depicting how PET works. ^{11}C decays by emitting a positron particle, which possesses a maximum kinetic energy of 0.960 MeV. This positron can travel several millimeters in living tissue before it annihilates with another electron, giving off two coincident gamma rays at 511 keV energy. Scintillation detectors connected in coincidence can register this decay event, and when many of these coincidence detectors are arranged in a ring around the sample, the instrument is capable of reconstructing 2-D and 3-D images

Biosystems, Dublin, CA, USA). Image analysis can be carried out using ImageQuant™ TL 7.0 software (Cytiva Life Sciences, Inc., Marlborough, MA, USA) (*see* **Note 5**).

While autoradiography is static, it does offer a significantly higher spatial resolution than PET, typically 100 μm under normal operation (Fig. 5A). However, we found that when imaging very thin tissues such as roots, one can improve on the resolving power of autoradiography by scanning the phosphor plate while the tissue is still in contact with the plate (Fig. 4B). This can be accomplished by first exposing the tissue on a glass plate that can later be inserted into the instrument. Many radiographic imaging stations sold today have the capability to perform fluorescence imaging and come equipped

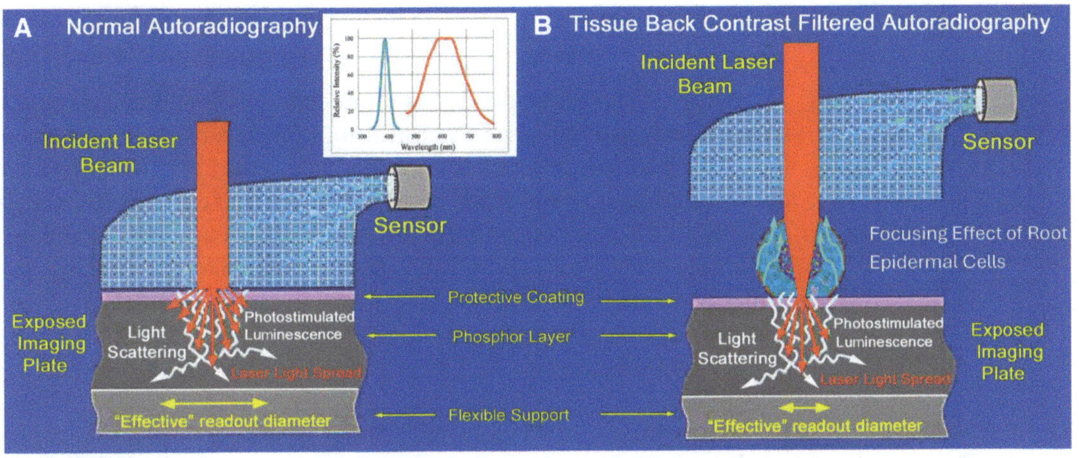

Fig. 4 Schematic drawings showing differences in how normal autoradiography and tissue back contrast autoradiography work. With normal autoradiography, one exposes a phosphor plate to tissue that contains radioactivity, then removes the plate for scanning (**A**). Image resolution for ^{11}C is typically around 100 μm and is dependent on the focusing capabilities of both the projected laser light and the emitted light. With tissue back contrast filtered autoradiography (**B**), the tissue is not removed during scanning. We suggest that the epidermal cells along the outer surface of a root may aid in focusing both the impinging neon laser light and the emitted blue light

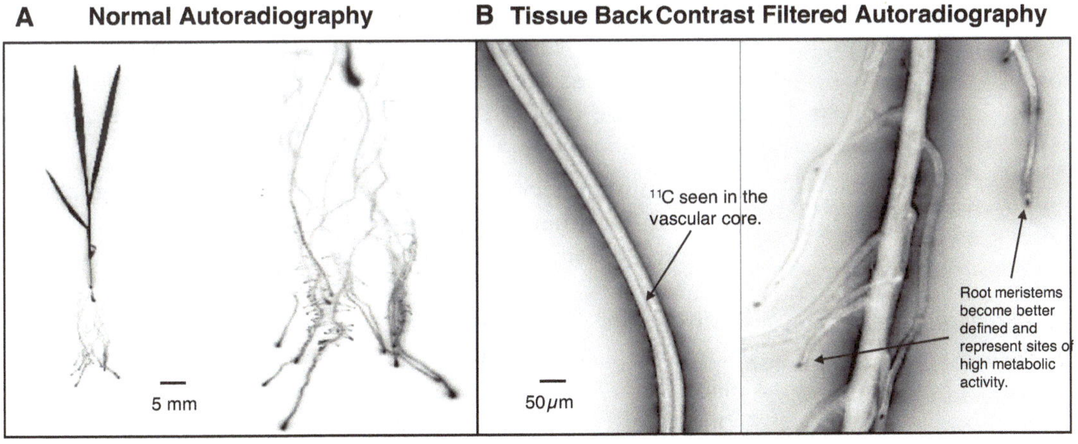

Fig. 5 Improving autoradiography resolution by tissue back contrast filtering. A normal autoradiograph of rice shoots and roots is shown in the left-side panel (**A**). With tissue back contrast filtering, significant improvement in image resolution can be obtained, providing greater detail on sink strength of the root's meristematic region, as well as being able to resolve the vascular core (**B**). The latter feature opens possibilities to examine the leakage of ^{11}C as it moves through the phloem. It should be noted that thicker tissues will interfere with both positron and light transmission

with suitable clear glass imaging plates for doing this. The resultant radiographic images (Fig. 5B) can provide greater contrast of root meristems and even show ^{11}C within the vascular core of the root.

3 Methods

In brief, $^{11}CO_2$ must be produced immediately before use. The $^{11}CO_2$ is administered as a pulse in a continuous air stream to an individual leaf, clamped inside a cuvette. Assimilation and export of ^{11}C are determined based on a detector affixed to the leaf cuvette. Detectors and/or a PET camera can be positioned on the basal part of the leaf blade or on the stem to measure transport dynamics. One to two hours after the $^{11}CO_2$ pulse, a snapshot can be taken of ^{11}C distribution throughout the plant using phosphor plate imaging for a 2-D image of thin tissues or a PET camera for 3-D imaging of thick tissues.

3.1 Preparation of Detection Equipment

1. Synchronize any clocks that will be used to time the experiment, including clocks of computers on which data will be recorded.
2. Detector calibration: Zero detector signals. Place an appropriate radioactive point source in front of each detector and record the signal. Long-lived point sources will need to be calibrated against actual ^{11}C radioactivity to do a true calibration, but this is not essential. Calibration here is primarily intended to account for day-to-day variability and to match the signal of the two plant detectors.
3. Erase any phosphor plates that may be needed for imaging. It is helpful to have extra plates ready in case multiple images are needed. For example, in the case of overexposure, a second plate can be exposed for a shorter time to capture a usable image.
4. PET scanner checks: Perform any pre-run checks required for your PET scanner. Calibrate the PET scanner using the manufacturer's recommended protocols.

3.2 Preparation of the Plant

1. Position the plant in the plant tracer chamber well in advance of the experiment. Take any desired measurements of the plants (e.g., height and leaf length or area), or environmental conditions (e.g., temperature and light intensity).
2. Check airflows from the trap to the leaf cuvette and check the plant tracer chamber exhaust for appropriate flow.
3. Turn on and adjust lighting (50–2000 $\mu mol\, m^{-2}\, s^{-1}$ PAR) and airflow to the leaf cuvette to desired levels (200–600 cc/min), as appropriate to the experiment.
4. Arrange the plant in the chamber and clamp the leaf into the leaf cuvette at least 30 min prior to $^{11}CO_2$ feeding to allow the leaf time to stabilize. If the experiment entails comparing multiple individual plants, the leaf position used should be kept uniform amongst plants to control for developmental differences (*see* **Note 6**).

5. Position transport detectors along the plant stem, such that the collimator hole is not facing toward the leaf cuvette (Fig. 2A, B). Detectors can be placed to measure acropetal or basipetal transport, but at least two detectors must be placed in each desired direction to determine transport speed. For long-leaved plants, such as maize, these detectors may be placed along the leaf blade basal to the leaf cuvette to measure transport speed through the leaf midrib (*see* **Note 7**). The distance between the two detectors should be at least 10 cm apart for the best accuracy of the transport speed measurement and should be kept uniform within a study.
6. Position the PET scanner field of view on the plant as desired (Fig. 2C). Note that cleaner images will result if the PET scanner is not immediately adjacent to a higher radiation source than is inside the field of view. If necessary, background noise from nearby radiation sources (e.g., a load leaf near the position where imaging is desired) can be reduced by placing shielding between the source and the PET scanner. Depending on the geometry of your system, the coincidence counting used by PET scanners may reduce or eliminate the need for such shielding.
7. Measure the distance between all the transport detectors and between the detectors and the PET scanner. Measure the distance between the leaf cuvette and the first detector or PET scanner along the path the radiotracer is most likely to follow when transported within the plant (e.g., along the midrib of the leaf, petiole, and stem).

3.3 Production and Administration of $^{11}CO_2$

1. Condition $^{11}CO_2$ trap: Heat trap to 250°C while flowing non-reactive gas (e.g., N_2, helium, or argon) through the molecular sieve column for about 5 min. Stop gas flow and cool trap to 50°C or cooler, preferably room temperature, if time permits.
2. Generate $^{11}CO_2$: Direct a proton beam at N_2 target for 0.5–2 min, depending on the amount of $^{11}CO_2$ desired, beam energy, and beam current.
3. Trap $^{11}CO_2$: Switch trap valves such that airflow to the leaf cuvette bypasses the trap. Use remote-controlled valves to flow $^{11}CO_2$ from the target through tubing to the trap (Fig. 1). Local radiation safety laws and regulations may require additional containment before air is vented. The gas which has passed through the CO_2 trap will still contain $^{14}O_2$, $^{15}O_2$, and likely a trace of ^{11}CO, which are not captured by the molecular sieves. As the amount of time required to capture $^{11}CO_2$ on the trap depends on the rate of airflow and the diameter and length of tubing, it is best to use the readout of a radioactivity detector placed on the $^{11}CO_2$ trap to determine when enough $^{11}CO_2$ has been trapped and stop airflow from the target to the trap

at that time. It should be noted that during this process, there is a continuous flow of air through the leaf cuvette to maintain the photosynthetic machinery.

4. If a portable trap is used to transport the tracer to a remote plant chamber, disconnect the trap and transport it to the plant using appropriate containment and shielding to minimize personnel exposure to radioactivity. Connect tubing such that compressed air will flow through the portable trap to the leaf cuvette and attach the thermocouple and furnace power connectors.
5. Heat trap: Heat the trap to 250°C.
6. Pulse $^{11}CO_2$ to the leaf cuvette: As soon as the temperature reaches 250°C, switch the trap valves to redirect the flow of air passing through the leaf cuvette to first pass through the $^{11}CO_2$ trap and then through the leaf cuvette. Typically, a flow of 200 mL min^{-1} flow rate is sufficient to generate a 30-s-wide pulse of $^{11}CO_2$ within this stream. Once the tracer pulse passes through the cuvette, as determined by the PIN diode detector response, the air stream is again switched to bypass the trap but continually feed air to the cuvette for the duration of the tracer study.
7. Record data: Start recording data from detectors and the PET scanner at the exact time when the $^{11}CO_2$ is pulsed to the leaf. Alternatively, note the time of the $^{11}CO_2$ pulse to use later for decay correction.

3.4 Determining Whole-Plant Partitioning of ^{11}C

1. PET scan of whole plant: If the entire plant fits within the field of view of the PET camera (as when using a large animal PET), partitioning of ^{11}C throughout the plant can be viewed for any time point during the experiment. If the part of the plant extends outside of the field of view, a whole-plant snapshot of ^{11}C distribution can be captured after a fixed time, if the PET camera is equipped to scan along the body of the plant and reconstruct the scanned image (Fig. 2D). Image reconstruction software should be offered by the PET camera manufacturer (*see* **Note 8**).
2. Phosphor plate imaging: For small plants or thin tissues (e.g., thin stems, roots, or leaves), lay the plant or tissue on a flat surface after a fixed time of plant incubation. Incubation time should be predetermined by conducting several preliminary test runs. Expose the phosphor plate by pressing the plate tightly against the surface of the plant. Depending on the amount of radioactivity in the plant, as little as 30 s exposure may be sufficient, or as much as 20 min exposure may be required (Fig. 5A). Since the phosphor plates can be erased by light and reused, minimize incident light on the phosphor plates

following exposure to ^{11}C until the plates have been scanned by a phosphor plate reader (*see* **Note 9**).

3. Determine the percent of ^{11}C partitioned to regions of interest (ROI) within the plant using image analysis software.

$$\text{Partitioning } (\%) = (\text{signal}_{\text{ROI}} / \text{signal}_{\text{whole-plant}} \times 100\%)$$

These data do not require decay correction, since they are normalized to whole-plant radioactivity and are unitless.

3.5 Decay Correction and Data Processing

1. Decay correction: All dynamic data must be decay corrected to a common time point, such as the time of the $^{11}CO_2$ pulse. Carbon-11 has a highly predictable exponential decay, such that decay correction can be used to compare radioactivity measurements from different time points.

$$R_0 = R_t \times e^{\lambda t}$$

where R_0 and R_t represent radioactivity levels at time zero and time "t" respectively; λ is the decay constant, $(\ln 2)/t_{1/2}$, where $t_{1/2}$ is the half-life of the isotope. For ^{11}C ($t_{1/2} = 20.4$ min), the formula can be simplified such that only R_t and t are needed to determine R_0:

$$R_0 = R_t \times 2^{t/20.4}$$

Each radiation measurement at each time point for each detector must be decay corrected prior to any other analysis.

2. ^{11}C fixation and export: Data from a PIN diode detector attached to the leaf cuvette is used to determine the total amount of ^{11}C pulsed to the leaf, and the amount of ^{11}C assimilated by the leaf based on the amount of ^{11}C remaining in the leaf immediately after the pulse (Fig. 6A). Fixation of ^{11}C should be presented as a proportion or percent of the pulse height to enable comparison between runs with slightly different amounts of ^{11}C delivered to the leaf cuvette. If the portion of the curve indicating ^{11}C export from the leaf is linear, then export can be calculated as the slope of the regression line that fits the export portion of the curve. Alternatively, if the export is nonlinear (e.g., exponential), the export can be determined as the decrease in radioactivity between two fixed time points, divided by the time (Fig. 6A). Either of these methods is sufficient for relative comparisons, as long as one method is used consistently within a study.

3. Time of arrival: To determine the time when ^{11}C radioactivity arrives in the field of view of a detector, identify at what time the maximum slope of the radioactivity curve occurs (Fig. 6B). This can be determined by calculus after fitting a 3rd order

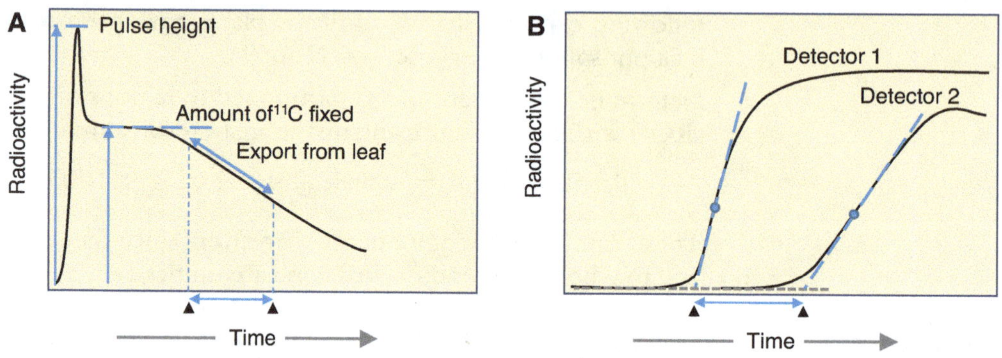

Fig. 6 Diagram of hypothetical radioactivity data graphed as a function of time. (**A**) Radioactivity in the leaf cuvette, measured by a PIN diode detector. Pulse height indicates the amount of ^{11}C pulsed to the cuvette. The radioactivity drops after the pulse and then levels out transiently, which indicates the amount of ^{11}C assimilated by the leaf, and then the decay-corrected radioactivity in the leaf declines gradually, due to ^{11}C export from the leaf. (**B**) Black curves indicate radioactivity measured by two detectors placed sequentially along a stem or leaf midrib. Blue dot indicates the point of the maximum slope of the radioactivity curve. The dotted gray line indicates baseline radioactivity. Dashed blue lines approximate the tangent line where the slopes of the curves are maximal. Time of arrival is indicated by black arrowheads where the tangent lines meet the baseline radioactivity. Transit time between the two detectors (double-headed arrow) is the difference between the time of arrival at detector 1 and the time of arrival at detector 2

polynomial to the curve. Often, a 3rd order polynomial is not a good fit to the curve, in which case the maximum slope can be determined in a spreadsheet, using an approach that is only slightly more computationally intensive. Calculate the best fit lines for the radioactivity curve over a fixed short period of time (e.g., 5 min), moving the time period forward by 30 or 60 s intervals for every consecutive 5 min period of the curve [10, 16]. Using the best-fit line with the greatest slope, project the best-fit line back toward the *x*-axis (Fig. 6B). The point where this line intersects the baseline level of radioactivity is the time of arrival (*see* **Note 10**). This approach can be used for data from transport detectors, or from a voxel or region of interest (ROI) that includes multiple voxels selected from dynamic PET data.

4. Transport speed: Divide the distance between the field of view of the two detectors or ROIs by the transit time between the two detectors or ROIs.
   Transit time = time of arrival_2 − time of arrival_1
   Transport speed = Distance between the two detectors/transit time
5. Loading time: The time elapsed between when $^{11}CO_2$ was pulse-delivered to the leaf and when the assimilated ^{11}C is first loaded into the phloem can be estimated. This loading time includes the time required for $^{11}CO_2$ to diffuse into the

leaf through stomata into the chloroplasts in the mesophyll, capture of the ^{11}C by photosynthesis, transport of triose phosphates from the chloroplasts into the cytosol, conversion to transportable biochemical forms, and transit from mesophyll cells to the phloem. To estimate the time it took for $^{11}CO_2$ to first enter the phloem, calculate the transport time between the load zone and the first transport detector:

$\text{Transport time}_{\text{LZ-to-TD1}} = d/\text{TS}$,

where d is the distance between the leaf cuvette and the first transport detector, and TS is the transport speed between 2 detectors, as determined above. Then subtract the transport time from the time of arrival at detector 1.

$\text{Loading time} = \text{time of arrival}_1 - \text{Transport time}_{\text{LZ-to-TD1}}$

4 Case Studies

1. Example of ^{11}C-radiotracer to examine ^{11}C-photosynthate movement across a graft junction. Grafting has been used in agriculture and horticulture for millennia to combine two genetically distinct crop varieties to create a plant with maximally beneficial shoot and root traits. Previously, it was thought that graft compatibility was an exclusive trait of dicotyledonous species; however, a recent study has reported a method by which it is possible to graft both inter and intraspecific monocot species by performing the grafting in pre-germinated seeds and kernels [17–19]. While the method had shown the ability of *Zea mays* to form symplastic connections between primary roots and shoots in both homo- and heterografts, CFDA dye assays were inconclusive whether vascular connections were established at the junctions, or whether the dye simply diffused through tissue [17]. Using a simple experimental setup relying on minimal dedicated radiation detectors, carbon-11 radiotracing was applied to determine whether vascular connectivity was established when the grafting technique was applied to differing genotypes of maize. The underlying hypothesis was that vascular connectivity would manifest in different degrees of ^{11}C partitioning to graft tissue, whereas a simple diffusion model would show no differences between genotypes.
2. Using a single transverse cut through the mesocotyl of an ungerminated maize kernel, embryos of bisected kernels from differing genotypes were realigned and held together with a small amount of adhesive, allowing normal germination and growth of plants possessing two shoots. Plants were grown in nutrient fortified hydrogel in glass cells that had a center partition according to published procedures [12] that fixed the

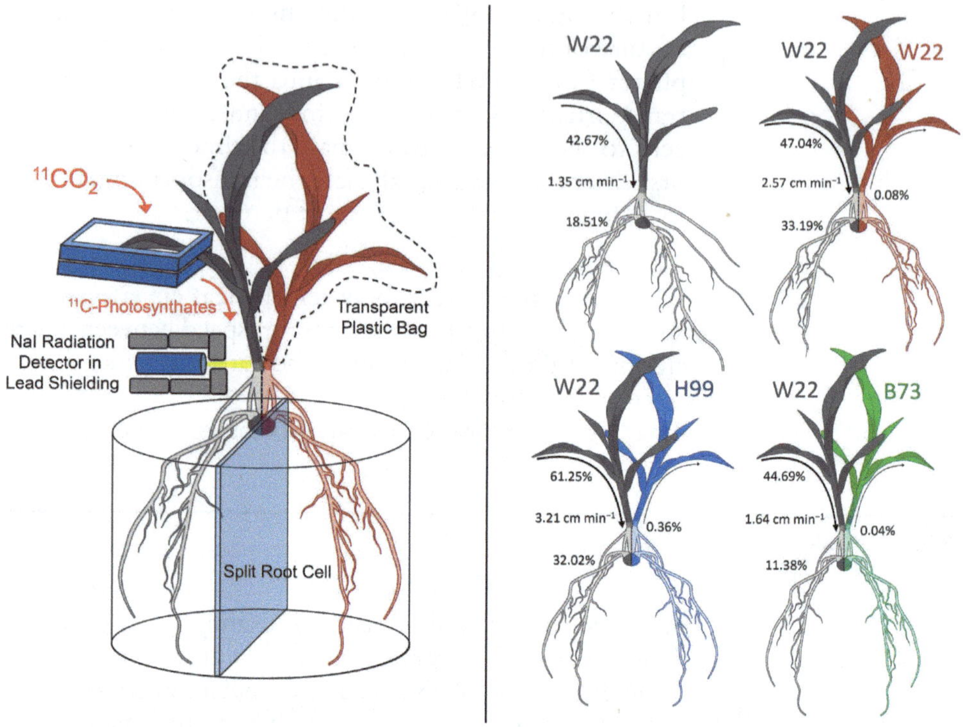

Fig. 7 The schematic drawing on the left side shows the experimental setup used to carry out ^{11}C radiotracing studies examining shoot:shoot maize grafts and the movement of tracer across the graft junction. The right-side panel shows schematic drawings of the various cultivar graft configurations that were examined. W22 isolates were used as controls for comparing leaf export, phloem transport speed, and root partitioning to values when grafts were attached: W22 homografts were highlighted in red; H99 heterografts were highlighted in blue; and B73 heterografts were highlighted in green. Means of leaf ^{11}C export, allocation of ^{11}C-photosynthates, and transport speeds are indicated on the drawings

kernel within the gel and provided a physical barrier between graft roots (Fig. 7). Four maize genotypes were examined; the W22 line was used as the isolate control, and shoot:shoot grafts were created generating a W22:W22 homograft, and W22:H99 and W22:B73 heterografts. Phloem–xylem connectivity was examined by administering radioactive $^{11}CO_2$ to a source leaf on one shoot and measuring partitioning of ^{11}C-photosynthate down that shoot to the roots, and back up the second shoot via the xylem (Fig. 7). As an added precaution against contamination, the adjacent graft shoot was encased in a clear plastic bag during tracer administration. A distance-normalized speed of ^{11}C-photosynthate arrival at the base of the W22 shoot was measured by placing a single shielded NaI scintillation detector at the base of the stem. Because speed was based on the difference in time between the $^{11}CO_2$ pulse maximum and the time-of-arrival of radioactivity at the lower detector, speed values can reflect effects of both phloem loading and phloem transport.

3. After 90 min, tissues were harvested, and radioactivity measured on a gamma counter providing quantitative data on source leaf export, partitioning of ^{11}C-photosynthates to W22 roots, and cumulative partitioning to graft tissues, including roots and shoots (Fig. 7). Grafts using the H99 maize line exerted a stronger sink strength resulting in increased leaf export, increased distance-normalized speed of arrival and increased ^{11}C partitioning to graft tissues. These observations support a vascular connectivity model rather than simple diffusion.

4. Example using autoradiography to examine changes in vascular sectoriality. Many higher plants possess a physiological organization that is based upon the carbon economy of their parts. While photosynthates are partitioned according to the relative strength of the sink tissues, there exists in many plant species a close relationship between partitioning, phyllotaxy, and vascular connectivity, giving rise to sectorial patterns of resource allocation that have evolved from selective ecological pressures over time [20]. In a recent study [21], the effects of smoke infiltration into soil on phloem sectoriality were examined using carbon-11 autoradiography in common sunflower (*Helianthus annuus* L.) as a model sectorial plant. Plants were grown in rhizoboxes (20 cm wide × 25 cm tall × 1 cm deep) either in soil or hydrogel to create a shallow 2-dimensional pattern in the root architecture (Fig. 8). Ninety minutes after $^{11}CO_2$ administration, radiographic imaging was applied to quantify the partitioning pattern of the ^{11}C-photosynthates within the sink tissues (Fig. 8). Smoke treatment significantly reduced phloem sectoriality. Catechol and resorcinol, chemical constituents of smoke, were also bioactive in causing reduced phloem sectoriality, but treatment with hydroquinone had no effect. The implications of these findings for plant growth and fitness were discussed in prior published work [21, 22].

5. Example using PET Imaging to examine stem sugar storage and its impact on sink partitioning. Studies of large plants at later developmental stages necessitate using a large PET camera [10]. We used a Siemens HR+ human PET scanner (~0.56 m bore diameter) to study reproductive stage sorghum [23]. A portable $^{11}CO_2$ trap [15] enabled carrying tracer safely to the scanner and administering the dose to an intact leaf that was affixed within a leaf cuvette. As above, the cuvette held a PIN diode detector for registering ^{11}C pulse size, fixation, and export from the leaf. That PIN diode data was coupled with PET images, giving insight into whole-plant carbon partitioning (Fig. 9). Comparisons were made between dwarf sweet sorghum and grain sorghum lines spanning from vegetative to

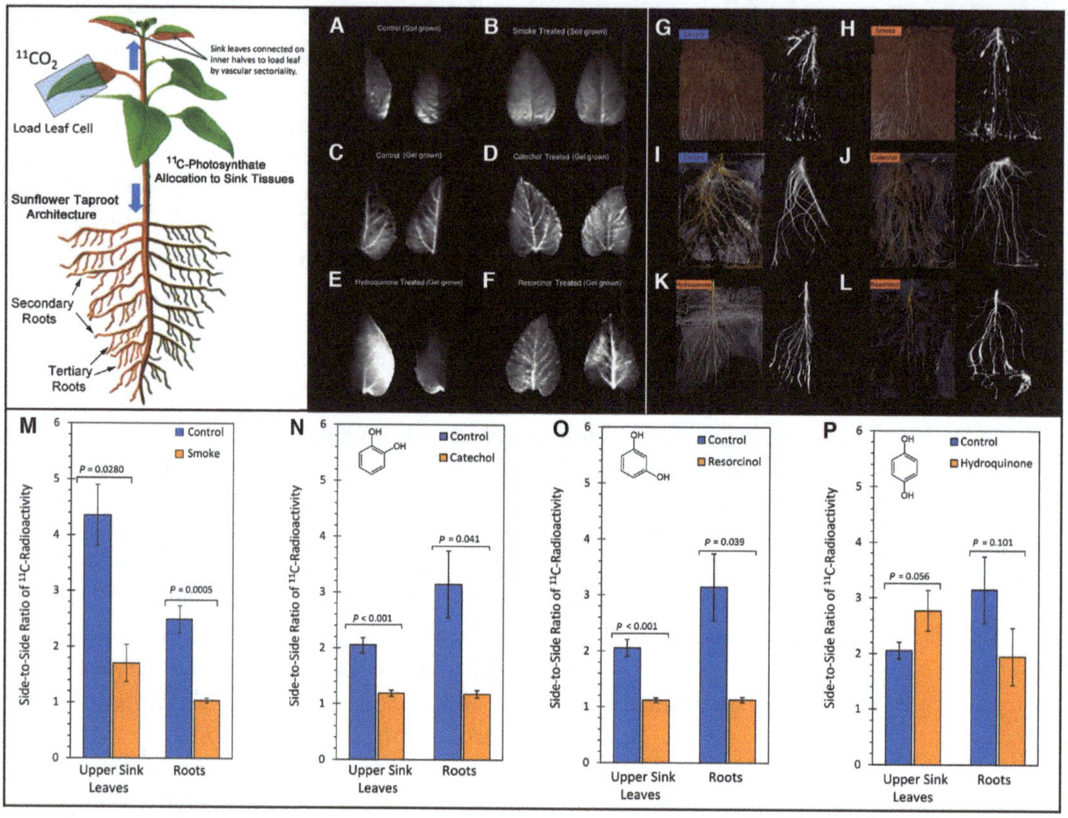

Fig. 8 The upper left schematic drawing shows the experimental setup used for administering $^{11}CO_2$ to a mature source leaf on a live sunflower plant. Patterns of partitioning of ^{11}C-photosynthates to sink tissues (both leaves and roots) are shown in red to depict the natural path of connectivity determined by the sectorial vascular architecture (i.e., half of two younger sink leaves directly above, and half of the root system directly below the source leaf). (**A–F**) Radiographic images showing patterns of ^{11}C-photosynthate allocation to sink leaves. (**A**) Control sink leaves of plants grown in soil show the characteristic pattern of sectorial vascular connectivity. (**B**) Sink leaves from plants grown in liquid smoke-treated soil show loss of sectoriality. (**C**) Control sink leaves of plants grown in a hydrogel matrix show similar sectorial behavior as soil-grown plants. (**D**) Sink leaves of plants grown in a hydrogel matrix treated with 150 μM catechol show loss of sectorial connectivity. (**E**) Sink leaves of plants grown in a hydrogel matrix treated with 150 μM hydroquinone show similar connectivity patterns as control plants. (**F**) Sink leaves of plants grown in a hydrogel matrix treated with 150 μM resorcinol showed loss of sectorial connectivity. Root photographs with matching radiographic images are shown in panels (**G–I**), showing patterns of ^{11}C-photosynthate partitioning to sink roots. (**G**) Control sink roots of plants grown in soil show the characteristic pattern of sectorial vascular connectivity. (**H**) Sink roots from plants grown in liquid smoke-treated soil show loss of sectorial connectivity. (**I**) Control sink roots of plants grown in a hydrogel matrix show similar sectorial behavior as soil-grown plants. (**J**) Sink roots of plants grown in a hydrogel matrix treated with 150 μM catechol show loss of sectoriality. (**K**) Sink roots of plants grown in a hydrogel matrix treated with 150 μM hydroquinone show similar sectorial patterns as control plants. (**L**) Sink roots of plants grown in a hydrogel matrix treated with 150 μM resorcinol show loss of sectoriality. (**M–P**) Changes in vascular sectoriality as a function of treatment type are depicted as the side-to-side ratio of ^{11}C-radioactivity measured from the radiographic images of leaves and roots. (**M**) Comparison of liquid smoke treatment to control, (**N**) comparison of catechol treatment to control, (**O**) comparison of resorcinol treatment to control, and (**P**) the comparison of hydroquinone treatment to control. The midrib of the leaf and the taproot were used as the dividing lines for radioactivity ratios. A value of 1 in the graph reflects no sectoriality. Data represent means ± SE for $N = 4$ replicates where $p < 0.05$ was statistically significant. (This figure was recreated from [21] with permission of MDPI Publishing)

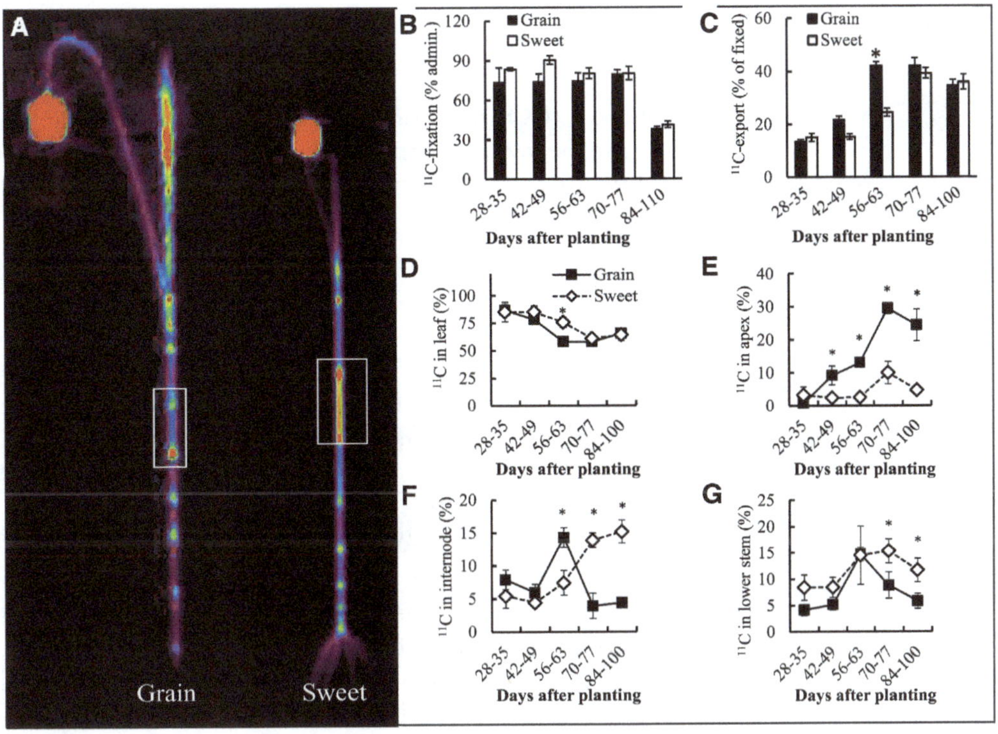

Fig. 9 Partitioning of ^{11}C-photosynthates to different plant tissues as a percentage of the total ^{11}C fixed by the labeled leaf. Representative PET images (**A**) of ^{11}C-photosynthate partitioning 2 h after $^{11}CO_2$ was fed to a single leaf of grain and sweet dwarf sorghum plants. The internode below each leaf sheath vascular connection to the stem is highlighted by a white box. Red colored regions indicate high radioactivity from ^{11}C, while blue and black indicate regions of low radioactivity. ^{11}C fixation (**B**) and export (**C**) changed with plant development. Carbon-11 partitioning is shown for the leaf that was fed $^{11}CO_2$ (**D**), the stem apex, above where the labeled leaf attached to the stem node (**E**), the subtending internode directly connected to the load leaf (**F**), and lower stem internodes (**G**). Mean ± SE ($n = 3$–4). Asterisks indicate time points at which differences between grain and sweet sorghum were statistically significant. (This figure was recreated from [23] with permission of Springer Publishing)

reproductive stages. Photosynthetic rates were found to be similar between sweet and grain sorghum, suggesting that stem sugar accumulation was not driven by differences in the photosynthetic machinery. However, sugar accumulation in sweet sorghum internodes appeared to depend on processes that occurred after export from leaves, which favored partitioning of photosynthates to storage reserves in stems over other sink tissues [23].

5 Notes

1. Optional $^{11}CO_2$ trap cooling: a fan or compressed air stream that can be switched on and directed at the molecular sieve column of the $^{11}CO_2$ trap to rapidly cool the column and speed turn-around time for subsequent experiments.
2. Optional radiation detection on trap: a radiation detector affixed to the $^{11}CO_2$ trap is helpful to monitor how much ^{11}C is produced by the cyclotron, and how much is delivered to the plant during each run.
3. A miniaturized $^{11}CO_2$ trap has been designed to fit inside a hand-held lead pig [15], allowing transport of up to 3.7 GBq, depending on the thickness of the lead. This portable design offers the potential to transport $^{11}CO_2$ short distances, e.g., across campus or a 10–20 min drive from a cyclotron to a plant tracer lab, using proper radiological transport procedures.
4. Transport detectors can provide an inexpensive means to measure transport speed relative to a PET camera. A PET camera is required if 3-D imaging of tissues is desired. A cheaper, albeit destructive, alternative to obtain whole-plant ^{11}C partitioning is to dissect the plant after the incubation period, and immediately measure the radioactivity in each of the plant parts using a calibrated gamma radiation detector, equipped with a well-type ion chamber (e.g., well counter and Capintec).
5. Phosphor plate image analysis software: In theory, phosphor plate images could be analyzed with freeware (e.g., NIH ImageJ). However, use of the manufacturer's software (e.g., ImageQuant) is advisable due to several features designed for use with the instrument. For example, rapid detection of plate overexposure can alert the user to the need to reimage the plant using a shorter exposure time and prevent associated artifacts from being propagated through the data stream.
6. If the load leaf does not fill the leaf cuvette, take a photo of the leaf in place with a ruler looking directly down at the cuvette. With image analysis software, the photo can be used to determine the actual leaf area inside the cuvette and normalize ^{11}C fixation values for differences in leaf area.
7. Allow sufficient distance between the leaf cuvette and the detectors so that background radiation from the cuvette will be minimal relative to the signal detected through the collimation. This distance must be determined empirically in initial trial experiments, because it will depend on how much ^{11}C is administered to the leaf, and what proportion of the ^{11}C is exported from the leaf to the rest of the plant. If export

is low (e.g., mutants defective in sucrose export), additional lead shielding may be positioned between the load zone and the gamma detectors to achieve sufficient signal-to-noise.

8. In plants less than 1 m tall, 1–1.5 h incubation time is usually sufficient for ^{11}C to be distributed throughout the plant for comparison of whole-plant ^{11}C partitioning amongst different plants, or comparison between repeated measures of the same plant (e.g., before and after a treatment).

9. While PET is ideal for nondestructive imaging of thick tissues, such as thick stems or roots potted in soil, phosphor plate imaging is ideal for thin tissues. In thick tissues, positrons emitted by ^{11}C travel a short distance (<1 mm on average) before striking an electron, annihilating, and emitting two gamma rays in opposite directions. This phenomenon allows noninvasive imaging of thick tissues based on coincidence detection of the two gamma rays. The distance of positron travel prior to annihilation, however, limits the theoretical spatial resolution of PET imaging. In thin tissues, positrons may escape from the tissue without encountering an electron, allowing direct imaging of the positrons. Phosphor plate imaging exploits this phenomenon to generate a high-resolution image. Thin tissues can be imaged by PET, but the image may not be quantitative since positrons that escape from the tissue before annihilation will not be seen as radioactivity within the tissue [24]. A simple workaround, if quantitation of thin tissues is desired, is to place material that will trap the positrons against the surface of the thin tissue. For example, aluminum foil or plexiglass can be placed against both sides of a leaf.

10. Several methods have been employed in the past to determine the time when ^{11}C radioactivity arrives in the field of view of a detector or an ROI of PET data. Determining the time when radioactivity rises significantly above baseline can be problematic because it is based on setting an arbitrary threshold above which the radioactivity signal must rise. This method is somewhat subjective because the time of arrival can be influenced by the magnitude at which the threshold is set. Additionally, noise in the data could make the time of arrival appear erroneously earlier or later than the actual arrival of the radioactivity. Using the time at which radioactivity reaches a maximum can be problematic, because data are sometimes not recorded long enough to capture the maximum. The inflection point, or the point where the rate of rise in radioactivity is greatest, can be determined objectively. However, using the time of the inflection point or maximum slope as an average time of arrival is not ideal, because the radioactivity curve may often "spread" as it passes through the stem (Fig. 6B), which

would result in an overestimate of the time of arrival and an underestimation of the transport speed. This spread might be due to leak and retrieval of radiolabeled photosynthates [25, 26], and the anatomical deviations of phloem from a straight pipeline.

Acknowledgments

Preparation of this paper was supported in part by the USDA National Institute of Food and Agriculture, McIntire Stennis project 1009319, 1025767, and 7010544.

Conflict of Interest The authors have no conflicts of interest to declare.

References

1. Babst BA, Coleman GD (2018) Seasonal nitrogen cycling in temperate trees: transport and regulatory mechanisms are key missing links. Plant Sci 270:268–277. https://doi.org/10.1016/j.plantsci.2018.02.021
2. Braun DM, Wang L, Ruan Y-L (2014) Understanding and manipulating sucrose phloem loading, unloading, metabolism, and signalling to enhance crop yield and food security. J Exp Bot 65(7):1713–1735. https://doi.org/10.1093/jxb/ert416
3. Ainsworth EA, Bush DR (2011) Carbohydrate export from the leaf: a highly regulated process and target to enhance photosynthesis and productivity. Plant Physiol 155:64–69. https://doi.org/10.1104/pp.110.167684
4. Sulpice R, Pyl E-T, Ishihara H, Trenkamp S, Steinfath M, Witucka-Wall H, Gibon Y, Usadel B, Poree F, Piques MC, Von Korff M, Steinhauser MC, Keurentjes JJB, Guenther M, Hoehne M, Selbig J, Fernie AR, Altmann T, Stitt M (2009) Starch as a major integrator in the regulation of plant growth. Proc Natl Acad Sci U S A 106(25):10348–10353. https://doi.org/10.1073/pnas.0903478106
5. Babst BA, Karve AA, Judt T (2013) Radiometabolite analysis of carbon-11 biochemical partitioning to nonstructural carbohydrates for integrated metabolism and transport studies. Plant Cell Physiol 54(6):1016–1025. https://doi.org/10.1093/pcp/pct045
6. Hanik N, Gomez S, Best M, Schueller MJ, Orians CM, Ferrieri RA (2010) Partitioning of new carbon as ^{11}C in *Nicotiana tabacum* reveals new insight into methyl jasmonate induced changes in metabolism. J Chem Ecol 36:1058–1067
7. Best M, Koenig K, McDonald K, Schueller MJ, Rogers A, Ferrieri RA (2011) Inhibition of trehalose breakdown increases new carbon partitioning into cellulosic biomass in *Nicotiana tabacum*. Carbohydr Res 346(5):595–601. https://doi.org/10.1016/j.carres.2011.01.018
8. Ferrieri AP, Agtuca B, Appel HM, Ferrieri RA, Schultz JC (2013) Temporal changes in allocation and partitioning of new carbon as ^{11}C elicited by simulated herbivory suggest that roots shape aboveground responses in *Arabidopsis thaliana*. Plant Physiol 161(2):692–704. https://doi.org/10.1104/pp.112.208868
9. Babst BA, Ferrieri RA, Gray DW, Lerdau M, Schlyer DJ, Schueller MJ, Thorpe MR, Orians CM (2005) Jasmonic acid induces rapid changes in carbon transport and partitioning in *Populus*. New Phytol 167(1):63–72
10. Karve A, Alexoff D, Kim D, Schueller M, Ferrieri R, Babst BA (2015) In vivo quantitative imaging of photoassimilate transport dynamics and allocation in large plants using a commercial positron emission tomography (PET) scanner. BMC Plant Biol 15(1):273. https://doi.org/10.1186/s12870-015-0658-3
11. Knoblauch M, Knoblauch J, Mullendore DL, Savage JA, Babst BA, Beecher SD, Dodgen AC, Jensen KH, Holbrook NM (2016) Testing the Münch hypothesis of long distance phloem transport in plants. eLife 5:e15341. https://doi.org/10.7554/eLife.15341
12. Qu W, Robert CAM, Erb M, Hibbard BE, Paven M, Gleede T, Riehl B, Kersting L, Cankaya AS, Kunert AT, Xu Y, Schueller MJ, Shea C, Alexoff D, Lee SJ, Fowler JS, Ferrieri RA (2016) Dynamic precision phenotyping

reveals mechanism of crop tolerance to root herbivory. Plant Physiol 172(2):776–788. https://doi.org/10.1104/pp.16.00735
13. Savage JA, Holbrook NM, Zwieniecki MA (2013) Phloem transport velocity varies over time and among vascular bundles during early cucumber seedling development. Plant Physiol 163(3):1409–1418. https://doi.org/10.1104/pp.113.225359
14. Ross-Elliott TJ, Jensen KH, Haaning KS, Wager BM, Knoblauch J, Howell AH, Mullendore DL, Monteith AG, Paultre D, Yan D, Otero-Perez S, Bourdon M, Sager R, Lee J-Y, Helariutta Y, Knoblauch M, Oparka KJ (2017) Phloem unloading in *Arabidopsis* roots is convective and regulated by the phloem-pole pericycle. eLife 6:e24125. https://doi.org/10.7554/eLife.24125
15. Kim D, Alexoff DL, Schueller M, Babst B, Ferrieri R, Fowler JS, Schlyer DJ (2014) The design and performance of a portable handheld $^{11}CO_2$ delivery system. Appl Radiat Isot 94(0):338–343. https://doi.org/10.1016/j.apradiso.2014.09.008
16. Suwa R, Fujimaki S, Suzui N, Kawachi N, Ishii S, Sakamoto K, Nguyen NT, Saneoka H, Mohapatra PK, Moghaie RE, Matsuhashi S, Fujita K (2008) Use of positron-emitting tracer imaging system for measuring the effect of salinity on temporal and spatial distribution of ^{11}C tracer and coupling between source and sink organs. Plant Sci 175:210–216. https://doi.org/10.1016/j.plantsci.2008.03.022
17. Reeves G, Tripathi A, Singh P, Jones MRW, Nanda AK, Musseau C, Craze M, Bowden S, Walker JF, Bentley AR, Melnyk CW, Hibberd JM (2022) Monocotyledonous plants graft at the embryonic root–shoot interface. Nature 602:280–286
18 Zimmerman DJ, Studer AJ (2023) Maize kernel grafting: a simplified method. In: 65th annual maize genetics meeting, St. Louis, MO. https://documents.maizegdb.org/maizemeeting/abstracts/2023Program.pdf. Accessed 10 Apr 2025
19. Zimmerman DJ, Ferrieri RA, Schueller MJ, Studer AJ (2024) Vasculature connection in maize twin-grafts. In: 66th annual maize genetics meeting, Raleigh, NC. https://documents.maizegdb.org/maizemeeting/abstracts/2024Program.pdf. Accessed 10 Apr 2025
20. Orians CM, Babst B, Zanne AE (2005) Vascular constraints and long-distance transport in dicots. In: Vascular transport in plants. Elsevier Academic Press, Burlington, MA, USA, pp 355–371
21. Noel R, Benoit M, Wilder SL, Waller S, Schueller MJ, Ferrieri RA (2022) Treatments with liquid smoke and certain chemical constituents prevalent in smoke reduce phloem vascular sectoriality in the sunflower with improvement to growth. Int J Mol Sci 23:12468
22. Noel R, Schueller MJ, Ferrieri RA (2024) Radiocarbon flux measurements provide insight into why a pyroligneous acid product stimulates plant growth. Int J Mol Sci 25:4207
23. Babst BA, Karve A, Sementilli A, Dweikat I, Braun DM (2021) Physiology and whole-plant carbon partitioning during stem sugar accumulation in sweet dwarf sorghum. Planta 254:80
24. Alexoff DL, Dewey SL, Vaska P, Krishnamoorthy S, Ferrieri RA, Schueller MJ, Schlyer DJ, Fowler JS (2011) PET imaging of thin objects: measuring the effects of positron range and partial-volume averaging in the leaf of *Nicotiana tabacum*. Nucl Med Biol 38(2):191–200. https://doi.org/10.1016/j.nucmedbio.2010.08.004
25. Ayre BG, Keller F, Turgeon R (2003) Symplastic continuity between companion cells and the translocation stream: long-distance transport is controlled by retention and retrieval mechanisms in the phloem. Plant Physiol 131(4):1518–1528. https://doi.org/10.1104/pp.012054
26. Gould N, Thorpe MR, Pritchard J, Christeller JT, Williams LE, Roeb G, Schurr U, Minchin PEH (2012) *AtSUC2* has a role for sucrose retrieval along the phloem pathway: evidence from carbon-11 tracer studies. Plant Sci 188-189:97–101. https://doi.org/10.1016/j.plantsci.2011.12.018

Check for updates

Chapter 18

Using ^{13}C to Quantify Phloem Transport on Tall Plants in the Field

Daniel Epron, Masako Dannoura and Caroline Plain

Abstract

Phloem transport dynamics play a crucial role in understanding the movement of photoassimilates within plants, particularly in tall species, where tracing carbon allocation and monitoring transport velocities present unique challenges. The difference in time lags between a labeling pulse of $^{13}CO_2$ of the foliage and the appearance of labeled C in the respiration at different locations along the stem of a tall plant is used to estimate at which velocities the isotope tracer, i.e., the labeled carbohydrates, are transported in the phloem sap. Here, we describe a method for pulse labeling tall plants in the field and subsequently tracing ^{13}C in the respiratory efflux of CO_2.

Keywords Carbon transfer, Carbon allocation, Laser absorption spectroscopy, Phloem, pulse labelling, Stable carbon isotope, Stem CO_2 efflux

1 Introduction

Pulse-labeling plants with $^{13}CO_2$ allows tracing of labeled carbon during its transfer from the source leaves, where labeled carbohydrates are produced by photosynthesis, via the phloem, to the sink organs, where these carbohydrates are partly used for respiration, releasing labeled CO_2, whose amount can be measured [1]. For security reasons (no radiation emission) and practical reasons (no decay, no legal restriction for use in the field), stable isotope-based labeling experiments are easier to perform in the field, especially on tall plants like trees (Fig. 1), than similar experiments using radioactive tracers like ^{11}C and ^{14}C [2–4].

Infrared laser absorption spectroscopy-based gas analyzers, such as tuneable diode laser absorption spectroscopy (TDLAS), quantum cascade laser (QCL) spectroscopy, cavity ring down spectroscopy (CRDS), off-axis integrated cavity output spectroscopy (OA-ICOS), or Fourier-transform infrared (FTIR) spectroscopy, enable continuous monitoring of the isotopic signature of respired CO_2, offering

Johannes Liesche and Sanna Sevanto (eds.), *Phloem*, Methods in Molecular Biology, vol. 3066,
https://doi.org/10.1007/978-1-0716-5488-0_18,

Fig. 1 Left: example of a crown labeling chamber installed on a pine tree. Right: example of a stem respiration chamber

the opportunity of a high measurement frequency at relatively low costs [5, 6].

The appearance of the tracer in the respiration of any organ is delayed relative to the labeling pulse, and this time lag depends both on the distance between the source leaves and the sink organ, and the velocity at which carbohydrates are transferred and the rate at which the transported sugars are catabolized to release $^{13}CO_2$ [1]. The difference in time lags observed in respired $^{13}CO_2$ at different positions along the stem (Fig. 2) is used to estimate the rate of carbon transfer and thus the phloem sap velocity. Since our first attempt in 2008 on beeches 10 m high [3], we have successfully used this approach in temperate forests on both broadleaved and coniferous species and in tropical tree plantations [7, 8].

2 Materials

2.1 Whole Crown Labeling Chamber

1. Labeling chamber made of plastic film (Fig. 1, left). The plastic film should exhibit a high transmittance of photosynthetic active radiation and a good tear resistance (e.g., 200-μm polyethylene) (*see* **Note 1**).
2. Two half-plates with semicircular openings form the base of the chamber. The opening should accommodate the stem (no special recommendation regarding the material).

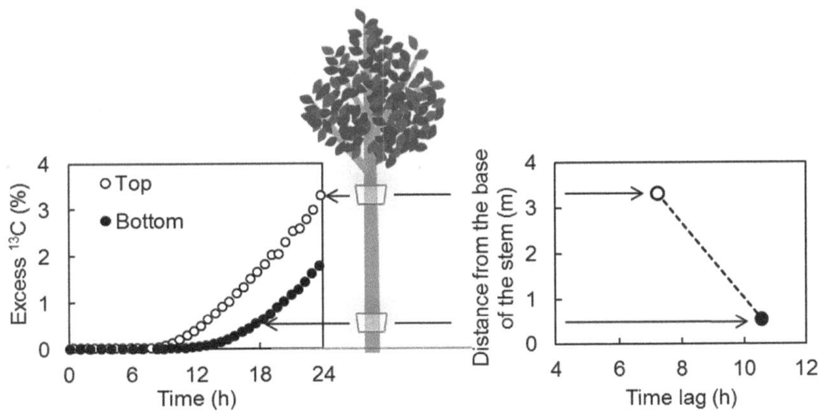

Fig. 2 Schematic diagram showing respiration chambers at two positions along the trunk. Left: time course of the excess ^{13}C of stem respiratory CO_2 efflux after pulse labeling showing the appearance of the tracer in the chamber setup at the base of the crown (top) and at the base of the stem (bottom). Right: relationship between time lag from the start of the labeling to the first appearance of ^{13}C in CO_2 efflux and distance of the chamber from the base of the stem

3. Putty to seal the base plate of the chamber to the stem below the foliage.
4. Scaffolding to fix the chamber above the plant (*see* **Note 2**).
5. Ropes to firmly secure the chamber to the scaffolding.
6. Fans to ensure a good mixing of the air within the labeling chamber (*see* **Note 3**). The power and the number of fans depend on the size of the chamber (*see* **Note 1**).
7. A CO_2 infrared gas analyzer to monitor the decrease in CO_2 concentration after closing the labeling chamber and before starting the labeling.

2.2 Labeling

1. Pure $^{13}CO_2$ (>99% of ^{13}C) in a pressurized gas cylinder (*see* **Note 4**). Around 10–20 l are needed to label trees 8–12 m high for 60 min.
2. A mass flow controller to deliver $^{13}CO_2$ at a fixed rate in the labeling chamber.

2.3 Stem Respiration Chamber

1. Two half-cylinders of typically 20 cm in length form the stem chamber. Made of any convenient material (e.g., polymethyl methacrylate), with semicircular openings to accommodate the stem (Fig. 1, right, *see* **Note 5**).
2. Putty to seal the stem chambers to the stem.
3. Solenoid valves to monitor sequentially two (or more) respiration chambers.
4. A membrane pump to circulate air between the chamber and the analyzer.

2.4 Laser-Based Gas Analyzer

1. A laser-based analyzer able to measure the two CO_2 isotopologues ($^{12}CO_2$ and $^{13}CO_2$), coupled to the respiration chambers either in a flow-through mode or closed-circuit mode, depending on the type of analyzer (*see* **Note 6**).
2. Working standards for calibration of the analyzer: synthetic air with known concentration in [$^{12}CO_2$] and [$^{13}CO_2$] (*see* **Note 7**).

3 Methods

3.1 Experimental Setup

1. Install stem chambers on the stem of the plant, typically one at the base of the crown and one at the base of the stem, i.e., at maximum distance from each other (Fig. 2, *see* **Note 8**), and seal them with putty.

2. Connect the stem chamber to the analyzer using tubing (e.g., in polyethylene/aluminum composite) and start recording $^{12}CO_2$ and $^{13}CO_2$ concentrations.
3. Assemble the scaffold around the plant.
4. Install the base plate of the crown chamber and seal both half-plates with putty.
5. Cover the crown with the labeling chamber.
6. Fix the chamber to the scaffolding.
7. Seal the chamber to the base plate with putty.

3.2 Pulse Labeling

1. Close the chamber and wait for the internal CO_2 concentration to decrease due to photosynthesis (*see* **Note 9**). The lower the amount of $^{12}CO_2$ remains in the chamber, the more CO_2 will be enriched in the chamber after $^{13}CO_2$ injection. Therefore, the amount of $^{13}CO_2$ absorbed will be high, and the signal will be strong.
2. Use the recorded decrease in CO_2 concentration to estimate the whole plant photosynthesis and calculate the flow rate at which $^{13}CO_2$ will have to be delivered.
3. Inject $^{13}CO_2$ into the labeling chamber using a mass flow controller (*see* **Note 10**). First, set the flow rate to a high value to quickly reach the total concentration of CO_2 initially present in the labeling chamber. The air inside the chamber will be greatly enriched in $^{13}CO_2$ without increasing the total CO_2 concentration too much above that of the ambient air (*see* **Note 11**). Then set the flow rate to the value that makes it possible to compensate for the amount of CO_2 absorbed by the crown (typically between 150 and 350 ml/min for 8–12 m tall trees).
4. Label for 30–60 min. The duration of the pulse should be as short as possible for a precise determination of time lags between the start of the labeling and the first appearance of labeled compounds in the respiratory efflux.

3.3 Tracing

1. Measure the isotope signature of the respiratory CO_2 efflux sequentially for the two chambers (*see* **Note 12**). The measurement frequency depends on the type of analyzer and on how many chambers have to be measured, in case more than one plant has been labeled.
2. Flush the respiration chamber with air or open it between two successive measurements.

3.4 Time Lags and Velocity

1. Visually estimate (Fig. 2) or compute the time lag (L) between the start of the labeling and the appearance of ^{13}C in respiratory efflux. The time lag can be computed by fitting a quadratic

function to the relationships between excess ^{13}C (x^E) and the time after labeling (t):

$$x^E = 0 \quad \text{if } t < L$$

$$x^E = a(t - L) + b(t - L)^2 \text{ if } t \geq L$$

2. Calculate the velocity of carbon transfer in the trunk (V) with d, the distance between the two chambers, and L_B and L_T, the time lags for the chamber at the base of the stem (bottom) and at the base of the crown (top) (Fig. 2, *see* **Notes 13 and 14**):

$$V = \frac{d}{L_B - L_T}$$

4 Notes

1. The dimensions of the chamber need to be adapted to the size of the crown of the plant. Chambers of 20–40 m^3 are generally suitable for 8–12 m high trees.
2. The scaffolding must at least be as high as the labeled tree. Please pay attention to local safety rules for building and climbing scaffolding.
3. It is recommended to monitor air temperature, air humidity, and photosynthetic active radiation both inside and outside the labeling chamber. It is recommended to keep the air temperature inside the labeling chamber at a temperature close to that recorded outside, using an air conditioner. Cooling will also help keep relative humidity (RH) below the dew point, preventing the chamber from being covered with water droplets. However, effective cooling can cause a substantial drop in RH, which can also exhibit rapid, large-amplitude variations due to intermittent operation of the air conditioner [9]. Recent tests have shown that adding a humidity-sensor-controlled mist generator to the labeling chamber would solve the problem, as the droplets evaporate more quickly when the humidity is low than when it is high.
4. ^{13}C-enriched CO_2 can also be produced within the labeling chamber by slowly mixing acid (e.g., H_2SO_4 or H_3PO_4) and highly ^{13}C-enriched carbonate (e.g., $Ca^{13}CO_3$ or $Na_2{}^{13}CO_3$) [10].

5. The respiration chamber can be covered by an insulated aluminum sheet to avoid light and an increase in temperature. This will prevent increased respiration and suppress bark photosynthesis, if any. Air temperature within the chamber can be monitored with a thermocouple.
6. The isotope composition of the respiratory efflux can be measured with an isotope ratio mass spectrometer (IRMS). But, unless using a mobile IRMS connected to flow-through respiration chambers [11], the main limitation in the field is the low frequency of sampling and the cost of analysis.
7. Wide ranges of $^{12}CO_2$ and $^{13}CO_2$ concentration are expected: the working standards should therefore bracket the expected concentration of both isotopologues.
8. The approach is based on the measurements of ^{13}C in respiration at two positions along a stem, and the distance between the two positions needs to be large enough so that a difference in time lag can be observed.
9. The foliage should not be kept for too long at low CO_2 concentration to prevent photoinhibition and excessive loss of water due to opening of stomata.
10. Instead of using a mass flow controller, ^{13}C-labeled gas can alternatively be pumped periodically into the chamber through a septum using a syringe connected to a foil bag prefilled with $^{13}CO_2$.
11. Exposing foliage to CO_2 concentrations well above ambient levels should be avoided because it increases photosynthesis, which would affect the amount of carbohydrates available for export, and thus phloem loading and transport.
12. The duration of a measurement is about 5–10 min for one chamber, depending on the type of analyzer.
13. Our calculation assumes that, if transferred carbon is not respired immediately (transient storage after the labeled carbohydrates leave the phloem sap) or if the diffusion time of CO_2 across the bark is not negligible, these delays are similar at all positions along the stem, so they are canceled by the subtraction of the two time lags [7].
14. The precision mainly depends on the measurement frequency (how often flux from a given chamber is measured), and therefore on how many chambers are connected to the analyzer [8].

Acknowledgments

We thank the many colleagues from Brazil, France, Japan, and Thailand who contributed to the development and improvement of this method.

Conflict of Interest The authors have no conflicts of interest to declare.

References

1. Epron D, Bahn M, Derrien D et al. (2012) Pulse-labelling trees to study carbon allocation dynamics: a review of methods, current knowledge and future prospects. Tree Physiol 32:776–798
2. Carbone MS, Trumbore SE (2007) Contribution of new photosynthetic assimilates to respiration by perennial grasses and shrubs: residence times and allocation patterns. New Phytol 176:124–135
3. Plain C, Gérant D, Maillard P et al. (2009) Tracing of recently assimilated carbon in respiration at high temporal resolution in the field with a tuneable diode laser absorption spectrometer after *in situ*$^{13}CO_2$ pulse labelling of 20-year-old beech trees. Tree Physiol 29:1433–1447
4. Högberg P, Högberg MN, Gottlicher SG et al. (2008) High temporal resolution tracing of photosynthate carbon from the tree canopy to forest soil microorganisms. New Phytol 177:220–228
5. Marron N, Plain C, Longdoz B et al. (2009) Seasonal and daily time course of the ^{13}C composition in soil CO_2 efflux recorded with a tunable diode laser spectrophotometer (TDLS). Plant Soil 318:137–151
6. Epron D, Ngao J, Dannoura M et al. (2011) Seasonal variations of belowground carbon transfer assessed by *in situ*$^{13}CO_2$ pulse labelling of trees. Biogeosciences 8:1153–1168
7. Dannoura M, Maillard P, Fresneau C et al. (2011) In situ assessment of the velocity of carbon transfer by tracing ^{13}C in trunk CO_2 efflux after pulse labelling: variations among tree species and season. New Phytol 190:181–192
8. Epron D, Cabral OMR, Laclau J-P et al. (2016) In situ $^{13}CO_2$ pulse labelling of field-grown eucalypt trees revealed the effects of potassium nutrition and throughfall exclusion on phloem transport of photosynthetic carbon. Tree Physiol 36:6–21
9. Duangngam O, Sathornkich J, Thaler P et al. (2023) Measuring photosynthesis of entire tree crowns and pulse label trees in large closed chamber with $^{13}CO_2$ in the field: design and testing. Trends Sci 20:6926
10. Kagawa A, Sugimoto A, Yamashita K et al. (2005) Temporal photosynthetic carbon isotope signatures revealed in a tree ring through $^{13}CO_2$ pulse-labelling. Plant Cell Environ 28:906–915
11. Subke J-A, Vallack HW, Tord M et al. (2009) Short-term dynamics of abiotic and biotic soil $^{13}CO_2$ effluxes after in situ $^{13}CO_2$ pulse labelling of a boreal pine forest. New Phytol 183:349–357

Chapter 19

Non-Invasive Determination of Phloem Transport Speed with ^{14}C

Christopher Vincent, Chiara Bernardini, Peter Minchin and Johannes Liesche

Abstract

Studying the phloem, the tissue responsible for distributing organic substances between plant organs, presents significant challenges due to its location deep within the plant body and its sensitivity to manipulation. One effective approach to investigate phloem transport is the use of tracers to measure transport speed with a noninvasive approach. This chapter provides a detailed protocol for using ^{14}C-labeled photoassimilates to study phloem transport velocity. A key advantage of this method is its noninvasive nature, as the isotope is supplied in the form of $^{14}CO_2$, which is assimilated by source leaves and converted into ^{14}C-labeled sugars. The movement of these labeled sugars is then tracked using photomultiplier-based X-ray detectors positioned near the stem. This method is particularly valuable for researchers without access to ^{11}C generation facilities. It can also be used to study partitioning of photoassimilates among sink organs over time, as well as rates of export from source tissues. Its relatively straightforward methodology makes it suitable for medium-throughput experiments under controlled conditions.

Keywords Phloem transport, Carbon allocation, Sieve element, Isotope, Carbon tracing, Bremsstrahlung, Transport velocity

1 Introduction

The phloem is the transport tissue of higher plants through which sugars, as the primary photoassimilates, and other organic molecules are distributed between leaves, roots, fruits, and other organs. The flow speed of the phloem sap is an important parameter for quantitative understanding of carbon allocation. However, functional data is scarce because this transport pathway is difficult to access experimentally. The phloem sieve elements are buried below several tissue layers and are under high pressure and, thus, are very sensitive to manipulation [1]. Not knowing how fast carbohydrates are transported from leaves to roots in forest trees is seen as an obstacle for anticipating the effects of climate change [2, 3]. Similarly, data on phloem transport speed in herbaceous plants, particularly crops,

Johannes Liesche and Sanna Sevanto (eds.), *Phloem*, Methods in Molecular Biology, vol. 3066,
https://doi.org/10.1007/978-1-0716-5488-0_19,

is crucial for optimizing resource allocation, improving yield, and developing strategies to enhance resilience to environmental stresses.

The method of choice for non-invasive measurements of phloem transport in the field is the tracing of ^{13}C-labeled photoassimilates using trace gas analyzers installed along the stem [4]. However, since the isotopic signature is measured in the respired gas instead of the phloem sap, uncertainty is introduced. Furthermore, the relatively large sample volumes required for measurements with an adequate signal-to-noise ratio prevent the application of this method to small plants. This can be overcome by labeling photoassimilates radioactively. Methods that investigate phloem translocation with radioactive tracers generally rely on destructive sampling [4, 5] or the short-lived ^{11}C-isotope, which emits strong gamma radiation upon decay that can be easily detected from outside the plant after application in the form of $^{11}CO_2$ [6]. Working with ^{11}C has several advantages, including the possibility of three-dimensional imaging over time, however it requires a synchrotron facility near the experimental site due to the 20.3-min half-life of ^{11}C [7].

Here, we provide a protocol to determine phloem sap flow speed by monitoring ^{14}C in vivo. This setup was originally presented by Black et al. [8] to measure carbon partitioning between fruit and roots of kiwifruit and later adapted for measurement of phloem flow speed [9, 10]. Instead of measuring radioactive decay in the form of $ß^-$ particles directly, we measure low-energy X-rays, called Bremsstrahlung radiation, which is produced by collision and attenuation of $ß^-$ particles by surrounding matter [11]. These X-rays penetrate much thicker tissue than the $ß^-$ particles. The long half-life of ^{14}C (5730 years) means it can be used without the need to have a generation source nearby, as for ^{11}C. Instead, ^{14}C is readily available from common suppliers.

With adequate control over environmental conditions, this setup allows for medium-throughput experiments that are essential to identify statistically significant differences between phloem translocation speed in herbaceous plants and young trees. We have used it successfully on Arabidopsis floret spikes [12], long-leaf pine needles [13], and citrus stems and leaves, in laboratory and greenhouse conditions [14, 15]. It can be used to test whether predictions from theoretical models apply [e.g.,_14]. Compared to field experiments, effort and time requirements are very modest. One disadvantage of this method is the distance limit between the signal source and detector, which restricts analysis to plants with relatively thin stems, unless the outer bark is removed. While time resolution can be very high (typically 30 s), spatial resolution is limited by the size of the scintillation detectors (5 cm diameter). Resolution can be increased by masking areas of the detector with lead to limit detection to a smaller area if the signal intensity is sufficiently high.

The intended result of this method is a rate of linear distance over time (speed = Δ distance/Δ time). The experimental procedure involves pulse-labeling of a source leaf with $^{14}CO_2$ supplied to a leaf chamber (Fig. 1). The signal of ^{14}C-labeled substances in the phloem, in most plants primarily sucrose, is then traced by time-series recordings from at least two scintillation detectors positioned along the stem (Fig. 2). From the intervals between arrivals of the tracer pulse at the different detectors, values for phloem transport speed are calculated (Figs. 2 and 3). There are various approaches to measuring the time between arrivals of a labeled pulse. Each has weaknesses and strengths, which are briefly discussed in the following.

Method 1: Direct first-time of arrival detection—First detection could be considered an approach to measuring maximum speed. The speed would be the distance traversed by the earliest measurable portion of the pulse. This can be calculated by identifying the baseline (background radiation) and calculating confidence intervals

Fig. 1 Illustration of the basic experimental setup for the determination of phloem transport speed with ^{14}C. A photosynthetically active leaf is exposed to $^{14}CO_2$ in a leaf chamber for a certain time period (pulse-labeling). The phloem transport of isotopic carbon, built into photoassimilates, is then traced by X-ray detectors positioned along the stem. The setup can be adapted to herbaceous plants such as Nicotiana benthamiana (shown here) or trees (shown in Fig. 2)

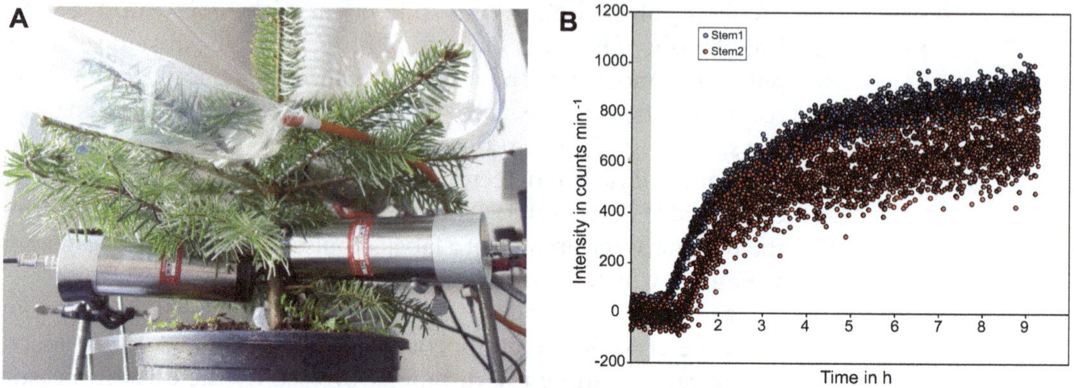

Fig. 2 Example of phloem transport speed measurement using ^{14}C carried out on a 2-year-old *Abies nordmanniana* tree. (**A**) The experimental setup with a leaf chamber (plastic bag) and two detectors positioned along the stem. Note that for broad-leafed plants, it is recommended to position detectors on the same side of the stem. Adequate ventilation as well as lighting is important. (**B**) Signal registered at the two detectors, with the upper detector (blue) showing an earlier and stronger signal increase than the lower detector (red). The difference in amplitude is due to photoassimilates being unloaded from the phloem along the stem. Gray area indicates the labeling period

for this baseline (Fig. 3A). The time of the first detections above the confidence intervals (in Fig. 3A, first data points above the red line) would be considered the time of arrival, and the distance between the times of arrival between two detectors would be the speed based on first detection. *Advantages*: Slope is not considered (see descriptions below of why a linear slope can complicate analysis). Calculation is simple after the variance parameters are determined. *Disadvantages*: Noise can make the exact point of first arrival difficult to determine, and measurement time depends on the sensitivity of both detectors (Figs. 2B and 3A). Errors primarily occur with low signal-to-noise ratios, which affect the detectors further away from the source the most. Summarizing count rates over 5–10-min blocks is usually not a solution because of the loss of time resolution, which can have a large impact on speed estimates if the detectors are close together. The detection of the first arrival can be used when all detectors show a high signal-to-noise ratio.

Method 2: Slope-based first arrival estimation—This approach also seeks to simplify the variability in detections by seeking the point of first detection. This approach uses the early linear portion of the signal increase as the pulse passes a detector (Fig. 3A, black, diagonal line). The slope of this line is regressed to the baseline to determine the time of first arrival. *Advantage*: This approach is likely closer to the true time of arrival, which is earlier than the sensitivity allowed by noise in many cases. *Disadvantages*: This approach introduces subjectivity in selecting what portion of the pulse slope is "linear." A different researcher might select a slightly different time segment over which to determine the slope. The different slope that emerged

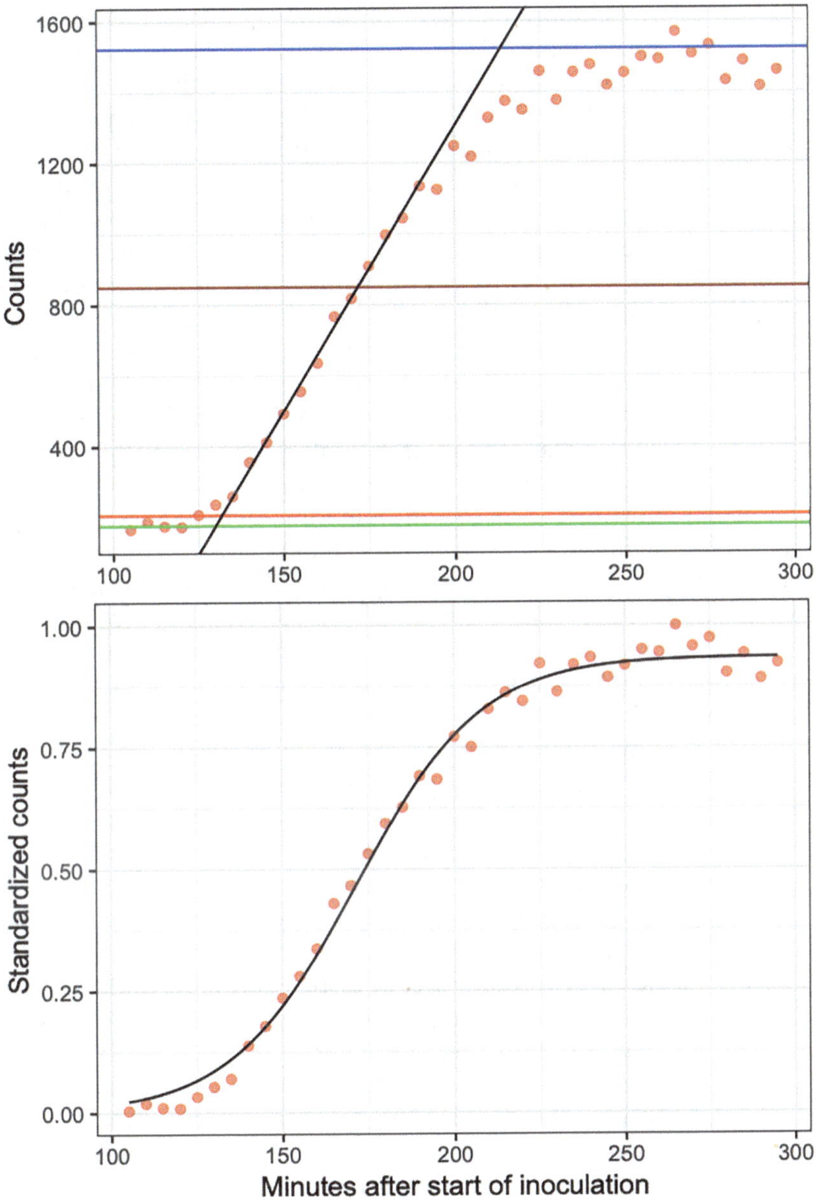

Fig. 3 (**A**) X-ray detector counts of the arrival of a ^{14}C pulse along the stem. Points represent 5-min sums of counts. Lines represent value estimates in the different approaches to estimating pulse arrival: baseline (green), 95% confidence interval of baseline (red), slope of linear portion of pulse (black), pulse height (blue), pulse half-height (brown). (**B**) Standardized values of the same counts to simplify fitting the logistic curve and make it comparable among detectors with different maxima and baseline levels, and includes a greater portion of the pre-pulse period. The black line represents the logistic curve. Note that the half-height estimates vary between the linear and the logistic approach

would result in a different estimate of the time of first arrival. Additionally, the assumption of a linear slope of pulse arrival is likely false, though useful for the sake of calculation. Over an extended time, the rise is approximately exponential, with the early part being approximately linear; however, a small part of any curve is approximately linear, making this approximation valuable.

Method 3: Time of half-height—This approach estimates the average time of arrival across the pulse, rather than the time of first arrival. *Advantages*: Rather than extrapolating outside the linear region, as Method 2 does, this approach is based much more on the tracer pulse and is quite independent of the tracer measurement sensitivity. This method, being based on a larger portion of the tracer pulse, gives a speed estimate incorporating more of the tracer profile and hence will be closer to an average speed. An advantage of this method is that it does not depend on the early profile shape, which is not accurately measured, and the timing as to when the tracer profile reaches half of its peak height is usually well defined. But it does depend on the tracer profile reaching a peak, which requires the measurements to continue for a considerably longer time to allow defining the peak height. *Disadvantages*: This method introduces a third variable (pulse height; Fig. 3, line d). Estimation of this variable depends on the timeframe selected, as in some experiments, the signal does not reach a clear peak within the time of the experiment (see Figs. 2B and 3). This may introduce additional subjectivity to the estimate, and estimates may vary between researchers, with no simple way a priori to ensure consistency.

Method 4: Logistic regression—This method requires more complicated calculations for data analysis than the previous three methods. However, it removes much of the subjectivity from the arrival estimation process. It can be used to estimate the time (point on the x-axis) of arrival at half height, or at any designated proportional height within the curve. The method consists of fitting a logistic curve to the pulse. Logistic curves follow the same pattern as is used in a logistic regression, but they do not require binary responses (Fig. 3B). Other nonlinear curves may be used to similar or improved effect, however, logistic regression inherently estimates the x value at half-height, greatly simplifying time of half-height calculation, even when not all portions of the line follow the pulse closely (Fig. 3B). *Advantages*: This method does not have the problems of subjective selections of timespans to calculate slopes or pulse heights. It also provides the additional variable of slope. *Disadvantages*: Requires statistical software or a programming language of some kind. Some curves may not fit the logistic pattern well.

Method 5: Whole-pulse speed distribution: The approach is agnostic about the shape of the curve, including exactly where it begins and ends. Rather, it uses the distance along the x-axis (time) at many points over the curve to generate many estimates of speed for each detector pair. Briefly, cumulative counts for each detector are summed, and standardized relative to the total at the end of the relevant time period (e.g., 24 h). Then, a high-resolution spline is fit to the relative cumulative counts for each detector, and the time point for each relative count point is calculated (essentially switching *x*- and *y*-axes). Finally, for each of these relative count points for each detector, a time difference can be calculated. The result is thousands of speed estimates across a pulse arrival curve, the distribution of which can be assessed. *Advantages*: This method does not require a mathematical function that accurately describes the shape of the pulse, and it provides more robust estimates than Methods 1–4 because its estimates include the entire pulse, rather than attempting to summarize each pulse with a single value. For the same reason, asymmetry between the shapes of two pulses is less likely to skew the results of this approach than those of Methods 1–4. *Disadvantages*: This also requires some statistical software and may still be susceptible to judgment about the starting and end points of curves.

2 Materials

1. Plant material: A plant generally has to be taller than 10 cm to arrange two detectors along the stem. In the case of Arabidopsis, we have used the floret spike in lieu of stem tissue for this reason. The maximum height depends on where the experiment is carried out, e.g., in a fume hood.
2. Growth lights: LED or halogen lamps that provide light intensity at adequate levels for the respective plant species.
3. Sodium Bicarbonate [14C]: Specific Activity: 40–60 mCi (1.48–2.22 GBq)/mmol.
4. Saturated citric acid solution.
5. Leaf chamber: For example, a zip-lock plastic bag. 3D-printed boxes with plexiglass tops or other transparent designs can be used. For optimal labeling efficiency, the chamber should leave as little air around the leaf as possible. Ideally, an air pump with tubing produces circulatory air movement to improve radiolabel exposure of the leaf. In some cases, we have used chambers with circulating pumps with a separate chamber with a septum to replace tape as seals and provide greater volume of gas to reduce [CO_2] fluctuations.
6. Transparent tape.
7. Syringe, 1 ml.
8. Scintillation detectors with a thin beryllium window (50 μm) and NaI crystal (1 mm thickness).
9. BNC cables of 1–3 m length.
10. Single-channel analyzer. Models that combine several single-channel analyzers in one housing and provide a USB

connection for control of all channels via software are most convenient.

11. Clamps and stands.

3 Method

^{14}C carbon is radioactive and must be handled accordingly. In this protocol, an isotope with an activity of 75 μCi (2.75 MBq) is used for one measurement. This level allows performing the experiment in standard laboratory or greenhouse conditions in some countries, while it is limited to specialized laboratories in others.

3.1 Setting Up the Experiment

1. Install growth lights in a fume hood or on a lab bench with adequate ventilation (*see* **Note 1**).
2. Set up the radiation detection equipment by connecting the scintillation detectors with single-channel analyzers and the analyzers with a computer.
3. Transfer the plants to the experimental site and arrange the scintillation detectors with clamps and stands. At least two detectors should be positioned at the stem, in line along the expected direction of transport, which is usually below the source leaf (*see* **Note 2**).
4. Measure the distance between detectors (*see* **Note 3**).
5. Loosely fit the leaf chamber or bag on the target leaf without restricting air flow toward the leaf, or on the rosette in case of small plants, i.e., Arabidopsis.
6. Let the plant acclimate to experimental conditions for 24 h (*see* **Note 4**).
7. Start signal recording to measure the background noise.

3.2 ^{14}C Application and Tracing

1. Close the leaf chamber with tape
2. Start signal recording. Use 1 min count rate integration time (*see* **Note 5**).
3. Inject 150 μl ^{14}C-bicarbonate to form a drop in one corner of the leaf chamber.
4. Inject 150 μl citric acid at the same location as the ^{14}C-bicarbonate. Mix quickly by pulling and pushing liquid with the syringe.
5. Seal the hole with tape if not using a septum.
6. Wait for pulse-labeling period, typically 30 min (*see* **Note 6**).
7. Open the leaf chamber by cutting a corner of the plastic bag. Use an air pump to flush air containing radiolabeled CO_2 from the chamber.

8. Wait until the signal front has passed all detectors and the curve of the pulse reaches the plateaux.
9. Discard plant according to radiological safety regulations.

3.3 Data Analysis

See the evaluation of different approaches to data analysis in the introduction to choose the most appropriate (*see* **Note 7**).

Direct First Arrival Detection

1. Determine noise level from baseline by calculating the 95% confidence interval (Fig. 3A, red line)
2. Determine the time at which the signal increases over the noise level for each detector.
3. Calculate speed with

$$\text{Speed} = \frac{\text{distance}_{2-1}}{\text{time}_2 - \text{time}_1}.$$

Slope-Based First Arrival Estimation

1. Calculate the time of arrival x for each detector with $x_1 = \frac{b_{\text{baseline1}} + b_1}{m_1}$

   Which is derived from the fact that the baseline of detector 1 (Fig. 3, line a) is:

$$y_{\text{baseline } 1} = b_{\text{baseline } 1}$$

   And the linear slope of the pulse (Fig. 3A, black line) is:

$$y_{\text{slope } 1} = x_1 \times m_{\ 1} + b_1$$

2. Calculate speed with:

$$\text{Speed} = \frac{\text{distance}_{2-1}}{x_2 - x_1}.$$

Time of Half-Height

At a given detector, this method consists of (1) estimating the baseline as in Methods 1 and 2, (2) estimating the slope of the linear portion of the pulse as in Method 2, and (3) estimating the pulse height. Estimating the pulse height requires isolating the time period in which counts are not increasing or decreasing significantly (*h;* Fig. 3A, purple horizontal line). The half height (*hh*; Fig. 3A, brown horizontal line) can be estimated at Detector 1 as:

$$hh_1 = \frac{h_1 - y_{\text{baseline1}}}{2} + y_{\text{baseline1}}$$

Thus, the time of arrival at the half-height is the intersection of the slope of the pulse and the half-height.

$$x_1 = \frac{hh_1 + b_1}{m_1}$$

Logistic Regression

Logistic curve fitting can be done in a variety of software packages. Because the R programing language is open-source and free, code using Base R Stats including the nlme package [16] is presented here. Although the logistic curve fits most cases we have observed well, it should not be used in cases that do not fit it.

1. Although not strictly necessary, for ease of visualization, it may be useful to normalize counts (Fig. 3B). We use a simple minimum–maximum standardization, for each channel:

$$y_{\text{standardized}} = \frac{y - \min}{\max - \min}$$

2. Create the logistic function with:

```
logis<-function(x, Asym,xmid,scal) Asym/(1+exp
(-(x-xmid)/scal))
```

In this example, the expression following the word function names the parameters to be estimated. The following expression defines the function. In this case, Asym represents the asymptote of the curve, xmid is the point along x where the half-height is reached, and scal is a slope parameter (*see* **Note 8**).

3. Fit the model to the data with

```
logis.nls<-nls(standard.sum~logis(time,Asym,
xmid,scal)
data=data
start=c(Asym=100,xmid=350,scal=50))
```

In this expression, the nls prompts fitting of the function to the data, where "time" becomes x. nls requires estimated starting values for each parameter to be estimated.

3. Call the parameter estimates for the fit with *summary (logis.nls)*

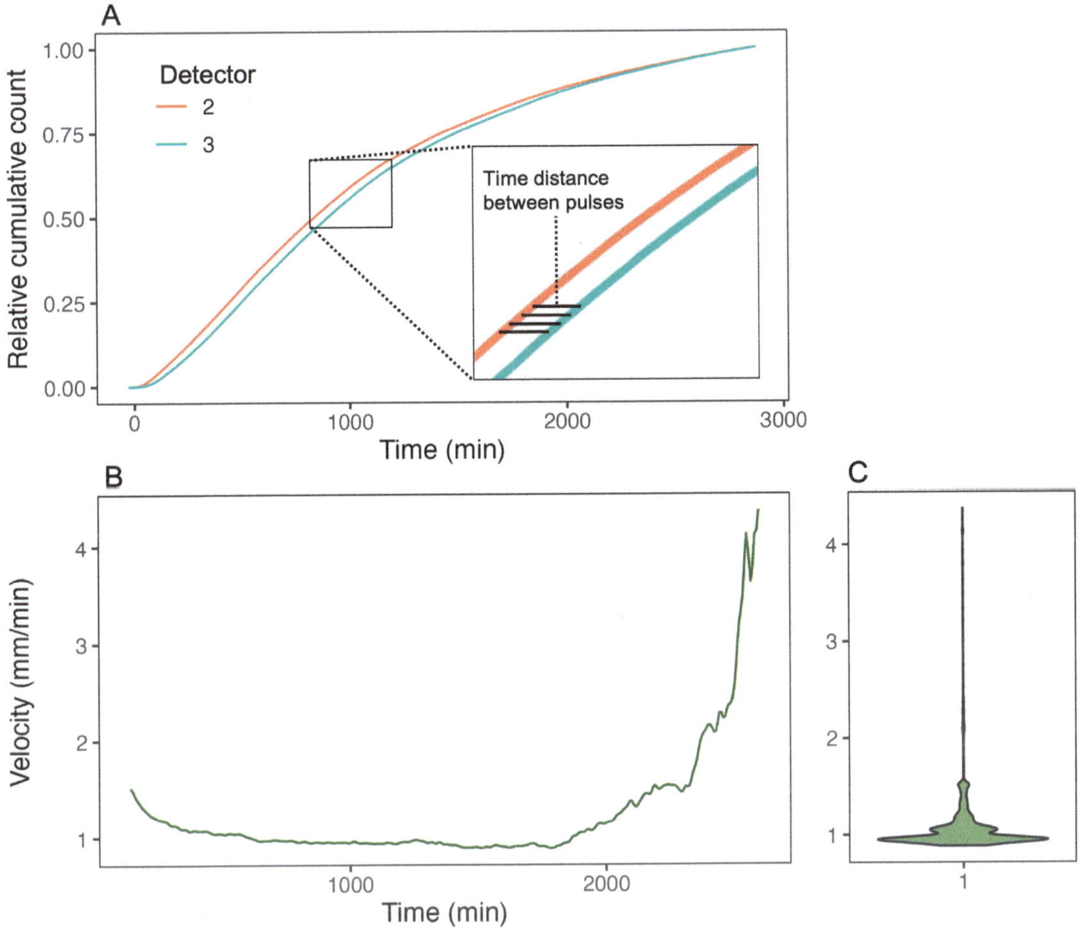

Fig. 4 Illustration of analysis of a single pulse in a longleaf pine (*Pinus pallustris*) needle using the whole-pulse method. (**A**) The arrival of the relative cumulative counts at two X-ray detectors over 24 h after labeling with ^{14}C. (**B**) The velocity estimates over the same time period. (**C**) The distribution of estimates from the pulse. Here the distribution of speed ranges from less than 0.5 mm min^{-1} to greater than 4 mm min^{-1}. However, the velocity estimates are highly concentrated around the 1 mm min^{-1} speed

Whole-Pulse Speed Distribution

Here we will also present an R-based implementation of this approach, though it could be performed using other data management software.

1. Fit a spline to predict 1000 values of time based on relative cumulative sum (cum_sum) counts of each detector (essentially, now time is the dependent variable and summed count is the independent variable):

```
spline.pred <- smooth.spline(cum_sum, time, df = 1000)
```

2. Join the data frames for each detector, such that there are parallel columns for the predicted time for the same relative cumulative count for each detector:

```
cbind(detector_1,detector_2)
```

3. Create a new variable to calculate speed (here, our example uses a "pipe" requiring the tidyverse package.

```
t_1_dark%>%
mutate(delta=time_2-time_1,#calculate time
difference between detectors
distance=63,#input distance between detectors in cm
speed=distance/delta)#calculate velocity
```

The result is now approximately 1000 estimates of speed (Fig. 4). Some portion of these, especially at the beginning or end of pulses of detectors that are close together, may have extremely high or low values, but these can be eliminated by selecting out the first and last 1%–3% of values. The result is a distribution of single pulse speeds across two detectors.

4 Notes

1. Experimental conditions, such as the light intensity, have strong influence on carbon export and presumably also speed [17, 18].
2. An additional detector can be directed at another part of the stem to test if/how much photoassimilate is transported toward the apex. Furthermore, one detector can be directed at the source leaf or at the rosette to follow carbon export dynamics.
3. Detectors are treated as measuring a single point. This is done knowing that it is not accurate. Detectors measure a region from which they may detect radiation. Typical detector diameter is approximately 5 cm. Thus, the detected area is approximately 5 cm along the stem. Additionally, detectors are not equally sensitive along the entire face, but rather, sensitivity decreases with distance from the center. However, because it is impossible to distinguish different origins of excitations along the detector, the area is summarized in an average distance from the adjacent detector. In this case, it is ideal to consider distances between the centers of adjacent detectors as representative of the average distance between them. One approach to limiting the source region of the signal for any detector is to wrap the detector in lead sheeting. This prevents detection of radiation coming from the sides or near the edges of the detector face.
4. Handling can influence phloem transport. The time to recover normal function can be different for different species. Some might be fine after 2 h.

5. Basic data formats can take many forms, and the specific arrangement of data depends on what is required for use by the intended analysis software. Most counters allow for counting over a range of intervals. As counts are typically expressed in counts per minute, we recommend this setting, which can be summarized if the 1-min time resolution is not required. The most advisable data arrangement is in "long-format," in which each line in the data represents a single observation (total counts in one minute) on a single counter. This arrangement is easy to tabulate and manipulate in most data management software packages or languages.
6. The duration of the labeling period can be adapted according to the plant species. Conifers, for example, usually need a 1 h pulse to generate a sufficiently strong signal, while plants with more photosynthetic surface area may require less time.
7. A few factors contribute to discrepancies among approaches. One is that not all labeled carbohydrates move at the same speed. The result is a slope of increasing detections from the baseline to some maximum. The maximum may be determined by the combination of detector sensitivity, distance from the stem, and the maximum possible signal from that portion of the stem. The slope may be determined by the rate with which the labeled pulse mixes with non-labeled carbohydrates loaded into the phloem and the number of phloem sieve elements and the varying rates of phloem transport among them.
8. It is possible to find the points at heights other than half-height by finding the intersection with the chosen height, but the combined estimates of xmid and scal are robust in this case to determine the time of arrival at half height and the slope of the pulse arrival [19].

Conflict of Interest

The authors have no conflicts of interest to declare.

References

1. Liesche J, Patrick J (2017) An update on phloem transport: a simple bulk flow under complex regulation. F1000Res 6:2096
2. Bonan GB (2008) Forests and climate change: forcings, feedbacks, and the climate benefits of forests. Science 320:1444–1449
3. Litton CM, Raich JW, Ryan MG (2007) Carbon allocation in forest ecosystems. Glob Change Biol 13:2089–2109
4. Blechschmidt-Schneider S (1990) Phloem transport in *Picea abies* (L.) Karst in mid-winter. Trees 4:179–186.

5. Dickson RE, Isebrands JS, Tomlinson PT (1990) Distribution and metabolism of current photosynthate by single-flush northern red oak seedlings. Tree Physiol 7:65–77
6. Hubeau M, Steppe K (2015) Plant-PET scans: in vivo mapping of xylem and phloem functioning. Trends Plant Sci 20:676–685
7. Minchin P, Thorpe M (2003) Using the short-lived isotope 11C in mechanistic studies of photosynthate transport. Funct Plant Biol 30:831–841
8. Black MZ, Minchin P, Gould N, Patterson K, Clearwater MJ (2012) Measurement of Bremsstrahlung radiation for in vivo monitoring of C-14 tracer distribution between fruit and roots of kiwifruit (*Actinidia arguta*) cuttings. Planta 236:1327–1337
9. Liesche J, Jensen KH, Minchin P, Bohr T, Schulz A (2013) Theoretical and experimental determination of phloem translocation speeds in gymnosperm and angiosperm trees. Acta Hort 991:45–52
10. Liesche J, Windt C, Bohr T, Schulz A, Jensen KH (2015) Slower phloem transport in gymnosperm trees can be attributed to higher sieve element resistance. Tree Physiol 35:376–386
11. Sowinski P, Bednarek B, Jelen K, Kowalski TZ, Ostrowski KW (1990) An in vivo method for the transport study of assimilated substances using C-14 isotope and X-ray proportional-counters. Acta Physiol Plant 12:139–148
12. Bernardini C, Santi S, Mian G et al. (2022) Increased susceptibility to Chrysanthemum Yellows phytoplasma infection in *Atcals7ko* plants is accompanied by enhanced expression of carbohydrate transporters. Planta 256:43
13. Liesche J, Vincent C, Han X, Zwieniecki M, Schulz A, Gao C, Bravard R, Marker S, Bohr T (2021) The mechanism of sugar export from long conifer needles. New Phytol 230(5):1911–1924
14. Li SY, Hussain SB, Vincent C (2024) Response of carbon fixation, allocation, and growth to source-sink manipulation by defoliation in vegetative citrus trees. Physiol Plant 176(3): 14304
15. Welker S, Pierre M, Santiago JP, Dutt M, Vincent C, Levy A (2022) Phloem transport limitation in Huanglongbing-affected sweet orange is dependent on phloem-limited bacteria and callose. Tree Physiol 42(2):379–390
16. Pinheiro J, Bates D, DebRoy S, Sarkar D, Core Team R (2018) nlme: linear and nonlinear mixed effects models. R package version 3.1-137. https://CRAN.R-project.org/package=nlme
17. Xu Q, Chen S, Yunjuan R, Chen S, Liesche J (2018) Regulation of sucrose transporters and phloem loading in response to environmental cues. Plant Physiol 176:930–945
18. Knox K, Paterlini A, Thomson S, Oparka K (2018) the coumarin glucoside, esculin, reveals rapid changes in phloem-transport velocity in response to environmental cues. Plant Physiol 178:795–807
19. Epron D, Bahn M, Derrien D, Lattanzi FA, Pumpanen J, Gessler A, Hogberg P, Maillard P, Dannoura M, Gerant D, Buchmann N (2012) Pulse-labelling trees to study carbon allocation dynamics: a review of methods, current knowledge and future prospects. Tree Physiol 32:776–798

Check for updates

Chapter 20

Measuring Phloem Transport Velocity on a Tissue-Level Using a Phloem-Mobile Dye

Jessica A. Savage and Maciej A. Zwieniecki

Abstract

Here, we describe an in vivo dye-tracking method for measuring phloem transport velocity in seedlings, leaves, and petioles and potentially other translucent plant tissues. The method requires measurement of the fluorescent signal of a phloem-mobile dye using sensitive photo-sensors placed externally to the plant. Following dye application, velocity is determined by either following a dye pulse or using laser fluorescence bleaching. Velocity is estimated by dividing the distance traveled by the dye by the time it takes to travel. This method can be used to measure phloem transport velocity on intact plants with minimal disturbance and has the potential to be used under a variety of growth conditions. Because there are large differences among species in their anatomy, this method should be optimized for individual plants and tissue types.

Keywords Transport velocity, Phloem transport, Carboxyfluorescein diacetate, Esculin, Fluorescence, Phloem-mobile dyes, Fluorescence bleaching, FRAP, Carbon allocation

1 Introduction

Phloem-mobile dyes have been used extensively for imaging translocation [1, 2], studying symplastic continuity between cells [3, 4], and estimating phloem transport velocity in individual sieve tubes [5, 6]. The majority of research that utilizes these dyes has examined phloem on a cellular level using confocal and fluorescent microscopy. However, many phloem-mobile dyes produce a strong fluorescent signal that is detectable with the naked eye or with a photo-sensor (e.g., a high-gain photodiode or photomultiplier) and can be imaged or measured without cutting into or manipulating the tissue of intact plants [7, 8]. Here, we outline a method for estimating phloem velocity on a tissue level by tracking the movement of a laser-bleached front or dye pulse of a phloem-mobile fluorescent dye using photo-sensors placed external to the plant. This technique has been used to estimate linear transport velocity in intact seedlings and leaves of woody and herbaceous plants [7, 9, 10]. It needs to be optimized for different

Johannes Liesche and Sanna Sevanto (eds.), *Phloem*, Methods in Molecular Biology, vol. 3066,
https://doi.org/10.1007/978-1-0716-5488-0_20,

species and tissue types because of variation in how light penetrates the tissue and differences in vascular architecture.

When bleaching is involved, the method is similar to FRAP (fluorescence recovery after photobleaching) techniques used in microscopy and involves tracking a dip in the fluorescence signal after the dye is temporarily bleached with a laser [5, 6, 11]. In the simplest form, a single photo-sensor is placed downstream of the laser, and phloem transport velocity can be calculated based on the time it takes the bleach front to travel to the sensor (Fig. 1A). Bleaching is required when using dyes like carboxyfluorescein diacetate (CFDA) that do not produce a clear front or pulse of fluorescence [5, 10, 12]. For dyes that do produce a clear pulse, like esculin (6,7-dihydroxy-coumarin 6-b-D-glucopyranoside), this technique can be applied without bleaching. When using this dye, two photo-sensors are placed along the sample (Fig. 1B), and phloem transport velocity is estimated by dividing the distance between the two photo-sensors by the time it takes the pulse to travel between them.

There are multiple dyes that have been used as phloem tracers [13, 14] that can be used for this technique. Each dye requires adjustment to the setup to account for differences in its excitation and emission wavelengths. Here we focus on two dyes, CFDA and esculin. CFDA is one of the most common dyes used for studying phloem translocation [1]. It is membrane-permeable when it is applied in the ester form, but becomes membrane-impermeable once it enters a living cell and is cleaved by esterases. The resulting

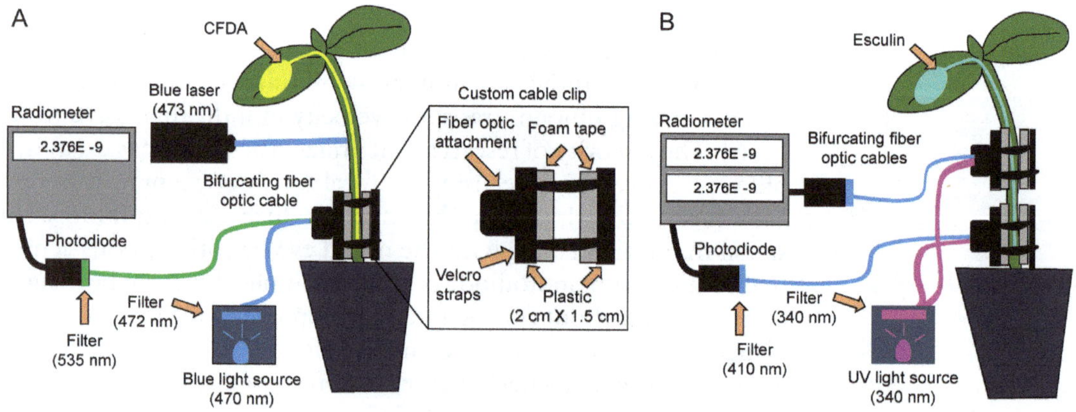

Fig. 1 (**A**) Equipment set-up for measuring phloem transport velocity in seedlings with CFDA. A custom cable clip is attached to the plant that holds a bifurcating fiber optic cable against the hypocotyl. One end of the fiber optic cable is attached to a light source, and the other is connected to a photo-sensor (e.g., a photodiode). A laser is targeted upstream of the clip on the hypocotyl. The fluorescent tracer is applied to a cotyledon and travels through the phloem. It is excited by the light source, and the intensity is recorded with a radiometer attached to the photodiode. (**B**) Equipment set-up for measuring phloem transport velocity with esculin. Two custom cable clips are attached to the hypocotyl with bifurcating fiber optic cables attached to a light source and separate photo-sensors. Note that the filters used for the two dyes are at different wavelengths

carboxyfluorescein (CF) remains in the symplast and readily moves in the phloem. The CF form is pH sensitive with low fluorescence activity in the range of apoplastic pH and high in the range of cellular pH, thus allowing for phloem-specific observations. Because CF is sequestered in sieve elements, it does not create a clean pulse of dye that can be tracked easily at a tissue level. As a result, it requires bleaching to create a clear front (Fig. 1A).

More recently, esculin has become a popular tracer used for tracking sap movement in the phloem [15–17]. Different than CFDA, which is initially membrane-permeable and can enter the phloem directly, esculin is transported into sieve elements by the sucrose transporter AtSUC2 [18, 19]. In species that contain this transporter, the dye only needs to get into the apoplast of a leaf to be loaded into the phloem. Movement of this dye is thought to be a good proxy for sugar movement in the phloem because transport occurs at the same rate as radiolabeled sucrose [15]. This dye creates a clear dye front that can be easily tracked with photo-sensors or with a fluorescent microscope (Fig. 1B).

2 Materials

Set up the equipment for measuring fluorescence under desired growth conditions (e.g., under a grow light and appropriate environmental conditions for the plant). If using a variation of the method that requires a laser, the equipment must be set up in a space that meets appropriate safety requirements. All work with the laser must be done following appropriate safety protocols and using laser safety goggles.

2.1 Equipment for Measuring Fluorescence with Photo-Sensors

1. Radiometer with one or two photo-sensors like high-gain photodiodes or photomultipliers (*see* **Note 1**). Two are needed if you are not using a laser. One is needed if you are using a laser.
2. A computer (datalogger) that can connect to a radiometer and log data.
3. LED source (blue, 470 nm for CFDA and ultraviolet, 340 nm for esculin).
4. Optional narrow band-pass filter for excitation: 472, 30 nm bandwidth for CFDA and 340 nm, 30 nm bandwidth or short-pass filter for esculin (*see* **Note 2**).
5. Narrow band-pass filter/s for emission: 535, 30 nm bandwidth for CFDA and 410, 30 nm bandwidth for esculin. You need one filter for each photodiode/photomultiplier.
6. Diode-pumped solid-state laser: 473 nm, 20 mW required when using CFDA but not esculin.

7. Bifurcating fiber optic cable: one cable per photo-sensor.
8. Custom cable clip (see Fig. 1 and **Note 3**): one clip per fiber optic cable.
9. Breadboard with optomechanical components (as needed, *see* **Note 4**).
10. Laser safety goggles if using a laser.

2.2 Dye Application

1. CFDA phloem tracer: Dissolve 100 mg of carboxyfluorescein diacetate (CFDA) in 1 mL of acetone and keep at 4 °C. When ready to apply to the plant, mix 5 μL of stock solution with 95 μL of water. The mixture will create a white precipitate (*see* **Note 5**).
2. Esculin phloem tracer: Dissolve 10 mg of esculin (6,7-dihydroxy-coumarin 6-b-D-glucopyranoside) in 1 mL of water (*see* **Note 5**).
3. Soft tip marker.
4. Sandpaper: fine-grain, 500-grit for application to the surface of soft leaves.
5. Plastic film to cover the site of dye application on the leaf surface.
6. Small syringe without a needle or dermal roller for infiltration of dye into hard leaves.

2.3 Anatomical Analysis

1. 1 cc syringe with hypodermic needle.
2. 0.1% aqueous safranin O.
3. Razor blades.
4. Microscope slides.
5. Compound fluorescent microscope with GFP filter with low magnification objectives (e.g., X4 or X10, depending on sample size).

3 Methods

3.1 Set Up

1. Allow the plant to acclimate to the lab conditions for several hours before starting measurements (*see* **Note 6**).
2. Assemble components of the system (see Fig. 1A, B).
3. Place custom cable clip/s on the plant (*see* **Note 4**). If using two clips, position them at least 10 mm apart.
4. Turn on the radiometer and check the background fluorescence (*see* **Note 7**).

5. If using a laser, align it to strike at least 10 mm above the custom clip. Mark the location where it hits the plant with a soft-tipped marker (*see* **Note 8**).

3.2 Dye Application (see Note 9)

Surface Application Method

1. Abrade a source leaf gently with sandpaper or a dermal roller. Be careful to minimize damage to the leaf or cotyledon. If a leaf is being measured, the dye should be applied at least 3 cm away from the clip to avoid accidental dye transfer.
2. Add a drop (<50 μL) of dye on the abraded portion of the leaf, followed by a small piece of plastic to minimize evaporation.
3. If using esculin, dye can be left on the leaf for 10–30 min and then rinsed with water to create a pulse. Plastic should remain on the abraded leaf to avoid desiccation of the application site.

Infiltration Method

1. Fill the syringe tip with dye while a large volume of the syringe remains air-filled and gently press the tip against the leaf abaxial surface (or side with stomata present).
2. Increase pressure in the syringe and wait a few minutes for the dye to penetrate the internal leaf volume.

3.3 Measurement of Bleach Front Using CFDA

1. Monitor the radiometer readout for a large gain in radiation above the background (see Fig. 2A). This increase reflects the arrival of the fluorescent dye to clip (usually, the signal will slowly increase before reaching saturation).

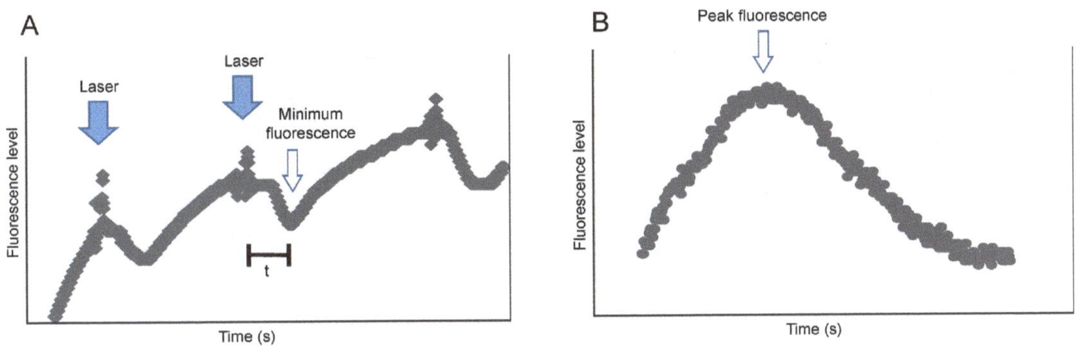

Fig. 2 (**A**) Output graph from photodiode during dye tracing experiment using CFDA and bleaching. Note that when the laser is on, there is a spike in the fluorescence level. The distance between this spike (blue arrow) and the base of the dip (white arrow) is the time it takes for the bleached fluid to move between the laser and the photodiode. (**B**) Output graph from photodiode during dye tracing experiment with esculin. Note the clear pulse of dye detected with the sensor. By comparing when peak fluorescence (white arrow) is detected at two different points, it is possible to estimate dye velocity without bleaching

2. Once the fluorescent dye arrives at the sensors, turn on the laser for 6–8 s. Make sure that it is still aligned with the mark made on the plant.
3. If the bleaching is successful and the sensor and laser are aligned on the same vascular bundle, a dip in radiation will appear in the reading after a short delay (see Fig. 2A). If no dip is observed within a reasonable time, you may need to troubleshoot the set up (*see* **Note 10**).
4. After the bleach front has passed, turn on the laser again to get a second measurement. Multiple measurements can be successfully made prior to saturation of the fluorescence signal (*see* **Note 11**).

3.4 Measurement of Pulse Using Esculin

1. Monitor the radiometer readout for a large gain in radiation above the background (see Fig. 2B). This increase reflects the arrival of the fluorescent dye to clip.
2. Continue to monitor the dye until the pulse of fluorescence passes by each of the two sensors (see Fig. 2B).

3.5 Anatomical Analysis

1. Remove the fiber optic cable/s from the custom clip/s and use a syringe to inject safranin O dye into the opening for the fiber optic cable. The dye will mark the location where the sensor was positioned.
2. Remove the plant from the clip/s and measure the distance between the laser mark and the injection site or the distance between the two clips. This is the length that the bleach front traveled (D).

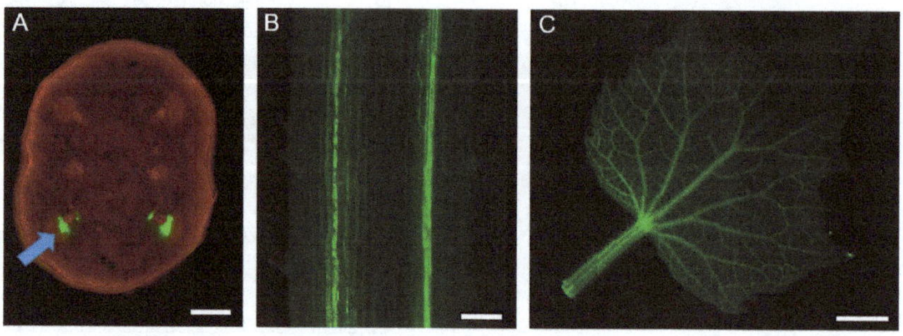

Fig. 3 Visualizing a fluorescence tracer in the phloem. (**A**) A cross-section of a cucumber seedling with CF (green) present in two vascular bundles. By altering the alignment of the sensor and laser on the stem, it is possible to target specific vascular bundles. The blue arrow indicates the orientation required to measure phloem transport velocity in the lower left bundle. (**B**) A longitudinal section of a similar seedling clearly shows two distinct traces with CF. (**C**) Visualization of a cucumber sink leaf using a fluorescent stereomicroscope. This micrograph clearly shows the strength of the fluorescent signal in completely intact tissue. The white bars in (**A**) and (**B**) are 400 μm and in **C** is 5 mm

3. Section the plant and view under a fluorescent microscope to confirm which vascular bundle contains the fluorescent dye and which bundles were measured during the experiment (see Fig. 3).

3.6 Data Analysis

1. Download data from the experiment and graph light intensity over time.
2. If you used a laser, determine the time when the laser was turned on based on the first observed spike and the time when the bleach front traveled in front of the laser based on the observed dip in fluorescence (see Fig. 2). The difference between these time points is the time it took the bleach front to travel from the laser to the sensor (t).
3. If you did not use a laser, determine the time of peak fluorescence for each photodiode/photomultiplier. The difference between these time points is the time it took for the pulse to travel between the sensors (t).
4. Calculate velocity as D/t, where D is the distance the bleach front/dye pulse traveled, and t is the time.

4 Notes

1. Select a radiation sensor that has a large dynamic range of detection, for example, $1e^{-11}$ to $1e^{-4}$ W cm^{-2} with higher gains for more opaque plant tissues. A large range minimizes the potential for sensor saturation and allows for work with different tissue types. A stereomicroscope that has a monochrome digital camera designed for fluorescence imaging can also be used to measure changes in the fluorescent signal of the tissue.
2. Excitation filters are optional, depending on the spectrum of the source light. If the source has a broad spectrum, these filters can help reduce noise. If a microscope is being used, standard light/filter sets for GFP and DAPI can be used for CFDA and esculin, respectively.
3. The custom clip holding the fiber optic should be made of completely opaque material with black closed-cell foam or foam tape that is arranged to seal around the plant (see Fig. 1). It is important that no ambient light penetrates the area of fluorescence signal detection. One can test the light seal by changing ambient light levels to see if the radiation sensor picks up the light change. The fiber optic cable should be securely attached to the clip, but also be easily removed without disturbing the clip's position on the plant. This can be

done by incorporating a connector specific to the fiber optic cable in the clip, for example, a fiber bundle ferrule with an SMA fiber optic cable.

4. With seedlings, the clip can be placed on the hypocotyl and oriented towards specific vascular bundles (see Figs. 1 and 3). With leaves, the clip should be situated on the abaxial side of the leaf on a large vein (see Fig. 4) or on the petiole. The leaf clip must be stable so that it will not shift during the span of the measurement. The clip should not place strain on the plant. If the plant is large, the clip can be supported with optomechanical components (e.g., posts, rails, and mounts) attached to a breadboard.
5. Dimethyl sulfoxide (DMSO) at a 0.1% concentration can be added to the CFDA or esculin if the dye does not readily enter the phloem.
6. Phloem transport will temporarily stop in many plants after they are disturbed and/or moved to a new environment. If the plant does not translocate the dye, give it longer to acclimate to new conditions.
7. The background fluorescence level is often species-specific and wavelength-specific. Some species will autofluoresce at a similar wavelength as the dye due to lignin content (excitation at ~488 nm, emission at ~530 nm) or presence of flavins (excitation 380–490 nm, emission 520–560 nm). This can lead to a high background signal. If the autofluorescence can be bleached, it will decrease over time due to the LED light source. It is important to determine the background fluorescence of the sample before taking measurements.
8. An adequate distance from the laser strike to the clip is important as "fiber optic" properties of the plant tissue can provide an excess of light to photomultipliers, causing permanent

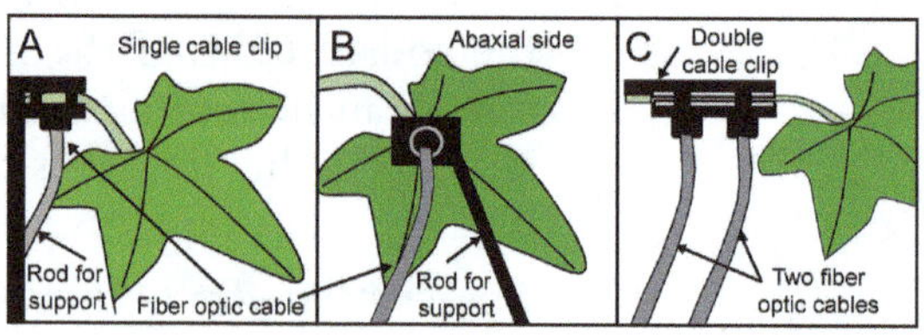

Fig. 4 Variation in cable clip design. A clip can be customized for use on (**A**) petioles, (**B**) midveins, and also (**C**) made to incorporate more than one fiber optic cable. Note that for (**C**), each cable is bifurcating, with one end going into the leaf clip and the other end split between the excitation light and the sensor. This two-cable design could be used without a laser if tracking a dye that does not accumulate in the tissue. When working on leaves, it is important to secure the clip to an optomechanical post or arm that can be adjusted to reduce strain on the plant

damage. Often, there will be a measurable spike in light intensity when the laser is on (see Fig. 2).

9. There are species-specific differences in how readily CFDA and esculin enter the phloem. Some of these variations are due to leaf structure, loading type, and the symplastic continuity of the cells in the leaf. Several techniques can be used to get dye into the phloem, including surface application and infiltration. Which technique is preferable is often species-specific.

10. If there is no dip in fluorescence after using the laser, readjust the alignment of the clip and laser and try again. If it still is not visible, stop taking measurements, mark where the sensor is positioned (see Sect. 3.5), and section the sample. Look under a microscope and note whether there was dye present and whether the sensor and laser are oriented in front of the same vascular bundle. Also note the depth of the phloem and how much dye is present. If only a small amount of dye is present and/or it is deep in the tissue or blocked by a large fiber cap, it might not be possible to bleach the dye and/or detect changes in fluorescence without modifying the method. Outer layers of cells can be trimmed off to reduce the light path's length, if necessary. This is similar to what is done when observing intact sieve tubes on a cellular level [2]. However, if the tissue is cut, it is important to wait until the system has recovered before trying additional measurements.

11. Over time, CFDA will build up in the phloem tissue and will cause the fluorescence level to saturate [10, 5, 13]. Typically, it is only possible to see the bleach before the fluorescence signal saturates. As a result, it is important to start taking measurements shortly after the dye first moves in front of the sensor. Depending on the plant, it may only be possible to take one to three measurements in sequence.

Acknowledgments

This work was partially supported by the National Science Foundation Integrative Organismal Systems 1942916 (JAS).

Conflict of Interest The authors have no conflicts of interest to declare.

References

1. Grignon N, Touraine B, Durand M (1989) 6(5) carboxyfluorescein as a tracer of phloem sap translocation. Am J Bot 76(6):871–877
2. Knoblauch M, van Bel AJE (1998) Sieve tubes in action. Plant Cell 10:35–50
3. Liesche J, Martens HJ, Schulz A (2011) Symplasmic transport and phloem loading in gymnosperm leaves. Protoplasma 248:181–190
4. Martens HJ, Roberts AG, Oparka KJ, Schulz A (2006) Quantification of plasmodesmatal

endoplasmic reticulum coupling between sieve elements and companion cells using fluorescence redistribution after photobleaching. Plant Physiol 142:471–480. https://doi.org/10.1104/pp.106.085803
5. Froelich DR, Mullendore DL, Jensen KH et al. (2011) Phloem ultrastructure and pressure flow: sieve-element-occlusion-related agglomerations do not affect translocation. Plant Cell 23:4428–4445. https://doi.org/10.1105/tpc.111.093179
6. Oparka KJ, Read ND (1994) The use of fluorescent probes for studies of living plant cells. In: Harris N, Oparka KJ (eds) Practical approach series; plant cell biology. Oxford University Press, Inc., pp 27–50
7. Savage JA, Haines DF, Holbrook NM (2015) The making of giant pumpkins: How selective breeding changed the phloem of *Cucurbita maxima* from source to sink. Plant Cell Environ 38:1543–1554
8. Roberts AG, Cruz SS, Roberts IM et al. (1997) Phloem unloading in sink leaves of Nicotiana benthamiana: comparison of a fluorescent solute with a fluorescent virus. Plant Cell 9:1381–1396. https://doi.org/10.1105/tpc.9.8.1381
9. Jensen KH, Lee J, Bohr T et al. (2011) Optimality of the Munch mechanism for translocation of sugars in plants. J R Soc Interface 8:1155–1165. https://doi.org/10.1098/rsif.2010.0578
10. Savage JA, Zwieniecki MA, Holbrook NM (2013) Phloem transport velocity varies over time and among vascular bundles during early cucumber seedling development. Plant Physiol 163:1409–1418
11. Carroll NJ, Jensen KH, Parsa S et al. (2014) Measurement of flow velocity and inference of liquid viscosity in a microfluidic channel by fluorescence photobleaching. Langmuir 30:4868–4874. https://doi.org/10.1021/la404891g
12. Wright KM, Oparka KJ (1996) The fluorescent probe HPTS as a phloem-mobile, symplastic tracer: an evaluation using confocal laser scanning microscopy. J Exp Bot 47:439–445
13. Knoblauch M, Vendrell M, de Leau E et al. (2015) Multispectral phloem-mobile probes: properties and applications. Plant Physiol 167:1211–1220. https://doi.org/10.1104/pp.114.255414
14. Knox K, Oparka K (2018) Illuminating the translocation stream. Curr Opin Plant Biol 43:113–118. https://doi.org/10.1016/j.pbi.2018.03.009
15. Knox K, Paterlini A, Thomson S, Oparka K (2018) The Coumarin glucoside, esculin, reveals rapid changes in phloem-transport velocity in response to environmental cues. Plant Physiol 178:795–807. https://doi.org/10.1104/pp.18.00574
16. Ross-Elliott TJ, Jensen KH, Haaning KS et al. (2017) Phloem unloading in Arabidopsis roots is convective and regulated by the phloem-pole pericycle. eLife 6:e24125. https://doi.org/10.7554/eLife.24125
17. Xu J, Wei X, Xiong M et al. (2022) A method for simultaneously monitoring phloem and xylem reconnections in grafted watermelon seedlings. Sci Horticult 299:111058. https://doi.org/10.1016/j.scienta.2022.111058
18. Gora PJ, Reinders A, Ward JM (2012) A novel fluorescent assay for sucrose transporters. Plant Methods 8:13. https://doi.org/10.1186/1746-4811-8-13
19. Rottmann TM, Fritz C, Lauter A et al. (2018) Protoplast-esculin assay as a new method to assay plant sucrose transporters: characterization of AtSUC6 and AtSUC7 sucrose uptake activity in Arabidopsis Col-0 ecotype. Front Plant Sci 9:430. https://doi.org/10.3389/fpls.2018.00430

Chapter 21

Determination of the Phloem Mass Flow Velocity and Calculation of the Volumetric Flow Rate

Alexandra C. U. Furch, Matthias R. Zimmermann and Jonas Nötzold

Abstract

Here, we describe a microscopic approach to measure the mass flow velocity and to calculate the volumetric flow rate in the midrib of leaves using a phloem-mobile fluorescent dye. The method requires measurements of the carboxyfluorescein (CF) fluorescence using a fluorescent microscope and the determination of the sieve element (SE) cross-sectional area. The translocated CF within the vascular system is tracked in the leaf using cross-sections and the mass flow velocity is deduced. Additionally, SE cross-sectional areas are measured in the center of the leaf at a high resolution to calculate the volumetric flow rate per SE. This method can be applied to all higher plants. However, as the velocity depends on various parameters like plant species, age, and developmental stage, all data needs to be collected and analyzed in each experiment.

Keywords Transport velocity, Volumetric flow rate, Carboxyfluorescein, Fluorescence microscope, Sieve elements

1 Introduction

The phloem of higher plants has diverse functions in plant development, reproduction, signaling, and growth [1] and to do so, all the different metabolites have to be translocated from one site to another within the plant. The force behind the phloem mass flow is an osmotically generated and driven pressure gradient within the sieve elements (SE) [2]. The phloem mass flow is influenced by various developmental and environmental factors. Abiotic factors such as cold [3] or heat [4] can lead to a reduction in the mass flow by a reversible sieve tube occlusion, but biotic factors such as bacteria [5, 6] also influence the flow speed. Furthermore, the flow properties change rapidly during plant development and growth [7]. Therefore, the basic understanding of phloem flow is of great importance and interest.

Here, we present a simple method for measuring the phloem mass flow velocity and the volumetric flow rate by tracking a phloem

Johannes Liesche and Sanna Sevanto (eds.), *Phloem*, Methods in Molecular Biology, vol. 3066,
https://doi.org/10.1007/978-1-0716-5488-0_21,

mobile fluorescent dye using light- and fluorescence microscopy. This method was developed for the use of the fluorescent dye 5,6-carboxyfluorescein diacetate (CFDA). CFDA permeates the plasma membrane in the nonfluorescent acetate form and is cleaved in the cytosol to the membrane-impermeable fluorescent compound CF. The trapped CF is transported inside SEs by mass flow, which can be observed by widefield fluorescence microscopy or confocal laser scanning microscopy (CLSM [5]).

2 Materials

2.1 Phloem Mass Flow Velocity Measurement

1. Intact plant (*see* **Note 1**).
2. Buffer solution: Mix 2 mM KCL, 1 mM $CaCl_2$, 1 mM $MgCl_2$, 50 mM Mannitol, 2.5 mM MES. Adjust to pH 5.7 with NaOH.
3. 5,6-carboxyfluorescein diacetate (CFDA): Dissolve 100 mg CFDA in 1 mL Dimethylsulfoxide (DMSO) to produce a stock solution. Mix 1 μL stock solution with 999 μL buffer solution to produce the working solution. Store the stock solution at 4°C. Prepare the working solution freshly on the day of the experiment.
4. Razor blades.
5. Black marker pen.
6. Objective slides and cover glass.
7. Pipettes: 1 mL and 1 μL.
8. Fluorescence microscope: with a 10× objective and illumination and detection properties suitable for CF. The optimal detection band for CF fluorescence detection is 510–580 nm.
9. Ruler.
10. Stopwatch.

2.2 Volumetric Flow Rate Calculation

1. Leaf sample: about 1 × 1 cm from the exact center of the same leaf from 2.1.
2. Ruler.
3. Razor blade.
4. Fixation solution: 2.5% (w/v) glutaraldehyde, 2% (v/v) paraformaldehyde in 0.1 M sodium-potassium phosphate buffer (pH 7.4).
5. Aluminum foil.
6. Cryofixation medium: e.g., Epredia™ Richard-Allan Scientific™ Neg-50™.
7. 2 mL tubes.

8. Cryomicrotome.
9. Sample holder.
10. Objective slides and cover glass.
11. Nail polish.
12. Light microscope equipped with a 40× objective.
13. Computer with image analysis software.

3 Methods

3.1 Phloem Mass Flow Velocity Measurement

1. Crop a leaf tip (~0.2 cm) with a fresh, sharp razor blade (Fig. 1A).
2. Apply the CFDA working solution at the cut leaf tip (Fig. 1B; *see* **Note 2**).
3. After incubation for 10–30 min (depending on the plant species) remove the leaf from the plant using a razor blade (*see* **Note 3**).
4. Mark every centimeter with a black marker pen, starting at the leaf tip (Fig. 1C).
5. Prepare cross-sections by hand every cm using a sharp razor blade, starting at the base of the leaf (*see* **Note 4**).
6. Place the cross-sections in order on a microscope slide (Fig. 1D).
7. Cover the sections with a cover glass without adding water (*see* **Note 5**).
8. Check the CF-fluorescence of each cross-section using the fluorescence microscope (Fig. 1E–H).
9. Calculate the velocity (V_a) up to the section that no longer shows fluorescence by dividing the distance the CF moved by the exact inoculation time.

3.2 Volumetric Flow Rate Calculation

1. Cut pieces of about 1 × 1 cm from the exact center of the leaf (Fig. 2A).
2. Fix the pieces in tubes filled with fixation solution for 2 days (Fig. 2B).
3. Form a cylinder from aluminum foil (1 cm wide, 1.5 cm high).
4. Fill the aluminum cylinder to the top with cryosectioning medium.
5. Place the leaf piece vertically into the cryosectioning medium (Fig. 2C).
6. Place the aluminum cylinder on the quick freeze platform of the cryomicrotome at −60°C and let it freeze through for 10 min.

Fig. 1 Experimental setup for measuring the phloem mass flow in *Beta vulgaris*. The leaf tip (max. 2 mm) was excised with a fresh razor blade (**A**). The leaf tip was placed in a 2 ml tube and incubated in CFDA working solution (**B**). Starting from the cut, positions were marked at 1 cm intervals (**C**), and cross sections were made at these positions using a razor blade (**D**). These sections were examined under a fluorescence microscope for CF-specific fluorescence (**E–H**)

7. Remove the aluminum foil and attach the frozen sample to the cryoform at −60°C, applying additional cryosectioning medium around it (Fig. 2D).
8. Mount the sample holder with the sample into the cryomicrotome (Fig. 2E).
9. Cut sections at a thickness of 20 μm using the cryomicrotome with a specimen head temperature of −60°C.
10. Place the cross-sections on a microscope slide, add a drop of water, and cover the sections with a cover glass.

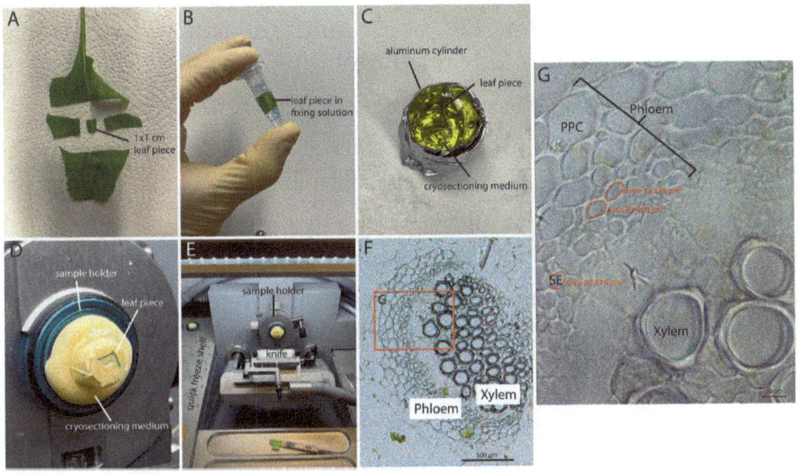

Fig. 2 Experimental setup for measuring the area of sieve elements (vasculature) in *Beta vulgaris*. A leaf segment (1 × 1 cm) containing the midrib was excised from the central region of the leaf using a razor blade (**A**). The segment was fixed in the fixing solution for 2 days (**B**). Subsequently, the sample was vertically embedded in an aluminum cylinder (1 × 1.5 cm) filled with cryosectioning medium and quick-frozen at −60 °C using a cryoform (**C**, **D**). The frozen cryoform was mounted onto the sample holder and positioned within the cryomicrotome (**D–F**). Light microscopy was used to image the resulting cryosections and to measure the size of the sieve elements (**F**, **G**)

11. Imagine the cross-sections using a light microscope equipped with a 40× objective.
12. Measure the area of SEs ($\tilde{A}_{SE}$) using image analysis software (Fig. 2F).
13. Calculate the volumetric flow rate (J_v) per SE by multiplying the measured phloem mass flow velocity (V_a) by the median area of ten measured SEs per cross section ($\tilde{A}_{SE}$) (*see* **Note 6**):

$$J_v\left[\frac{cm^3}{h}\right] = V_a\left[\frac{cm}{h}\right] * \tilde{A}_{SE}\left[cm^2\right]$$

4 Notes

1. In principle, all plants can be used for this method. Depending on the size of the leaves, only the application technique of the fluorescent dye needs to be adapted. Here, we have used the sugar beet as an example for the application of the method.

2. The tip of the leaf must be permanently dipped into the CFDA solution. The best results are achieved by immersing the leaf tip in a fully filled 2 ml tube.
3. Stop the exact time to calculate the speed later.
4. It is important to start at the base of the leaf so that no CF is carried over to the following cuts.
5. Water would wash the dye out of the tissue.
6. Since all sections originate from the center, you only get the flow rates for this position. Of course, you can also create sections from all other leaf positions and obtain the corresponding flow rates for each leaf position.

Acknowledgments

J.N. and A.C.U.F. are supported by Bavarian State Ministry of Food, Agriculture, Forestry, and Tourism (StMELF) and Federal Ministry of Agriculture, Food and Regional Identity. We thank Leon Friemert (FSU, Jena, Germany) for excellent assistance in the laboratory.

Conflict of Interest The authors have no conflicts of interest to declare.

References

1. Schulz A, Thompson GA (2009) Phloem structure and function. In: Encyclopedia of life sciences. John Wiley & Sons, Ltd., Chichester. https://doi.org/10.1002/9780470015902.a0001290.pub2
2. Knoblauch M, Knoblauch J, Mullendore DL, Savage JA, Babst BA, Beecher SD, Dodgen AC, Jensen KH, Holbrook NM (2016) Testing the Münch hypothesis of long distance ohloem transport in plants. eLife 5:e15341
3. Peuke AD, Windt C, van As H (2006) Effects of cold-girdling on flows in the transport phloem in *Ricinus communis*: is mass flow inhibited?. Plant Cell Env 29:15–25
4. Furch ACU, Zimmermann MR, Will T, Hafke JB, van Bel AJE (2010) Remote-controlled stop of phloem mass flow by biphasic occlusion in *Cucurbita maxima*. J Exp Bot 61:3697–3708
5. Gallinger J, Zikeli K, Zimmermann MR, Görg LM, Mithöfer A, Reichelt M, Seemüller E, Gross J, Furch ACU (2021) Specialized 16SrX phytoplasmas induce diverse morphological and physiological changes in their respective fruit crops. PLoS Pathog 17:e1009459
6. Furch ACU, Zimmermann MR, Noll GA, Wrobel LS, Scholz SS, Buxa-Kleeberg SV, Hafke JB, Fliegmann J, Mithöfer A, Ehlers K, Haufschild T, Nötzold J, Koch AM, Grabe V, Teutemacher F, Maaß J-P, Prüfer D, Oelmüller R, Peiter E, Kogel K-H, van Bel AJE (2025) Transformation of flg22 perception into electrical signals decodes in vasculature leads to sieve tube blockage and pathogen resistance. Sci Adv 11:eads6417
7. De Schepper V, Swaef TD, Bauweraets I, Steppe K (2013) Phloem transport: a review of mechanisms and controls. J Exp Bot 64:4839–4850

Part IV

Characterizing and Measuring Phloem Loading

Chapter 22

Measuring Esculin Export from Leaves as Proxy for Sucrose Loading Rates in the Phloem

Jianqiu Li and Johannes Liesche

Abstract

Phloem loading of sucrose and the transport from source leaves to sink tissues is vital for plant growth and carbon allocation. Traditional methods to measure phloem loading are often time-consuming or require specialized equipment. Here, we introduce a rapid, cost-effective esculin-based fluorometric assay as a reliable proxy for sucrose loading. Esculin, a fluorescent coumarin glucoside, is specifically transported by sucrose transporters in plants with active apoplastic phloem loading. After application to source leaves, esculin fluorescence is measured in the extracted leaf sap, providing a sensitive and relatively high-throughput method for analyzing phloem loading dynamics. Validated against established techniques, the assay is accessible to nonspecialized laboratories and enables investigations into environmental and developmental regulation of phloem loading, offering insights into plant growth and stress responses.

Keywords Carbon allocation, Carbon partitioning, Phloem loading, Source-sink, Sugar transport

1 Introduction

Phloem loading, the process of transporting sucrose from the site of production to the phloem sieve elements in minor veins, is a critical component of plant carbon allocation and energy distribution, influencing overall plant growth and productivity [1, 2]. Traditional methods for measuring phloem loading, such as isotopic carbon tracing or EDTA-facilitated phloem exudate analysis, often require specialized equipment, are time-consuming and artifact-prone, or involve handling radioactive materials [3–5]. To address these limitations, we developed a novel esculin-based fluorometric assay that provides a rapid, cost-effective, and reliable method for measuring phloem loading activity in plants. Esculin, a fluorescent coumarin glucoside, is specifically transported by sucrose transporters (SUTs), making it a good proxy for sucrose in phloem loading studies [6, 7]. After application to source leaves, esculin is taken up by sucrose transporters and transported in the phloem. By extracting leaf sap

Johannes Liesche and Sanna Sevanto (eds.), *Phloem*, Methods in Molecular Biology, vol. 3066,
https://doi.org/10.1007/978-1-0716-5488-0_22,

and measuring esculin fluorescence using a fluorometer, the rate of exported esculin can be inferred based on the intensity of the signal at two time points.

However, it is important to note that the esculin assay is limited to plants with esculin-transporting SUTs and an active apoplastic phloem loading strategy. Monocot SUTs do not seem to transport esculin, and plants with symplastic phloem loading strategies, such as Cucurbits and many tree species, do not utilize SUTs for phloem loading.

The esculin is applied to the abaxial side of the source leaves through infiltration. After a short incubation period, the fluorescence of esculin in leaf extracts is measured using a fluorometer, providing a relative quantification of phloem loading activity. This method has been validated against established techniques, such as isotopic carbon export and phloem exudate sucrose analysis, demonstrating consistent trends and the expected responsiveness to changes in light levels and other environmental conditions [8]. Furthermore, the esculin assay is particularly advantageous for high-throughput studies, as it requires minimal sample preparation time and does not necessitate specialized equipment, making it accessible to nonspecialized laboratories.

This protocol outlines the step-by-step workflow for the esculin assay, including the preparation of esculin solution, leaf infiltration, sample collection, and fluorescence measurement. By providing a straightforward and efficient approach, this method enables researchers to investigate the dynamics of phloem loading and its regulation by environmental factors and signals, contributing to a deeper understanding of whole-plant carbon allocation and its implications for plant growth and stress responses.

2 Materials

2.1 Plant Material

1. Dicot plants with an active apoplastic phloem loading type and leaves of at least 1 cm^2 area. For example, 3- to 4-week-old Arabidopsis plants. Prepare enough plants for unlabeled control and measurements at timepoints 0 and 1.

2.2 Chemicals

1. Esculin solution: Dissolve 17 mg esculin in 50 mL sterile double-distilled H_2O (ddH_2O) to make a 1 mM solution. Heat the solution to 45 °C for 5 min to ensure complete dissolving. Make the solution fresh when you use it at a volume appropriate for the planned experiments.
2. Methanol–water–chloroform (MWC) solution: Prepare under the fume hood by adding 30 mL methanol to 7.5 mL sterile ddH_2O and then adding 12.5 mL chloroform. Prepare the solution fresh before use. Adjust the volume according to

experimental requirements, keeping the 15:60:25 volume ratio of water, methanol, and chloroform.

2.3 Tools

1. Growth chamber or other facility for plant growth.
2. Analytical balance.
3. Measurement cylinder: 100 mL volume.
4. Syringe without needle: 1 or 5 mL volume.
5. Liquid nitrogen in a Dewar flask. Handle with care.
6. Fume hood.
7. High-flux tissue grinder.
8. Hotplate magnetic stirrer.
9. Centrifuge.
10. Pipette with autoclave-sterilized tips.
11. Water bath.
12. Autoclave-sterilized centrifuge tubes: 1.5 and 2 mL volume.
13. Disposable microcuvettes: 500 µL volume (*see* **Note 1**).
14. Fluorometer with excitation in the range of 335 nm and emission detection in the range of 460 nm (*see* **Note 1**).

3 Methods

3.1 Infiltration with Esculin

1. Prepare plants for processing. Mark the leaves to be treated at the petiole (*see* **Note 2**).
2. Using a 1 or 5 mL syringe without a needle, infiltrate the esculin solution into the abaxial side of the marked leaves (Fig. 1, *see* **Note 3**).
3. Blot dry any esculin solution splashed onto the surface of the leaves.
4. Place the infiltrated plants under standard growth conditions.
5. After 15 min (timepoint 0) excise a first batch of marked leaves, record their area, and transfer them into 2 mL centrifuge tubes containing steel beads.
6. Flash-freeze these samples in liquid nitrogen.
7. After 2 h (timepoint 1; *see* **Note 4**), excise the second batch of marked leaves, record their area, and transfer them into 2 mL centrifuge tubes containing steel beads.
8. Flash-freeze the samples in liquid nitrogen.
9. Homogenize all samples into powder using a high-throughput tissue grinder. Set the grinder to 40 Hz for 30 s per cycle and perform two grinding cycles (*see* **Note 5**).

Fig. 1 Arabidopsis leaves before and after esculin infiltration. Infiltrated leaves appear darker on both the abaxial and adaxial surfaces than those without infiltration. The leaves 2 h post-esculin infiltration resemble those without infiltration on both abaxial and adaxial surfaces

3.2 Measurement of the Esculin Fluorescence

1. Add 1 mL of MWC solution to the homogenized samples.
2. Vortex thoroughly and incubate the mixture in a water bath at 50°C for 30 min.
3. Centrifuge the samples at 12,000 *g* for 20 min. Carefully transfer the supernatant into a new 1.5 mL sterile centrifuge tube (*see* **Note 6**).
4. Add sterile ddH_2O equivalent to 60% of the supernatant volume. Mix thoroughly and incubate the mixture in the dark for 2 h (*see* **Note 7**).
5. Transfer 200 μL of the supernatant to a disposable microcuvette with a 500 μL capacity.

6. Measure the fluorescence intensity of the supernatant using a fluorometer. Set the excitation wavelength to 335 nm and the emission detection range to 460 ± 5 nm.

3.3 Calculation of Esculin Export

1. Record the fluorescence intensity of samples taken at time-points 0 and 1 as I_0 and I_1, as well as the control I_C. The esculin export E was with the following formula containing values for leaf area A and t, the time between t_0 and t_1:

$$E = \frac{(I_1 - I_C) - (I_0 - I_C)}{A \times t}$$

4 Notes

1. The protocol can also be carried out using a microplate reader with 96-well plates instead of microcuvettes if such an instrument is available. This setup allows for higher throughput and simultaneous measurement of multiple samples.
2. For plants with small leaves, several leaves may be collected from the same plant for sampling. The selected leaves should be uniform in size and consistent in growth. For *Arabidopsis thaliana*, typically two to four leaves are used for each datapoint.
3. Ensure that the infiltration area covers >95% of the leaf lamina. Infiltrated leaves will appear wet and darkened in color. Perform infiltration consistently to minimize mechanical injury and reduce intersample variation.
4. The sampling time can be adjusted according to species and experiment conditions. We recommend periods between time-points of 1–2 h for Arabidopsis. For *Nicotiana tabacum* even shorter periods (30 min) can provide reliable results.
5. Homogenization should be performed promptly after flash-freezing to prevent sample degradation due to freeze–thaw cycles.
6. When collecting the supernatant after centrifugation, exercise caution to avoid aspirating the pellet.
7. For some high-intensity samples or in case of limited dynamic range of the fluorometer, further dilute the supernatant with sterile ddH_2O to ensure the fluorescence measurements fall within the detection range.

Conflict of Interest

The authors have no conflicts of interest to declare.

References

1. Smith AM, Stitt M (2007) Coordination of carbon supply and plant growth. Plant Cell Env 30(9):1126–1149. https://doi.org/10.1111/j.1365-3040.2007.01708.x
2. Liesche J, Patrick JW (2017) An update on phloem transport: a simple bulk flow under complex regulation. F1000Research 6:2096. https://doi.org/10.12688/f1000research.12233.1
3. Christy AL, Fisher DB (1978) Kinetics of C-photosynthate translocation in morning glory vines. Plant Physiol 61(2):283–290. https://doi.org/10.1104/pp.61.2.283
4. Minchin PEH, Thorpe MR (2003) Using the short-lived isotope 11C in mechanistic studies of photosynthate transport. Funct Plant Biol 30(9):831–841. https://doi.org/10.1071/FP03032
5. Xu Q, Ren Y, Liesche J (2019) Studying phloem loading with EDTA-facilitated phloem exudate collection and analysis. Methods Mol Biol 2014:125–133. https://doi.org/10.1007/978-1-4939-9550-2_10
6. Knoblauch M, Vendrell M, de Leau E, Paterlini A, Knox K, Ross-Elliot T et al (2015) Multispectral phloem-mobile probes: Properties and applications. Plant Physiol 167(4):1211–1220. https://doi.org/10.1104/pp.114.255216
7. Sivitz AB, Reinders A, Johnson ME, Krentz AD, Grof CP, Perroux JM et al (2007) Arabidopsis sucrose transporter AtSUC9. High-affinity transport activity, intragenic control of expression, and early flowering mutant phenotype. Plant Physiol 143(1):188–198. https://doi.org/10.1104/pp.106.089003
8. Ren Y, Zhang Z, Zhanakhmetova D, Li W, Chen S, Werner T, Liesche J (2024) Fast and simple fluorometric measurement of phloem loading exposes auxin-dependent regulation of Arabidopsis sucrose transporter AtSUC2. Plant J 120(5):2305–2318. https://doi.org/10.1111/tpj.17110

Check for updates

Chapter 23

Relative Quantification of Phloem Loading with $^{13}CO_2$ in Grass Leaves

Jiamin Hao and Johannes Liesche

Abstract

A central challenge in plant physiology research is tracking the movement of photoassimilates, especially following their allocation via the phloem. To obtain better insight into movement of photoassimilates, we present a stable isotope-based approach to measure photoassimilate export from leaves. In this simple procedure, $^{13}CO_2$ pulse label is supplied to whole leaves or defined sections of the leaves, sugars are extracted at two timepoints and the reduction of ^{13}C -signal quantified. The method uses the stable ^{13}C isotope of carbon, is safe to use, and has been successfully applied in both monocot and dicot species.

Keywords Sieve tubes, Carbon-13, Sucrose transport, Carbohydrate tracking

1 Introduction

The phloem plays a central role in the long-distance transport of photoassimilates. By directing and facilitating the flow of sugars from source organs (primarily mature leaves) to sink organs (such as roots, stems, fruits, and seeds), the phloem is not only vital for maintaining carbon partitioning and energy balance in plants but also serves as a key component in plant responses to environmental changes [1]. Under abiotic stresses such as drought and salinity, phloem function is often among the first impaired/affected physiological processes, and the decline in transport efficiency directly constrains overall plant growth and stress resilience [2, 3]. Therefore, systematically elucidating the dynamics of carbon transport within the phloem not only aids assessing the physiological status of plants, but can also contribute to the theoretical foundations and genetic resources for improving crop stress tolerance and resource use efficiency.

The dynamic analysis of sugar loading into the phloem and export from the leaf is a crucial step for elucidating source–sink interactions and carbon allocation patterns. The application of stable isotope labeling techniques, such as leaf-feeding with $^{13}CO_2$, coupled

Johannes Liesche and Sanna Sevanto (eds.), *Phloem*, Methods in Molecular Biology, vol. 3066,
https://doi.org/10.1007/978-1-0716-5488-0_23,

with physiological and molecular approaches, has enabled precise tracing of photoassimilate transport routes, rates, and regulatory nodes. Notably, it has been widely employed to elucidate cellular mechanisms of sucrose loading such as distinguishing between active transport and passive diffusion pathways and to identify associated transporters and key genes [4, 5]. $^{13}CO_2$ feeding experiments have enabled systematic insights into the spatiotemporal dynamics and regulatory networks of carbon flux at whole-plant and even ecosystem levels [6–8].

Although various techniques have been employed to study carbon transport in the phloem, each has notable limitations. For instance, while ^{11}C, ^{14}C, and ^{18}F labeling offer high sensitivity and real-time monitoring capabilities, their radioactive nature poses substantial operational challenges, restricting their broader application in practical research [9]. Another common tracer, esculin, has proven effective in dicots [10], but not in monocots. Moreover, its application can interfere with experimental outcomes [11]. Among available methods, stable isotope ^{13}C pulse-labeling offers distinct advantages due to its environmental safety and excellent biocompatibility. However, ^{13}C pulse-labeling has limitations, including lower detection sensitivity compared to ^{11}C. Moreover, it requires destructive sampling as in this protocol, or tracing in the respiratory efflux as described in Chapter 18.

The following protocol describes ^{13}C pulse-labeling to track the export of photoassimilates from grass leaves. We have typically applied this to maize plants at the three-leaf stage and flag leaves of mature wheat plants. Instead of whole leaves, $^{13}CO_2$ can also be supplied to specific leaf regions to compare the export rates and transport directions of photoassimilates from different segments. The destructive sampling requires control and experimental group samples to be taken from different plants (Fig. 1). This means that

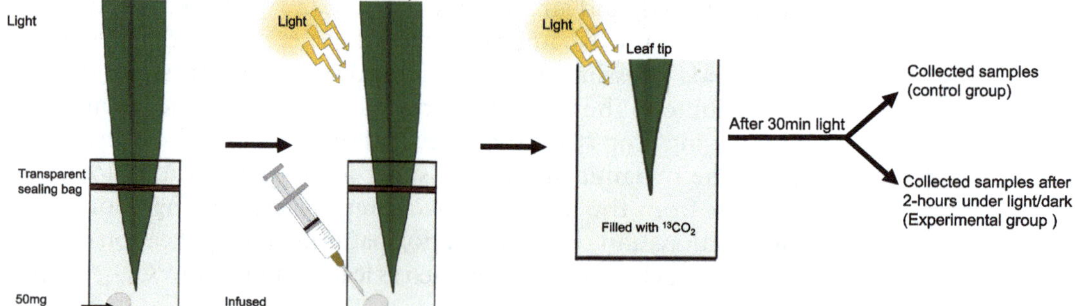

Fig. 1 ^{13}C pulse-labeling experiment. Due to the destructive sampling, control and experimental group samples are taken from different plants

two comparable plants are needed for one datapoint. In practice, at least six plants are needed to minimize the influence of plant-to-plant variation on results.

2 Materials

2.1 Plant Material

1. Plants grown in an incubator, greenhouse or in the field at the conditions adequate for your experiment (*see* **Notes 1and 2**). At least six plants are needed to have three replicates of measurements at the t0 and t1 time points. Select leaves or leaf areas for experiments that can be compared across plants(*see* **Note 3**).

2.2 Chemicals and Solutions

1. Citric acid solution ($C_4H_2O_7{\cdot}H_2O$): dissolve 5 g $C_4H_2O_7{\cdot}H_2O$ in 10 mL double distilled water (ddH_2O). This solution should be freshly prepared.
2. $NaH^{13}CO_3$ powder.

2.3 Tools and Instruments

1. Growth chamber.
2. Syringe without needle, 5 mL volume.
3. Transparent plastic bag (6 cm * 8 cm).
4. 65°C oven.
5. Laboratory balance.
6. Tin foil cups.
7. Elemental Analyzer (e.g., MACRO Cube Elemental Analyzer, calibrated with sulphate). Recommended accuracy C%: 41.81% ± 0.5%.
8. Isotope Ratio Mass Spectrometer (*see* **Note 4**), calibrated with international standard Pee Dee Belemnite. Recommended accuracy: $\delta^{13}C$: ± <0.1 ‰.

3 Methods

3.1 CO_2 Application

1. Place 50 mg $NaH^{13}CO_3$ powder at the bottom of the sealed transparent plastic bag (*see* **Note 5**) (Fig. 1).
2. Secure the transparent plastic bag containing $NaH^{13}CO_3$ powder to a leaf portion or a whole leaf (*see* **Note 6**). Seal any gaps between bag and leaf with transparent tape (*see* **Note 7**).
3. Inject an excess volume (~1.5 mL) of citric acid solution into the sealed transparent bags using a 5 mL syringe to ensure complete reaction of $NaH^{13}CO_3$ (*see* **Note 8**).

4. Upon cessation of gas evolution in the sealed transparent bag (indicating completion of the reaction) after 15–20 s, immediately commence timing the 30-min pulse-labeling period. Pulse labeling can be extended for leaves with low photosynthetic activity.
5. After the labeling period, carefully detach the transparent bags using scissors to fully expose the leaf tissue to the environmental atmosphere (*see* **Note 9**).

- **T0 group:** Immediately excise the treated leaf/leaf portion of the T0-group of plants, place in light-impermeable paper bags, and dry in an oven at 65°C for 72 h.
- **T1 group:** After an additional incubation of typically between 0.5 and 2 h, harvest the treated leaf/leaf portions of the T1-group of plants, transfer to light-impermeable paper bags, and dry in a 65°C oven for 72 h.

6. Measure and record the dry weight of the desiccated leaf samples.

3.2 Calculation of carbon exporting

1. Homogenize the desiccated leaf tissues to a fine powder, sieve through a 100-mesh screen, and quantify $\delta^{13}C$ and total organic carbon (TOC) using elemental analyzer and Isotope Ratio Mass Spectrometer.
2. Convert ^{13}C atomic percentage ($AT\%$, from isotope ratio mass spectrometry) to mass percentage ($m\%$) using the following formula:

$$m\% = \frac{13 \times AT\%}{(13 \times AT\%) + (12 \times (1 - AT\%))}$$

3. Calculate the dry-weight-based ^{13}C content (C_{13C}, in mg/g), with C% as the total organic carbon content measured by elemental analysis, using the formula:

$$C_{13C} = m\% \times C\% \times 10$$

4. Calculate the export rate v:

$$v = -\frac{C_{13c(30\,\text{min})} - C_{13c(2\,\text{h})}}{C_{13c(30\,\text{min})}}$$

4 Notes

1. The plants used in the experiment should have the same height, developmental stage, and leaf length (±1 cm). Because different leaf length led to different distance among tip, middle and basal zones. Importantly, the plants selected should be healthy, with no biotic or abiotic stress (unless this is the purpose of the experiment).
2. Typical conditions for maize seedlings in an incubator are constant temperature (25°C), light (300 $\mu mol \cdot photons \cdot m^{-2} s^{-1}$), and diurnal rhythm (16 h light/8 h dark).
3. If $^{13}CO_2$ is applied to a specific area of the leaf, make sure to apply it to the same area on all leaves as the distance from the leaf base influences phloem loading rates.
4. Laser-based carbon isotope analysis instruments could be used for this purpose. However, instead of gasification of extracts, one could prefer the option direct detection of ^{13}C in the respired air with these instruments (see Chap. 18).
5. $NaH^{13}CO_3$ powder requires a microbalance to be accurately weighed (±0.05 mg). Furthermore, when adding powder to a transparent plastic sealable bag, concentrate the powder as much as possible and avoid adherence to other parts of the bag.
6. This protocol uses a fully sealed plastic bag as leaf chamber, in which $^{13}CO_2$ is released. As a consequence, the ^{13}C concentration as well as the total CO_2 concentration inside the leaf chamber is not fully controlled. Some experimental questions might require a more precise setup in which a leaf chamber is connected to an inlet and outlet to create a gas exchange circuit comprising a CO_2 meter. $^{13}CO_2$ can then be supplied to the tubing from a gas canister/cylinder equipped with a flow controller at defined concentration with concomitant reduction in $^{12}CO_2$ as described for example in Chap. 18.
7. If the depression along the midrib compromises the seal integrity, gently fill the recess with cotton or soft tissue paper before securing it with transparent tape. At the leaf tip, the transparent sealable plastic bag requires no modification and can be directly placed over the tip (Fig. 1). For other leaf portions, create an incision approximately 1 cm from the bag's base (adjust incision width according to leaf width) to allow the leaf to pass through the incision.
8. The introduction of citric acid solution induces vigorous CO_2 evolution. Therefore, inject slowly to prevent direct contact of bubbles with leaf tissues. Exercise caution to avoid accidental

needle punctures, and immediately seal the injection site with transparent tape after the injection.

9. Use scissors carefully to avoid injuring leaves or hands.

Conflict of Interest

The authors have no conflicts of interest to declare that are relevant to the content of this chapter.

References

1. Geiger DR, Saunders MA, Cataldo DA (1969) Translocation and accumulation of translocate in the sugar beet petiole. Plant Physiol 65:1657–1665
2. Sevanto S (2014) Phloem transport and drought. J Exp Bot 65:1751–1759
3. Park SH, Lee B-R, La VH, Mamun MA, Bae D-W, Kim T-H (2021) Drought intensity-responsive salicylic acid and abscisic acid crosstalk with the sugar signaling and metabolic pathway in Brassica napus. Plants 10:610
4. Li G, Zhou C, Yang Z, Zhang C, Dai Q, Huo Z, Xu K (2022) Low nitrogen enhances apoplastic phloem loading and improves the translocation of photoassimilates in rice leaves and stems. Plant Cell Physiol 63:991–1007
5. Huang SR, Ai Y, Du J, Yu L, Wang X, Yang W, Sun X (2022) Photosynthetic compensation of maize in heterogeneous light is impaired by restricted photosynthate export. Plant Physiol Biochem 182:50–56
6. Hobbie EA, Werner RA (2004) Intramolecular, compound-specific, and bulk carbon isotope patterns in C3 and C4 plants: a review and synthesis. New Phytol 161:371–385
7. Pace ML, Cole JJ, Carpenter SR, Kitchell JF, Hodgson JR, Van de Bogert MC, Bade DL, Kritzberg ES, Bastviken D (2004) Whole-lake carbon-13 additions reveal terrestrial support of aquatic food webs. Nature 427:240–243
8. Gessler A, Brandes E, Buchmann N, Helle G, Rennenberg H, Barnard RL (2009) Tracing carbon and oxygen isotope signals from newly assimilated sugars in the leaves to the tree-ring archive. Plant Cell Environ 32: 780–795
9. Tran TM, Hampton CS, Brossard TW, Harmata M, Robertson JD, Jurisson SS, Braun DM (2017) In vivo transport of three radioactive [18F]-fluorinated deoxysucrose analogs by the maize sucrose transporter ZmSUT1. Plant Physiol Biochem 115:1–11
10. Ren YJ, Zhang ZY, Zhanakhmetova D, Li WH, Chen SL, Werner T, Liesche J (2024) Fast and simple fluorometric measurement of phloem loading exposes auxin-dependent regulation of Arabidopsis sucrose transporter ATSUC2. Plant J 117:2305–2318
11. Knoblauch M, Vendrell M, Leau ED, Paterlini A, Knox K, Ross-Elliot T, Reinders A, Brockman SA, Ward J, Oparka K (2015) Multispectral phloem-mobile probes: properties and applications. Plant Physiol 167:1211–1220

Check for updates

Chapter 24

Predicting Phloem Loading Types with Electron Microscopy

Johannes Liesche

Abstract

Phloem loading is a critical step in the transport of photosynthetically produced sugars from leaves to sink tissues. The type of phloem loading—active apoplasmic, active symplasmic, or passive symplasmic—can be inferred from the density and structure of plasmodesmata (PD) along the phloem-loading pathway. This protocol describes a method to predict phloem loading types using transmission electron microscopy (TEM) to quantify PD density. The method follows established standards for sample preparation, imaging, and data analysis, enabling comparative studies across species.

Keywords Carbon allocation, Plant anatomy, Plasmodesmata, Sugar transport, Vasculature

1 Introduction

Phloem loading is a fundamental process in plants that facilitates the transport of photosynthetically produced sugars from source tissues, such as leaves, to sink tissues, including roots, fruits, and seeds. Photosynthetically produced sugars are loaded into the phloem via three principal loading mechanisms: active apoplasmic, active symplasmic, and passive symplasmic loading. These mechanisms differ in their energy requirements and the role of plasmodesmata (PD) in facilitating symplastic connectivity between cells [1, 2]. Although not all details of the mechanisms have been elucidated, for example, regarding potential contributions of active processes in the passive loading type, a comparative analysis clearly showed the three types to constitute principally different mechanisms [3].

Understanding the phloem loading type of a plant is critical for several reasons. First, it provides insights into the plant's physiological adaptations to its environment. For example, active apoplasmic loading is common in herbaceous plants, including many crop species, and is associated with high rates of sugar transport to support rapid growth and reproduction. In contrast, passive symplasmic loading is often found in trees, which typically have slower growth rates [4, 5].

Johannes Liesche and Sanna Sevanto (eds.), *Phloem*, Methods in Molecular Biology, vol. 3066,
https://doi.org/10.1007/978-1-0716-5488-0_24,

These differences in loading mechanisms reflect evolutionary adaptations to diverse ecological niches and growth forms [6].

Second, knowledge of phloem loading types can inform agricultural practices and crop improvement strategies. For instance, active apoplasmic loaders rely on sugar transporters to move sugars into the phloem, making them more susceptible to environmental stresses that affect energy availability. In contrast, passive symplasmic loaders may be better suited to low-energy environments due to their reliance on diffusion. By understanding the phloem loading type, researchers and agronomists can develop targeted strategies to enhance crop resilience, optimize sugar transport, and improve yield under varying environmental conditions [1, 7, 8].

Third, the phloem loading type has implications for plant–pathogen interactions. Plasmodesmata, which are central to symplasmic loading, can serve as conduits for the spread of viruses and other pathogens within the plant [9]. Plants with high plasmodesmatal density may be more vulnerable to systemic infections, highlighting the importance of understanding PD structure and function in relation to phloem loading.

Despite its importance, the phloem loading type of many plant species remains unknown. The classification of phloem loading types has traditionally been challenging and required the combination of results from different experiments [10]. Our method enables the classification of phloem loading types within a single experiment. The protocol aims to provide a standardized method for predicting phloem loading types using transmission electron microscopy (TEM).

It should be noted that for researchers without access to a transmission electron microscope, a combination of other methods, such as sugar concentration measurements, ^{14}C-uptake studies, and fluorescence-based techniques, would be a more practical option to investigate phloem loading mechanisms. Moreover, the protocol uses a standard TEM. If an instrument for volume EM (e.g., Serial Block-Face Scanning EM) is available, employing this will increase the accuracy of the classification.

2 Materials

1. Plant Material: Source leaves from species of interest.
2. Sodium cacodylate buffer: 0.1 M, pH 7.2.
3. Karnovsky's fixative: prepare freshly by mixing 2.5% glutaraldehyde and 2% formaldehyde in a 0.1 M sodium cacodylate buffer, adjusted to pH 7.2.
4. Osmium tetroxide: prepare freshly by dissolving to 1% in 0.1 M sodium cacodylate buffer, pH 7.2.

5. Acetone.
6. Embedding Medium: Spurr's resin or similar.
7. Flat embedding molds.
8. Ventilated oven: for polymerizing the resin blocks.
9. Uranyl acetate: 1%.
10. Lead citrate: 0.5%.
11. Vacuum chamber.
12. Tissue rotator.
13. Ultramicrotome.
14. Glass or diamond knife for ultramicrotome.
15. Transmission electron microscope (*see* **Note 1**).
16. Image analysis software (e.g., ImageJ).

3 Methods

3.1 Sampling and Imaging

1. Collect leaf pieces (~2 × 4 mm) from healthy source leaves, avoiding larger veins.
2. Immerse samples in Karnovsky's fixative under vacuum for 2 h.
3. Replace fixative with 0.1 M sodium cacodylate buffer and rotate for 20 min.
4. Post-fix with 1% osmium tetroxide for 1 h, followed by rinsing in buffer and water.
5. Dehydrate samples through an acetone series (50%, 70%, 90%, 100%), 20–30 min per step.
6. Infiltrate samples with Spurr's resin in a graded series (1:3, 1:1, 3:1 resin:acetone) for 30 min each.
7. Embed samples in flat molds and polymerize at 70°C for 8 h.
8. Use an ultramicrotome to cut ultrathin sections (50 nm thickness).
9. Collect sections on grids coated with pioloform film.
10. Stain sections with 1% uranyl acetate for 5 min.
11. Rinse and stain with 0.5% lead citrate for 3 min.
12. Capture images of minor veins, including some of the surrounding mesophyll, using a TEM. Take images at different magnifications between 2600× and 10,500× (*see* **Notes 2 and 3**).Image sets of at least 10 minor veins are needed for the following analysis.

3.2 Image Analysis

1. Load images into image analysis software.
2. Determine cell types on lower resolution images: mesophyll, bundle sheath, phloem parenchyma, companion cells (Fig. 1; *see* **Note 4**).
3. Count PD per minor vein per cell type-interfaces: mesophyll to bundle sheath cell, bundle sheath cell to phloem parenchyma, bundle sheath cell to companion cell, and phloem parenchyma to companion cell. Count any portion of a PD visible in the TEM micrograph (*see* **Note 5**).
4. Extrapolate the number of PD to a 1 μm thickness of the minor vein by multiplying mean PD counts per section by 20 (20×50 nm = 1 μm). With a section thickness of 50 nm and an average PD diameter of 39 nm, it is unlikely that any PD is missed. This provides the number of PD per 1 μm minor vein length, which is used as the measure of PD density.
5. Normalize PD density to the highest value along the phloem-loading pathway (Fig. 2).
6. Classify phloem loading type according to the following scale:

   - Passive symplastic: Normalized counts at mesophyll cell (MC):MC, MC:bundle sheath cell (BSC), BSC:vascular parenchyma (VP), and VP:companion cell (CC) interfaces vary by less than 50%

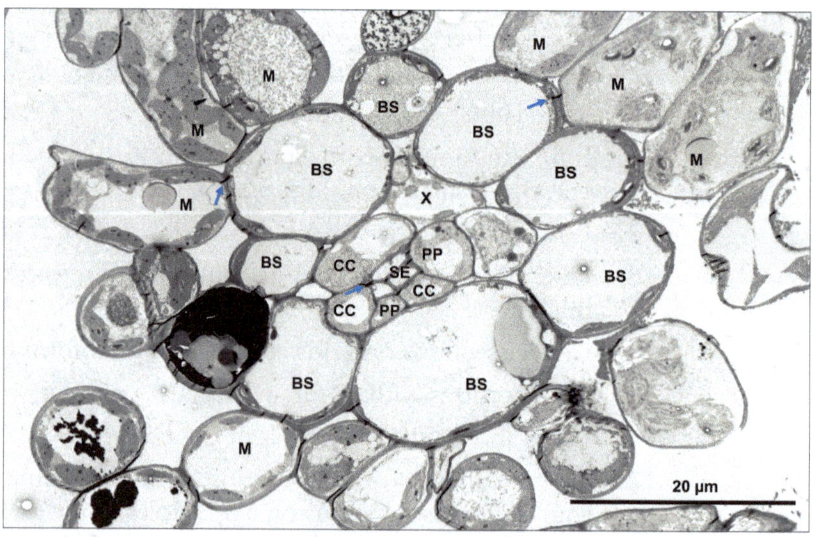

Fig. 1 Overview image of a minor vein in a *Betula pubescens* source leaf. Some mesophyll cells (M), bundle sheath cells (BS), phloem parenchyma cells (PP), companion cells (CC), and sieve elements (SE) are marked. Sample preparation artifacts (blue arrows) should be minimized. If they interfere with PD counting, then the section should not be used for analysis

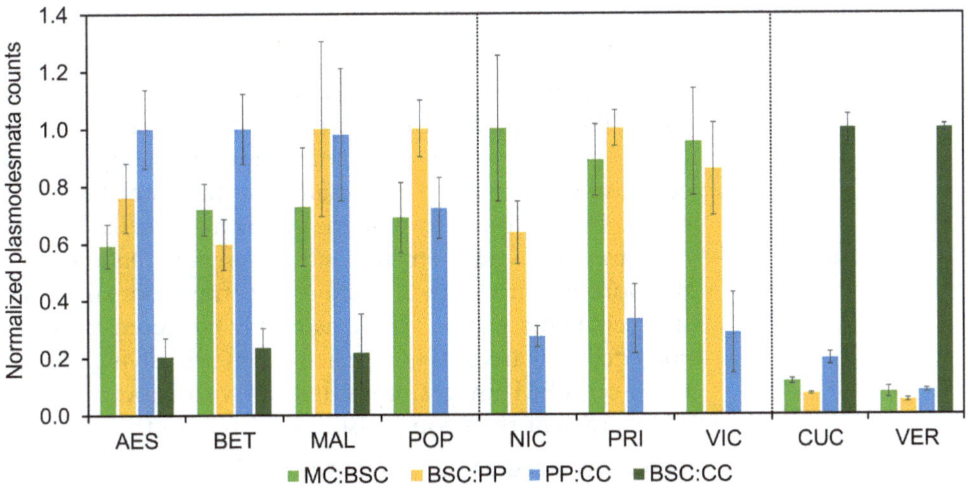

Fig. 2 Normalized PD counts in the passive symplasmic phloem loaders *Aesculus hippocastanum* (AES), *Betula pubescens* (BET), *Malus domestica* (MAL), and *Populus x canescens* (POP), the active symplasmic phloem loaders *Nicotiana benthamiana* (NIC), (PRI), and *Vicia faba* (VIC), and the active symplastic phloem loaders *Cucurbita pepo* (CUC) and *Verbena bonarensis* (VER). Note the relative similarity of plasmodesmata density along the whole phloem loading pathway in the passive loaders from mesophyll (MC) via bundle sheath cell (BSC) and phloem parenchyma (PP) to companion cell (CC). The active symplastic loaders are easily identified by the massive number of plasmodesmata at the BSC:CC interface. Active apoplastic loaders have much fewer plasmodesmata at the BSC:PP and the BSC:CC interfaces than at the upstream interfaces. (Data from Liesche et al. [11])

- Active apoplastic: normalized counts at PP:CC interface below 50% of counts at MC:BSC and BSC:PP interfaces.
- Active symplastic: normalized counts at BSC:CC interface several times higher than all other interfaces.

4 Notes

1. Any TEM for Life Science research, even one with low specifications (e.g., 80 kV voltage), will be adequate. However, measurements can be more accurate when three-dimensional volumes can be recorded, for example, with TEM tomography or, even better, serial blockface scanning EM or automatic tape collecting ultramicrotome–EM. But note that these methods will require more expertise and computing power for image analysis.
2. The use of consistent imaging settings is recommended to ensure comparability across samples. However, since PDs only need to be counted, optimal image quality is not necessarily required.

3. Avoid double-counting PD by imaging nonconsecutive sections.
4. Phloem parenchyma cells and companion cells can have similar sizes. The companion cells typically have a larger fraction of cytosol, which also appears darker (more protein). Companion cells also often have several small vacuoles instead of one larger one, typically found in phloem parenchyma cells. Companion cells are always adjacent to a sieve element. On some images, the pore plasmodesmata units connecting companion cells to sieve elements will be visible as highly clustered PDs at thickened cell wall areas. If inexperienced in identifying phloem cell types on TEM images, consultation with experts is generally more efficient than comparison to images in the literature.
5. To ensure accuracy and avoid bias, PD counting can be performed independently by different scientists.

Conflict of Interest

The authors have no conflicts of interest.

References

1. Liesche J, Patrick J (2017) An update on phloem transport: a simple bulk flow under complex regulation. F1000Res 6:2096. https://doi.org/10.12688/f1000research.12233.1
2. Zhang C, Turgeon R (2018) Mechanisms of phloem loading. Curr Opin Plant Biol 43:71–75. https://doi.org/10.1016/j.pbi.2017.12.007
3. Liesche J, Gao C, Binczycki P, Andersen SR, Rademaker H, Schulz A, Martens HJ (2019) Direct comparison of leaf plasmodesma structure and function in relation to phloem-loading type. Plant Physiol 179(4):1768–1778. https://doi.org/10.1104/pp.18.01353
4. Fu Q, Cheng L, Guo Y, Turgeon R (2011) Phloem loading strategies and water relations in trees and herbaceous plants. Plant Physiol 157:1518–1527. https://doi.org/10.1104/pp.111.186262
5. Schulz A (2015) Diffusion or bulk flow: how plasmodesmata facilitate pre-phloem transport of assimilates. J Plant Res 128:49–61. https://doi.org/10.1007/s10265-014-0673-2
6. Davidson A, Keller F, Turgeon R (2011) Phloem loading, plant growth form, and climate. Protoplasma 248:153–163. https://doi.org/10.1007/s00709-010-0237-6
7. Liesche J (2017) Sucrose transporters and plasmodesmal regulation in passive phloem loading. J Integr Plant Biol 59:311–321. https://doi.org/10.1111/jipb.12526
8. Fink D, Dobbelstein E, Barbian A, Lohaus G (2018) Ratio of sugar concentrations in the phloem sap and the cytosol of mesophyll cells in different tree species as an indicator of the phloem loading mechanism. Planta 248:661–676. https://doi.org/10.1007/s00425-018-2933-7
9. Harries P, Ding B (2011) Cellular factors in plant virus movement: at the leading edge of macromolecular trafficking in plants. Virology 411:237–243. https://doi.org/10.1016/j.virol.2010.12.041
10. Rennie EA, Turgeon R (2009) A comprehensive picture of phloem loading strategies. Proc Natl Acad Sci U S A 106:14162–14167. https://doi.org/10.1073/pnas.0902279106
11. Liesche J, Gao C, Binczycki P, Andersen SR, Rademaker H, Schulz A, Martens HJ (2019) Direct comparison of leaf plasmodesma structure and function in relation to phloem-loading type. Plant Physiol 179:1768–1778. https://doi.org/10.1104/pp.18.01353

Chapter 25

Measurement of Subcellular Metabolite Concentrations in Relation to Phloem Loading

Gertrud Lohaus

Abstract

The key step of carbon export from green leaves is the loading of sugars into the phloem. To fully understand and quantify this process, it is essential to know the concentration of sugars in the different compartments of the cells along the phloem loading pathway. However, determining subcellular metabolite concentrations has been technically challenging. This paper describes a technique to measure metabolite levels in the chloroplast, the cytosol, and the vacuole of mesophyll cells with high accuracy. The nonaqueous fractionation (NAF) technique is arguably the method of choice to analyze the subcellular metabolite distributions, as it minimizes the risk of metabolite interconversions or redistribution during the process. The principle of NAF is the separation of small subcellular particles, which are obtained by homogenization, lyophilization, and sonication, in a nonaqueous density gradient. Due to the varying composition-dependent density of the fragments, their segregation reflects compartmental distributions throughout the gradient. By determining marker enzymes for chloroplast stroma, cytosol, and vacuole in gradient fractions, the proportions of each subcellular compartment in each gradient fraction can be analyzed. The measured distribution of marker enzymes and of metabolites in each fraction of the gradient can be used to calculate the subcellular distribution of the metabolites.

Keywords Nonaqueous fractionation, Subcellular metabolite concentration, Sugar, Amino acid, Cell metabolism

1 Introduction

In order to understand carbon partitioning, photosynthesis regulation, source–sink interactions, or phloem loading, it is important to know how photoassimilates are distributed between the subcellular compartments of mesophyll cells in source leaves. To gain insight into in vivo subcellular metabolite concentrations, it is essential that any biological activity of the plant material is stopped to avoid metabolite redistribution during the separation procedure. Nonaqueous fractionation (NAF) is a powerful technique that allows the determination of metabolite levels in the vacuolar, chloroplastic, and cytosolic compartments of plant tissue under conditions that preserve the

Johannes Liesche and Sanna Sevanto (eds.), *Phloem*, Methods in Molecular Biology, vol. 3066,
https://doi.org/10.1007/978-1-0716-5488-0_25,

physiological state during photosynthesis [1]. Fragments of subcellular compartments are separated in the frozen state, under nonaqueous conditions in which biological activities, such as metabolite leakage, conversion, and redistribution, are arrested. Figure 1 shows an overview of the general procedure of NAF. Leaves are frozen in liquid nitrogen and then broken into small fragments, which are generated during lyophilization and ultrasonication of ground material. The central assumption is that the metabolites and proteins in a particular region of the cell aggregate as the plant material is lyophilized. The

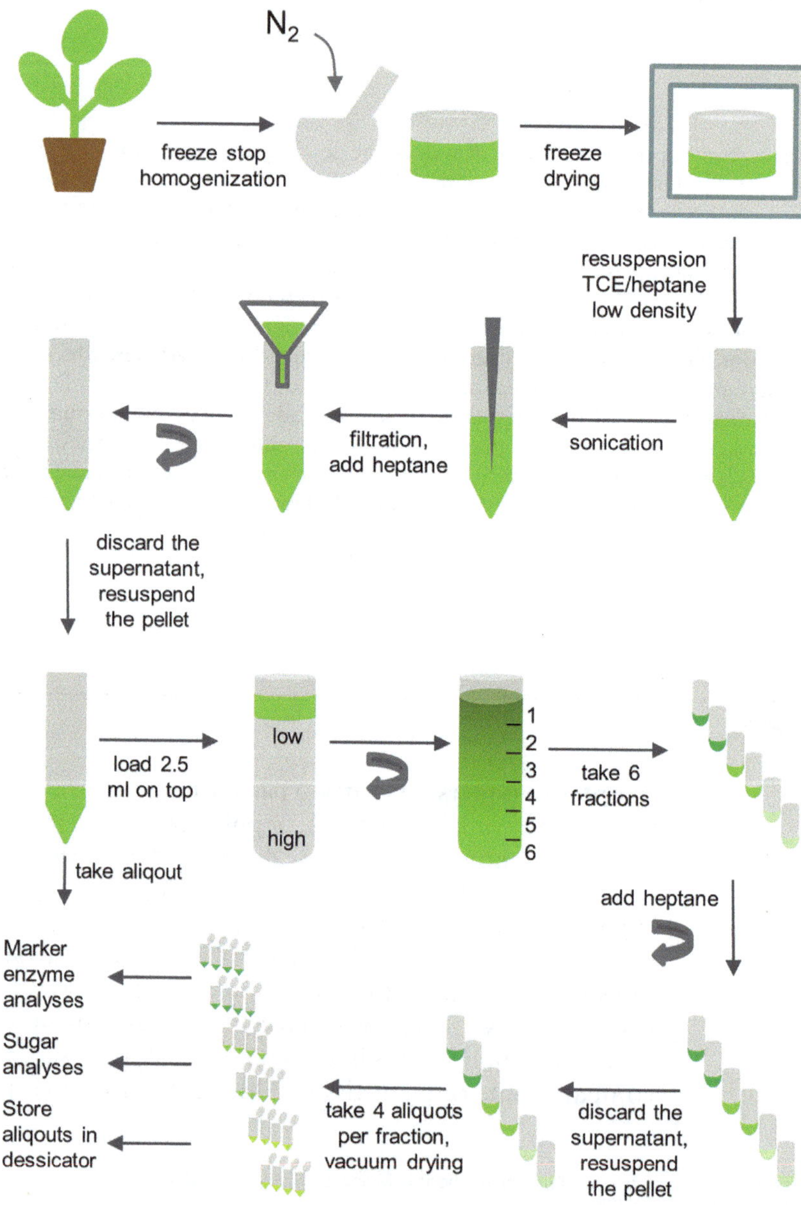

Fig. 1 Overview of the nonaqueous fractionation procedure

subcellular fragments are separated by their composition-dependent density using equilibrium centrifugation in a gradient that consists of two nonaqueous solvents with different densities (heptane and tetrachloroethylene). This prevents redistribution of the water-soluble metabolites and enzymes within the gradient during the separation of the plant tissue and therefore allows for the attribution of the amounts of metabolites to the subcellular compartments, which are identified via the activities of specific marker enzymes. Different densities of cell compartment membranes lead to a spatial separation of subcellular fragments within the density gradient [1]. The amount of metabolites and compartment-specific markers, which are also used as references to estimate subcellular metabolite distributions, are analyzed in all collected gradient fractions [2].

NAF was first applied to study high-energy phosphates in mammalian cells [3]. Later, it was mostly used in plant sciences to determine the partitioning of photosynthetic assimilates in leaves of herbaceous plants [1, 2, 4–11], and it was also adapted to leaves of woody species [9, 12] or to heterotrophic tissues like potato tubers [13]. In most cases, the subcellular concentrations of sugars (e.g., hexoses and sucrose), sugar alcohols, organic acids, or amino acids were determined. Additionally, nonaqueous fractionation in combination with GC/MS and LC/MS was used in metabolite profiling of the subcellular distribution of a large number of phosphorylated metabolites and other known and unknown compounds [13, 14].

Phloem loading is an important step for the export of sugars from leaves, and different types of phloem loading have been described. Active phloem loading leads to elevated sugar concentrations in the phloem in comparison to surrounding mesophyll cells [2]. For passive phloem loading, where sugars are assumed to diffuse into sieve elements, a higher sugar concentration in the cytosol of mesophyll cells compared to the phloem is required [15]. Therefore, knowledge of metabolite concentrations in subcellular compartments of mesophyll cells and in the phloem is important to understand possible modes of phloem loading.

2 Materials

2.1 Homogenization and Freeze–Drying of Leaf Material

1. Ball mill or mortar and pestle.
2. Liquid nitrogen: about 2 L.
3. Small plastic box and elastic bands.
4. Cleaning tissue: laboratory-grade (e.g., Kimwipes).
5. Lyophilizer.
6. Silica gel containing a desiccator.

2.2 Preparation of a Density Gradient

1. Solution A: For 100 mL, measure the volume of 67 mL tetrachloroethane (TCE) and 33 mL heptane, both analytical grade, and mix (v/v). This gives a low-density solution of $\rho = 1.31\ g\,cm^{-3}$ (*see* **Note 1**). Add a water adsorbent (molecular sieve beads; 0.4 µm) at least 2 days before the experiment and store the solution at 4°C (*see* **Note 2**) in brown bottles or bottles wrapped with aluminum foil.
2. Solution B: For 100 mL Solution B, measure 83 mL TCE and 17 mL heptane, both analytical grade, and mix (v/v). This gives the high-density solution of $\rho = 1.46\ g\,cm^{-3}$ (*see* **Note 1**). Add a water adsorbent (molecular sieve beads; 0.4 µm) at least 2 days before the experiment and store the solution at 4°C (*see* **Note 2**) in brown bottles or bottles wrapped with aluminum foil.
3. Brown glass bottle or glass bottle wrapped with aluminum foil: 250 mL volume.
4. Gradient mixer, glass, 15 mL, for the preparation of the density gradient.
5. Falcon centrifuge tube: 15 mL volume.

2.3 Fractionation of the Tissue

1. Falcon centrifuge tubes: 50 mL volume.
2. Vortex mixer.
3. Ultrasonic homogenizer (sonicator) with a titanium tip.
4. Nylon net: 200 cm², <40 µm pore size.
5. Centrifuge tubes: 1.5 and 2 mL volumes.
6. Pasteur pipette.
7. Cooling centrifuge for 50 mL Falcon tubes that can reach 3200 × *g*.
8. Cooling centrifuge for 15 mL Falcon tubes with swing-out rotor that can reach 5000 × *g*.
9. Cooling centrifuge for 1.5 and 2 mL centrifuge tubes that can reach 13,000 × *g*.
10. Vacuum concentrator (e.g., Speedvac).

2.4 Marker Enzyme Analyses

1. Extraction buffer: 50 mM Hepes–KOH (pH 7.4), 5 mM $MgCl_2$, 1 mM Na_2–EDTA, 0.1% Triton, 10% glycerol. Weigh 2.98 g Hepes, 0.119 g $MgCl_2$ (or 0.254 g $MgCl_2 \times 6H_2O$), and 0.084 g Na_2–EDTA and transfer to a 250 mL graduated cylinder. Add water to a volume of 200 mL. Mix and adjust pH with KOH to pH 7.4. Add 0.25 mL Triton and 25 mL glycerol and mix. Fill up to 250 mL with water. Store at 4°C.
2. Cooling centrifuge for 2 mL centrifuge tubes that can reach 13,000 × *g*.

3. Centrifuge tubes: 2 mL volume.
4. Plastic cuvettes: 0.6 mL volume.
5. Photometer.
6. GAP-DH buffer: 100 mM Hepes–KOH (pH 8.0), 30 mM $MgCl_2$, 20 mM KCl, 2 mM Na_2–EDTA. Prepare a 100 mL solution. Store at 4°C.
7. Dithiothreitol (DTT): 200 mM; dissolve in 1 mL water, prepare fresh each time.
8. ATP: 200 mM, dissolve in 1 mL water, can be stored at −20°C.
9. NADPH: 10 mM, dissolve in 1 mL GAP-DH buffer, prepare fresh each time.
10. Phosphoglycerate kinase: 3.7 U μL^{-1}; dissolve in 1 mL GAP-DH buffer, prepare fresh each time.
11. 3-phosphoglycerate: 200 mM; dissolve in 1 mL water, can be stored at −20°C.
12. UGPase buffer: 100 mM Tris HCl (pH 8.0), 2 mM $MgCl_2$. Prepare a 100 mL solution. Store at 4 °C.
13. $NADP^+$: 18.5 mM; dissolve in 1 mL water, can be stored at −20 °C.
14. NaF: 150 mM, dissolve in 1 mL water, can be stored at −20 °C.
15. UDP–Glucose: 150 mM; dissolve in 1 mL water, can be stored at −20 °C.
16. Glucose-1,6-bisphosphate: 1.5 mM; dissolve in 1 mL UGPase buffer, can be stored at −20 °C.
17. Phosphoglucomutase: 0.3 U μL^{-1}; dissolve in 1 mL UGPase buffer, prepare fresh each time.
18. Glucose-6-phosphate-dehydrogenase: 0.1 U μL^{-1}; dissolve in 1 mL UGPase buffer, prepare fresh each time.
19. $Na_2P_2O_7$: 148 mM; Dissolve in 1 mL UGPase buffer, prepare fresh each time.
20. Sodium citrate (pH 4.5): 100 mM; Prepare 50 mL solution, adjust pH with KOH, and store at 4 °C.
21. P-nitrophenyl-α-D-mannopyranoside: 10 mM; Dissolve 0.030 g in 10 mL water (prepare fresh).
22. Boric acid (pH 9.8): 0.8 M; Prepare a 50 mL solution, adjust pH with KOH, and store at 4 °C. Borate starts to dissolve after adjustment to pH 9.8.

2.5 Extraction of Soluble Metabolites

1. Chloroform:methanol mixture (1.5:3.5): For 100 mL, mix 30 mL chloroform and 70 mL methanol in a glass bottle.

2. Plastic tubes: 15 mL volume.
3. Round bottom flask (glass bulb): 100 mL volume.
4. Centrifuge for 15 mL plastic tubes that can reach 400 *g*.
5. Rotary evaporator.
6. Syringe-filter: 1 mL, 0.20 μm nylon.

2.6 Sugar Quantification and Data Analysis

1. HPLC with pulse amperometrical detector: pulse settings are 50, 700, and −800 mV for 500, 540, and 540 ms. Alternatively, a refractive index detector (RID) or an evaporative light scattering detector (ELSD) can also be used.
2. Anion exchange column (typically CarboPac PA10).
3. NaOH: 2 L of 80 mM.
4. Computer with relevant software (e.g., Peaknet, Excel, and Bestfit).

3 Methods

Carry out all procedures at room temperature (RT) unless specified otherwise (*see* **Note 2**)

3.1 Sample Preparation and Freezing

1. Precool the lyophilizer to −30°C.
2. Cut the leaves off the plants and remove the midrib (*see* **Note 4**). Homogenize a total of about 4 g leaf material to a very fine powder in liquid nitrogen using a precooled mortar and pestle or a ball mill (*see* **Note 5**).
3. Put the leaf tissue powder in a plastic box and cover it with cleaning tissues, fixed with an elastic band. After the nitrogen has evaporated, but before the samples have begun to thaw, transfer the box into a lyophilizer. Dry the powder at 4 Pa for about 4 days.
4. Open the lyophilizer under a nitrogen atmosphere. Close the plastic box with the leaf tissue powder quickly, place it in a desiccator, and store it at −20 °C until fractionation on a density gradient. Dried material can be stored in a desiccator protected from light and humidity for up to 3–4 weeks.

3.2 Preparation of the Density Gradient

1. Prepare solutions A and B at least 1 day before the fractionation procedure.
2. Prepare the gradient mixer. Lock the valve between container A and B and the outlet valve of the gradient mixer. Give 6 mL of solution A (low density) in the left container and 5 mL of solution B (high density) in the right container of the gradient mixer (Fig. 2). Stir the solution in container B.

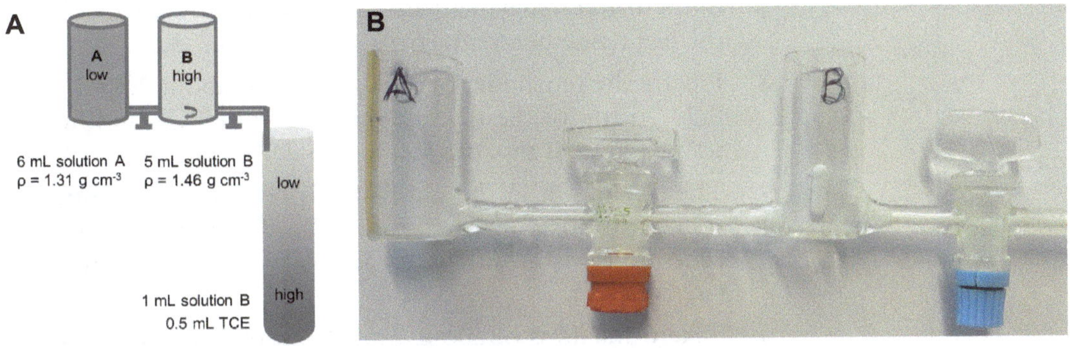

Fig. 2 Preparing the density gradient. (**A**) Drawing indicating volumes and densities of solutions used for the preparation of the density gradient. (**B**) Image of the gradient mixer

3. Prepare the density gradient immediately before fractionation of the plant tissue (see Sect. 3.3) by pipetting 1 mL of solution B into the 15 mL Falcon tube and underlaying it carefully with 0.5 mL of TCE. Use a Pasteur pipette for this.
4. Overlay this with 10 mL of a mixture of solutions A and B with exponentially changing density (high to low density) from the gradient mixer.
5. Fill the Falcon tube with the hose from container B (open the outlet valve and the valve between container A and B).
6. Fix the end of the hose to the inside of the tube wall, about 2 cm beneath the upper edge.

3.3 Fractionation of the Tissue

1. Resuspend the dry leaf powder (about 200 mg) in 20 mL of solution A in a 50 mL plastic tube and vortex.
2. Ultrasonicate the powder for a total of 95 s, with 5 s pulses and 5 s breaks. To prevent overheating during ultrasonication, place the plastic tube on ice.
3. Pour the suspension through a nylon net (<40 μm pore size), then pour three times 10 mL heptane through the net. Collect the discharge in a 50 mL plastic tube.
4. Centrifuge the filtrate for 10 min at 3200 × *g*.
5. Discard the supernatant and resuspend the pellet in 3 mL solution A.
6. Withdraw four aliquots of 50 μL each from the unfractionated material and transfer them into separate 2 mL Eppendorf tubes.
7. Dry the samples in a vacuum concentrator for about 30 min at RT and keep them in a silica-gel containing desiccator at RT in the dark. These aliquots are used to determine the activities

of the marker enzymes (see Sect. 3.4) and the metabolites in the leaf tissue before fractionation (see Sect. 3.5).

8. Pipette the remainder of the suspension (about 2.5 mL) carefully on top of the density gradient from TCE and heptane using a Pasteur pipette (see Sect. 3.2).
9. The separation in the gradient is performed by centrifugation at 5000 × *g* for 50 min. Use a swing-out rotor and centrifuge with a slow start and without any brakes at 4°C (*see* **Note 6**).
10. Take 6 fractions of about 2 mL from the top to the bottom of the gradient and transfer each fraction into 15 mL Falcon tubes. Note the exact volumes.
11. Add 10 mL heptane to each fraction in the Falcon tubes.
12. Centrifuge at 3200 × *g* for 10 min (4 °C).
13. Discard the supernatants and resuspend each pellet in 3 mL heptane.
14. Pipette four aliquots of 500 μL of each suspension (fraction) into 1.5 mL Eppendorf tubes.
15. Dry the samples in a vacuum concentrator (speed vac) for about 30 min at RT and keep them in a silica-gel containing desiccator at RT in the dark. One of the aliquots is used to measure the marker enzyme activity in a given fraction (see Sect. 3.5), one to measure the concentration of sugars or other metabolites (see Sect. 3.5), and the remaining two aliquots serve as a reserve.

3.4 Measurements of Compartment-Specific Markers

To unambiguously assign a specific compartment to the distributions in the density gradient, the amounts of compartment-specific markers within the six collected gradient fractions are determined (Fig. 3). These marker distributions, which must be sufficiently distinct from one another, are then used to evaluate the compartmental enrichment and separation of distinct organelles or subcellular spaces. In most experiments with leaves, NADP-glycerin aldehyde phosphate-dehydrogenase (NADP-GAPDH), UDP–glucose pyrophosphorylase (UGPase), and α-mannosidase are taken as marker enzymes for chloroplast stroma, cytosol, and vacuole, respectively (*see* **Note 7**).

1. Resuspend the dried aliquots (one of the unfractionated material and one of each fraction; see Sect. 3.3; Steps 7 and 15) in 500 μL of extraction buffer.
2. Vortex for 1 min, keep on ice for 5 min, and vortex again. Centrifuge the samples at 13,000 × *g* for 2 min at 4°C.
3. Use the supernatants to measure the marker enzyme activities.

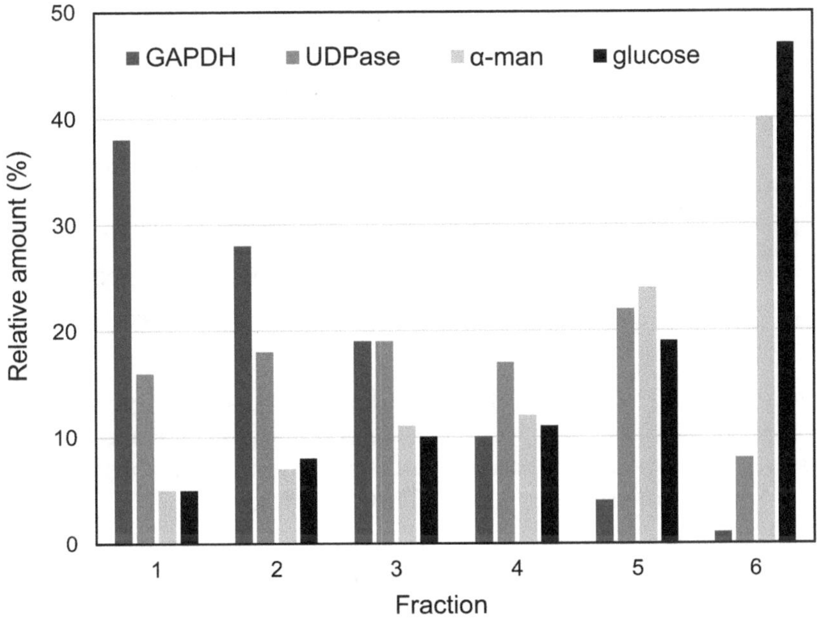

Fig. 3 Typical distribution of compartment-specific markers in a nonaqueous gradient from *Brassica napus* leaves. The distribution of vacuolar (α-mannosidase), cytosolic (UDPase), and plastidic (GAPDH) markers is shown, as well as the distribution of glucose. The values of marker enzyme activities (GAPDH, UDPase, α-mannosidase) or of glucose concentration in each fraction are depicted as a percentage of the total. Fraction 1 represents the upper side of the gradient with the lowest density, and fraction 6 the lower side with the highest density. The proportions of the marker enzymes and the glucose concentrations are used for the calculation procedure (best fit). The calculation program evaluated which of the possible distributions yields the best fit (agreement) with the experimental result on the basis of the linear least squares method. The example results in the distribution of glucose between the subcellular compartments: 3% of glucose is located in the chloroplast, 1% in the cytosol, and 96% in the vacuole

NADP-Dependent Glyceraldehyde 3-Phosphate-Dehydrogenase (GAPDH)

The method is based on a decrease in the absorbance at 340 nm due to the oxidation of NADPH over time in two coupled reactions [16]:

$$3-\text{phosphoglycerate}+\text{ATP} \xrightarrow{\text{Phosphoglycerate kinase}} 1,3-\text{bis}-\text{phosphoglycerate}$$

$$1,3-\text{bis}-\text{phosphoglycerate}+\text{NADPH} \xrightarrow{\text{GAPDH}} \text{NADP}^{+}+3-\text{phosphoglycerol aldehyde}$$

1. Pipette in a 1 mL plastic cuvette:
   580 μL GAP-DH buffer.
   20 μL 200 mM DTT.
   20 μL 200 mM ATP.
   20 μL 10 mM NADPH.

20 μL 3.7 U μL^{-1} phosphoglycerate kinase.
20 μL sample.

2. Mix carefully and measure the extinction at 340 nm.
3. After about 5 min, start the reaction with 20 μL 200 mM 3-phosphoglycerate and measure the continuous decrease of the extinction per time (about 10 min).
4. Calculate the activity using Beer–Lambert's law and the molar extinction coefficient for NADPH of 6.3 mM^{-1} cm^{-1} at 340 nm.
5. Repeat the procedure at least once for each sample.

Uridine Diphosphate (UDP)-Glucose-Pyrophosphorylase

The activity of uridine diphosphate (UDP)-glucose-pyrophosphorylase is assayed in the direction of Glc-1-P according to [17]. The method is based on an increase in the absorbance at 334 nm due to the reduction of NADP$^+$ over time

$$\mathrm{UDP-glucose + Na_4P_2O_7 + UGP} \xrightarrow{\mathrm{UDP-glucose-pyrophosphorylase}} \mathrm{UTP + Glucose-1-P}$$

$$\mathrm{Glucose-1-P} \xrightarrow{\mathrm{Phosphoglucomutase}} \mathrm{Glucose-6-P}$$

$$\mathrm{Glucose-6-P + NADP^+} \xrightarrow{\mathrm{Glucose-6-P-DH}} \mathrm{6-P-gluconolactone + NADPH}$$

1. Pipette in a 1 mL plastic cuvette:
   600 μL 100 mM UGPase buffer
   10 μL 18.5 mM NADP$^+$.
   10 μl 150 mM NaF.
   10 μL 150 mM UDP-Glucose.
   10 μL 1.5 mM Glucose-1,6-bisphosphate.
   10 μL 0.3 U μL^{-1} Phosphoglucomutase.
   10 μL 0.1 U μL^{-1} Glucose-6-phosphate-dehydrogenase.
   70 μL sample.
2. Mix carefully and measure the extinction at 334 nm.
3. After about 3 min, start the reaction with 10 μL 148 mM $Na_2P_2O_7$ and measure the continuous decrease of the extinction over time for about 10 min.
4. Calculate the activity by using Beer–Lambert's law and the molar extinction coefficient for NADPH of 6.18 mM^{-1} cm^{-1} at 334 nm.
5. Repeat the procedure at least once for each sample.

α-Mannosidase

The activity of α-mannosidase is determined according to Ye–The [18] based on an increase in the absorbance at 405 nm due to the formation of p-nitrophenol in the reaction

$$\text{p} - \text{nitrophenyl} - \alpha - \text{D} - \text{mannopyranoside} + \text{H}_2\text{O} \xrightarrow{\alpha-\text{mannosidase}} \text{D} - \text{mannose} + \text{p} - \text{nitrophenol}$$

1. Pipette in a 2 mL Eppendorf tube:
   50 μL sample.
   500 μL 100 mM sodium citrate.
   500 μL 10 mM p-nitrophenyl-α-D-mannopyranoside.
2. Mix the solutions by pipetting up and down.
3. Incubate the assay mixture at 37°C for 60 min.
4. Stop the reaction by adding 500 μL 0.8 M boric acid and mix the solutions by pipetting up and down.
5. Centrifuge the samples for 5 min at 13,000 × *g*.
6. Transfer the supernatant into plastic cuvettes and measure the samples in the photometer at 405 nm.
7. Prepare individual blanks by pipetting in a 2 mL Eppendof tube:
   500 μL 100 mM sodium citrate
   500 μL 10 mM p-nitrophenyl-α-D-mannopyranoside
   500 μL 0.8 M boric acid.
8. Mix the solutions by pipetting up and down.
9. Add 50 μL sample.
10. Centrifuge the samples for 5 min at 13,000 × *g*.
11. Transfer the supernatant into plastic cuvettes and measure extinction at 405 nm using a photometer
12. Calculate the enzyme activity by using a molar extinction coefficient for p-nitrophenol of 18.5 mM^{-1} cm^{-1} at 405 nm against an individual blank.
13. Repeat the procedure at least once for each sample.

3.5 Extraction of Soluble Metabolites

1. Aliquots of each dried fraction (1–6) of the gradient and an aliquot of the unfractionated dry material (see Sect. 3.3; Steps 7 and 15) are used for the extraction of sugars, amino acids, and other metabolites.
2. Add 5 mL of a chloroform:methanol mixture (1.5:3.5) to the dried material in a 15 mL plastic tube.
3. Homogenize the samples by vortexing them and keep them on ice for 30 min.
4. Add 3 mL of water to the homogenate, shake for 30 s.

5. Centrifuge at 400 × *g* for 3 min. Three phases become visible: the upper aqueous phase, the white interface (proteins), and the lower organic phase.
6. Transfer the aqueous supernatant (avoiding protein from the interphase) into a round-bottom flask.
7. Repeat Steps 2–5.
8. Add 2 mL of water to the lower phase in the plastic tube and shake for 30 s.
9. Centrifuge at 400 × *g* for 3 min.
10. Transfer the aqueous supernatant into the glass bulb.
11. Evaporate the combined aqueous phases in a rotary evaporator.
12. Redissolve the vacuum-dried residue in 1 mL of ultrapure water.
13. Filter the solution through a syringe filter (0.20 μm nylon).
14. Store the samples at −80 °C until analyses of sugars, amino acids, or other metabolites.

3.6 Sugar Quantification and Data Analysis

1. Analyze sugars using high-performance liquid chromatography (HPLC) according to the instrument manufacturer's instructions.
2. Calculate the distribution of the marker enzymes in relative proportion per fraction, e.g., 50% of the total GAP-DH activity (chloroplast) is in fraction 1, 10% each in fractions 2–6. The same for UDPase (cytosol) and α-mannosidase (vacuole).
3. Calculate the distribution of a specific sugar, e.g., glucose, in relative proportion per fraction as well.
4. For the evaluation of the subcellular distribution of metabolites between the three main compartments, a calculation procedure according to [2] is used. Enter the data from Steps 2 and 3 into the calculation program, which is based on a best-fit algorithm (available upon request, *see* **Note 8**).
5. After calculation, it is evaluated which case yields the best fit (agreement) with the experimental result on the basis of the linear least squares method. As a criterion of the best fit, a parameter Q is used.

$$Q = \sqrt{\sum \Delta_i^2 / n - 1}$$

Δ_i represents the difference between the calculated amount and the measured amount in each fraction and in the number of fractions (*see* **Note 9**). To avoid results being falsified by analytical errors, the calculations are usually based on mean values obtained from measurements of at least three density gradient fractionations of the same leaf material.

6. Determine the distribution of sugars between the subcellular compartments. The percentage distribution of metabolites between the subcellular compartments and the metabolite contents in leaves allows for the evaluation of average values for the metabolite contents in the subcellular compartments (e.g., in $\mu mol\,g^{-1}$ FW) (*see* **Note 10**).
7. Convert subcellular contents into concentrations. The further conversion of subcellular contents based on gram fresh weight (FW) into concentrations requires the knowledge of the relative volumes of the subcellular compartments. A determination can be done by morphometric analysis of light and electron micrographs [8, 9, 12, 19, 20].

4 Notes

1. The densities of the gradients depend on the plant species and the tissue. The densities for leaves of *Brassica napus*, for example, are $1.31\,g\,cm^{-3}$ and $1.46\,g\,cm^{-3}$. It might be necessary to determine the most suitable densities and density differences for each plant species. Gradients with density differences between 1.33 and $1.50\,g\,mL^{-1}$ to 1.35 and $1.53\,g\,mL^{-1}$ were used for leaves from woody species, and density differences between 1.27 and $1.45\,g\,mL^{-1}$ to 1.34 and $1.58\,g\,mL^{-1}$ were used for leaves from herbaceous species
2. It is important to prepare the solutions A and B from fresh tetrachloroethylene and to protect tetrachloroethylene and the solutions from light because the formation of Cl radicals during the storage results in the complete inhibition of the activities of the marker enzymes in the fractions. Particularly, the solution with the lower density (Solution A) can change with time because heptane evaporates.
3. In the protocol upon which this work is based [1, 2], the majority of the fractionation procedure was carried out at 4°C. Because it is difficult to work water-free during the whole experimental procedure, Farré et al. [13] tested whether metabolites were stable during fractionation at room temperature, where water-free handling can be guaranteed to a higher degree. They found that metabolite levels per unit of frozen material were comparable to levels found after fractionation at room temperature.
4. It is important to consider the presence or absence of light when harvesting leaves or tissues: the content of photoassimilates like sugars, starch, and malate increases during the light period and decreases during the dark period.

5. The particles applied to the nonaqueous gradient should ideally be as small as possible to reduce the potential of including material from different organelles. The use of a ball mill followed by sonication for 2 min leads to an average particle size of 2–3 μm [13].
6. After testing several centrifugation velocities and durations [14], we concluded that equilibrium distribution was already achieved at 5000 *g* and 50 min instead of 25,000 × *g* and 130 min [2]. This considerably shortens the exposure time of the sample material to the nonaqueous solvents.
7. Chlorophyll (chloroplasts), PEP-carboxylase (cytosol), or acid phosphatase and nitrate (vacuole) can be used as alternative markers [14]. The optimal markers are to be tested for each plant species and tissue. It must be noted that the suspension of the dried material in nonaqueous solvents and the homogenization by sonication alter the activities of the employed marker enzymes to different extents. However, this change of activity appears to be uniform for the corresponding enzymes in all fractions and does not affect the suitability of these enzymes as marker enzymes [1]. As far as the other compartments in the mesophyll cells are concerned, it was shown that mitochondria, for example, can also be measured by the activity of citrate synthase. However, in most cases, there is a cofractionation of the mitochondrial marker with cytosolic markers [13]. Besides, it is also not yet possible to distinguish between vacuole and apoplast in the nonaqueous gradients due to cofractionation of apoplastic and vacuolar markers, for example, α-mannosidase [21].
8. This procedure is based on the assumption that the metabolites are confined to the stromal, cytosolic, and vacuolar compartments, indicated by the corresponding marker enzymes. The evaluation is done by a computer program which tests all possible cases for the distribution of a certain metabolite between the three compartments, using increments of 1%; for example:

(1) stroma 100%, cytosol 0%, vacuole 0%;

(2) stroma 99%, cytosol I %, vacuole 0%;

(3) stroma 99%, cytosol 0%, vacuole 1%

In steps of 1%, there are 5151 possibilities for the distribution of a metabolite between the three compartments, and it is calculated which of these possible cases yields the best agreement with the experimental results. The total amount of metabolites thus calculated for each fraction of the gradient is compared to the amount actually measured in these fractions.

This method provides highly reproducible results for metabolites that are confined to a single compartment only. A higher variation was found for metabolites located in more than one compartment, and the variability was highest when the proportion found in a particular compartment was low. Farré et al. [13] estimated that the limit of detection of a compound in a particular compartment is around 5% of the total amount in the tissue.

It should be pointed out that the nonaqueous fractionation method was developed for the determination of those subcellular metabolite concentrations that are exclusively located in the mesophyll, for example, Calvin cycle intermediates [1]. For metabolites also present in tissues other than the mesophyll, such as sucrose also present in the phloem, the method tends to somewhat overestimate the concentrations in the mesophyll cell compartments, because the amounts of metabolites in other tissues are assumed to be negligible. This is also the case for raffinose-family-oligosaccharides (RFOs), because RFOs are synthesized in intermediary cells of the phloem and the concentration in the phloem is at least one order of magnitude higher than in the mesophyll cells. It is to be expected that a considerable amount of RFOs associated with the mesophyll cells arises from the high content in the phloem.

9. Krüger et al. [14] used two different algorithms: the best fit algorithm described here and the nonnegative least squares algorithm. Both approaches solve a system of linear equations defined by the subcellular compartments (determined by the amount of relative marker in the fractions) to find the relative assignment of a metabolite to the compartments by minimizing the discrepancy between the measured and the fitted fraction abundances. Application of both algorithms revealed that the results of the subcellular distributions are similar. The best fit method by Riens et al. [2] was further developed, and bestFit (v1.2) is available from CSB.DB website [22, 23] at http://csbdb.mpimp-golm-mpg.de/bestfit.html.
10. The nonaqueous fractionation technique enables the determination of metabolites in the vacuolar, stromal, and cytoplasmic compartments under conditions that preserve the physiological state during photosynthesis. For this analysis, a leaf was treated as a uniform metabolic compartment representing an average mesophyll cell, subdivided into the three compartments listed above. This is obviously a simplification of a more complex pattern, but until now, it is one of the very few ways to obtain information about subcellular metabolite levels in vivo.

Conflict of Interest

The authors have no conflicts of interest to declare.

References

1. Gerhardt R, Heldt HW (1984) Measurement of subcellular metabolite levels in leaves by fractionation of freeze-stopped material in non-aqueous media. Plant Physiol 75:542–547
2. Riens B, Lohaus G, Heineke D, Heldt HW (1991) Amino acid and sucrose content determined in the cytosolic, chloroplastic, and vacuolar compartments and in the phloem sap of spinach leaves. Plant Physiol 97:227–233
3. Elbers R, Heldt HW, Schmucker P, Soboll S, Wiese H (1974) Measurement of the ATP/ADP ratio in mitochondria and in the extramitochondrial compartment by fractionation of freeze-stopped liver tissue in non-aqueous media. Z Physiol Chem 355:378–393
4. Stitt M, McLilley R, Gerhardt R, Heldt HW (1989) Metabolite levels in specific cells and subcellular compartments of plant leaves. Methods Enzymol 174:518–552
5. Heineke D, Wildenberger K, Sonnewald U, Willmitzer L, Heldt HW (1994) Accumulation of hexoses in leaf vacuoles: studies with transgenic tobacco plants expressing yeast-derived invertases in cytosol, vacuole or apoplasm. Planta 194:29–33
6. Moore Bd, Palmquist DE, Seemann JR (1997) Influence of plant growth at high CO2 concentrations on leaf content of ribulose-1,5-bisphosphate carboxylase/oxygenase and intracellular distribution of soluble carbohydrates in tobacco, snapdragon, and parsley. Plant Physiol 115:241–248
7. Tilsner J, Kassner N, Struck C, Lohaus G (2005) Amino acid contents and transport in oilseed rape (*Brassica napus* L.) under different nitrogen conditions. Planta 221:328–338
8. Voitsekhovskaja OV, Koroleva OA, Batashev DR, Knop C, Tomos AD, Gamalei YV, Heldt HW, Lohaus G (2006) Phloem loading in two Scrophulariaceae species. What can drive symplastic flow via plasmodesmata?. Plant Physiol 140:383–395
9. Nadwodnik J, Lohaus G (2008) Subcellular concentrations of sugar alcohols and sugars in relation to phloem translocation in *Plantago major*, *Plantago maritima*, *Prunus persica*, and *Apium graveolens*. Planta 227:1079–1089
10. Knaupp M, Mishra KB, Nedbal L, Heyer AG (2011) Evidence for a role of raffinose in stabilizing photosystem II during freeze-thaw cycles. Planta 234:477–486
11. Findling S, Zanger K, Krueger S, Lohaus G (2015) Subcellular distribution of raffinose oligosaccharides and other metabolites in summer and winter leaves of *Ajuga reptans* (Lamiaceae). Planta 241:229–241
12. Öner-Sieben S, Lohaus G (2014) Apoplastic and symplastic phloem loading in *Quercus robur* and *Fraxinus excelsior*. J Exp Bot 65:1905–1916
13. Farré E, Tiessen A, Roessner U, Geigenberger P, Trethewey RN, Willmitzer L (2001) Analysis of the compartmentation of glycolytic intermediates, nucleotides, sugars, organic acids, amino acids, and sugar alcohols in potato tubers using a nonaqueous fractionation method. Plant Physiol 127:685–700
14. Krüger S, Giavalisco P, Krall L, Steinhauser MC, Büssis D, Usadel B, Flügge UI, Fernie AR, Willmitzer L, Steinhauser D (2011) A topological map of the compartmentalized *Arabidopsis thaliana* leaf metabolome. PLoS ONE 6:e17806. https://doi.org/10.1371/journal.pone.0017806
15. Rennie EA, Turgeon R (2009) A comprehensive picture of phloem loading strategies. Proc Natl Acad Sci U S A 106:14162–14167
16. Wirtz W, Stitt M, Heldt HW (1980) Enzymatic determination of metabolites in the subcellular compartments of spinach chloroplasts. Plant Physiol 106:661–671
17. Zrenner R, Willmitzer L, Sonnewald U (1993) Analysis of the expression of potato uridinediphosphate-glucose pyrophosphorylase and its inhibition by antisense RNA. Planta 190:247–252
18. Ye-The LI (1967) Studies on the glycosidase in Jack bean meal. I. Isolation and properties of α -mannosidase. J Biol Chem 242:5474–5480
19. Winter H, Robinson DG, Heldt HW (1993) Subcellular volumes and metabolite concentrations in barley leaves. Planta 191:180–190
20. Leidreiter K, Kruse A, Heineke D, Robinson DG, Heldt HW (1995) Subcellular volumes and metabolite concentrations in potato (*Solanum tuberosum* cv. Désirée) leaves. Bot Acta 108:439–444
21. Fettke J, Poeste S, Eckermann N, Tiessen A, Pauly M, Geigenberger P, Steup M (2005) Analysis of cytosolic heteroglycans from leaves of transgenic potato (*Solanum tuberosum* L.) plants that under- or overexpress the Pho 2 phosphorylase isozyme. Plant Cell Physiol 46:1987–2004
22. Steinhauser D, Usadel B, Luedemann A, Thimm O, Kopka J (2004) CSB.DB: a comprehensive systems-biology database. Bioinformatics 20:3647–3651
23. Klie S, Krueger S, Krall L, Giavalisco P, Flügge UI, Willmitzer L, Steinhauser D (2011) Analysis of the compartmentalized metabolome—a validation of the non-aqueous fractionation technique. Front Plant Sci 2:55

Check for updates

Chapter 26

Methods to Estimate Carbon Export from Leaves

Jess Gersony, Maria Park and Uri Hochberg

Abstract

In most plant species, the majority of carbon assimilation takes place in leaves. The export of carbon from leaves to the rest of the plant's organs is thus a fundamental process in plant biology. At the same time, only a few studies have estimated the rate of this process and its dynamics. In the current chapter, we will describe two simple methods for evaluating the rate of carbon export from leaves. Both methods calculate the export as the difference between carbon assimilation and carbon storage. The first method evaluates the carbon storage based on the leaf mass per area (LMA) difference between two time points, while the second method evaluates the carbon storage based on the difference in non-structural carbohydrate (NSC) concentration. We will also explore the differences between the methods and the assumptions that could reduce the accuracy of each method.

Keywords Phloem export, Carbon, Nonstructural carbohydrates, Leaf mass per area, Physiology

1 Introduction

Phloem export is the process that enables the translocation of photosynthates from the leaves into the carbon sinks (stems, roots, fruits), facilitating their respiration and growth. Understanding the rate of this translocation is critical for our basic understanding of plant function and for modeling their responses to the environment. Environmental conditions that typically harm plants' growth, such as cold nights and drought, are likely to reduce carbon export from leaves [1, 2]. Similarly, carbon export from leaves can be reduced when direct limitations on carbon sinks are imposed [3]. Also, when it comes to daily changes in environmental conditions, carbon export shows considerable variability. A summary of results from studies that measured phloem export from leaves through a diurnal period (Fig. 1) shows that, while there is substantial variability between species and methods, overall the available data suggest that daytime phloem export is much higher than during the dark period. Despite its fundamental role in plant biology and its high dynamics, there are relatively few measurements of phloem export rate, especially

Johannes Liesche and Sanna Sevanto (eds.), *Phloem*, Methods in Molecular Biology, vol. 3066,
https://doi.org/10.1007/978-1-0716-5488-0_26,

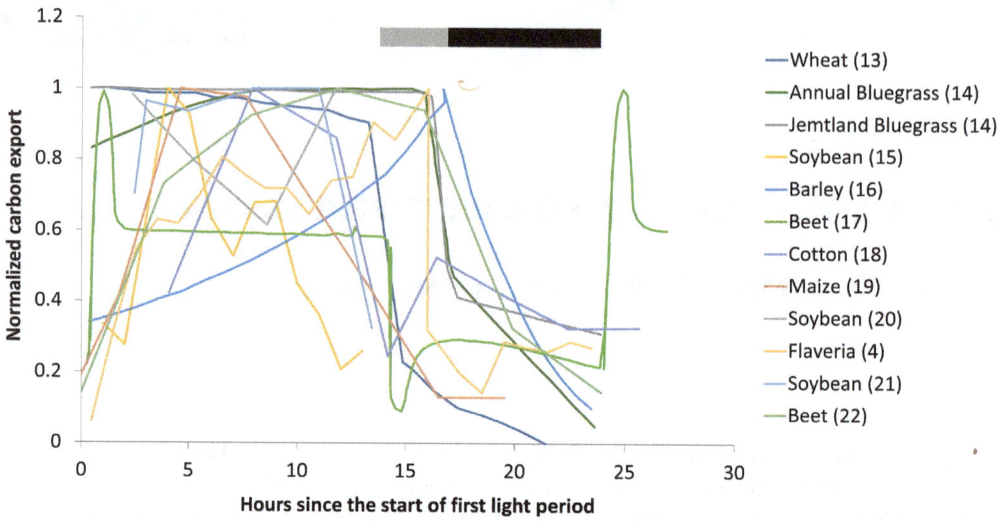

Fig. 1 Summary figure of carbon export data in the literature. Each line shows export values from a different study. These numbers are normalized to the maximum export rate in each study. The gray/black bar represents the dark period. Not all studies had the same hours of daylight, so the gray bar shows when some studies were dark, but not all. Each number after the common species name is the citation from which the data came

when compared to other fluxes from leaves such as photosynthesis or transpiration. Part of the reason for the lack of data is that the measurement protocol has not made it into the standard method protocols of ecophysiology laboratories. The current chapter aims to introduce the relevant protocols for phloem export measurements in a simple, easily adaptable manner.

To date, carbon export has been measured in three different ways: leaf weight dynamics combined with gas exchange measurements, leaf non-structural carbohydrate dynamics combined with gas exchange measurements, and C_{14} labeling. The latter was suggested to be less accurate as unlabeled carbon was exported out of the leaf [4]. Accordingly, this chapter will describe the methods that use leaf weight dynamics (Method 1) and NSC dynamics (Method 2) and compare their results. To show a practical usage of the method, we will describe how we used it in mature forest trees for the first time. Specifically, we studied five mature red oak trees at Harvard Forest in Petersham, MA, in late summer using both methods.

2 Materials

2.1 Method 1

- Hole puncher.
- Analytical balance accurate to 1×10^{-5} g.
- Freeze drier.

- Gas exchange instrument.
- Cooler with dry ice.

2.2 Method 2

- Nonstructural carbohydrate measurement.
  - For example, see [5, 6].
  - Example instruments and supplies needed for the above method include: Megazyme starch and Sigma fructose assay kits, freeze drier, plate reader (SpectraMax i3), Sigma Aldrich glucose assay reagent, phosphoglucose isomerase and invertase, peach standards for additional validation (NIST SRM 1547, Sigma-Aldrich), and a cuvette reader (100 UV–Vis Spectrometer, Cary).
- Gas exchange instrument.
- Cooler with dry ice.

3 Methods

Both methods rely on the premise that exported carbon can be calculated by measuring the carbon coming into the leaf over a certain amount of time and the change in carbon in the leaf over that same time period. In other words, the amount of carbon that is being exported out of the leaf can be calculated by a mass balance approach. Exported carbon (C_{EX}) is the sum of the dynamics of carbon storage ($C_{T\text{-}1} - C_T$; carbon at prior time point – carbon at time point) and net carbon that has come in via gas exchange (C_A):

$$C_{EX} = (C_{T-1}) - (C_T) + (C_A) \tag{1}$$

The two methods differ in the way in which they calculate the carbon storage dynamics. The first uses the change in LMA (leaf mass per area) while the other uses the change in NSC concentration. A large part of the description below is the method to convert the units so that the ($C_{T\text{-}1} - C_T$) can be added to the (C_A), as in Eq. (1).

3.1 Method 1

1. Take leaf disks from the leaves of interest multiple times in 1 day. It is necessary to have more than one sampling time point in order to compare at least two weights. For this hole punching, you can use a standard desktop hole–puncher.
   a. We typically limit ourselves to 12 leaf disks out of a *Quercus rubra* or *Acer rubrum* leaf to limit the damage to the leaves. E.g., one scenario would be to sample 2–3 holes at each time point, for 4–6 time points over the course of a day.

   b. We also avoid sampling from major or secondary veins.
   c. We typically have 3–4 leaves per plant of interest, depending on the plant age and size/leaf area.

2. Once these leaf disks are sampled, place them in a small, labeled paper coin-envelope and place them on dry ice to immediately freeze the samples.
   a. Store in a −80°C freezer or on dry ice until they can be freeze-dried.
3. At the same time points as decided upon for hole punching, take measurements of gas exchange using an infrared gas analyzer (IRGA), e.g., Li-Cor 6800, ideally on the same leaves from which you are hole-punching. Take the IRGA measurement prior to the hole punching, and avoid areas where there are holes. The lamina should fill the chamber of the IRGA.
4. Freeze dry the leaf disks for 48 h.
5. Once dry, store the leaf disks in a desiccator while waiting to be weighed.
6. Weigh the leaf disks together for each leaf for each time point (e.g., 2–3 disks at a time) on an analytical balance accurate to 1×10^{-5} g. Record the weight and number of disks that were measured for each leaf for each time point.
7. Once the field measurements and leaf disk weighing are complete, aggregate the data into one spreadsheet to calculate export.
8. To calculate export:
   a. To calculate the net assimilation input between time points:
      i. Average the net assimilation rate between the two time points (e.g., what is the average amount of photosynthesis occurring between time point 1 and time point 2).
   b. To calculate the change in weight between the two time points:
      i. Calculate the dry weight of each individual leaf disk by dividing the average dry weight of the leaf disks at a time point by the number of leaf disks.
      ii. Subtract the weight of time point 2 from time point 1 and divide it by the leaf disk area (m^{-2}) and the time difference (in seconds) between the two sampling points. This number represents the change in weight per time per area (grams of NSC m^{-2} s^{-1}).
         1. Here we are making the assumption that all of the dry weight change is due to NSC.

iii. Multiply this number by 41.5%, which is the ~ percentage of sucrose, glucose, and fructose, as well as starch, that is carbon by weight. This number is now in the units of grams of carbon $m^{-2} s^{-1}$.

1. Sucrose is 41.1% carbon by weight; glucose and fructose are both 40% carbon by weight. Starch is 44% carbon by weight. So here it would help to know the exact ratio of suc:glu:fruc:st in your leaf. But, barring that, one could use the number 41.5% as an approximation.

iv. To create the same units as the ones of assimilation, multiply this number by 1,000,000 (conversion from g to μg) and then divide by the molar weight of carbon (12.01 g mol^{-1}). This number is now in the units of "μ mol of carbon m^{-2} s^{-1}".

v. Add together carbon storage change (result of 8.b.iv.) and the assimilation (from 8.a.i.) to calculate the export rate (carbon $m^{-2} s^{-1}$).

3.2 Method 2

1. Take measurements of gas exchange using an IRGA, e.g., Li-Cor 6800, on leaves of your plant of interest at multiple times over the day. It is necessary to have at least two sampling points in order to compare the two carbon contents.
2. At the same time points of the gas exchange measurements, collect samples of leaves to test for NSCs.
3. Proceed to measure your samples for NSCs (soluble sugars and starches). We will not describe NSC methods here. Preferably, use an NSC method that quantifies sucrose, glucose, and fructose independently from one another.
   a. We typically quantify NSCs using a modified protocol of the Megazyme starch and Sigma Fru assay kits used by [5] and further described in [6].
   b. It is important to record the dry weight and fresh area of your leaf, for estimating the LMA. This will be important in calculating export.
4. Once the field measurements and NSC methods are complete, aggregate the data into one spreadsheet to calculate export.
5. To calculate export:
   a. To calculate the net assimilation input between time points:
      i. Average the net assimilation rate between the two time points (e.g., what is the average amount of photosynthesis occurring between time point 1 and time point 2). This number is in μmol $m^{-2} s^{-1}$.
   b. To calculate the change in carbon in NSCs between time points:

i. Multiply the concentration ($mg\,g^{-1}$) of starch, glucose, fructose, and sucrose by the amount of carbon, by weight, that is in each molecule (44%, 40%, 40% and 42%)

ii. Add these numbers together for each sample. Now this number is the concentration of carbon in NSCs in $mg\,g^{-1}$.

iii. Multiply this number by the LMA ($g\,m^{-2}$) calculated from the dry weight and area of your leaf. The result is in mg carbon m^{-2}.

iv. Do Steps i to iii for both time points.

v. Subtract the carbon weight ($mg\,m^{-2}$) of time point 2 from time point 1 and divide it by the time difference (in seconds) between the two sampling points. This number represents the change in weight per time per area (milligrams of NSC $m^{-2}\,s^{-1}$).

vi. Convert this number into $\mu mol\ m^{-2}\,s^{-1}$ by multiplying by 1000 (converting mg to μg) and dividing it by the molar mass of carbon ($12.01\ g\,mol^{-1}$) to shift from μg to μmol.

6. To calculate export, add the carbon storage dynamics (5. b. vi.) to the carbon assimilation (5. a. i).

4 Methods Comparison

To compare these methods on mature oak trees, we implemented both methods in August 2017. For Method 2, the data is replicated from [6]. For Method 1, regarding replication, we studied the same 5 trees as in [6]. We studied 4 leaves per tree, and for each leaf, we collected 4 hole punches per time period. The results highlight a similar magnitude and a similar pattern for each of the methods (Fig. 2). We would highlight that Method 1 in this comparison was a much more rapid measurement, taking only days to process the samples, whereas Method 2 took many weeks.

5 Notes

In the current experiment, we found a reasonable agreement between methods, but there are still several potential artifacts associated with the methods that should be considered. Below, we explore some considerations to take into account when engaging with these methods.

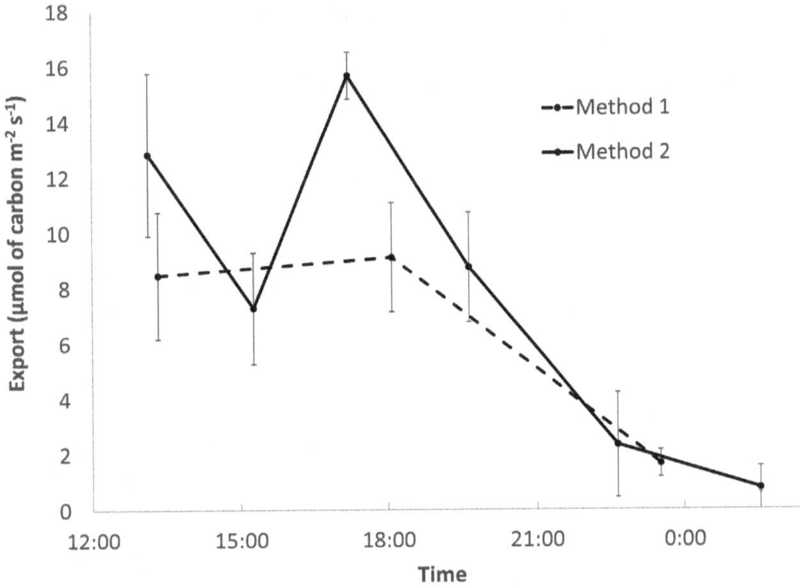

Fig. 2 A comparison across methods for five mature oak trees. Each dot represents the average and standard error across five trees. Method 1 (LMA method) is shown with the dashed line and Method 2 (NSC method) is shown with the solid line

5.1 Leaf Weight Dynamics

The conversion from leaf weight changes into changes in the carbon pool assumes that all the changes in the leaf mass per area (LMA) result from changes in the leaf carbon pool.

In most species, leaves tend to lose water during the day, resulting in shrinkage of both thickness and area. That shrinkage results in a higher concentration of NSCs per area, even without the accumulation of carbon. A survey of 15 species [7] found that leaf area shrinkage from saturation to turgor loss ranged from 0.5% to 14% (with a 4% average). In our experiment, the daily increase in LMA was up to 5% in one tree (with an average of 3% for the five trees), and so even 1% leaf shrinkage could explain a large proportion of the daily LMA changes we reported. As red oaks have relatively stiff leaves (less shrinkage) and, in the current experiment, they did not experience drought, the effect was probably much smaller. In contrast, in tomatoes and grapevines (which have soft leaves), the accumulation of NSCs during the day explained only part of the increase in dry mass [8, 9]. Accordingly, we recommend that prior to utilization of the method, the daily leaf area shrinkage should be evaluated, and in cases where the change is 1% or more, the leaf shrinkage would be accounted for in the carbon export calculation.

The daily accumulation of mass does not necessarily arrive from carbon. Minerals are also known to accumulate in the leaf on a daily basis [10, 11]. In particular, potassium and magnesium were found to increase by ~10% during the light period. However, considering that measuring the changes in minerals would seriously complicate the method and that the overall minerals concentration is typically less than 10% of the leaf dry weight, we do not recommend accounting for the error.

5.2 NSCs Dynamics

Carbon can be stored in many molecules, including sugars, sugar–alcohols, organic acids, amino acids, proteins, and more. In the NSCs method, we assume that our analyses capture all of the compounds that contain carbon and have a substantial daily pattern. In red oaks, this assumption was tested by Gersony et al. [6], who checked for the daily change of all the major carbon osmolytes and found them to be negligible compared with starch and sucrose. However, other species might accumulate other compounds during the day. For example, species from the *Oleaceae* family are known to accumulate about 200 mmol kg^{-1} of malate and mannitol during the day [12]. Before the NSCs method is applied, direct measurements or a literature survey of the daily pattern of the major carbon pools is recommended [13–22].

Conflict of Interest

The authors have no conflicts of interest to declare.

References

1. Tombesi S, Cincera I, Frioni T, Ughini V, Gatti M, Palliotti A, Poni S (2019) Relationship among night temperature, carbohydrate translocation and inhibition of grapevine leaf photosynthesis. Env Exp Bot 157:293–298
2. Perry A, Sperling O, Rachmilevitch S, Hochberg U (2025) Carbon dynamics under drought and recovery in grapevine's leaves. Plant Cell Env 48(5):3379–3390
3. Fondy BR, Geiger DR (1980) Effect of rapid changes in sink-source ratio on export and distribution of products of photosynthesis in leaves of *Beta vulgaris* L. and *Phaseolus vulgaris* L. Plant Physiol 66(5):945–949
4. Leonardos ED, Micallef BJ, Micallef MC, Grodzinski B (2006) Diel patterns of leaf C export and of main shoot growth for Flaveria linearis with altered leaf sucrose–starch partitioning. J Exp Bot 57(4):801–814
5. Gauthier PP, Crous KY, Ayub G, Duan H, Weerasinghe LK, Ellsworth DS, Tjoelker MG, Evans JR, Tissue DT, Atkin OK (2014) Drought increases heat tolerance of leaf respiration in Eucalyptus globulus saplings grown under both ambient and elevated atmospheric [CO2] and temperature. J Exp Bot 65(22):6471–6485
6. Gersony JT, Hochberg U, Rockwell FE, Park M, Gauthier PP, Holbrook NM (2020) Leaf carbon export and nonstructural carbohydrates in relation to diurnal water dynamics in mature oak trees. Plant Physiol 183(4):1612–1621
7. Scoffoni C, Vuong C, Diep S, Cochard H, Sack L (2014) Leaf shrinkage with dehydration: coordination with hydraulic vulnerability and drought tolerance. Plant Physiol 164(4):1772–1788
8. Bertin N, Tchamitchian M, Baldet P, Devaux C, Brunel B, Gary C (1999) Contribution of carbohydrate pools to the variations in leaf mass per area within a tomato plant. New Phytol 143(1):53–61
9. Perry A, Sperling O, Ben-Gal A, Holbrook NM, Rachmilevitch S, Hochberg U (2026) Do Daily and Seasonal Changes in Non-Structural Carbohydrates in Grapevine Leaves Contribute to Osmotic Adjustment and Regulation of Photosynthesis? Physiologia Plantarum 178(1), e70683.
10. Phillis E, Mason TG (1942) On diurnal variations in the mineral content of the leaf of the cotton plant. Ann Bot 6(23):437–442

11. Hocking PJ, Pate JS, Atkins CA, Sharkey PJ (1978) Diurnal patterns of transport and accumulation of minerals in fruiting plants of *Lupinus angustifolius* L. Ann Bot 42(6):1277–1290
12. Peltier JP, Marigo D, Marigo G (1997) Involvement of malate and mannitol in the diurnal regulation of the water status in members of Oleaceae. Trees 12(1):27–34
13. Balaguer L, Barnes JD, Panicucci A, Borland AM (1995) Production and utilization of assimilates in wheat (*Triticum aestivum* L.) leaves exposed to elevated O_3 and/or CO_2. New Phytol 129(4):557–568
14. Borland AM, Farrar JF (1985) Diel patterns of carbohydrate metabolism in leaf blades and leaf sheaths of Poa annua L. and Poa× Jemtlandica (Almq.) Richt. New Phytol 100(4):519–531
15. Cure JD, Rufty Jr TW, Israel DW (1991) Assimilate relations in source and sink leaves during acclimation to a CO_2-enriched atmosphere. Physiol Plant 83(4):687–695
16. Farrar SC, Farrar JF (1985) Carbon fluxes in leaf blades of barley. New Phytol 100(3):271–283
17. Fondy BR, Geiger DR (1982) Diurnal pattern of translocation and carbohydrate metabolism in source leaves of *Beta vulgaris* L. Plant Physiol 70(3):671–676
18. Hendrix DL, Huber SC (1986) Diurnal fluctuations in cotton leaf carbon export, carbohydrate content, and sucrose synthesizing enzymes. Plant Physiol 81(2):584–586
19. Kalt-Torres W, Kerr PS, Usuda H, Huber SC (1987) Diurnal changes in maize leaf photosynthesis: I. Carbon exchange rate, assimilate export rate, and enzyme activities. Plant Physiol 83(2):283–288
20. Kerr PS, Rufty Jr TW, Huber SC (1985) Endogenous rhythms in photosynthesis, sucrose phosphate synthase activity, and stomatal resistance in leaves of soybean (Glycine max [L.] Merr.). Plant Physiol 77(2):275–280
21. Qiu J, Israel DW (1992) Diurnal starch accumulation and utilization in phosphorus-deficient soybean plants. Plant Physiol 98(1):316–323
22. Rao IM, Fredeen AL, Terry N (1990) Leaf phosphate status, photosynthesis, and carbon partitioning in sugar beet: III. Diurnal changes in carbon partitioning and carbon export. Plant Physiol 92(1):29–36

Part V

Connecting Phloem Function and Plant Physiology

Chapter 27

Neutron Imaging to Reveal Phloem Function in Trees

Sanna Sevanto and Dominique Derome

Abstract

Understanding phloem function is key to understanding the regulation of plant function. The phloem distributes the main building blocks that support plant function and growth from the production sites to sites of consumption and storage. It also distributes important signaling molecules and interacts with the adjacent water-transport tissue, the xylem, to support hydraulic integrity of plants. Yet, studying phloem function is challenging because of the sensitivity of this tissue to mechanical stimulation that can lead to blockage of flow. Furthermore, its small size, adjacent location, and interactions with the much larger xylem tissue that operates under an opposite pressure gradient to the phloem make experiments of phloem function in opaque, woody tissues challenging. Here, we present a method to measure water transport rates within the phloem and between the xylem and the phloem using neutron imaging with heavy water, D_2O, as a tracer. The method is based on differences in neutron attenuation by hydrogen and deuterium atoms that create a contrast when imaging plant tissues or soil that are relatively transparent to neutrons. This method allows nondestructive visualization of water flow within the xylem and phloem tissues and calculation of flow rates at the spatial and temporal resolution determined by the experimental setup.

Keywords Deuterium, Heavy water, Hydrogen, Phloem transport, Transport velocity, Xylem, Xylem–phloem water exchange

1 Introduction

Neutron imaging can be used to track water movement in plants using heavy water as a tracer [1]. In contrast to X-rays that interact with electrons to a degree that is proportional to atom number, neutron imaging is based on neutron interactions with the nuclei of atoms and does not follow a clear pattern related to atom size. This leads to unique attenuation patterns of neutrons of relevance to us here us the high attenuation of hydrogen. This makes neutron imaging suitable for many organic materials that contain hydrogen, particularly when containing water, which is hard to quantify with X-ray imaging [2]. Neutron interactions with the imaged material depend on the energy of the neutrons used in imaging. Thermal neutrons (energy up to 0.025 eV), with their slower speeds, exhibit higher

Johannes Liesche and Sanna Sevanto (eds.), *Phloem*, Methods in Molecular Biology, vol. 3066,
https://doi.org/10.1007/978-1-0716-5488-0_27,

contrast and are effective at imaging small objects or objects with a small amount of water, while fast neutrons (energy 0.5–20 MeV), with their higher energies, offer greater penetration and are suitable for imaging larger, thicker objects.

For imaging, the materials must have structures that differ in neutron attenuation to create contrast. Generally, plant cell walls are relatively transparent to neutrons. Cellulose, hemicelluloses, and lignin that make up the wood cell wall are overall quite transparent to neutrons. Similarly, sandy soils low in organic matter (low carbon content) are relatively transparent to neutrons, but hydrogen atoms scatter neutrons readily. This creates contrast that allows imaging or measuring plant tissue or soil water (H_2O) content with neutrons (Fig. 1). Imaging varying water content in wood has been successful with liquid uptake in dry samples, but also with water vapor adsorption [3, 4]. However, when the sample already contains water, adding a contrast agent that moves with water allows time-lapse imaging of water movement from which flow rates can be calculated. An often-used contrast agent is heavy water (D_2O). The additional neutron in the deuterium nucleus compared to hydrogen changes both the spin and the attenuation cross-section of the deuteron, making it quite transparent to neutrons [5]. D_2O is also transported through plant tissues like normal water, even if it is toxic to living tissues at high concentrations [6, 7]. This allows utilizing the contrast between H_2O and D_2O for high-resolution imaging of D_2O translocation with both nuclear magnetic resonance (NMR) and neutron imaging methods [7–9]. One interesting application has been tracking water uptake from the soil and through the roots [8, 10–12].

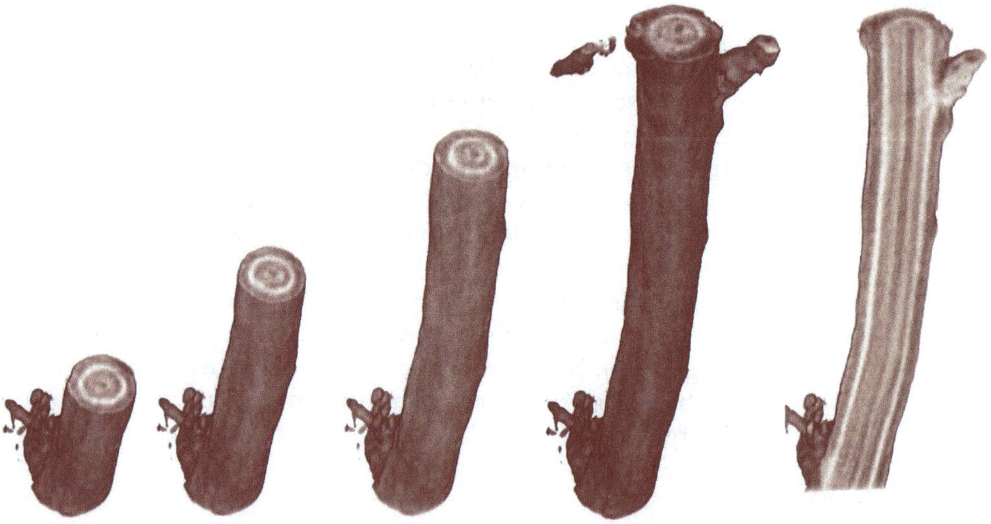

Fig. 1 Neutron tomography image of a pine root. The lighter areas are higher in water content than the darker areas, following the growth ring pattern inside the root. (Image courtesy Adrian Losko)

Applying these methods for measuring phloem function relies on detailed anatomical knowledge for separating the contributions of the xylem and storage tissues from the contribution of the conductive phloem. At the moment, neither of these methods provides a means of detecting flow in single phloem conduits due to a lack of spatial resolution. Therefore, the use of D_2O as a tracer is very well suited for studying bulk flow patterns and their dynamics under plant stress rather than defining flow in single phloem conduits.

Neutron imaging can be conducted in radiography or tomography modes. In both cases, the neutron beam passes through the sample to a neutron detector behind the sample (Fig. 2), and the image is created from the output of this detector. In radiography, the sample is stationary, and the attenuation of each pixel of the image results from the superposition of the respective attenuation of all the elements the beam passes through. In tomography, the sample is rotated so that a three-dimensional image of the internal structures of the object is created by the observation and recording of the differences in the effects of the object on the detected neutron beam. In both modes, obtaining a high-enough quality image requires an exposure time that depends on the qualities of the neutron beam and the detector. With the systems we have used, the exposure time is such that measuring flow rates in the tomography mode is not feasible, mostly because the time needed to obtain enough rotated images to make a three-dimensional image is too long compared to the flow velocities in the xylem and the phloem. Therefore, we focus here on radiographic imaging. The techniques presented here can be applied to tomographic imaging once the imaging technology of the neutron beam fluxes allows for shorter timesteps for image acquisition. While this chapter describes methods for imaging water flow using D_2O as a tracer, it is possible to develop other tracers from, e.g., chlorine [13] or other higher attenuation atoms that, in the

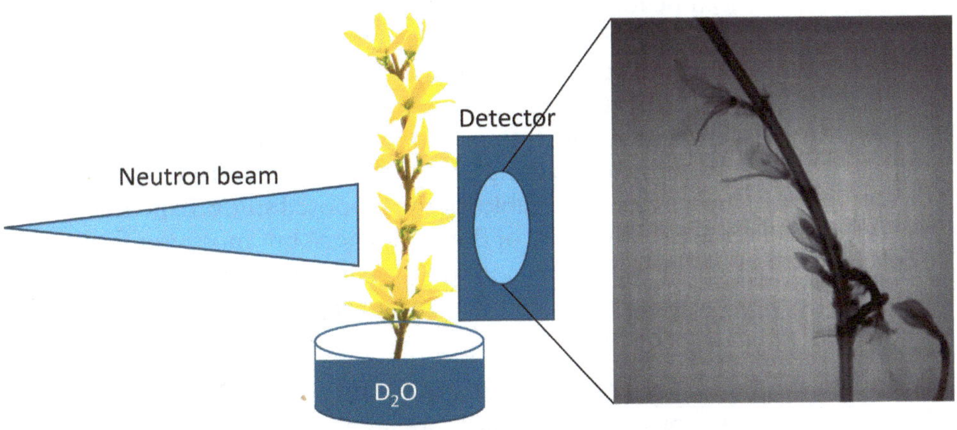

Fig. 2 A schematic presentation of the sample and detector setup in a neutron beam for imaging a Forsythia branch with flowers

right environment and with high enough spatial resolution, could allow tracing the movement of these materials within plant tissues with neutron imaging.

2 Materials

2.1 Imaging Systems

Neutron imaging requires a neutron source that produces a coherent neutron beam and specialized equipment for neutron detection and image capture. Generally, neutrons are produced through various nuclear reactions, including nuclear fission, nuclear fusion, and spallation reactions, where high-energy protons or electrons bombard a heavy metal target, causing the target nucleus to disintegrate and release neutrons. Neutron detection and image capture are facilitated by scintillators or X-ray films with conversion screens that convert neutrons to visible light to be imaged with digital imaging cameras. While portable high-yield neutron generators have been developed recently [14], these instruments still lack the contrast and spatial resolution needed for imaging plants. Therefore, the easiest way to use neutron imaging to study plants is to apply for measurement time at an imaging facility. There are several facilities around the world that accept research applications to use their instruments. These include the Atomic Agency (IAEA) user facility, the European Spallation Source (ESS) in Sweden, Paul Scherrer Institute (PSI) in Switzerland, and the National Institute of Standards and Technology (NIST) Neutron Imaging Facility (NNIF), the Spallation Neutron Source (SNS) at Oak Ridge National Laboratory, and the North Carolina State University's Neutron Imaging Facility (NIF) in the USA. Other institutions that offer neutron imaging services include the JPARC in Japan and ISIS in the UK. Each facility has its own application time window and review processes, instructions for which can be found on its web page. If a proposal is selected, the institute typically provides technical support and training for the researchers to conduct the imaging. Therefore, it is not necessary to be an expert in neutron imaging to use it as a tool. Each facility also has its own instruments with their strengths and weaknesses. Therefore, when planning an imaging study, it is important to evaluate whether the instrument is suitable for the proposed study. Experimental design is the responsibility of the researchers, but most user facilities encourage proposers to contact their instrument leaders to discuss their proposals and help design the experiments for success with their instruments.

2.2 Imaging Materials

1. The plant materials to be imaged (*see* **Note 1**).
2. Plant holders, tape, and putty that can keep the sample stationary during imaging (Fig. 3) (*see* **Note 2**).

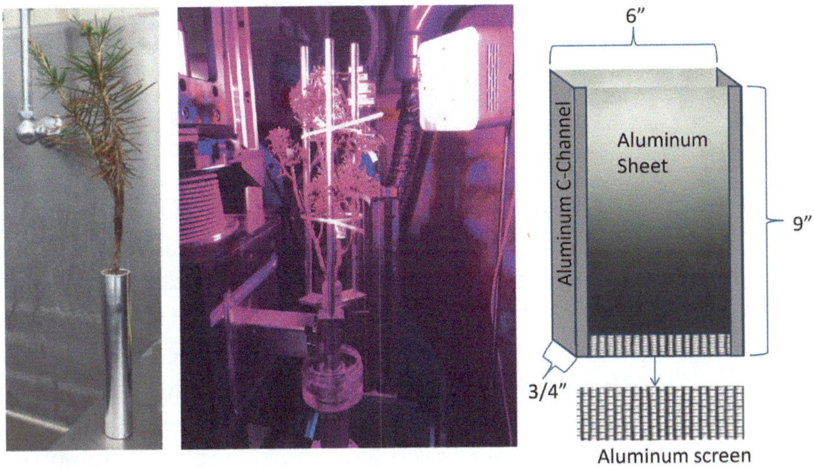

Fig. 3 Photographs of examples for plant holders. A pinion pine seedling in a cylindrical aluminum container for neutron tomographic imaging of roots (left), and a setup at the measurement chamber to measure flow rate in a juniper branch (middle). Here, the window of view is determined by the square tube on the left, and the plant holder consists of a cylindrical aluminum tube attached to a plastic water container with an adjustable orifice attached to aluminum rods secured by zip ties to hold the branch stationary. The pink color is due to the growth light on the right. An example of the design of a rectangular, custom-made plant container for imaging plant root water uptake with a screen at the bottom (right)

3. Heavy water (D_2O). The amount needed can be estimated based on sample size (stem diameter and leaf area) and typical transpiration rate.
4. Two growth lamps to induce stomatal opening. One is to be placed above the sample during the measurement, and the other is to illuminate the samples while waiting for their turn to be measured.
5. Syringes, tubing, or containers that can hold water and heavy water to allow for easy exchange of H_2O to D_2O (*see* **Note 3**).
6. Scalpels, razor blades, or clippers to trim the water-uptake surfaces as needed.
7. Teflon tape, if measuring xylem-phloem water exchange.

2.3 Data Processing Tools

1. Portable hard drives that can store TBs of data, depending on the size and duration of your experiment (number of replicates/plants measured), or access to a server supported by the imaging facility.
2. An image processing software that can handle image stacks and perform simple mathematical operations on images.
3. A computer with a relatively high-frequency processor and good working memory (*see* **Note 4**).

3 Methods

3.1 Measurement Chamber Setup

1. Place the growth light in the measurement chamber above the sample position. If comparing different measurements, it is good to measure chamber temperature and relative humidity during the measurements. We have also used thermocouples attached to the leaves to measure leaf temperature and calculate transpiration rates during the measurements to account for differences in comparisons [15, 16] (*see* **Note 5**).
2. Set up and secure the sample holder system, and test that it is stable and changing samples is relatively easy. If working with a team, it is good to practice changing samples with all team members and even write down a step-by-step protocol.

3.2 Sample Preparation

1. Place samples under a growth light to ensure stomatal opening. Select a sample appropriately to match the dimensions of the beam flux.
2. Select the plant parts to be imaged. The window of view will depend on the instrument and detector used. Generally, the larger the window of view, the lower the spatial resolution that can be obtained. If the goal is to compare the effects of different treatments on water movement in different plants, try to image the samples that you are most interested in comparing in a single run. This eliminates artifacts related to the measurement setup and changes in the environment (temperature, water vapor deficit) from affecting the differences detected. Choose healthy plants. If imaging leaves, try to select samples of a similar developmental stage.
3. Prior to imaging, make sure that the plants are functional by measuring leaf gas exchange. This can be done a day or two before if that is more practical.
4. Prepare the samples for imaging. Leaf samples can be cut and placed in a water container right before imaging. If imaging branches, recut the branches under water before placing them into a water container. If imaging plants with intact root systems in soil, we recommend withholding water for some time before imaging to ensure that the soil is not saturated, to allow plant uptake to change quickly from H_2O to D_2O (*see* **Note 6**).

   a. If measuring water exchange between the xylem and the phloem, peel the bark and the phloem from the end of the branch with a razor blade and seal the interface between the cut bark-phloem surface and the xylem and enough of the bark with Teflon tape (Fig. 4) so that the branch part in water is tape-covered. This forces the water uptake via

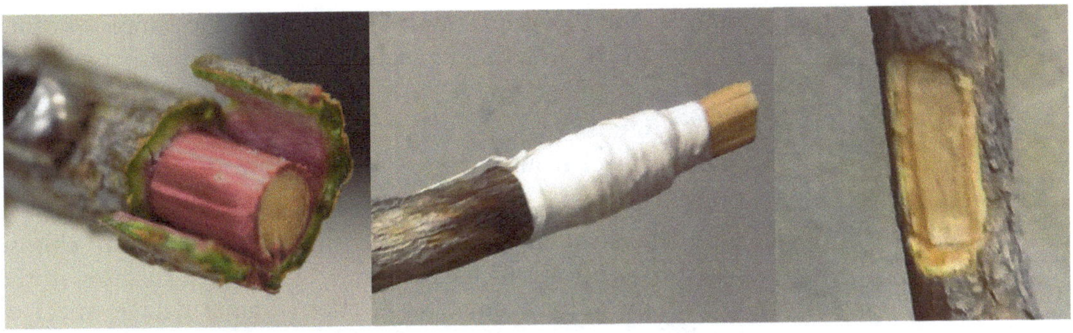

Fig. 4 A photograph showing how to peel the bark and phloem on a pinion pine branch (left), a juniper branch with Teflon tape sealing the bark-phloem cut so that water uptake will occur only via the xylem (middle), and a juniper branch with exposed xylem at the window of view to allow subtracting xylem from the whole stem ROIs to calculate phloem flow rate

the xylem and allows for detecting water movement from the xylem to the phloem.

b. If measuring water flow in the phloem, slice a small piece of bark and phloem out from the middle of the branch on opposite sides of the branch in the middle of the window of view (Fig. 4). This will allow for subtracting xylem transport from phloem transport in the data analysis.

3.3 Imaging

1. Measure the open beam without the sample.
2. Set up the sample and test that the ROI is visible.
3. Image the sample with H_2O to obtain at least ~10 time-lapse images to create an H_2O baseline (*see* **Note 7**).
4. Change the H_2O to D_2O (*see* **Note 8**).
5. Run time-lapse imaging until the D_2O front has passed the ROI.
6. Close the shutter; do not remove the sample, but take the dark-current image to allow for neutron-scattering corrections in the data.
7. Finish the measurement and remove the sample.
8. Repeat for the next sample.

3.4 Data Processing

1. Correct the images for neutron-scattering effects by subtracting dark-current images from the data.
2. Correct the images for non-uniform illumination (vignetting) and pixel-sensitivity variations (such as dust on optics or sensor imperfections): divide the dark current-subtracted data by the open-beam measurement.
3. Clean the image stack for beam shutdowns or other anomalies (*see* **Note 9**). Scattering neutrons can bounce back from the

samples and the water source at the base of the sample or other objects in the hutch, and specialized software can correct for this.

Optional steps

4. Subtract background from your images to increase signal-to-noise ratio.
5. Normalize the maximum intensity to 1. You can choose whether you want to have your D_2O interface to show in the images as white against black (the original data format) or black against white (reversed). Choose what is easiest and be consistent.
6. Clean images in any other way to make the D_2O–H_2O interface clearer. Just be consistent so that the same procedures are conducted on all images.

3.5 Calculating Flow Rates

1. Select at least two regions of interest (ROIs) within the tissues of interest that are large enough to give some statistics for the pixel intensity value to provide a reasonable average intensity at each time step (*see* **Notes 10 and 11**).
2. Produce a time series of average intensity for each ROI.
3. Fit the Weibull curve to the intensity (I) data as a function of time (t)

$$I\,(\%\text{ of }I_{\max}) = 100 - \frac{100}{1 + e^{a(t - t_{I50})}} \tag{1}$$

where a and t_{I50} are the fitted parameters (*see* **Note 12**).

1. Calculate the time of D_2O arrival by determining the time at the first bend of the Weibull curve. In plant physiological studies, this is often done by determining the time at which $I =$ 12% (or 88%) of $I_{\max}$ (*see* **Note 13**).
2. Set the scale in the images correctly and measure the distance between the ROIs along which the flow rate is to be calculated.
3. Calculate the flow rate by dividing the distance between the ROIs by the time of D_2O arrival (Fig. 5) (*see* **Note 14**).

4 Notes

1. Neutron imaging can be used to image water movement in any plant tissue. Leaves and branches can be placed directly in water containers for imaging. When imaging water transport through soil, it is important to use a soil and plant container material that is transparent to neutrons and adjust the sample

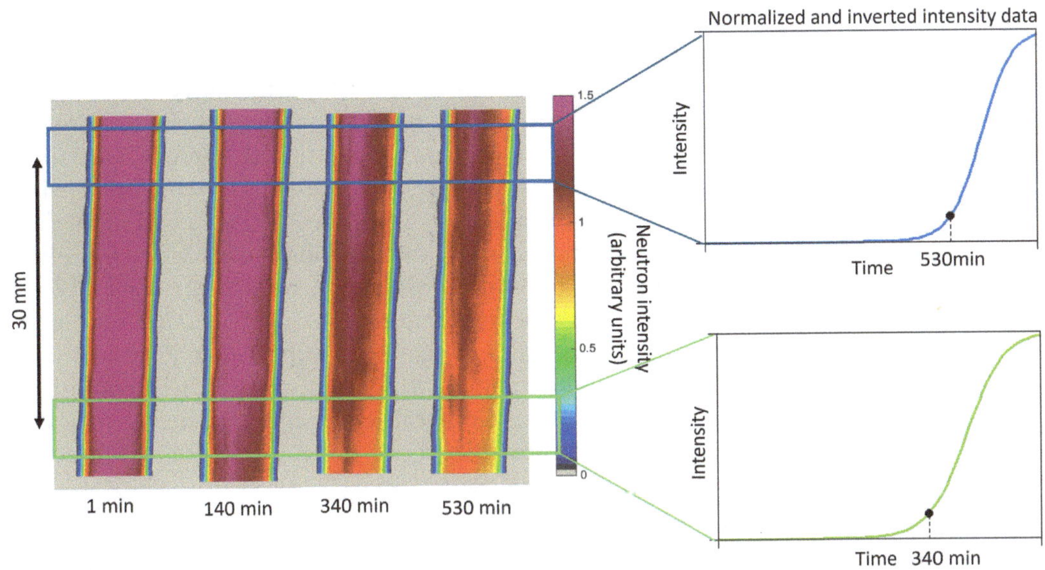

Fig. 5 A schematic illustration of flow rate calculation inside a juniper branch xylem. The replacement of H_2O by D_2O is shown as decreasing neutron intensity (left). A blue and green ROI is selected 30 mm apart. The average intensity in each ROI is plotted against time, and a Weibull curve (shown in the figure) is fitted to the data. The time for D_2O arrival is determined from the first curve of the fit, and the flow velocity is calculated as distance over time (here, 30 mm/290 min). Note that the intensity change is reverted in the graphs on the right compared to the image on the left

thickness to the imaging beam energy and other qualities. Too large a sample thickness makes the soil opaque to neutrons and prevents imaging roots. We have used silica sand as a growing medium by itself or mixed with 10% of silty loam soil without affecting the visibility of roots [16]. Plant container materials that are transparent to neutrons include pure aluminum and boron-free glass. Note that plastic (even transparent plastic) is not transparent to neutrons. We have used rectangular aluminum containers for radiography and cylindrical ones for tomography. When planning the container design, it is important to consider the material thickness in relation to the beam attenuation to ensure visibility through the material. Discuss this with the beamline technologist/scientist before building your containers. When designing the containers, plan also how the contrast agent (D_2O) will be delivered to the plant. When delivering the contrast agent from the bottom of the container, a mesh bottom has worked well.

2. It is important that the part of the sample that is in the view of the window for imaging remains as stationary as possible during imaging. Movement of the plant will make data

analysis challenging. This is because contrast changes that indicate water movement in an image are created by changing the amount of hydrogen atoms in the ROI. This means that movement of a drier plant part to the ROI or removal of plant from the ROI can be misinterpreted as water movement (D_2O replacing H_2O) because a drier plant part and air are more transparent to neutrons than H_2O. The design of the plant holder will depend on which tissues are imaged. For leaves, aluminum wire grids have been used [17]; for branches, it might be enough to stabilize the sample with laboratory stands and clamps or other structures both above and below the imaging window of view. When imaging root water uptake through soil, ensuring stationarity of the soil container might suffice.

3. The water containers can be plastic as long as their walls do not interfere with the imaged area in the window of view. The size of these and the length of the tubing will depend on the measurement configuration. If opening and closing and securing the door to the measurement chamber is relatively fast, one can simply change an H_2O container to a D_2O container below the plant. If opening and closing/securing the door is slow, we have used a small-diameter PVC tube attached to the plant container that allows sucking the H_2O out and adding D_2O (or vice versa) from outside the measurement chamber without stopping the measurement and opening the door.
4. We have processed neutron image stacks with normal desktop and laptop computers, but this often requires processing the data piecewise, which slows the work down.
5. Any tool, instrument, or sample that is in the measurement chamber when the neutron beam is on, potentially gets activated. Therefore, they are not allowed to be removed from the chamber before "cooling off." Do not leave any materials that you would like to use outside the chamber in the measurement chamber. Have duplicates.
6. Measurement time at the imaging facilities is allocated on an hourly basis, and it is usually continuous hours, independent of the time of day. Make sure that you have a large enough team to help so that everyone can take breaks or sleep. The duration of each experiment can be estimated based on expected water movement velocities. Water transport in small leaves can be imaged in 3–4 h. Larger branches will require 6–10 h depending on the distance of the water surface from the ROI and the size of the window of view. Generally, a shorter distance means faster measurements. When imaging small leaves, and if changing the water container rather than using tubing to change H_2O to D_2O, it is good to have some

distance for the D_2O to travel before it reaches the window or view to create a time buffer for closing the measurement chamber and restarting the measurements.

7. The exposure time will depend on the measurement system. Lower exposure gives grainier images but better temporal resolution. Discuss and test your needs with a test sample with the imaging personnel at the facility.
8. If you imaged a plant with D_2O and want to replicate imaging on the same plant, it is possible to reverse the tracer back to H_2O as long as the D_2O toxicity has not affected the functionality of the plant. In our experience, 8 h of imaging in pine samples could still be reversed [16], but aspen leaves saturated for 4 h with D_2O showed toxicity effects (necrosis around veins).
9. In many facilities, the neutron beam production fluctuates either with a known schedule and/or can have abrupt cuts. This results in black images (i.e., missing data) when the beam is down. It is possible to obtain beam-intensity information from the facility to scale the brightness of your images.
10. What is a reasonable size can be tested by how much the intensity varies by small changes in ROI size. Generally, you want to have at least >20 pixels in the area to reduce the effects of random noise.
11. Selecting the ROIs in the right tissue is the most important step for obtaining representative flow rates. When calculating phloem flow rate, draw the ROI within the phloem tissue. Saving the samples and checking them on a light microscope to identify and measure the phloem size will help selecting the ROI correctly. If you cannot select a large enough ROI that includes only phloem, you need to select representative ROIs of the other tissues in the image, estimate their relative contributions to the beam intensity, and subtract those to obtain values for the phloem. See [16] for an example of subtracting soil effects from water flow in the roots).
12. The Weibull curve is a customary curve used in plant physiological measurements (e.g., [18]) and usually fits the intensity data well, provided the initial state before D_2O exposure at the ROI is imaged for long enough. If an exponential curve fits your data better, you can also use that.
13. You can also choose some other location on the curve to represent the arrival of D_2O. Just be consistent. The first curve tells the first arrival time of D_2O at the location. The second curve tells when the location in question starts to be saturated with D_2O. Both have plant physiological meaning and can be used for calculating either the arrival time or the saturation time. And the steepness of the Weibull curve can be used to obtain information about the tissue saturation rate.

14. The flow rates from this calculation come in units of length per time. If you want to convert these to mass or volume per time, you will need to measure the tissue dimensions as well as make estimates for relative water content compared to dry matter. This can be done from the samples after imaging or by using known controls during imaging [2].

References

1. Matsushima U, Kardjilov N, Hilger A, Graf W, Herpich WB (2008) Application potential of cold neutron radiography in plant science research. J Appl Bot Food Qual 82:90–98
2. https://www.psi.ch/en/lns-applied-materials/comparison-to-x-ray
3. Sedighi GM, Vontobel P, Lehmann E, Carmeliet J, Derome D (2014) Liquid uptake in Scots pine sapwood and heartwood visualized and quantified by neutron radiography. Mater Struct47:1083–1096
4. Zhou X, Desmarais G, Carl S, Mannes D, Derome D, Carmeliet J (2022) Investigation of coupled vapor and heat transport in hygroscopic material during adsorption and desorption. Build Environ 214:108845
5. Manke I, Hartnig C, Kardjilov N, Messerschmidt M, Hilger A, Strobl M, Lehnert W, Banhart J (2008) Characterization of water exchange and two-phase flow in porous gas diffusion materials by hydrogen-deuterium contrast neutron radiography. Appl Phys Lett 92:244101
6. Czajka DM, Finkel AJ, Fischer CS, Katz JJ (1961) Physiological effects of deuterium on dogs. Am J Physiol Legacy Content 201:357–362
7. Malone MW, Yoder J, Hunter JF, Espy MA, Dickman LT, Nelson RO, Vogel SC, Sandin H, Sevanto S (2016) In vivo observation of tree drought response with low-field NMR and neutron imaging. Front Plant Sci 7:564. https://doi.org/10.3389/fpls.2016.00564
8. Zarebanadkouki M, Kim YX, Carminati A (2013) Where do roots take up water? Neutron radiography of water flow into the roots of transpiring plants growing in soil. New Phytol 199:1034–1044
9. Nelson RO, Vogel SC, Hunter J, Watkins EB, Losko AS, Tremsin AS, Borges NP, Cutler TE, Dickman LT, Espy M, Gautier C, Madden AC, Majewski J, Malone MW, Mayo DR, McClellan KJ, Montgomery D, Mosby S, Nelson AT, Ramos K, Schirato RC, Schroeder K, Sevanto S, Swift AL, Vo L, Williamson T, Winch N (2018) Neutron imaging at LANSCE –from cold to ultrafast. J Imaging 4:45
10. Zarebanadkouki M, Kroener E, Kaestner A, Carminati A (2014) Visualization of root water uptake: quantification of deuterated water transport in roots using neutron radiography and numerical modeling. Plant Physiol 166:487–499
11. Kroener E, Zarebanadkouki M, Kaestner A, Carminati A (2014) Nonequilibrium water dynamics in the rhizosphere: how mucilage affects water flow in soils. Water Resour Res 50:6479–6495. https://doi.org/10.1002/2013WR014756
12. Warren JM, Bilheux H, Kang M, Voisin S, Cheng C-L, Horita J, Perfect E (2013) Neutron imaging reveals internal plant water dynamics. Plant Soil 366:683–693
13. Derluyn H, Vontobel P, Mannes D, Derome D, Lehmann E, Carmeliet J (2019) Saline water evaporation and crystallization-induced deformations in building stone: insights from high-resolution neutron radiography. Transp Porous Media 128:895–913
14. Williams DL, Brown CM, Tong D, Sulyman A, Gary CK (2020) A fast neutron radiography system using a high yield portable DT neutron source. J Imaging 6:128. https://doi.org/10.3390/jimaging6120128
15. Leigh A, Sevanto S, Ball MC, Close JD, Ellsworth DS, Knight CA, Nicotra AB, Vogel S (2012) Do thick leaves avoid thermal damage in critically low wind speeds?. New Phytol 194:477–487
16. Sevanto S, Gehring CA, Ryan MG, Patterson A, Losko A, Vogel S, Carter K, Dickman LT, Espy MA, Kuske CR (2023) Ectomycorrhizal root colonization differentially affects water uptake and drought tolerance in two pinyon pine genotypes. Sci Rep 13:14424. https://doi.org/10.1038/s41598-023-41191-5
17. Defraeye T, Derome D, Aregawi W, Cantre A, Hartmann S, Lehmann E, Carmeliet J, Voisard F, Verboven P, Nicolai B (2014) Quantitative neutron imaging of water distribution, venation network and sap flow in leaves. Planta 240:423–436. https://doi.org/10.1007/s00425-014-2093-3
18. Skelton RP, West AG, Dawson TE (2015) Predicting plant vulnerability to drought in biodiverse regions using functional traits. PNAS 112:5744–5749. https://doi.org/10.1073/pnas.150337611

Check for updates

Chapter 28

A Dual-Reporter System for the Analysis of Phloem Structural and Signaling Responses In Vivo

Lisa S. Wrobel and Gundula A. Noll

Abstract

Plants have evolved an effective defense mechanism in the phloem to prevent the spread of pathogens and minimize the loss of phloem sap following injury. Specific structural phloem proteins known as P-proteins rapidly seal affected sieve elements by plugging the sieve plates, a phenomenon defined as sieve element occlusion. This chapter describes a live cell imaging method for the analysis of P-protein responses and signal propagation in vivo without tissue sectioning or mechanical manipulation. It is based on an *Arabidopsis thaliana* dual-reporter line where P-proteins are labeled with fluorescent tags in a complementation background, allowing real-time visualization of the parietal protein network during sieve element occlusion. The calcium sensor Yellow Cameleon 3.6 is specifically expressed and anchored in the sieve elements, enabling the detection of calcium waves and their effects on P-protein structure over longer distances in vivo. Using this protocol, a wide range of external triggers—including chemical treatments, buffers, and wounding—can be applied with precision, allowing the analysis of P-protein functions and long-distance signaling in the phloem. The method is readily adaptable to other genetically encoded sensors and can be used to investigate P-protein-independent processes as well as the diverse signaling and structural responses of additional sieve element components.

Keywords Sieve element occlusion, P-protein, Genetically encoded sensors, Calcium signaling, Phloem wounding

Abbreviations

CC	Companion cell
CFDA	Carboxyfluorescein diacetate
CFP	Cyan fluorescent protein
CLSM	Confocal laser scanning microscopy
CFP	Cyan fluorescent protein
DMSO	Dimethylsulfoxide
FRET	Förster resonance energy transfer
mRFP	Monomeric red fluorescent protein
SE	Sieve element
SEOR	Sieve element occlusion-related

Johannes Liesche and Sanna Sevanto (eds.), *Phloem*, Methods in Molecular Biology, vol. 3066,
https://doi.org/10.1007/978-1-0716-5488-0_28,

1 Introduction

Phloem injury is a risk to plant health and survival because the loss of phloem sap disrupts the transport of photoassimilates, signaling molecules, and other essential compounds, while enabling the ingression and systemic spread of pathogens [1]. To counteract this, plants have evolved multiple protective mechanisms against both mechanical damage and pathogens. In most angiosperms, structural phloem proteins (P-proteins) are arranged in a parietal network along the plasma membrane of the sieve elements (SEs). P-proteins are encoded by members of the *SIEVE ELEMENT OCCLUSION-RELATED* (*SEOR*) gene family and are almost exclusively expressed in immature SEs. They assemble into large protein bodies that, in mature SEs, disperse into the P-protein network [2].

In response to wounding, P-proteins detach from their parietal position and form protein plugs on the downstream sieve plates, potentially triggered by an influx of external Ca^{2+} ions. The occlusion helps to preserve turgor within the phloem system, inhibits the loss of photoassimilates, and blocks the spread of pathogens [3, 4]. The sensitivity of P-proteins means that analysis by microscopy is challenging. Any kind of handling and sectioning can mimic the effects of wounding and inadvertently trigger the response, resulting in artifacts.

The laboratory model *Arabidopsis thaliana* (Arabidopsis) can be used to overcome these issues because the roots are transparent, providing an opportunity to visualize the sensitive phloem system in vivo and in real time. Another advantage of Arabidopsis is that only two functional *SEOR* genes are present (*AtSEOR1* and *AtSEOR2*), which simplifies the genetic manipulation of P-protein functions [5, 6]. The P-protein network can be visualized by fusing the AtSEOR1 protein to a fluorescent tag such as monomeric red fluorescent protein (mRFP), which can be monitored by confocal laser scanning microscopy (CLSM) [1]. To avoid AtSEOR1 overexpression, leading to artifacts in the form of fluorescent mobile protein agglomerates [6], the AtSEOR1-mRFP construct is introduced into an *atseor1* knockout background [7], generating a complementation line in which the tagged AtSEOR1 integrates seamlessly into the parietal P-protein network.

The phloem also functions as a conduit for long-distance signaling. In response to mechanical or biological stimuli, it can transmit electropotential waves (EPWs) featuring characteristic patterns of membrane depolarization and repolarization [8–11]. EPWs can lead to the activation of Ca^{2+} channels and initiate calcium fluxes, which are considered key components of systemic signaling and may act in concert with local occlusion responses. Ca^{2+} and reactive oxygen species (ROS) contribute to signal propagation depending on signal duration, intensity, and spatial dynamics. These so-called signaling signatures are decoded by target cells and activate various downstream responses [12, 13]. Genetically encoded sensors have been developed to monitor these signals in vivo. One example is the calcium sensor Yellow Cameleon 3.6 (YC3.6). When Ca^{2+} binds to its calmodulin domain, YC3.6 undergoes a conformational change that brings two fluorophores—cyan fluorescent protein (CFP) and cpVenus—into closer proximity, enabling Förster resonance energy transfer (FRET). Excitation of CFP (440 nm) then results in energy transfer to Venus, producing a detectable shift in emission from blue to yellow [14]. For the simultaneous investigation of calcium dynamics and the conventional P-protein response, a dual-reporter line has been generated combining mRFP-labeled AtSEOR1 and the YC3.6 sensor expressed under the control of the SE-specific *AtSEOR2* promoter. A plasma membrane localization tag was added to prevent sensor translocation with the phloem mass flow and ensure colocalization with the P-protein network. Phloem mass flow can be visualized using fluorescent dyes such as esculin and carboxyfluorescein diacetate (CFDA). Both are phloem-mobile dyes that are specific for living cells and restricted to the phloem tissue. They can be used to verify the active transport of the phloem in root tissues [15–17].

The method described in this chapter combines the principles discussed above and enables the simultaneous investigation of EPWs, molecular signaling (e.g., calcium and ROS), and the P-protein response in intact transporting SEs. It is designed to image phloem tissue over longer distances in vivo without mechanical manipulation or sectioning. This allows detailed studies of the dynamics of P-protein reactions and offers further insight into the underlying mechanisms of SE occlusion and long-distance signaling. Furthermore, it provides broad potential for microscopic analysis, including in vivo observation during chemical treatments, tracking of phloem mass flow, and visualization of fluorescence-labeled structures within an intact, functioning phloem network (Fig. 1).

2 Materials

2.1 Plant Material

1. The Arabidopsis mutant *atseor1* in the Col-0 wild-type background (*see* **Note 1**).
2. The transgenic Arabidopsis line P*AtSEOR1*:mRFP::AtSEOR1 (*atseor1*) (*see* **Note 2**).
3. The transgenic Arabidopsis line P*AtSEOR2*:$YC3.6_{PM}$ (*see* **Note 3**).
4. The transgenic Arabidopsis line mRFP::AtSEOR1 - $YC3.6_{SE}$ (*see* **Note 4**).

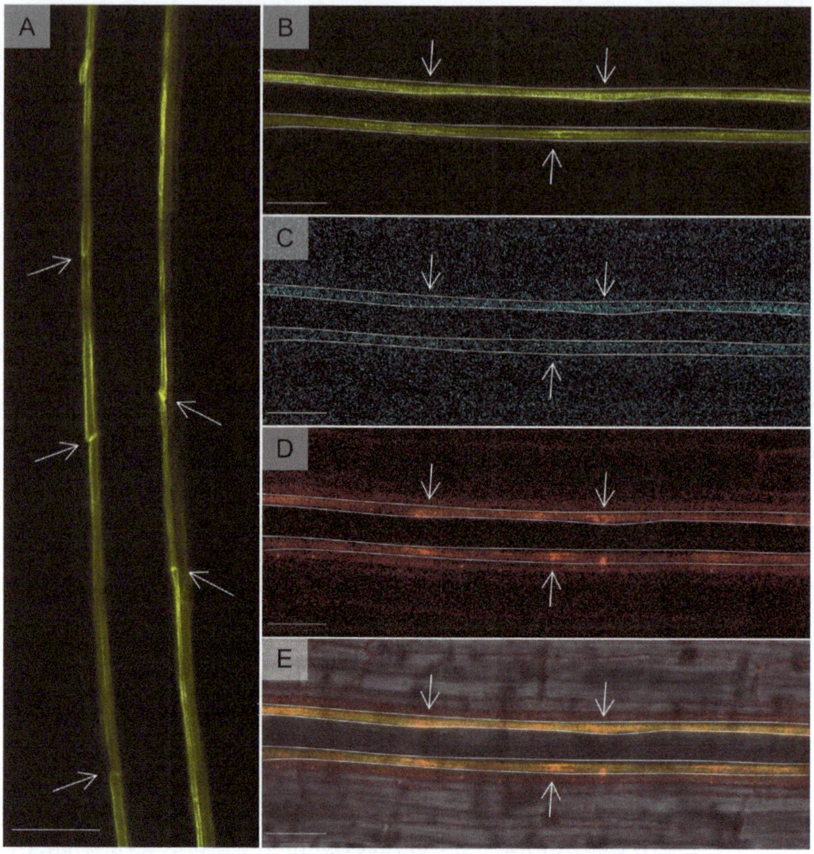

Fig. 1 P-protein structures and the Ca^{2+} detection system Yellow Cameleon (YC3.6) in transgenic roots of the dual-reporter line. (**A**) Sieve elements (SEs) of the dual-reporter line expressing YC3.6 under the control of the *AtSEOR2* promoter. The Venus signal is detected solely at the plasma membrane of the SEs. (**B**) Venus signal (yellow). (**C**) CFP signal (blue) of the YC3 sensor. (**D**) Signal of mRFP-labeled P-proteins. The pictures were used for the recording of Ca^{2+} waves and P-protein reactions and are low resolution to increase the recording speed. (**E**) Overlay of the mRFP and YC3.6 Venus signals combined with transmitted light. Sieve plates are indicated by arrows. Scale bar = 25 μm

2.2 Consumables and Equipment

1. Kits and reagents for genomic DNA and mRNA extraction, cDNA synthesis, and PCR (*see* **Note 5**).
2. Binary vector pLab12 and molecular cloning reagents (*see* **Note 5**).
3. *Agrobacterium tumefaciens* strain EHA105 and equipment for its transformation and cultivation (*see* **Note 5**).
4. Climate-controlled greenhouse/growth room with standard plant cultivation equipment, including sowing trays/pots and soil.
5. Fine forceps.
6. Magnifying glasses or a standard binocular microscope.

7. Air-tight plastic containers with close-fitting sealable lids.
8. Fume hood.
9. Standard 9-cm Petri dishes for sterile seed propagation.
10. Glass-bottomed Petri dishes with lids for microscopy (*see* **Note 6**).
11. Inverted confocal microscope (*see* **Note 7**).
12. Razor blades.
13. Fine sandpaper (*see* **Note 8**).
14. Scalpel.
15. Cauterization wire.

2.3 Chemicals

1. 13% sodium hypochlorite solution.
2. Half-strength Murashige and Skoog (½MS) medium [18] (*see* **Note 9**).
3. ½MS agar plates (*see* **Note 9**).
4. Sterile water.
5. Lanthanum solution: 10 mM $LaCl_3$, 10 mM KCl, and 5 mM NaCl [19].
6. Saline: 10 mM $CaCl_2$, 10 mM KCl, and 5 mM NaCl [19].
7. Hyperpolarization buffer: 0.1 mM KCl, 10 mM $CaCl_2$, 10 mM MES–Tris pH 6.15 (*see* **Note 10**) [20].
8. Depolarization buffer: 0.1 mM KCl, 10 mM MES–Tris pH 6.15 (*see* **Note 10**) [20].
9. Apoplast buffer: 1 mM $CaCl_2$, 2 mM KCl, 50 mM mannitol, 25 mM MES, 1 mM $MgCl_2$ pH 5.7 (*see***Note 11**) [21].
10. Sodium phosphate buffer (25 mM, pH 4) (*see* **Note 12**).
11. Esculin solution: 5 g/L in sodium phosphate buffer (*see* **Note 13**).
12. CFDA solution (*see* **Note 14**).

3 Methods

3.1 Generation of Transgenic Arabidopsis Lines for P-Protein Localization and Calcium Signaling

Labeling of P-Proteins in Arabidopsis Using mRFP-Tagged AtSEOR1

1. Isolate genomic DNA from Arabidopsis Col-0 wild-type plants (*see* **Note 5**) and amplify 1 kb of the *AtSEOR1* promoter (*see* **Note 15**) to drive SE-specific transgene expression [22].

2. Prepare cDNA from mRNA isolated from Arabidopsis Col-0 wild-type leaves, amplify the *AtSEOR1* coding sequence, and add the mRFP tag to the N-terminus (*see* **Notes 5 and 16**).
3. Assemble the final construct (P*AtSEOR1*:mRFP::AtSEOR1) in the binary vector pLab12 [23] with the CaMV 35S terminator and introduce the vector into Agrobacterium strain EHA105 [24] (*see* **Notes 5 and 16**).
4. Use the floral dip method [25] to transform theArabidopsis knockout line atseorl [7].

Specific Expression of YC3.6$_{SE}$ in Sieve Elements

1. Isolate genomic DNA from Arabidopsis Col-0 wild-type plants (*see* **Note 5**) and amplify 1 kb of the *AtSEOR2* promoter (*see* **Note 17**) to drive SE-specific transgene expression [22].
2. Prevent translocation of the YC3.6 sensor along with phloem mass flow by fusing it to an Ltl6b tag plasma membrane localization signal (*see* **Note 18**).
3. Assemble the final construct (PAtSEOR2:YC3.6::LtI6b) in the binary vector pLab12 [23] and introduce the vector into Agrobacterium strain EHA105 [24] (*see* **Note 5**).
4. Use the floral dip method [25] to transform Arabidopsis Col-0 plants.

Generation of the Dual-Reporter Line Combining mRFP::AtSEOR1 and YC3.6$_{SE}$

1. Grow homozygous mRFP-AtSEOR1 and YC3.6$_{SE}$ plants in soil until the flowering stage under long-day growth conditions (22°C, 16-h photoperiod).
2. Select fully closed flower buds from YC3.6$_{SE}$ plants.
3. Remove all floral organs except the pistils using fine forceps and carefully expose the stigma. Magnifying glasses or a binocular microscope will be required to visualize the manipulation.
4. Transfer pollen from fully open mRFP::AtSEOR1 flowers onto the exposed stigma of YC3.6$_{SE}$ plants by direct contact.
5. Repeat the procedure once daily for ten consecutive days to increase the likelihood of successful fertilization.
6. After the final cross, remove all remaining unpollinated flowers as well as floral buds to promote seed set from successfully fertilized flowers.

3.2 Plant Cultivation

Sterilizing Arabidopsis Seeds in a Chlorine Atmosphere

1. Place seeds in a plastic container with a sealable lid.
2. Pour 50 ml of a 13% solution of sodium hypochlorite into a beaker and place it next to the seeds inside the container.
3. Carefully add 1.5 ml of concentrated HCl to the sodium hypochlorite to generate chlorine gas and immediately close the container and seal the lid.

4. Keep the seeds in the chlorine gas atmosphere for 3–4 h at room temperature.
5. After treatment, open the container under a fume hood and spread the sterile seeds on ½MS agar plates.

Sterile Cultivation Conditions

1. Germinate the seeds on ½MS agar plates under long-day conditions (22°C, 16-h photoperiod).
2. Grow seedlings for 5–7 days on plates positioned vertically to allow roots to grow along the agar in one accessible layer.
3. Transfer individual seedlings to separate plates to prevent root entanglement.
4. After 10–15 days, select seedlings suitable for imaging (*see* **Note 19**).

3.3 Imaging Setup

Preparation of the Seedling

The observation of phloem tissue in fragile Arabidopsis roots requires a carefully designed experimental setup to maintain structural integrity and minimize artifacts caused by handling or preparation (Fig. 2).

1. Place a small agar block (cut from the ½MS plate) in a glass-bottomed Petri dish, close to the edge.
2. Add a thin film of sterile water to the dish to maintain humidity throughout the experiment.
3. Carefully transfer the seedling from the ½MS plate and drape the distal roots over the small agar block, extending the main part of the seedling across the bottom of the Petri dish (Fig. 2A, B).
4. Fix the roots in position against the bottom of the Petri dish by placing a second agar block above the first one, on top of the roots. The region of interest for imaging lies beneath the second agar block, which keeps the root at a fixed focal level. The distal portion of the root should remain draped over the first agar block so it is exposed and accessible for manipulation or treatment (Fig. 2A, B)
5. The distance between the stimulus application site and the imaging region can be adjusted 0.5–3.5 cm by varying the position and size of the agar blocks. This flexibility allows the investigation of long-distance signaling through the roots.
6. Place a drop of water on the shoot to prevent dehydration. A Petri dish lid may also be used, depending on the experimental requirements.

Microscopy

1. Place the Petri dish containing the prepared seedling on the stage of an inverted confocal microscope and begin imaging with a 40× objective (*see* **Note 7**).

Fig. 2 Setup of the inverse confocal microscope for the observation of sieve elements in Arabidopsis roots and the application of stimuli. (**A**) A 14-day-old seedling on a glass-bottomed Petri dish. The root is positioned between two agar blocks, preventing movement during the application of stimuli. (**B**) Schematic overview of the experimental setup for in vivo observation. P-protein plugs on the sieve plates are shown in red. (**C**) Seedlings in a glass-bottomed Petri dish under the inverse confocal microscope. A glowing wire is used to wound the roots

2. Focus on a single phloem strand within a root region that is stably fixed beneath the second agar block.
3. Identify a region showing a clearly parietal P-protein network before adjusting the settings (Section: Settings).

Settings

1. Adjust the confocal microscope to the settings listed in Table 1, which are optimized for phloem imaging. Use the appropriate excitation and emission wavelengths for the fluorophore (Table 2). Capturing z-stack images allows stable tracking of the SE plane over time (*see* **Note 20**).
2. Begin scanning and recording 1–2 min before the application of the stimulus. Record continuously for 10–15 min to capture both baseline activity and stimulus-induced responses.

3.4 Verifying Phloem Integrity and Transport Using Mobile Tracers

1. Prepare a saturated solution of esculin or a working solution of CFDA (*see* **Notes 13 and 14**).
2. Gently abrade the upper surface of the leaf using fine sandpaper or make small incisions in the leaf tip using a scalpel.

Table 1
CLSM settings optimized for phloem imaging

Setting	Value
Pinhole	95 μm
Scan speed	400 Hz
Frame average	1
Line average	1
Resolution	Typically 1024×250 pixels, but adapt to root size and position
Acquisition mode	Use xyzt mode to generate both z-stack and time-lapse sequences

Table 2
Excitation and emission wavelengths for different tracers and fluorophores

Signal	Excitation	Emission	Reference
Mobile tracers			
CFDA	488 nm	505–530 nm	[26]
Esculin	405 nm	445–465 nm	[27]
Fluorophores			
CFP	440 nm	445–510 nm	
mRFP	543 nm	560–706 nm	
Venus	514 nm	520–552 nm	
Yellow Cameleon 3.6	440 nm (CFP)	445–510 nm (CFP) and 520–552 nm (Venus)	[20]

3. Apply 10 μl of the dye solution directly to the treated area of the leaf surface.
4. Monitor the roots for fluorescence, which typically appears in the phloem strands 60–90 min later, depending on the seedling size (Fig. 3).

3.5 Application of External Stimuli

Stimulus-induced responses in the phloem can be investigated using a variety of external triggers that are applied to the exposed root tip in the Petri dish, simulating natural environmental stressors or mechanical injury (Table 3). For example, natural changes in pH, salinity, or temperature can be mimicked by applying specific chemicals or buffers, whereas localized burning simulates injury and has consistently been shown to induce a strong and reliable P-protein response in SEs [4].

1. For chemical stimuli, prepare the hyperpolarization and depolarization buffers (*see* **Note 10**) or another appropriate solution

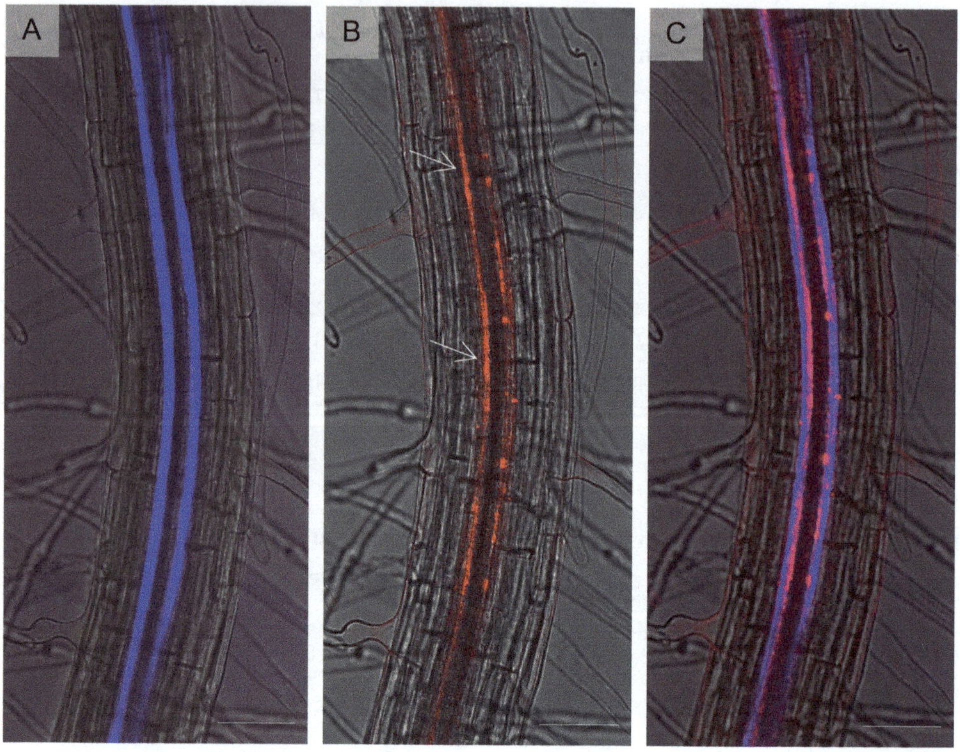

Fig. 3 Visualization of phloem mass flow in Arabidopsis roots using esculin. The dye is used to trace mass flow in the phloem of P*AtSEOR1*:mRFP::AtSEOR1 (*atseor1*) plants expressing mRFP-labeled P-proteins anchored in a parietal network. (**A**) Esculin is detected in the two phloem strands. (**B**) The mRFP-labeled P-proteins have assembled in the sieve elements (SEs). (**C**) Overlay showing the P-protein structure in the loaded SEs and the precise superimposition of the two signals

Table 3
Chemical and physical stimuli applied to roots for the analysis of phloem responses

Stimulus	Specific application	References
Ethanol	20%, 50%, or 70% solution	
DMSO	Undiluted	
Glutamic acid		[29]
Hyperpolarization and depolarization	Hyperpolarization/depolarization buffers (*see* **Note 7**)	[20]
Heat or cold	Saline at 40°C or 5°C	
Pathogen-associated molecular patterns	Examples include bacterial flagellin or fungal chitosan	[30, 31]
Burning	Brief touch with a glowing cautery wire	[32]

(Table 3); for burning, prepare the cautery wire (Fig. 2C) (*see* **Note 21**).

2. Position the seedling in the Petri dish as described in Section: Preparation of the Seedling and confirm active phloem transport as described in Sect. 3.4.
3. Start the time-lapse recording sequence.
4. For chemical stimulation, apply ~15 µl of the chemical solution (Table 3) directly to the exposed root region on top of the first agar block. Hyperpolarization and depolarization buffers can be applied sequentially to manipulate membrane potentials [20]. Hyperpolarization has been shown to activate specific Ca^{2+} channels, resulting in elevated cytosolic Ca^{2+} concentrations in guard cells and roots [20, 28]. For stimulation by burning, gently touch the tip of the cautery wire against the root region on top of the first agar block to induce a small injury (*see* **Note 21**).
5. Continue imaging the upstream SEs, which are located beneath the second agar block, to capture potential long-distance responses within the phloem.

3.6 Recording and Analysis

Measuring the Intracellular Calcium Signal

1. Use an overlay of the recorded image series to verify the continuous tracking of SEs.
2. Isolate the mRFP channel to visualize dynamic changes in the P-protein structures over the course of the experiment.
3. Define the regions of interest by manually framing individual SEs.
4. Quantify the fluorescence intensity of the two YC3.6 fluorophores (CFP and Venus) within each region of interest over time.
5. Calcium waves manifest as transient peaks in the FRET ratio, accompanied by inverse changes in CFP and Venus emission intensities (Fig. 4).

Inhibiting the Intracellular Calcium Signal

To investigate the regulation of P-protein responses, calcium signaling can be inhibited using trivalent lanthanum ions (La^{3+}) [19].

1. Position the seedling in the Petri dish as described above (Section: Labeling of P-Proteins in Arabidopsis Using mRFP-Tagged AtSEOR1).
2. Fill the dish with a 10 mM La^{3+} solution (or an appropriate saline control).
3. Incubate the roots in the solution for at least 15 min before applying the stimulus.

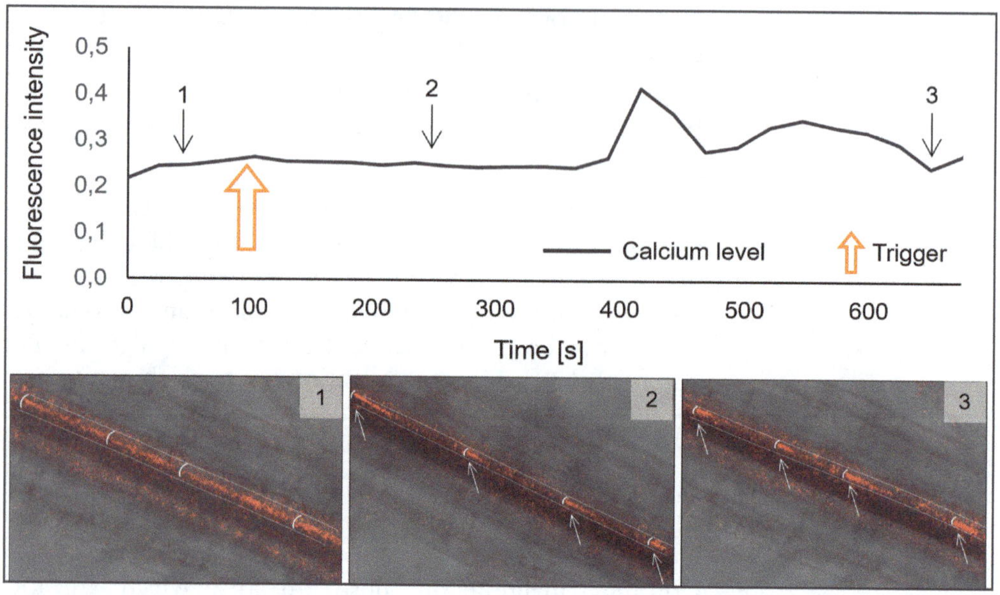

Fig. 4 Calcium spikes and the formation of protein plugs in response to wounding observed in multiple experiments ($n = 5$). The calcium wave patterns vary, but their temporal correlation with P-protein plug formation is consistent. The experimental timeline begins with the application of a stimulus (yellow arrow), followed by a calcium peak and then the formation of P-protein plugs. The CLSM images show the initial state of the network in image 1 and the newly formed protein plugs on the sieve plates in images 2 and 3

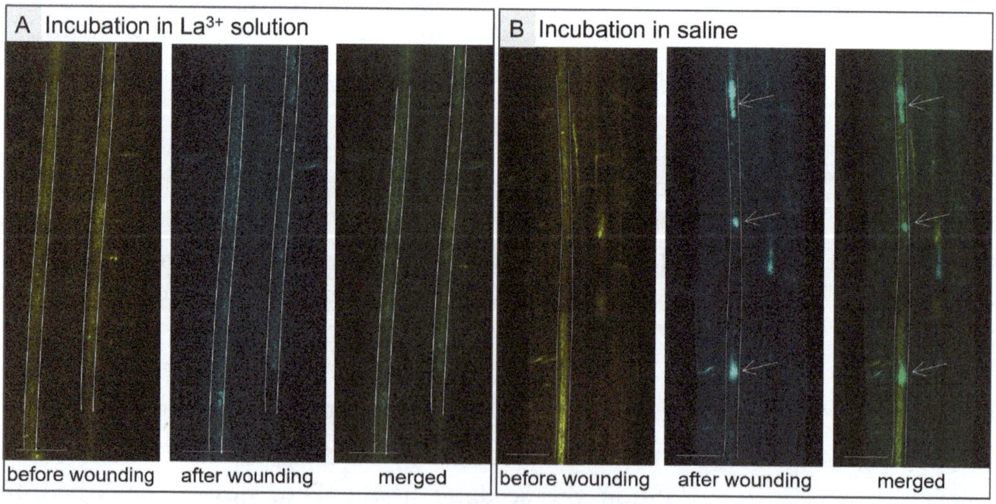

Fig. 5 Correlation between P-protein plug formation and calcium signaling. (**A**) No P-protein detachment or plug formation was observed in roots exposed to La^{3+}. The conventional P-protein network remained unchanged. (**B**) Control roots treated with saline responded to injury with clear plug formation at the sieve plates indicated by arrows

4. Apply stimulus to the exposed root tip (Sect. 3.5) while continuously recording.
5. A lack of response under these conditions suggests that the response is dependent on calcium signaling (Fig. 5).

4 Notes

1. The *atseor1* mutant (ID: SM_3.34780) is a knockout line produced by transposon mutagenesis in the Col-0 wild-type background. It has a dSpm insertion at locus At3g01680, which corresponds to the *AtSEO1* gene [7]. The mutant line is also resistant to the herbicide phosphinothricin (PPT) due to the presence of the *bar* gene in the dSpm insertion element.
2. The transgenic line P*AtSEOR1*:mRFP::AtSEOR1(*atseor1*) was created by transforming the *atseor1* mutant (*see* **Note 1**) with a construct comprising the AtSEOR1 sequence with an N-terminal mRFP fusion under the control of the *AtSEOR1* promoter. It therefore retains the PPT resistance of the original mutant. Transformation was carried out using the binary expression vector pLab12 (*see* **Note 5**). Arabidopsis plants were transformed using the floral dip method [25].
3. The transgenic line P*AtSEOR2*:YC3.6_{PM} was created by transforming Col-0 wild-type plants with a construct comprising the Yellow Cameleon system (YC3.6) fused to a PM tag under the control of *AtSEOR2* promoter. The construct also carried a *bar* gene conferring PPT resistance.
4. The transgenic Arabidopsis line mRFP AtSEOR1 - YC3.6_{SE} was created by crossing parental lines expressing mRFP-AtSEO1 (*see* **Note 2**) and YC3.6_{PM} (*see* **Note 3**) and recovering F1 hybrid offspring carrying both transgenes: [P*AtSEOR1*:mRFP::AtSEOR1(*atseor1*)] × [P*AtSEOR2*:YC3.6_{PM}].
5. Detailed protocols for the extraction of genomic DNA and leaf mRNA, cDNA synthesis, PCR, molecular cloning, and the introduction of binary vectors into Agrobacterium are beyond the scope of this chapter. We used commercially available kits for these steps, and detailed protocols can be found in the literature accompanying those kits. The pLab12 plasmid is based on the binary expression vector pFGC5941, which confers kanamycin resistance for selection in bacteria and PPT resistance for selection in plants [23].
6. We use dishes with a 50 mm overall diameter and a 30 mm diameter of clear glass (Mat Tek Europe).

7. An inverted confocal microscope equipped for CLSM allows free access above the sample for manipulation. We use the Stellaris 8 FALCON instrument (Leica Microsystems) fitted with 20×, 40×, or 63× objective using water or oil immersion as appropriate.
8. Grade P150–P220 is the most suitable.
9. MS medium is 4.4 g/L MS salts + vitamins and 40 g/L sucrose (pH 5.8). Add 7 g/L Gelrite or 8 g/L agar to make the solid medium. The components of the MS salts and vitamins are provided in ref. [18] but can be sourced as ready-mixed powders from many suppliers (e.g., Duchefa Biochemie, Sigma-Aldrich/Merck, Phytotechnology Laboratories). Half-strength medium is made by halving the concentration of all components, with the exception of the Gelrite or agar in the solid medium, which is required at full strength to achieve the appropriate stiffness.
10. Prepare a stock of 20 mM MES–Tris buffer, which is double the required concentration. To make the depolarization buffer, mix with an equal volume of 0.2 mM KCl to reach final concentrations of 10 mM MES–Tris and 0.1 mM KCl. To make the hyperpolarization buffer, mix 20 mM MES–Tris buffer with an equal volume of 0.2 mM KCl + 20 mM $CaCl_2$ to reach final concentrations of 10 mM MES–Tris, 0.1 mM KCl, and 10 mM $CaCl_2$. Adjust to pH 6.15 using HCl and KOH [20].
11. This buffer simulates the conditions in the apoplast and is optimized for transport studies [21]. Adjust to pH 5.7 using HCl and KOH.
12. Sodium phosphate buffer is usually made by mixing the monobasic and dibasic forms in defined ratios to get the correct pH, but pH 4 is at the limit of the buffering range, so only the monobasic form is needed in this case, and HCl is used to adjust the pH (*see* **Note 13**).
13. Esculin has been shown to provide more consistent results at pH 4 [33].
14. The CFDA stock solution is prepared in DMSO at a concentration of 500 mM and should be stored at −20°C. Working solutions of 1–50 μM are prepared in apoplast buffer (*see* **Note 11**).
15. The 1-kb *AtSEOR1* promoter can be amplified using primers 5′-AGA <u>GGT ACC</u> AAA ACA TGC ATA GAA TAA ACC-3′/5′-AGA <u>CTC GAG</u> TCT TGG TTC AGT TTG CTT TTG-3′ (restriction sites underlined) and is sufficient to drive SE-specific gene expression [34].

16. The *AtSEOR1* promoter sequence can be amplified using primers 5′-AAA AAG CTT AAA ACA TGC ATA GAA TAA ACC -3′/5′-AAA CCA TGG TCT TGG TTC AGT TTG CTT-3′. The fluorophore mRFP can be amplified using primers 5′-AAA CCA TGG CCT CCT CCG AGG AC-3′ /5′-AAA CCC GGG TAC TCG ACT TCC GGC-3′. The *AtSEOR1* coding sequence can be amplified using primers 5′ -AAA CCC GGG ATG GAG TCG CTG ATC A-3′/5′-AAA TCT AGA TTA GAA GTT GTA GTT CTC GTC T-3′. The cauliflower mosaic virus 35S terminator can be amplified using primers 5′-AAA TCT AGA GTC CGC AAA AAT CAC C-3′/5′-AAA GGT ACC GGA ATT AGA AAT TTT ATT GAT AGT-3′. These amplicons can then be digested with the corresponding restriction enzymes and ligated into pLab12.1.
17. The 1-kb *AtSEOR2* promoter can be amplified using primers 5′-AGA GGT ACC TCA GCC GGA GAT CAT CC-3′/5′ -AGA CTC GAG ATT GGC GAG GTT GAG AG-3′ (restriction sites underlined) and is sufficient to drive SE-specific gene expression [22].
18. The complete sequence is derived from the PM-YC3.6-LtI6b/pTkan vector, comprising YC3.6 fused to the Ltl6b tag and the *rbcsT* terminator sequence [20].
19. It is important to select seedlings of the appropriate age and developmental state. Younger seedlings will not have developed enough functional phloem tissue with supporting P-protein structures, whereas older seedlings have a larger root system that is more difficult to transfer undamaged to Petri dishes and has started to lose transparency.
20. The number of scans for the z-stack is important for the live recording. Choose a distance between the scans of 2.5 and 4 nm. It is important to cover the width of the SE, but try to use as few scans as possible to increase the speed of the recording (see Section: Settings).
21. Mechanical wounding by burning consistently induces reliable P-protein responses, whereas chemical treatments have often failed to induce detectable changes in either YC3.6 fluorescence or the P-protein network. This raises the possibility that chemical stimuli do not reach the phloem tissue effectively. Unlike the epidermis, the vascular cylinder lies beneath multiple cell layers and is protected by the Casparian strip. Externally applied substances may therefore remain confined to superficial tissues, failing to reach the SEs. However, it is important to ensure that the roots are not extensively damaged or severed with the cautery wire. A brief touch is sufficient to induce a small injury that will trigger phloem signaling.

5 Outlook—A Toolbox for the Analysis of Sieve Elements In Vivo

Investigating phloem physiology has long been hampered by technical limitations. The method presented here addresses several major challenges associated with the analysis of SEs in vivo, including (1) limited microscopic access to SEs, which are deeply embedded within the tissue; (2) avoidance of artifacts caused by mechanical manipulation of plant tissues and therefore of the P-protein network; (3) the controlled application of defined external stimuli under live imaging conditions; and (4) the absence of tools allowing the simultaneous visualization of tagged cellular structures and real-time signaling events in SEs. The latter is addressed by our method, which combines fluorescence-labeled P-proteins with the calcium sensor YC3.6 in a dual-reporter line, enabling functional analysis of the transport-active phloem in vivo. This setup enables the live imaging of P-protein reactions at distances of 5 mm or more from the wounding site while simultaneously recording calcium waves within the same SEs. Furthermore, we found that the inhibition of calcium channels inhibits calcium signaling and the P-protein response, strongly supporting a Ca^{2+}-dependent, regulated mechanism of SE occlusion by P-proteins.

Although the system was initially developed to investigate P-protein-mediated wound responses, it has the potential for broader applications. The YC3.6 sensor enables real-time imaging of calcium dynamics within SEs and facilitates the detection of stimulus-dependent Ca^{2+} signatures. These spatiotemporal patterns are known to be decoded by target cells and translated into distinct physiological responses [12]. Beyond calcium signaling, the method can be readily adapted to include other genetically encoded indicators. For example, ROS signals mediate plant stress responses and can be monitored using redox-sensitive sensors such as roGFP-Orp1, which responds to hydrogen peroxide (H_2O_2) via ratiometric changes in fluorescence [35]. Expressing such sensors under the control of SE-specific promoters and fused to a membrane anchor for parietal localization would enable the detection of ROS waves propagating in the phloem and the analysis of crosstalk between ROS and Ca^{2+} signaling. Another promising extension involves the monitoring of cytosolic pH changes. In Arabidopsis epidermal cells, cytosolic acidification of up to 0.25 pH units has been linked to Ca^{2+} signaling and can be visualized using pH-sensitive fluorescent sensors [36]. However, we lack comparable data for SEs or companion cells, highlighting a compelling direction for future research.

In summary, the method described here provides a flexible and expandable platform for functional phloem imaging under near-physiological conditions. It enables the dissection of both structural and signaling processes in SEs and offers broad potential for the investigation of diverse, P-protein-independent mechanisms in phloem biology.

Conflict of Interest

The authors have no conflicts of interest to declare that are relevant to the content of this chapter.

References

1. Noll GA, Furch ACU, Rose J, Visser F, Prüfer D (2022) Guardians of the phloem—forisomes and beyond. New Phytol 236(4):1245–1260. https://doi.org/10.1111/nph.18476
2. Ernst AM, Jekat SB, Zielonka S, Müller B, Neumann U, Rüping B, Twyman RM, Krzyzanek V, Prüfer D, Noll GA (2012) *Sieve element occlusion* (*SEO*) genes encode structural phloem proteins involved in wound sealing of the phloem. Proc Natl Acad Sci U S A 109(28):E1980–1989. https://doi.org/10.1073/pnas.1202999109
3. Furch ACU, Hafke JB, Schulz A, van Bel AJE (2007) Ca^{2+}-mediated remote control of reversible sieve tube occlusion in *Vicia faba*. J Exp Bot 58(11):2827–2838. https://doi.org/10.1093/jxb/erm143
4. Knoblauch M, van Bel AJE (1998) Sieve tubes in action. Plant Cell 10(1):35–50
5. Jekat SB, Ernst AM, von Bohl A, Zielonka S, Twyman RM, Noll GA, Prüfer D (2013) P-proteins in Arabidopsis are heteromeric structures involved in rapid sieve tube sealing. Front Plant Sci 4:225. https://doi.org/10.3389/fpls.2013.00225
6. Anstead JA, Froelich DR, Knoblauch M, Thompson GA (2012) Arabidopsis P-protein filament formation requires both AtSEOR1 and AtSEOR2. Plant Cell Physiol 53(6):1033–1042. https://doi.org/10.1093/pcp/pcs046
7. Tissier AF, Marillonnet S, Klimyuk V, Patel K, Torres MA, Murphy G, Jones JD (1999) Multiple independent defective suppressor-mutator transposon insertions in Arabidopsis: a tool for functional genomics. Plant Cell 11(10):1841–1852. https://doi.org/10.1105/tpc.11.10.1841
8. van Bel AJE, Furch ACU, Will T, Buxa SV, Musetti R, Hafke JB (2014) Spread the news: systemic dissemination and local impact of Ca^{2+} signals along the phloem pathway. J Exp Bot 65(7):1761–1787. https://doi.org/10.1093/jxb/ert425
9. Fromm J, Lautner S (2007) Electrical signals and their physiological significance in plants. Plant Cell Environ 30(3):249–257. https://doi.org/10.1111/j.1365-3040.2006.01614.x
10. Stahlberg RC, Cosgrove DJ (1997) The propagation of slow wave potentials in pea epicotyls. Plant Physiol 113(1):209–217. https://doi.org/10.1104/pp.113.1.209
11. Zimmermann MR, Maischak H, Mithöfer A, Boland W, Felle HH (2009) System potentials, a novel electrical long-distance apoplastic signal in plants, induced by wounding. Plant Physiol 149(3):1593–1600. https://doi.org/10.1104/pp.108.133884
12. Kudla J, Batistic O, Hashimoto K (2010) Calcium signals: the lead currency of plant information processing. Plant Cell 22(3):541–563. https://doi.org/10.1105/tpc.109.072686
13. Batistič O, Kudla J (2012) Analysis of calcium signaling pathways in plants. Biochim Biophys Acta 1820(8):1283–1293. https://doi.org/10.1016/j.bbagen.2011.10.012
14. Nagai T, Yamada S, Tominaga T, Ichikawa M, Miyawaki A (2004) Expanded dynamic range of fluorescent indicators for Ca^{2+} by circularly permuted yellow fluorescent proteins. Proc Natl Acad Sci U S A 101(29):10554–10559. https://doi.org/10.1073/pnas.0400417101
15. Ren Y, Zhang Z, Zhanakhmetova D, Li W, Chen S, Werner T, Liesche J (2024) Fast and simple fluorometric measurement of phloem loading exposes auxin-dependent regulation of Arabidopsis sucrose transporter AtSUC2. Plant J 120(5):2305–2318. https://doi.org/10.1111/tpj.17110
16. Knox K, Paterlini A, Thomson S, Oparka K (2018) The coumarin glucoside, esculin, reveals rapid changes in phloem-transport velocity in response to environmental cues. Plant Physiol 178(2):795–807. https://doi.org/10.1104/pp.18.00574
17. Oparka KJ, Duckett CM, Prior DAM, Fisher DB (1994) Real-time imaging of phloem unloading in the root tip of Arabidopsis. Plant J 6(5):759–766. https://doi.org/10.1046/j.1365-313X.1994.6050759.x
18. Murashige T, Skoog F (1962) A revised medium for rapid growth and bioassays with tobacco tissue cultures. Physiol Plant 15(3):473–497. https://doi.org/10.1111/j.1399-3054.1962.tb08052.x
19. Peng HC, Walker GP (2020) Sieve element occlusion provides resistance against *Aphis gossypii* in TGR-1551 melons. Insect Sci 27(1):33–48. https://doi.org/10.1111/1744-7917.12610

20. Krebs M, Held K, Binder A, Hashimoto K, Den Herder G, Parniske M, Kudla J, Schumacher K (2012) FRET-based genetically encoded sensors allow high-resolution live cell imaging of Ca^{2+} dynamics. Plant J 69(1):181–192. https://doi.org/10.1111/j.1365-313X.2011.04780.x
21. Musetti R, Buxa SV, De Marco F, Loschi A, Polizzotto R, Kogel KH, van Bel AJ (2013) Phytoplasma-triggered Ca^{2+} influx is involved in sieve-tube blockage. Mol Plant-Microbe Interact 26(4):379–386. https://doi.org/10.1094/MPMI-08-12-0207-R
22. Rüping B, Ernst AM, Jekat SB, Nordzieke S, Reineke AR, Müller B, Bornberg-Bauer E, Prüfer D, Noll GA (2010) Molecular and phylogenetic characterization of the *sieve element occlusion* gene family in Fabaceae and non-Fabaceae plants. BMC Plant Biol 10: 219. https://doi.org/10.1186/1471-2229-10-219
23. Post J, van Deenen N, Fricke J, Kowalski N, Wurbs D, Schaller H, Eisenreich W, Huber C, Twyman RM, Prüfer D, Gronover CS (2012) Laticifer-specific cis-prenyltransferase silencing affects the rubber, triterpene, and inulin content of *Taraxacum brevicorniculatum*. Plant Physiol 158(3):1406–1417. https://doi.org/10.1104/pp.111.187880
24. Hood EE, Gelvin SB, Melchers LS, Hoekema A (1993) New *Agrobacterium* helper plasmids for gene transfer to plants. Transgenic Res 2(4):208–218. https://doi.org/10.1007/BF01977351
25. Clough SJ, Bent AF (1998) Floral dip: a simplified method for Agrobacterium-mediated transformation of *Arabidopsis thaliana*. Plant J 16(6):735–743. https://doi.org/10.1046/j.1365-313x.1998.00343.x
26. Oparka KJ, Santa Cruz S, Prior DAM, Padgett HS, Beachy RN (1997) Gating of epidermal plasmodesmata is restricted to the leading edge of expanding infection sites of tobacco mosaic virus (TMV). Plant J 12(4):781–789. https://doi.org/10.1046/j.1365-313X.1997.12040781.x
27. Ross-Elliott TJ, Jensen KH, Haaning KS, Wager BM, Knoblauch J, Howell AH, Mullendore DL, Monteith AG, Paultre D, Yan D, Otero S, Bourdon M, Sager R, Lee JY, Helariutta Y, Knoblauch M, Oparka KJ (2017) Phloem unloading in Arabidopsis roots is convective and regulated by the phloem-pole pericycle. Elife 6:e24125. https://doi.org/10.7554/eLife.24125
28. Véry AA, Davies JM (2000) Hyperpolarization-activated calcium channels at the tip of Arabidopsis root hairs. Proc Natl Acad Sci USA 97(17):9801–9806. https://doi.org/10.1073/pnas.160250397
29. Toyota M, Spencer D, Sawai-Toyota S, Jiaqi W, Zhang T, Koo AJ, Howe GA, Gilroy S (2018) Glutamate triggers long-distance, calcium-based plant defense signaling. Science 361(6407):1112–1115. https://doi.org/10.1126/science.aat7744
30. Buxa SV (2014) Microscopic identification of plant immune responses in phloem tissue of higher plants relating to bacterial infection. [PhD thesis] Justus-Liebig-University, Giessen
31. Furch ACU, Zimmermann MR, Noll GA, Wrobel LS, Scholz SS, Buxa-Kleeberg SV, Hafke JB, Fliegmann J, Mithöfer A, Ehlers K, Haufschild T, Nötzold J, Koch AM, Grabe V, Teutemacher F, Maaß JP, Prüfer D, Oelmüller R, Peiter E, Kogel KH, van Bel AJE (2025) Transformation of flg22 perception into electrical signals decoded in vasculature leads to sieve tube blockage and pathogen resistance. Sci Adv 11(9):eads6417. https://doi.org/10.1126/sciadv.ads6417
32. Furch AC, Zimmermann MR, Will T, Hafke JB, van Bel AJ (2010) Remote-controlled stop of phloem mass flow by biphasic occlusion in *Cucurbita maxima*. J Exp Bot 61(13):3697–3708. https://doi.org/10.1093/jxb/erq181
33. Gora PJ, Reinders A, Ward JM (2012) A novel fluorescent assay for sucrose transporters. Plant Methods 8: 13. https://doi.org/10.1186/1746-4811-8-13
34. Jekat S (2012) Phylogenetische und molekularbiologische Charakterisierung der strukturellen P-Proteine in Pflanzen. [PhD thesis] University of Muenster, Muenster
35. Morgan B, Sobotta MC, Dick TP (2011) Measuring E_{GSH} and H_2O_2 with roGFP2-based redox probes. Free Radic Biol Med 51(11):1943–1951. https://doi.org/10.1016/j.freeradbiomed.2011.08.035
36. Moseyko N, Feldman LJ (2001) Expression of pH-sensitive green fluorescent protein in *Arabidopsis thaliana*. Plant Cell Environ 24(5):557–563. https://doi.org/10.1046/j.1365-3040.2001.00703.x

Chapter 29

Regulatory DNA Sequences Controlling Gene Expression in Leaf Phloem Cells

Chen Deng, Li-Qing Chen and Ji-Yun Kim

Abstract

This chapter discusses the regulatory DNA sequences that enable cell-type-specific gene expression in leaf phloem tissues, focusing on phloem parenchyma, companion cells, and sieve elements. Molecular biology techniques, including reporter gene analysis, single-cell RNA sequencing, and live imaging, especially when combined with bioinformatic analysis, are powerful tools for identifying key promoter regions and/or cis-regulatory elements responsible for targeted expression. However, identifying conserved, cell type-specific promoter sequences across diverse plant species for cell-specific regulation remains challenging in high-throughput contexts. Through this review, we aim to provide insights into the cellular functions of the phloem cell types and facilitate precise genetic modifications to manipulate agronomic traits. We anticipate that identifying phloem cell-type specific regulatory elements will have significant implications for biotechnological applications aimed at improving crop productivity and resilience, highlighting the potential for tailored gene expression strategies in plant vascular tissues.

Keywords Cell type, Gene expression, Genetic modification, Phloem parenchyma, Companion cell, Sieve element

1 Introduction

In vascular plants, leaves utilize conductive tissues to allocate various nutrients (such as photoassimilates) and relay diverse signals (e.g., florigen and mobile molecules) to coordinate development, growth, and environmental responses. Plants have two types of vascular tissues, one of which is the phloem tissue, responsible for loading and transporting photoassimilates (mainly as sucrose in many species). Given the importance of the leaf phloem, extensive research has been directed toward understanding the cellular biology of phloem cells that constitute the phloem tissue in leaves—namely companion cells, sieve elements, and to a lesser extent, phloem parenchyma.

Plant scientists have achieved precise regulation of gene expression in targeted cell types by utilizing regulatory DNA sequences [1]. The differential deployment of DNA regulatory elements, which

Johannes Liesche and Sanna Sevanto (eds.), *Phloem*, Methods in Molecular Biology, vol. 3066,
https://doi.org/10.1007/978-1-0716-5488-0_29,

are pivotal in gene transcription and silencing, underlies the cell type-specific utilization of the genome and facilitates cellular specialization. These regulatory elements, mediating cell-specific gene expression, are often situated within accessible chromatin regions (ACRs), where nucleosome positioning is altered due to interactions between transcription factors (TFs) and cis-regulatory elements (CREs). Genome-wide identification of ACRs can be accomplished using methods such as MNAse-seq (micrococcal nuclease digestion with deep sequencing) or ATAC-seq (Assay for Transposase-Accessible Chromatin with sequencing) [2]. Recent advances in single-cell genomics have provided more precise and comprehensive methods for identifying regulatory elements. The integration of single-cell or single-nucleus RNA sequencing (sc/snRNA-seq), single-nucleus ATAC-seq (snATAC-seq), and DNA affinity purification sequencing (DAP-seq) now offers more efficient approaches for the de novo elucidation of phloem-specific gene regulatory networks [3–5]. Moreover, in cases where chromatin accessibility changes do not align with gene expression alterations, something often observed during bacterial infection, profiling additional epigenetic features at the single-cell level, including three-dimensional chromatin architecture, DNA methylation, and histone modifications, can provide valuable insights [6]. Here, we summarize our current knowledge on phloem cell-specific promoters, the putative transcription factor binding motifs, and potential applications in agrobiotechnology.

1.1 Phloem Parenchyma

The phloem tissue in vascular plants comprises living cells known as phloem parenchyma cells, which primarily function in the storage of nutrients such as lipids and starch, as well as other substances, including resins and tannins. Additionally, phloem parenchyma cells play a crucial role in the export of sugars and various organic compounds, facilitating their loading into the companion cell-sieve element complex for long-distance translocation from source to sink tissues, as well as lateral transport. In some plant species employing apoplasmic phloem loading, including *Arabidopsis thaliana* (Arabidopsis) and *Senecio vulgaris*, the cell wall invaginations of phloem parenchyma cells, analogous to those observed in animal enterocytes, exhibit enhanced development following exposure to high light intensity, methyl jasmonate, or sucrose treatments [7, 8]. These specialized cells, characterized by the presence of cell wall ingrowths, are designated as phloem parenchyma transfer cells. From an anatomical perspective, the polarized development of cell wall ingrowths in phloem parenchyma cells is hypothesized to enhance their capacity for transmembrane transport [9].

The involvement of phloem parenchyma transfer cells in phloem loading is further supported by the detection of the sucrose uniporter AtSWEET11 in Arabidopsis, as well as the sugar accumulation phenotype observed in the *sweet11; sweet12* double mutants [10–13].

The promoter region of *AtSWEET11* (pAtSWEET11), encompassing nucleotides –2231 to –29 relative to the start codon, has been demonstrated to drive β-glucuronidase (GUS)-mCitrine reporter expression in cells located within the midvein of Arabidopsis leaves [14]. However, promoter activity alone is insufficient to define the phloem parenchyma-specific expression pattern of AtSWEET11 [12]. Consistent with this, specific coding sequences corresponding to transmembrane domains 3 (TM3, nucleotides +211 to +291) and 7 (TM7, nucleotides +565 to +642) of *AtSWEET11* are essential for directing phloem parenchyma-specific GUS/YELLOW FLUORESCENT PROTEIN (YFP) transcription in minor and major veins when a more extensive promoter fragment (nucleotides –2937 to +1) is utilized [12]. Intriguingly, the regulatory role of both the SWEET promoter and coding region sequences in conferring phloem parenchyma-specific gene expression is conserved across at least four vascular plant species, including *Selaginella moellendorffii*, *Azolla filiculoides*, *Populus trichocarpa*, and *Sorghum bicolor* [12]. These findings indicate that the phloem parenchyma-specific expression of SWEET genes is governed by mechanisms operating at both transcriptional and post-transcriptional levels.

ScRNA-seq analyses using mature leaf samples have facilitated the identification of multiple promoters specific to phloem parenchyma cells [4, 15]. Utilizing a protocol designed to enrich vascular cells from leaves, Kim et al. demonstrated that the mRNAs of *AtSWEET13*, a paralog of *AtSWEET11*, and multiple members of the UmamiT amino acid transporter family, including *AtUMAMIT20*, are preferentially enriched within the phloem parenchyma of mature Arabidopsis leaves [4]. The localization of fusion proteins driven by native promoters, specifically *AtSWEET13* fused to YFP and BASIC LEUCINE-ZIPPER9 (AtbZIP9) fused to GREEN FLUORESCENT PROTEIN (GFP), was confirmed in phloem parenchyma cells through live imaging using confocal scanning laser microscopy [4]. Moreover, promoter activity of *AtbZIP9* within the veins of mature leaves appears to depend on regulatory DNA elements situated between nucleotides –2289 and –1300 [16]. Histochemical analysis employing GUS staining revealed that *AtUMAMIT20*-GUS fusion proteins, driven by a native promoter region extending to nucleotide –3000, are localized in the leaf veins of Arabidopsis [17]. Similarly, promoter regions approximately 2000 nucleotides upstream of genes such as Arabidopsis *NDR1/HIN1-LIKE1* (*AtNHL1*), *CYCLING DOF FACTOR 4* (*AtCDF4*), *CYTOCHROME P450 79B2* (*AtCYP79B2*), *AtCYP79B3*, and an adenine nucleotide alpha hydrolase-like superfamily protein (*AT3G11930*) exhibit activity within leaf veins [15, 18].

To date, only a limited number of transcription factors involved in the regulation of gene expression within phloem parenchyma cells have been characterized. Notably, the C2H2-DNA-BINDING

FINGER (C2C2-DOF) family member AtCDF5 and the bZIP family protein ELONGATED HYPOCOTYL 5 (AtHY5) have been shown to upregulate *AtSWEET11* expression in Arabidopsis leaves [19, 20]. Similarly, in rice, the OsDOF11 transcription factor enhances the expression of *OsSWEET11* and *OsSWEET14* in phloem parenchyma cells [21]. It is anticipated that additional TFs governing phloem parenchyma-specific gene expression will be identified through the application of single-cell transcriptomics across diverse plant species [3–5].

The expression of genes specific to phloem parenchyma cells is likely dependent on CREs that are enriched in these cell types. Recent studies employing multiplexed DNA affinity purification sequencing (multiDAP-seq) and snRNA-seq have demonstrated that DNA binding motifs for several TFs—including TEOSINTE BRANCHED1, CYCLOIDEA, PROLIFERATING CELL FACTOR 24 (AtTCP24); BASIC HELIX-LOOP-HELIX (bHLH) family protein BES1-INTERACTING MYC-LIKE 1 (AtBIM1); BRASSINAZOLE-RESISTANT (BZR) family protein BES1/BZR1 HOMOLOG 2 (AtBEH2); BRI1-EMS-SUPPRESSOR 1 (BES1) family protein BETA-AMYLASE 2 (AtBMY2); AUXIN RESPONSE FACTOR 2 (AtARF2); APETALA2-ETHYLENE-RESPONSIVE ELEMENT BINDING PROTEINS (AP2/EREBP) family members such as ETHYLENE-RESPONSIVE TRANSCRIPTION FACTOR 040 (AtERF040), C-REPEAT/DRE BINDING FACTOR 2 (AtCBF2), COOPERATIVELY REGULATED BY ETHYLENE AND JASMONATE 1 (AtCEJ1); and C2C2-DOF family protein DOF AFFECTING GERMINATION 2 (AtDAG2)—are enriched within putative regulatory regions spanning 2000 base pairs upstream to 500 base pairs downstream of the translation start sites of phloem parenchyma-expressed genes in *Arabidopsis*, *Arabidopsis lyrata*, *Capsella rubella*, and *Brassica oleracea* [3]. These CREs that facilitate gene transcription in phloem parenchyma cells are hypothesized to reside within ACRs. For instance, DNA binding motifs of C2C2-DOF family proteins, including *At*DAG2 and OBF-BINDING PROTEIN 1 (*At*OBP1), have been found to be enriched in ACRs associated with phloem parenchyma [6]. Moreover, candidate TFs implicated in the regulation of phloem parenchyma-specific gene expression have been identified through analyses of gene co-expression networks and TF binding motifs at the single-cell level in leaves. These include *At*bHLH68, *At*bHLH112, *At*DOF2.2, *At*DOF5, PHLOEM EARLY DOF 2 (*At*PEAR2), and *At*OBP2 [5].

It is therefore plausible that the activity of the aforementioned phloem parenchyma-specific promoters depends on these TFs and their corresponding DNA binding motifs located within ACRs (see Table 1) [2, 6, 22]. Future investigations on these TFs and their DNA binding motifs are anticipated to elucidate the regulatory mechanisms governing gene expression specific to phloem parenchyma cells.

Table 1
Putative transcription factor families that bind to the accessible chromatin regions of phloem parenchyma-specific promoters

Gene name (Gene ID)	ACR	Transcription factor families
SWEET11(AT3G48740)	Chr3:18054723-18054897	DOF
	Chr3:18056286-18056465	bHLH, DOF
UMAMIT20 (AT4G08290)	Chr4:5238671-5238801	DOF
NHL1(AT3G11660)	Chr3:3679673-3679803	DOF
CDF4(AT2G34140)	Chr2:14415623-14415770	AP2/EREBP, bHLH, DOF
	Chr2:14416004-14416134	ARF, DOF
	Chr2:14416161-14416291	AP2/EREBP, DOF
CYP79B2 (AT4G39950)	Chr4:18524837-18525339	AP2/EREBP, ARF, DOF, TCP
	Chr4:18524071-18524327	bHLH, DOF
CYP79B3 (AT2G22330)	Chr2:9488349-9488479	DOF
	Chr2:9487212-9487603	AP2/EREBP, bHLH, DOF, TCP

ACR, Accessible chromatin regions

1.2 Companion Cells

Within the phloem, companion cells are closely associated with sieve elements, forming functional symplasmic units that facilitate nutrient and sugar translocation throughout the plant. The companion cells, interconnected to the sieve elements via plasmodesmata, support metabolic processes, mediate active loading of photosynthates into sieve tubes, and function as central regulatory hubs for the movement of substances within the vascular tissue. A distinct subset of companion cells, termed companion transfer cells, is characterized by the uniform deposition of wall ingrowths along their internal cell wall. These transfer cells are found in the minor leaf veins of *Pisum sativum* but are absent in Arabidopsis [23].

Companion cells, akin to phloem parenchyma cells, possess transmembrane nutrient transporters that facilitate the uptake of nitrogen and carbon. For instance, the import of sucrose into companion cells is mediated by *At*SUC2/SUT1, a well-characterized proton/sucrose symporter specific to companion cells and expressed in the source leaves of Arabidopsis [24]. The promoter region of *AtSUC2* (*pAtSUC2*), comprising a 945-nucleotide sequence upstream of the start codon, is sufficient to drive companion cell-specific gene expression [25]. Mutational analyses of *pAtSUC2* have identified putative binding sites for DOF (*CTTT* motif, nucleotides –569 to –566) and homeodomain-leucine zipper (HD-Zip) subfamily I (*AAT(A/T)ATT* motif, nucleotides –619 to –606) as critical for *pAtSUC2* activity [26]. Furthermore, the DOF binding motif is enriched in ACRs adjacent to companion cell-expressed genes across multiple species, including *Arabidopsis*, *S.bicolor*, *Zea mays*, *Panicum miliaceum*, and *Urochloa fusca* [6, 27]. In rice, the HD-Zip motif similarly shows enrichment in ACRs near companion cell-expressed genes [27]. Additionally, AUXIN RESPONSE FACTOR 1 (*At*ARF1) has been shown to activate *AtSUC2* expression by binding to the CRE *TGTCTC* located between nucleotides −1615 and −148 [28]. Conversely, upon sugar treatment, ETHYLENE-INSENSITIVE3 (*At*EIN3) represses *AtSUC2* expression through binding to its promoter region spanning nucleotides −960 to −800 [29]. The proton gradient necessary for *At*SUC2-mediated sucrose import is likely generated by ARABIDOPSIS H^+-ATPASE ISOFORM 3 (*At*AHA3), whose promoter region (~4000 nucleotides upstream) exhibits companion cell-specific expression [30].

Galactinol synthase is localized within the companion cells of plant leaves, particularly in the specialized intermediary cells (a type of companion cells) of minor veins. In the intermediary cells, galactinol synthase catalyzes the synthesis of galactinol, a key precursor in the biosynthesis of raffinose-family oligosaccharides (RFOs). The promoter region of *Cucumis melo GALACTINOL SYNTHASE 1* (*pCmGAS1*) drives companion cell-specific expression in the leaves of Arabidopsis, *Nicotiana tabacum* (*N. tabacum*), and *Populus tremula* × *alba*, although these species do not utilize the polymer trap mechanism for phloem loading [31]. Interestingly, pCmGAS1 is only active in a subset of Arabidopsis companion cells that also express *FLOWERING LOCUS T* (*AtFT*) [32]. snRNA-seq further identified eight promoter sequences that are active in companion cells without high pAtFT activity, namely *pPROTEIN BROTHER OF FT AND TFL 1* (*pAtBFT*, nucleotides −2396 to −1), *pAT1G24575* (nucleotides −2521 to −1), *pAT1G67865* (nucleotides −2488 to −1), *pMONOTHIOL GLUTAREDOXIN-S2* (pAtGRXS2, nucleotides −1939 to −1), *pAT2G26695* (nucleotides −2226 to −1), *pPHLOEM ASSOCIATED RNA CHAPERONE-LIKE* (*pAtPARCL*, nucleotides −1911 to −1), *pGRXS9* (nucleotides −2719 to −1), and *pCYSTEINE-RICH TRANSMEMBRANE MODULE 12* (*pAtCYSTM12*, nucleotides −2574 to −1) [32].

Companion cells in plants serve a pivotal function in the context of plant viral infections by acting as selective gateways to the vascular system, specifically the phloem. They facilitate the translocation of viruses from adjacent cells into sieve elements, thereby enabling systemic viral dissemination and subsequent unloading into distal tissues. In Arabidopsis ecotype Shahdara, the reprogramming of the *INDOLE-3-ACETIC ACID INDUCIBLE 26* gene enhanced the phloem loading efficiency of tobacco mosaic virus through transcriptional regulatory mechanisms [33].

Unlike *pAtSUC2*, the companion cell-specific promoter derived from commelina yellow mottle virus (*pCoYMV*) is upregulated in Arabidopsis in response to sugar availability in the growth medium [34]. The nucleotide region spanning –159 to –84 of *pCoYMV* harbors regulatory elements responsible for driving gene expression within the vascular tissues of *N. tabacum* leaves [35]. Furthermore, *pCoYMV* has been effectively utilized to direct vascular-specific gene expression in *Malus domestica* [36].

A 403-nucleotide bidirectional promoter from wheat dwarf virus strain F, which regulates the expression of the movement protein V1 and the DNA replication protein C1-1, confers companion cell-specific gene expression across multiple species, including *N. tabacum*, *Nicotiana benthamiana*, Arabidopsis, and *C. melo* [37]. Additionally, three cis-regulatory elements within the rice tungro bacilliform virus (RTBV) genome—namely the GATA motif (nt –143 to –135), ASL Box (nt –98 to –79), and BOX II (nt –53 to –39), relative to nucleotide 7354—function synergistically to mediate companion cell-specific gene expression in transgenic rice plants [38]. Although the bZIP motif is not enriched in accessible chromatin regions adjacent to companion cell-expressed genes in rice [27], the rice bZIP transcription factor OsRF2a is capable of activating the RTBV companion cell-specific promoter through direct binding to the BOX II element [39].

In Arabidopsis leaves, the specificity of companion cell expression within minor veins can be governed by a 1392-nucleotide DNA sequence termed *MATURE MINOR VEIN ELEMENT 1* (*AtMMVE1*), located between the genes encoding *ABC TRANSPORTER OF THE MITOCHONDRION 1* (*AtATM1*) and *INDOLE-3-ACETIC ACID INDUCIBLE 11* (*AtIAA11*) [40]. Within *AtMMVE1*, three distinct regulatory regions have been characterized, positioned upstream of the *AtIAA11* start codon: nucleotides –1108 to –917 facilitate expression across all veins; nucleotides –989 to –719 specifically confer expression in minor veins; and nucleotides –241 to –1 function to repress expression in larger veins. Treatment with auxin nearly doubles the activity of the *AtIAA11* promoter (*pAtIAA11*), a response potentially mediated by an ARF binding motif (*TGTCTC*) located between nucleotides –172 and –167 [40]. Furthermore, when three tandem copies of the sequence spanning nucleotides –1393 to –800 are positioned upstream of the minimal Cauliflower Mosaic Virus 35S promoter, they drive robust and specific gene expression within minor veins.

Putative regulatory regions of genes expressed in companion cells from *Arabidopsis*, *A. lyrata*, *C. rubella*, and *B. oleracea* exhibit significant enrichment of DNA-binding motifs corresponding to various TF families. These include the CYSTEINE-RICH POLYCOMB-LIKE PROTEIN (CPP) family member TESMIN/TSO1-LIKE CXC 2 (*At*TCX2) and the V-MYB AVIAN MYELOBLASTOSIS VIRAL

ONCOGENE HOMOLOG (MYB) family protein *At*MYB3R1, *At*TCP24, *At*BMY2, MYB-related proteins such as TELOMERE REPEAT-BINDING PROTEIN 2 (*At*TRP2), REVEILLE 8 (*At*RVE8), *At*HY5, *At*DAG2, GOLDEN2 (G2)-like family protein MYB-RELATED PROTEIN 2 (*At*MYR2), AT-RICH INTERACTION DOMAIN (ARID) 3 (*At*ARID3), SQUAMOSA PROMOTER BINDING PROTEINS (SBP) family protein SBP-LIKE 5 (*At*SPL5), the Cys2-Cys2 (C2C2)-YABBY family protein CRABS CLAW (AtCRC), and the Cys-Cys-His-His (C2H2) family protein WIP DOMAIN PROTEIN 5 (*At*WIP5) [3]. In Arabidopsis leaves, DNA-binding motifs for *At*DAG2, *At*OBP1, and *At*MYR1 are notably enriched within ACRs that demarcate companion cells [6]. Furthermore, these TF family motifs are present within the ACRs of several Arabidopsis companion cell-specific promoters previously identified (Table 2) [2, 6, 22]. Analyses of gene co-expression and TF binding motifs at the single-cell level in leaves have also implicated *At*DOF2, *At*DOF2.2, *At*DOF4.6, *At*DOF5, *At*PEAR2, *At*OBP2, INTERCELLULAR TRAFFICKING DOF 1 (*At*ITD1), *At*TCP5, *At*MYR1, MYB family proteins such as EARLY FLOWERING MYB PROTEIN (*At*EFM), HRS1 HOMOLOG 5 (*At*HHO5), PHR1-LIKE 1 (*At*PHL1), *At*PHL12, and ALTERED PHLOEM DEVELOPMENT (*At*APL) as putative regulators of genes expressed in companion cells [5]. In at least one of five Poaceae family members examined, motifs corresponding to NAC, MYB, WRKY, HD-ZIP, ERF/DREB, Trihelix, C4-GATA, TCP, DOF, MIKC, GARP_ARR-B, GARP_G2-like, RAV, and AP2 TF families are enriched within ACRs proximal to companion cell-expressed genes [27]. Consistent with phloem parenchyma-specific promoters, nearly all Arabidopsis companion cell-specific promoters described herein contain DOF binding motifs within their accessible chromatin regions (Tables 1 and 2). Recent evidence indicates that DOF proteins may serve as master regulators of phloem-expressed genes throughout the evolution of vascular plants [41]. Dissecting the interactions between DOF proteins and other phloem-enriched TFs will be critical for deciphering the regulatory network that confers cell type-specific gene expression in phloem parenchyma and companion cells.

Unlike the DOF transcription factors, bHLH and WRKY DNA binding motifs are not significantly enriched within the putative regulatory regions of companion cell-expressed genes. Yet several members of these families—including *At*bHLH48, *At*bHLH60, PHYTOCHROME-INTERACTING FACTOR 4 (*At*PIF4), *At*PIF5, *At*PIF7, *At*bHLH60, *At*bHLH65, *At*bHLH72, *At*WRKY71, and *At*WRKY75—were shown to enhance *AtFT* transcription through direct promoter interaction [42]. In contrast, transcription factors such as *At*MYR1, *At*MYR2, *At*MYB53, *At*MYB106, and *At*EFM, whose DNA-binding motifs are significantly enriched within the putative regulatory regions of companion cell-expressed genes, function as repressors of *AtFT* transcription [42]. snRNA-seq combined

Table 2
Putative transcription factor families that bind to the accessible chromatin regions of companion cells-specific promoters

Gene name (Gene ID)	ACR	Transcription factor families
AHA3 (AT5G57350)	Chr5:23236758-23236888	C2H2, DOF, MYB
	Chr5:23238412-23238542	bZIP, C2H2, DOF, SBP
	Chr5:23239185-23239365	ARID, bZIP, DOF, G2-like, MYB, MYB-related
SUC2 (AT1G22710)	Chr1:8032771-8033544	bZIP, C2H2, DOF, C2C2-YABBY, MYB, SBP, TCP
IAA26 (AT3G16500)	Chr3:5614974-5615104	bZIP, DOF, MYB, MYB-related
	Chr3:5615706-5615976	C2H2, CPP, MYB-related, SBP
IAA11 (AT4G28640)	Chr4:14141286-14141662	bZIP, C2H2, DOF, C2C2-YABBY, MYB, SBP, TCP
	Chr4:14140960-14141222	ARID, DOF, C2C2-YABBY, G2-like, MYB, SBP
PHL12 (AT3G12730)	Chr3:4049250-4049510	bZIP, C2H2, DOF, MYB
TET6 (AT3G12090)	Chr3:3853920-3854050	ARID, bZIP, C2H2, DOF
	Chr3:3854087-3854234	bZIP, C2H2, DOF
NAKR1 (AT5G02600)	Chr5:587367-587497	ARID, DOF
AT1G24575	Chr1:8711868-8711998	ARID, bZIP, DOF
	Chr1:8712031-8712178	DOF, MYB, SBP, TCP
AT2G26695	Chr2:11363407-11363537	C2H2, DOF, TCP
GRXS9 (AT2G30540)	Chr2:13013531-13013661	bZIP, DOF, MYB, SBP
	Chr2:13013822-13013952	bZIP, DOF, MYB
	Chr2:13014042-13014238	bZIP, C2H2, DOF, SBP

ACR, accessible chromatin region

with motif analysis further revealed that the DNA-binding motif of the transcriptional *AtFT* repressor NITRATE-INDUCIBLE GARP-TYPE TRANSCRIPTIONAL REPRESSOR 1.2 (*At*NIGT1.2) is enriched in the promoters of genes highly expressed in Arabidopsis leaf cells that exhibit pAtFT activity [32]. Notably, the mRNA levels of *AtNIGT1.2*, *AtNIGT1.4*, and other known transcriptional repressors of *AtFT* are comparatively low in these cells with active *pAtFT* [32]. Together, these findings suggest that transcriptional repression may serve a pivotal role in shaping gene expression patterns within the companion cells of Arabidopsis leaves.

Building on the view of companion cells as a transcriptionally specialized cell type, recent studies indicate that this specialization also intersects with hormone-responsive pathways. Recent studies have shown that despite the lack of significant enrichment of ARF binding motifs in the ACRs of companion cell-expressed genes in both Brassicaceae and Poaceae [3, 27], auxin signaling can still be effectively modulated using companion cell-specific promoters. For instance, auxin signaling in tomato was successfully manipulated by driving the companion cell-specific expression of the melon Aux/IAA auxin response factor gene *CmF-308* or the tomato *CYCLOPHILIN 1* gene using *pAtSUC2* [43, 44]. Together, these findings demonstrate the utility of companion cell-specific regulatory DNA sequences as effective tools for precise engineering in crops, with implications for agrobiotechnological applications.

1.3 Immature Sieve Elements

Mature sieve elements, characterized by the absence of most organelles, including nuclei, ribosomes, and rough endoplasmic reticulum, rely on imported (from the companion cells or immature sieve elements) or pre-existing functional proteins. Consequently, gene expression in sieve elements occurs predominantly during their immature stages. The synthesis of structural phloem proteins—including sieve element occlusion (SEO) proteins and other phloem-associated proteins essential for phloem functionality and sieve tube occlusion following injury—is confined to this limited developmental window.

Several promoters have been characterized that drive transcription specifically in immature sieve elements. In *Medicago truncatula*, the ~1000-nucleotide promoters *pMtSEO1*, *pMtSEO2*, and *pMtSEO3*, along with the ~1450-nucleotide *N. tabacum* promoter *pNtSEO1*, direct immature sieve element-specific expression [45–47]. Functional dissection of *pMtSEO1* revealed two critical regulatory regions: nucleotides −1021 to −136 confer immature sieve element expression, while nucleotides −136 to −41 repress expression in companion cells [48]. In Arabidopsis, RESTRICTED TEV MOVEMENT 1 (*At*RTM1) and RESTRICTED TEV MOVEMENT 2 (*At*RTM2) encode inhibitors of tobacco etch virus long-distance movement in sieve elements [49–51]. The regulatory regions associated with *At*RTM1 (1150 nucleotides upstream, 760 nucleotides downstream) and *At*RTM2 (1195 nucleotides upstream, 653 nucleotides downstream) drive immature sieve element-specific expression in Arabidopsis [52, 53].

Co-expression analysis using well-characterized sieve element-expressed genes—namely *EARLY NODULIN-LIKE PROTEIN 9* (*AtENODL9*), *SIEVE ELEMENT-LINING CHAPERONE1* (*AtSLI1*), *AtSEOR1*, and *AtSEOR2*—as query has facilitated the identification of regulatory DNA elements governing gene expression in immature sieve elements [54]. Furthermore, scRNA-seq of Arabidopsis leaves, combined with reporter gene assays, indicates that the promoters of *AtENODL9*, *ASCORBATE PEROXIDASE 5* (*AtAPX5*), *GLUTATHIONE S-TRANSFERASE F4* (*AtGSTF4*), AWPM-19-like 4 (*AtPM19L4*), *MILDEW RESISTANCE LOCUS O 13* (*AtMLO13*), RESTRICTED TEV MOVEMENT 3 (*AtRTM3*), and *RETICULON-LIKE PROTEIN B9* (*AtRTNLB9*) may be active within immature sieve elements of leaves [4, 54].

Integration of multiple single-cell expression atlases further accelerates the identification of immature sieve element-specific genes. For instance, the Arabidopsis *SIEVE ELEMENT CARBOHYDRATE-BINDING PROTEIN 1* (*AtSECB1*) promoter is active in developing phloem tissues, with the GFP fusion protein exhibiting discontinuous localization along sieve element peripheries [41]. Higher-order *AtSECB* knockout mutants (*secb1 secb2 secb5*, *secb1 secb2 secb5 secb6*, and *secb1 secb2 secb4 secb5 secb6*) display phloem transport defects, including delayed flowering, accumulation of purple pigmentation, and growth inhibition [41]. As single-cell genomic approaches expand across diverse plant species, additional genes and their upstream regulatory elements that are specific to immature sieve elements are anticipated to be identified.

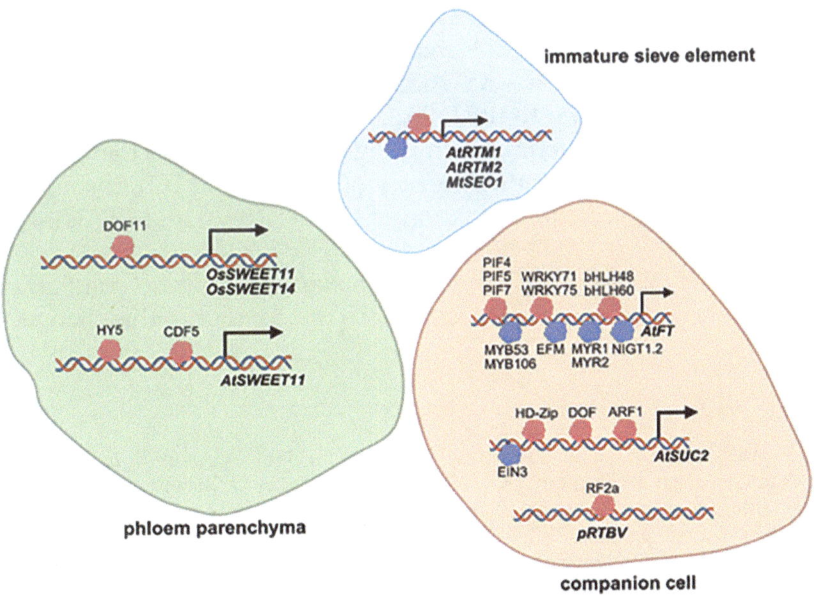

Fig. 1 Transcriptional regulation of gene expression within phloem cells. The expression of *SWEET* genes in phloem parenchyma cells is regulated by the transcription factors *Os*DOF11, *At*CDF5, and *At*HY5. In companion cells, the transcription of *AtFT* is activated by bHLH family members (*At*PIF4, *At*PIF5, *At*PIF7, *At*bHLH48, *At*bHLH60) as well as *At*WRKY71 and *At*WRKY75, while it is repressed by *At*MYR1, *At*MYR2, *At*MYB53, *At*MYB106, *At*EFM, and *At*NIGT1.2. The transcription of *AtSUC2* in companion cells is positively regulated by *At*ARF1, members of the HD-Zip family, and DOF family members, and negatively regulated by *At*EIN3. Additionally, *Os*RF2a functions as an activator of RTBV's promoter in companion cells. The expression of *AtRTM1*, *AtRTM2*, and *MtSEO1* is controlled by *cis*-regulatory elements situated within their respective promoter regions. Created in BioRender. Deng C (2025) https://BioRender.com/ksve74s

2 Conclusion

The regulatory elements governing gene expression across leaf phloem cell types—companion cells, phloem parenchyma, and immature sieve elements—are now being systematically characterized through advances in single-cell genomics, chromatin accessibility profiling, and comparative transcriptomics (Fig. 1). Such discoveries are revealing not only the molecular mechanisms underlying cell type identities but also the evolutionary innovations that enabled long-distance transport and signaling in vascular plants. Translating the expanding regulatory toolkit into precision biotechnological applications will enable targeted manipulation of phloem-mediated processes, from optimizing nutrient distribution and systemic signaling to engineering pathogen resistance and stress tolerance in crops facing climate change.

Acknowledgments

We thank Johannes Liesche for providing constructive suggestions on the manuscript. This work was supported by the National Research Foundation of Korea (NRF) grant funded by the Korea government (MSIT) (No. RS-2025-00560593), and by the Korea Polar Research Institute (KOPRI, PAP, PE26915) funded by the Ministry of Oceans and Fisheries. Chen Deng was supported by the Brain Pool program funded by the Ministry of Science and ICT through the National Research Foundation of Korea (RS-2025-15373144).

Conflict of Interest The authors have no conflicts of interest to declare that are relevant to the content of this chapter.

References

1. Brophy JAN, Magallon KJ, Duan L, Zhong V, Ramachandran P, Kniazev K, Dinneny JR (2022) Synthetic genetic circuits as a means of reprogramming plant roots. Science 377:747–751. https://doi.org/10.1126/science.abo4326
2. Lu Z, Marand AP, Ricci WA, Ethridge CL, Zhang X, Schmitz RJ (2019) The prevalence, evolution and chromatin signatures of plant regulatory elements. Nat Plants 5:1250–1259. https://doi.org/10.1038/s41477-019-0548-z
3. Baumgart LA, Greenblum SI, Morales-Cruz A, Wang P, Zhang Y, Yang L et al. (2025) Recruitment, rewiring and deep conservation in flowering plant gene regulation. Nat Plants 11:1514–1527. https://doi.org/10.1038/s41477-025-02047-0
4. Kim J-Y, Symeonidi E, Pang TY, Denyer T, Weidauer D, Bezrutczyk M et al. (2021) Distinct identities of leaf phloem cells revealed by single cell transcriptomics. Plant Cell 33:511–530. https://doi.org/10.1093/plcell/koaa060
5. Ferrari C, Manosalva Pérez N, Vandepoele K (2022) MINI-EX: integrative inference of single-cell gene regulatory networks in plants. Mol Plant 15:1807–1824. https://doi.org/10.1016/j.molp.2022.10.016
6. Nobori T, Monell A, Lee TA, Sakata Y, Shirahama S, Zhou J et al. (2025) A rare PRIMER cell state in plant immunity. Nature 638:197–205. https://doi.org/10.1038/s41586-024-08383-z
7. Amiard V, Demmig-Adams B, Mueh KE, Turgeon R, Combs AF, Adams WW (2007)

Role of light and jasmonic acid signaling in regulating foliar phloem cell wall ingrowth development. New Phytol 173:722–731. https://doi.org/10.1111/j.1469-8137.2006.01954.x
8. Wei X, Nguyen STT, Collings DA, McCurdy DW (2020) Sucrose regulates wall ingrowth deposition in phloem parenchyma transfer cells in Arabidopsis via affecting phloem loading activity. J Exp Bot 71:4690–4702. https://doi.org/10.1093/jxb/eraa246
9. Pate J, Gunning B (1972) Transfer cells. Annu Rev Plant Physiol 23:173–196. https://doi.org/10.1146/annurev.pp.23.060172.001133
10. Chen L-Q, Qu X-Q, Hou B-H, Sosso D, Osorio S, Fernie AR, Frommer WB (2012) Sucrose efflux mediated by SWEET proteins as a key step for phloem transport. Science 335:207–211. https://doi.org/10.1126/science.1213351
11. Wei X, Huang Y, Nguyen STT, Collings DA, McCurdy DW (2022) Asymmetric wall ingrowth deposition in Arabidopsis phloem parenchyma transfer cells is tightly associated with sieve elements. J Exp Bot 73:5414–5427. https://doi.org/10.1093/jxb/erac234
12. Zhang C, Li Y, Wang J, Xue X, Beuchat G, Chen L-Q (2021) Two evolutionarily duplicated domains individually and post-transcriptionally control SWEET expression for phloem transport. New Phytol 232:1793–1807. https://doi.org/10.1111/nph.17688
13. Kim J-Y, Symeonidi E, Pang TY, Denyer T, Weidauer D, Bezrutczyk M et al. (2021) Distinct identities of leaf phloem cells revealed by single cell transcriptomics. Plant Cell 33:511–530. https://doi.org/10.1093/plcell/koaa060
14. Procko C, Lee T, Borsuk A, Bargmann BOR, Dabi T, Nery JR et al. (2022) Leaf cell-specific and single-cell transcriptional profiling reveals a role for the palisade layer in UV light protection. Plant Cell 34:3261–3279. https://doi.org/10.1093/plcell/koac167
15. Liu Z, Wang J, Zhou Y, Zhang Y, Qin A, Yu X et al. (2022) Identification of novel regulators required for early development of vein pattern in the cotyledons by single-cell RNA-sequencing. Plant J 110:7–22. https://doi.org/10.1111/tpj.15719
16. Illescas-Miranda J, Llanos-Casado V, Contreras E, Carrillo-Barral N, Iglesias-Fernández R (2025) Temporal and spatial expression analysis of AtbZIP9 during seed and silique development in *Arabidopsis thaliana* (L.) Heynh. Plant Gene 44:100536. https://doi.org/10.1016/j.plgene.2025.100536
17. Prior MJ, Weidauer D, Locci F, Liao J-Y, Kuwata K, Deng C et al (2025) The Arabidopsis amino acid transporter UmamiT20 confers *Botrytis cinerea* susceptibility. J Exp Bot 77:eraf496. https://doi.org/10.1093/jxb/eraf496
18. Procko C, Lee T, Borsuk A, Bargmann BOR, Dabi T, Nery JR et al. (2022) Leaf cell-specific and single-cell transcriptional profiling reveals a role for the palisade layer in UV light protection. Plant Cell 34:3261–3279. https://doi.org/10.1093/plcell/koac167
19. Chen X, Yao Q, Gao X, Jiang C, Harberd NP, Fu X (2016) Shoot-to-root mobile transcription factor HY5 coordinates plant carbon and nitrogen acquisition. Curr Biol 26:640–646. https://doi.org/10.1016/j.cub.2015.12.066
20. Liu Z, Wang J, Zhou Y, Zhang Y, Qin A, Yu X et al. (2022) Identification of novel regulators required for early development of vein pattern in the cotyledons by single-cell RNA-sequencing. Plant J 110:7–22. https://doi.org/10.1111/tpj.15719
21. Wu Y, Lee S-K, Yoo Y, Wei J, Kwon S-Y, Lee S-W et al. (2018) Rice transcription factor OsDOF11 modulates sugar transport by promoting expression of sucrose transporter and SWEET genes. Mol Plant 11:833–845. https://doi.org/10.1016/j.molp.2018.04.002
22. Chow C-N, Yang C-W, Wu N-Y, Wang H-T, Tseng K-C, Chiu Y-H et al. (2024) PlantPAN 4.0: updated database for identifying conserved non-coding sequences and exploring dynamic transcriptional regulation in plant promoters. Nucleic Acids Res 52:D1569–D1578. https://doi.org/10.1093/nar/gkad945
23. Henry Y, Steer MW (1980) A re-examination of the induction of phloem transfer cell development in pea leaves (*Pisum sativum*). Plant Cell Environ 3:377–380. https://doi.org/10.1111/1365-3040.ep11581888
24. Stadler R, Sauer N (1996) The *Arabidopsis thaliana AtSUC2* gene is specifically expressed in companion cells. Bot Acta 109:299–306. https://doi.org/10.1111/j.1438-8677.1996.tb00577.x
25. Imlau A, Truernit E, Sauer N (1999) Cell-to-cell and long-distance trafficking of the green fluorescent protein in the phloem and symplastic unloading of the protein into sink tissues. Plant Cell 11:309–322. https://doi.org/10.1105/tpc.11.3.309
26. Schneidereit A, Imlau A, Sauer N (2008) Conserved cis-regulatory elements for DNA-binding-with-one-finger and homeo-domain-leucine-zipper transcription factors regulate companion cell-specific expression of the *Arabidopsis thaliana* SUCROSE TRANSPORTER 2 gene. Planta 228:651–662. https://doi.org/10.1007/s00425-008-0767-4
27. Yan H, Mendieta JP, Zhang X, Luo Z, Marand AP, Liang Y et al (2025) A single-cell rice atlas integrates multi-species data

to reveal cis-regulatory evolution. Nat Plants 11:2050–2071. https://doi.org/10.1038/s41477-025-02106-6
28. Ren Y, Zhang Z, Zhanakhmetova D, Li W, Chen S, Werner T, Liesche J (2024) Fast and simple fluorometric measurement of phloem loading exposes auxin-dependent regulation of Arabidopsis sucrose transporter AtSUC2. Plant J 120:2305–2318. https://doi.org/10.1111/tpj.17110
29. Tong C, Li C, Cao X-Y, Sun X-D, Bao Q-X, Mu X-R et al. (2022) Long-distance transport of sucrose in source leaves promotes sink root growth by the EIN3-SUC2 module. PLoS Genet 18:e1010424. https://doi.org/10.1371/journal.pgen.1010424
30. DeWitt ND, Harper JF, Sussman MR (1991) Evidence for a plasma membrane proton pump in phloem cells of higher plants. Plant J 1:121–128. https://doi.org/10.1111/j.1365-313X.1991.00121.x
31. Haritatos E, Ayre BG, Turgeon R (2000) Identification of phloem involved in assimilate loading in leaves by the activity of the galactinol synthase promoter. Plant Physiol 123:929–938. https://doi.org/10.1104/pp.123.3.929
32. Takagi H, Ito S, Shim JS, Kubota A, Hempton AK, Lee N et al. (2025) Companion cells with high florigen production express other small proteins and reveal a nitrogen-sensitive FT repressor. eLife 14:RP102529. https://doi.org/10.7554/eLife.102529
33. Collum TD, Padmanabhan MS, Hsieh Y-C, Culver JN (2016) Tobacco mosaic virus-directed reprogramming of auxin/indole acetic acid protein transcriptional responses enhances virus phloem loading. Proc Natl Acad Sci U S A 113:E2740–E2749. https://doi.org/10.1073/pnas.1524390113
34. Dasgupta K, Khadilkar AS, Sulpice R, Pant B, Scheible W-R, Fisahn J et al. (2014) Expression of sucrose transporter cDNAs specifically in companion cells enhances phloem loading and long-distance transport of sucrose but leads to an inhibition of growth and the perception of a phosphate limitation. Plant Physiol 165:715–731. https://doi.org/10.1104/pp.114.238410
35. Medberry SL, Olszewski NE (1993) Identification of cis elements involved in Commelina yellow mottle virus promoter activity. Plant J 3:619–626. https://doi.org/10.1046/j.1365-313X.1993.03040619.x
36. Gittins JR, Pellny TK, Biricolti S, Hiles ER, Passey AJ, James DJ (2003) Transgene expression in the vegetative tissues of apple driven by the vascular-specific *rolC* and CoYMV promoters. Transgenic Res 12:391–402. https://doi.org/10.1023/A:1024286405310
37. DinantS,RipollC,PieperM,DavidC(2004)Phloem specific expression driven by wheat dwarf geminivirus V-sense promoter in transgenic dicotyledonous species. Physiol Plant 121:108–116. https://doi.org/10.1111/j.0031-9317.2004.00296.x
38. Yin Y, Chen L, Beachy R (1997) Promoter elements required for phloem-specific gene expression from the RTBV promoter in rice. Plant J 12:1179–1188. https://doi.org/10.1046/j.1365-313X.1997.12051179.x
39. Yin Y, Zhu Q, Dai S, Lamb C, Beachy RN (1997) RF2a, a bZIP transcriptional activator of the phloem-specific rice tungro bacilliform virus promoter, functions in vascular-development. EMBO J 16:5247–5259. https://doi.org/10.1093/emboj/16.17.5247
40. McGarry RC, Ayre BG (2008) A DNA element between At4g28630 and At4g28640 confers companion-cell specific expression following the sink-to-source transition in mature minor vein phloem. Planta 228:839–849. https://doi.org/10.1007/s00425-008-0786-1
41. Xue H-C, Xu Z-G, Liu Y-J, Wang L, Ming X, Wu Z-Y et al (2025) A unified cell atlas of vascular plants reveals cell-type foundational genes and accelerates gene discovery. Cell 188:6370. https://doi.org/10.1016/j.cell.2025.07.036
42. Niu F, Rehmani MS, Yan J (2024) Multilayered regulation and implication of flowering time in plants. Plant Physiol Biochem 213:108842. https://doi.org/10.1016/j.plaphy.2024.108842
43. Golan G, Betzer R, Wolf S (2013) Phloem-specific expression of a melon Aux/IAA in tomato plants alters auxin sensitivity and plant development. Front Plant Sci 4:329. https://doi.org/10.3389/fpls.2013.00329
44. Spiegelman Z, Ham B-K, Zhang Z, Toal TW, Brady SM, Zheng Y et al. (2015) A tomato phloem-mobile protein regulates the shoot-to-root ratio by mediating the auxin response in distant organs. Plant J 83:853–863. https://doi.org/10.1111/tpj.12932
45. Noll GA, Fontanellaz ME, Rüping B, Ashoub A, van Bel AJE, Fischer R, Prüfer D (2007) Spatial and temporal regulation of the forisome gene *for1* in the phloem during plant development. Plant Mol Biol 65:285–294. https://doi.org/10.1007/s11103-007-9217-0
46. Noll GA, Rüping B, Ernst AM, Bucsenez M, Twyman RM, Fischer R, Prüfer D (2009) The promoters of forisome genes *MtSEO2* and *MtSEO3* direct gene expression to immature sieve elements in *Medicago truncatula* and *Nicotiana tabacum*. Plant Mol Biol Rep 27:526–533. https://doi.org/10.1007/s11105-009-0120-5
47. Ernst AM, Jekat SB, Zielonka S, Müller B, Neumann U, Rüping B et al. (2012) Sieve element occlusion (SEO) genes encode

structural phloem proteins involved in wound sealing of the phloem. Proc Natl Acad Sci U S A 109:E1980–E1989. https://doi.org/10.1073/pnas.1202999109

48. Bucsenez M, Rüping B, Behrens S, Twyman RM, Noll GA, Prüfer D (2012) Multiple cis-regulatory elements are involved in the complex regulation of the sieve element-specific *MtSEO-F1* promoter from *Medicago truncatula*. Plant Biol 14:714–724. https://doi.org/10.1111/j.1438-8677.2011.00556.x
49. Mahajan SK, Chisholm ST, Whitham SA, Carrington JC (1998) Identification and characterization of a locus (RTM1) that restricts long-distance movement of tobacco etch virus in *Arabidopsis thaliana*. Plant J 14:177–186. https://doi.org/10.1046/j.1365-313X.1998.00105.x
50. Whitham SA, Anderberg RJ, Chisholm ST, Carrington JC (2000) Arabidopsis *RTM2* gene is necessary for specific restriction of tobacco etch virus and encodes an unusual small heat shock–like protein. Plant Cell 12:569–582. https://doi.org/10.1105/tpc.12.4.569
51. Cayla T, Batailler B, Le Hir R, Revers F, Anstead JA, Thompson GA, Dinant S (2015) Live imaging of companion cells and sieve elements in Arabidopsis leaves. PLoS One 10:e0118122. https://doi.org/10.1371/journal.pone.0118122
52. Chisholm ST, Parra MA, Anderberg RJ, Carrington JC (2001) Arabidopsis *RTM1* and *RTM2* genes function in phloem to restrict long-distance movement of tobacco etch virus. Plant Physiol 127:1667–1675. https://doi.org/10.1104/pp.010479
53. Cayla T, Batailler B, Le Hir R, Revers F, Anstead JA, Thompson GA, Dinant S (2015) Live imaging of companion cells and sieve elements in Arabidopsis leaves. PLoS One 10:e0118122. https://doi.org/10.1371/journal.pone.0118122
54. Sanden NC, Schulz A (2022) Identification of new proteins in mature sieve elements. Physiol Plant 174:e13634. https://doi.org/10.1111/ppl.13634

Chapter 30

Quantifying Fire-Induced Phloem Damage

Raquel Partelli Feltrin

Abstract

Despite the phloem's crucial role in carbohydrate translocation and its high vulnerability due to its position just beneath the bark, studies directly addressing fire effects on phloem remain very limited. Thus, this chapter aims to provide a concise overview of the methodologies that have been used directly or indirectly to investigate the impacts of fire on the phloem, including changes in phloem compounds, transport velocity, and phloem cell damage.

Keywords Phloem, Fire effects, Phloem damage, Cambial death, Visual assessment, Phloem sap composition, Phloem velocity

1 Introduction

Phloem plays a critical role in transporting photosynthates from the foliage to growing tissues and storage organs. Any disruption to its function caused by disturbances, such as fires, can affect plant survival [1–3] and have significant implications for entire ecosystems [4]. Despite phloem's vital role and inherent vulnerability to heat due to its position just beneath the bark, the impacts of fire on phloem structure, function, and chemical composition have received limited attention. Thus, this chapter provides an overview of the methodologies used in the few studies that have directly or indirectly assessed the impact of fire on phloem tissue.

2 Methods Used to Study Fire Impacts on Phloem

2.1 Phloem Cell Viability

Phloem dysfunction caused by fire-induced damage can impact tree health and potentially lead to tree death due to the disruption of carbohydrate transport [1, 2]. However, permanent phloem transport disruption only occurs when cambial cells are killed around the entire circumference of the tree. One of the most important

Johannes Liesche and Sanna Sevanto (eds.), *Phloem*, Methods in Molecular Biology, vol. 3066,
https://doi.org/10.1007/978-1-0716-5488-0_30,

and well-studied factors in determining post-fire tree mortality is the extent of cambial cell damage in the tree stem. Two primary approaches are commonly used to evaluate fire-induced damage to cambium tissues: visual assessment based on natural coloration changes [1, 5, 6] and chemical staining tests [7, 8] (Table 1). Both methods rely on visible indicators of cell viability and generally offer a binary outcome, indicating whether the tissue cells are alive or dead, rather than detailed physiological or structural insights. Cambial cells cannot be distinguished by the naked eye, and when the visual assessment is performed, the inner bark (i.e., phloem and cambial cells) is assessed. Thus, based on the inner bark coloration, the viability of phloem and cambial cells is determined.

The natural coloration change consists of visually determining the changes in phloem coloration compared to healthy phloem tissue from unburned plants. For instance, healthy phloem tissue usually presents a light green coloration depending on the species, and dead tissue often appears darker, ranging in color from yellow to brown [1, 9]. In Fig. 1, healthy phloem can be visually identified by its light green coloration in panels A and B, indicating healthy cambial cells. In contrast, in panels C and D, the phloem shows a brownish coloration due to fire damage, indicating necrosis of phloem and cambial cells. To further assess these visual changes, samples can be collected by removing the outer bark and exposing the phloem tissue. Different tools can be used to collect samples, such as a sharp tool (e.g., a knife or an axe) or an increment borer [10]. Depending on the thickness of the inner bark, it may be possible to visually identify the depth of the cell necrosis when using the increment borer.

Table 1

Methods used for assessing fire impacts on phloem cell viability, sap composition, and transport velocity

Method	Parameter measured	Species	Conditions	Reference
EDTA phloem exudation	Sugar, amino acids, and proteins	*Pinus brutia*	Field and laboratory	[9]
Ethanol extraction and enzyme digestion	Sugar and starch	*Pinus ponderosa*	Laboratory	[2, 3]
^{15}N tracer	Transport velocity	*Pinus brutia*	Field and laboratory	[9]
Visual assessment—natural coloration	Cell viability	Various	Field and laboratory	[1, 5, 6]
Visual assessment—orthotolidine	Cell viability	Various	Field and laboratory	[10–12]
Visual assessment—TTC	Cell viability	Various	Laboratory	[7]

Fig. 1 Visual phloem cell assessment. (**A**) Healthy phloem (Ph) in mature *Tsuga heterophylla* exhibits a reddish coloration, while (**B**) in *Pinus ponderosa* sapling, healthy phloem appears light green. (**C**) Dark brown phloem coloration in a *Pinus ponderosa* sapling stem cross-section, indicating dead phloem tissue several days after fire exposure. (**D**) Cross-section of a *Pinus ponderosa* sapling stem showing both heat-damaged (brown) and healthy (light green) phloem tissue following fire exposure

Studies that used the visual assessment technique usually collect samples along the circumference of the tree to determine the extent of phloem–cambium tissue necrosis. For instance, sampling can be performed by dividing the tree trunk into four quadrants and sampling the tissue at the center of each quadrant [8]. Destructive sampling, in which the plant is harvested, allows phloem tissue to be assessed on the stem cross-section, providing a better evaluation of the extent of phloem and cambium necrosis around the circumference of the stem [1] (Fig. 1). However, if assessments are conducted immediately after fire exposure, visible signs of phloem damage may not yet be apparent. In such cases, chemical staining techniques can be used to detect cell viability.

Two chemical stains are commonly used: orthotolidine and triphenyl tetrazolium chloride (TTC). Orthotolidine tests for the presence of peroxidase enzyme activity can be performed in the field. The method involves applying a solution of orthotolidine to the exposed phloem tissue. Living cells oxidize the reagent, resulting in a blue coloration, while dead cells remain uncolored [10, 11]. The sampling and exposure of the phloem tissue can be performed using sharp tools, an increment borer, or a stem cross-section, as described in the phloem natural coloration method. TTC is a more sensitive laboratory-based method for assessing phloem–cambium viability. Samples can be collected by cutting a section of the tree stem using a chainsaw [7] or a drill fitted with a cutting tool. In the laboratory, small cross-sections of the phloem–cambium tissue (0.1–4 mm) are subjected to vacuum infiltration with TTC solution, followed by incubation of the samples in the dark for approximately 24 h. Viable cells reduce TTC to red formazan, turning the tissue red, while non-viable cells remain uncolored [7, 12, 13].

2.2 Phloem Compounds

When photosynthate transport disruption occurs from the leaves to the roots due to heat-induced stem girdling, tree mortality can occur. In addition to the demand for non-structural carbohydrates (NSC), such as soluble sugars and starch, for growth and maintenance of metabolism, the depletion of NSC in trees exposed to fire can potentially indicate a higher demand for tissue repair [1] and mortality [2, 3].

A protocol [14] commonly used in studies to assess changes in a tree's NSCs caused by drought (e.g., [15–17]) has been applied to determine changes in NSCs in trees' phloem exposed to fire [2, 3]. As described in refs. [2, 3], inner bark samples (phloem and cambium tissue) are collected in at least two different locations along the circumference of the stem using a phloem punch at 1–1.5 m height in the tree stem. As previously described in Sect. 2.1, in these studies, phloem was visually analyzed to determine that the tissue collected was alive. For studies aiming to determine carbohydrate depletion, particularly when periodic sampling is planned, it is important that the phloem tissue is functioning. Although there is no evidence, it is possible that high temperatures may cause phloem NSC degradation, resulting in an initial decline, followed by long-term stabilization of NSC concentration as the tissue becomes non-functional, which could lead to misleading results.

Briefly, the protocol [14] used to analyze NSCs in the studies [2, 3] involves the extraction of soluble sugars using ethanol and enzymatic digestion of starches. Although the protocol describes the use of high-performance liquid chromatography (HPLC) to determine the concentration of soluble sugars and starch, it is also possible to use a spectrophotometer with wavelengths of 340 nm [2, 3]. See [14] for more details on how to prepare samples to determine soluble sugars and starch.

2.3 Phloem Sap Composition and Velocity

Abiotic stresses such as drought and fire can alter phloem sap composition [18, 19], and these alterations can indicate how plants respond to such stresses. For instance, flame-induced accumulation of non-structural (NS) proteins, such as heat shock proteins (HSP), can signal to sinks that stress has occurred, triggering the cessation of growth and a reduction in carbon assimilation, as suggested in [19]. In addition, an increase in phloem velocity may occur due to the need to shift to stress metabolism in the sink, and sugar is manipulated, sending a signal for stomatal closure [19]. However, NS proteins and phloem velocity tend to decrease as trees acclimate. Although phloem sap composition and velocity can improve our understanding of plant responses to fire and recovery, only one study [19] has investigated this, likely due to the challenges of obtaining uncontaminated phloem sap [20] and measuring phloem velocity (see Part III).

Although the cited study [19] appears to be the only one that has examined fire impacts on phloem sap composition and velocity, it employed well-established techniques such as EDTA phloem exudation (see Chap. 13). For phloem exudation, small pieces of bark (outer and inner bark) are sampled from twigs near the burned location of the branch, i.e., unburned twigs. The study did not describe the size of the bark pieces collected, but for phloem exudation, bark samples weighing 150–200 mg were incubated in EDTA solution. It is important to note that phloem transport relies on living cells, such as sieve elements and companion cells [21], which are highly susceptible to heat damage during fire. Similar to assessing phloem compounds, if phloem sap composition is analyzed from tissue where cell death has occurred, the results may be misleading, since compound concentrations might show an initial decrease or little change due to no active transport and cell activity. Thus, branches and twigs showing signs of fire scorch, such as darkened, cracked, or charred bark, and damaged foliage (either grayish or brown in color or consumed) should be avoided. In addition, phloem cell viability can be used to detect if there are changes in phloem coloration in the samples collected for phloem exudation. Particularly in the study [19], spectrophotometry was used to determine the concentration of different compounds in the phloem sap (see [19] for solution preparations and wavelength used for each compound measurement). However, other analytical techniques can be used to quantify phloem sap composition, such as gas chromatography–mass spectrometry (GC–MS), HPLC, chromatography, and others (see [22]).

Furthermore, the ^{15}N tracer method can be used to assess phloem velocity in trees exposed to fire. The method consists of cutting thin bark flaps (<2 cm long) from unburned twigs (<20 cm long) and submerging them for 30 min in an aqueous solution composed of ^{15}N glutamine, sucrose, EDTA, and chloramphenicol [19]. If the exposure is performed in the field, the solution should be placed in capped vials and transported frozen. Subsequently, the exposed twig should be cut into smaller pieces (1 cm long), placed in different Eppendorf vials, and stored in liquid N. Each twig section should be carefully labeled to ensure that the distance traveled by the isotope can be accurately tracked. In the laboratory, the bark, including the phloem, of each twig section should be separated from the wood and used to determine ^{15}N enrichment using isotope ratio mass spectrometry (see ref. [19] for solution and sample preparation). The ^{15}N content in each bark section indicates how far the isotope traveled through the phloem during the exposure period, i.e., the phloem velocity.

3 Conclusions

The aim of this chapter was to provide a summary of the methods that have been used to assess fire effects on phloem. For more details on the methodology, the reader is referred to the references in Table 1. Despite the phloem's critical role in carbohydrate transport and its high susceptibility to heat damage, our understanding of how fire impacts its structure and function remains limited. However, different techniques for studying phloem structure and functioning, such as those described in this book (e.g., chapters in Parts I and III), can be effectively applied to plants exposed to fire. Applying these approaches will help advance our understanding of fire impacts on phloem structure, functioning, and composition, as well as the mechanisms underlying post-fire tree mortality, recovery, and survival.

Acknowledgments

I thank Sanna Sevanto for proofreading and valuable comments.

Conflict of Interest The authors have no conflicts of interest to declare that are relevant to the content of this chapter.

References

1. Partelli–Feltrin R, Smith AMS, Adams HD, Alex Thompson R, Kolden CA, Yedinak KM, Johnson DM (2022) Death from hunger or thirst? Phloem death, rather than xylem hydraulic failure, as a driver of fire–induced conifer mortality. New Phytol 237:1154–1163. https://doi.org/10.1111/nph.18454
2. Reed CC, Hood SH (2024) Non structural carbohydrates explain post-fire tree mortality and recovery patterns. Tree Physiol 44(2):1–12. https://doi.org/10.1093/treephys/tpad155
3. Reed CC, Hood SH, Ramirez AR, Sala A (2025) Fire directly affects tree carbon balance and indirectly affects hydraulic function: consequences for post–fire mortality in two conifers. New Phytol 247:595–611. https://doi.org/10.1111/nph.70212
4. Salmon Y, Dietrich L, Sevanto S, Hölttä T, Dannoura M, Epron D (2019) Drought impacts on tree phloem: from cell–level responses to ecological significance. Tree Physiol 9(2):173–191. https://doi.org/10.1093/treephys/tpy153
5. Partelli–Feltrin R, Johnson DM, Sparks AM, Adams HD, Kolden CA, Nelson AS, Smith AMS (2020) Drought increases vulnerability of *Pinus ponderosa* saplings to fire–induced mortality. Fire 3:56. https://doi.org/10.3390/fire3040056
6. Steady WD, Partelli–Feltrin R, Johnson DM, Sparks AM, Kolden CA, Talhelm AF, Lutz

JA, Boschetti L, Hudak AT, Nelson AS, Smith AMS (2019) The survival of *Pinus ponderosa* saplings subjected to increasing levels of fire behavior and impacts on post-fire growth. Fire 2:23. https://doi.org/10.3390/fire2020023
7. Bova AS, Dickinson MB (2005) Liking surface fire behavior, stem heating, and tissue necrosis. Can J For Res 35:814–822
8. Hood SM, Cluck DR, Smith SL, Ryan KC (2008) Using bark char codes to predict post-fire cambium mortality. Fire Ecol 4(1):57–73
9. Ryan KC (1982) Techniques for assessing fire damage to trees. In: Lotan J (ed) Proceedings of the symposium: fire, its field effects. Intermountain Fire Council, pp 1–11
10. Hare RC (1965) Chemical test for fire damage. J For 63(12):939
11. Ryan KC, Peterson DL, Reinhardt ED (1988) Modeling long-term fire-caused mortality of Douglas-fir. For Sci 34(1):190–199
12. Parker J (1953) Some applications and limitations of tetrazolium chloride. Science 118(3055):77–79
13. Lopez Del Egido L, Navarro-Miró D, Martinez-Heredia V, Toorop PE, Iannetta PPMA (2017) Spectrophotometric assay for robust viability testing of seed batches using 2,3,5-triphenyl tetrazolium chloride: using Hordeum vulgare L. as a model. Front Plant Sci 16(8):747. https://doi.org/10.3389/fpls.2017.00747
14. Landhäusser SM, Chow PS, Dickman LT, Furze ME, Kuhlman I, Schmid S, Wiesenbauer J, Wild B, Gleixner G, Hartmann H, Hoch G, McDowell NG, Richardson AD, Richter A, Adams HD (2018) Standardized protocols and procedures can precisely and accurately quantify non-structural carbohydrates. Tree Physiol 38(12):1764–1778
15. Arend M, Link RM, Zahnd C, Hoch G, Schuldt B, Kahmen A (2022) Lack of hydraulic recovery as a cause of post-drought foliage reduction and canopy decline in European beech. New Phytol 234:1195–1205. https://doi.org/10.1111/nph.18065
16. Peltier DMP, Carbone MS, McIntire CD, Robertson N, Thompson RA, Malone S, LeMoine J, Richardson AD, McDowell NG, Adams HD, Pockman WT, Trowbridge AM (2023) Carbon starvation following a decade of experimental drought consumes old reserves in *Pinus edulis*. New Phytol 240:92–104. https://doi.org/10.1111/nph.19119
17. Netherer S, Lehmanski L, Bachlehner A, Rosner S, Savi T, Schmidt A, Huang J, Paiva MR, Mateus E, Hartmann H, Gershenzon J (2024) Drought increases Norway spruce susceptibility to the Eurasian spruce bark beetle and its associated fungi. New Phytol 242:1000–1017. https://doi.org/10.1111/nph.19635
18. Blicharz S, Beemster GTS, Ragni L, De Diego N, Spíchal L, Hernandiz AE, Marczak Ł, Olszak M, Perlikowski D, Kosmala A, Malinowski R (2021) Phloem exudate metabolic content reflects the response to water-deficit stress in pea plants (*Pisum sativum* L.). Plant J 106:1338–1355. https://doi.org/10.1111/tpj.15240
19. Alexou M, Dimitrakopoulos AP (2014) Early physiological consequences of fire as an abiotic stressor in metabolic source and sink of young Brutian pine (*Pinus brutia* Ten.). Tree Physiol 34:1388–1398. https://doi.org/10.1093/treephys/tpu098
20. Killiny N (2019) Collection of phloem sap, pros and cons. Plant Signal Behav 14(8):1618181. https://doi.org/10.1080/15592324.2019.1618181
21. De Schepper V, De Swaef T, Bauweraerts I, Steppe K (2013) Phloem transport: a review of mechanisms and controls. J Exp Bot 64(16):4839–4850. https://doi.org/10.1093/jxb/ert302
22. Broussard L, Abadie C, Lalande J, Limami AM, Lothier J, Tcherkez G (2023) Phloem sap composition: what have we learnt from metabolomics. Int J Mol Sci 24:6917. https://doi.org/10.3390/ijms24086917

Chapter 31

Methods for Assessing the Role of Phloem Transport in Plant Stress Responses

Sanna Sevanto

Abstract

Delivery of carbohydrates to tissues under stress is important for plant defenses, recovery, and survival. Yet, little is known on how phloem function is altered under stress, and how that influences plant responses to stress. This chapter summarizes current and emerging methods for quantifying phloem transport and assessing how environmental stress affects phloem function at whole plant scale. The techniques discussed range from isotopic and fluorescent tracer studies to advanced non-invasive imaging (MRI, NMR, and PET), carbon balance approaches, anatomical analysis, and stem diameter monitoring. Emphasis is placed on the methodological principles, experimental design, and applicability of each technique to stress studies.

Keywords Anatomy, Carbon balance, Drought, Hydraulic conductivity, Isotope, NMR, PET, Pathogen, Phloem, Stress, Stem diameter variation, Tracer

1 Introduction

Stomatal regulation of water loss is a well-known first response of plants to environmental stressors such as drought, freezing, and pathogen or insect attack [1–7]. Stomatal closure limits water loss and can delay mortality, but it concurrently restricts carbon uptake, influencing phloem transport and carbon balance at plant and ecosystem scales [8, 9]. Plant species vary widely in stomatal sensitivity and hydraulic architecture, shaping their ecological distributions and competition [10–12]. Traditionally, stomatal control of water loss in trees has been linked to the anatomy and resilience or efficiency of the plant hydraulic system [2]. In recent years, increasing attention has been paid to how phloem transport interacts with the hydraulic system, and how plant carbon reserves and their redistribution affect plant survival under stress [13–15]. Answering whether, and under what conditions, phloem transport ceases under stress, and how that affects plant survival or recovery, requires precise measurements of various parameters. This chapter reviews techniques developed to

Johannes Liesche and Sanna Sevanto (eds.), *Phloem*, Methods in Molecular Biology, vol. 3066,
https://doi.org/10.1007/978-1-0716-5488-0_31,

assess phloem function at whole plant level, mainly in trees. Many of the techniques are described in more detail in other chapters of this book. Here, the focus is on their application under stress and outlining how these methods inform our understanding of plant carbon allocation and drought resilience. Before reviewing the methods, this chapter discusses current knowledge and hypotheses regarding responses of phloem transport to stress. Thereby, the chapter provides a basis for choosing the best method for different studies.

2 Phloem Transport Responses to Stress

Phloem transport responses to stress reflect both hydraulic limitations and metabolic adjustments. Under moderate drought, metabolic adjustments might slow transport down via reduced sink activity because growth ceases before photosynthesis [16, 17] although in Arabidopsis thaliana and other herbs, phloem loading might continue despite cessation of photosynthesis under moderate drought [18, 19]. Severe drought further constrains flow through hydraulic coupling between the xylem and the phloem. As xylem water potential decreases, phloem must adjust osmotically to maintain turgor [20–25]. Without adjustment, water would exit the phloem, causing loss of turgor in the living cells and possible transport failure. This osmoregulation occurs primarily via carbohydrate and ion concentration changes [26, 27]. Increasing carbohydrate concentration in the phloem can increase the flow resistance via increasing the viscosity of the phloem sap [28, 29]. This effect is particularly important in the context of sucrose, the dominant transport sugar [30]. Viscosity rises exponentially with increasing sucrose concentration and may impede flow at water potentials below –4 MPa when it reaches ~20 times that of the pure water [29]. Experimentally, an increase in phloem sap viscosity has been observed in response to prolonged drought but so far, only to an extent that has a limited impact on the axial hydraulic conductivity of the phloem [31, 32]. Plants could mitigate the viscosity buildup by enhancing membrane permeability (e.g., via aquaporins) [33–35]. However, higher permeability increases water exchange between phloem conduits and their surroundings, thereby increasing vulnerability to turgor loss and creating a trade-off between permeability and transport stability [22, 23] unless applied only very locally. High hydraulic permeability also leads to high sensitivity of the phloem tissue to changes in xylem water potential. Thus, plants with high membrane permeability need a high capacity for fast osmoregulation to avoid excessive water loss from the phloem [23]. One way of increasing transport stability could be a phloem relay structure [20, 24, 36], where carbohydrates are unloaded and reloaded to the conduits at regular intervals along the entire transport pathway. While hypothesized, this mechanism has not been experimentally demonstrated.

Plants can also use alternative osmolytes such as potassium or hexoses [20, 37, 38] to regulate water balance with lower viscosity costs. Indeed, the proportion of potassium in phloem sap has been found to increase during water stress [37]. Substitution of sugars with other osmolytes in the phloem transport stream requires sugar storage in the leaves while photosynthetic products are waiting to be exported. Supporting this as a possible coping mechanism, a difference in sugar storage capacity between trees that continue photosynthesis at low leaf water potentials (anisohydric) and trees that close stomata at high water potentials (isohydric) during drought has been observed [39]. Generally, the various observations supporting different strategies for maintaining phloem transport highlight the potential variability in functional strategies and operational modes under stress. Differences in these strategies between species could contribute to the diversity of observed drought tolerance [22, 39]. Ultimately, theoretical and empirical evidence suggest that phloem performance under stress depends on the interplay of viscosity, conduit permeability, and local carbohydrate storage capacity. These parameters underlie the development of experimental and modeling approaches reviewed in the following sections.

3 Experimental and Analytical Methods for Studying Phloem Transport at Whole Plant Scale

3.1 Tracer and Imaging Methods

Tracer Methods

Principle

Tracer methods provide the most direct means to determine whether and how phloem transport responds to stress (Fig. 1). In these methods, a tracer that is assumed to travel with phloem flow and pass cell walls and membranes or sieve plates/areas as phloem sap is applied into the phloem. The appearance or velocity of tracer signal reflects the functioning of the transport pathway [40–46] (Fig. 1). The tracers vary from isotopes of carbon and water, or other metabolically important substances, to fluorescent dyes, fluorescent viruses, heat, and nanoparticles [40–46]. The detection methods are adapted to the particular tracer and its detectable properties. When carefully designed, tracer experiments can reveal tracer travel times from which phloem flow velocities [42, 43, 47] as well as tissue metabolic activity can be calculated [48, 49]. Combined with measurements of cellular pressure and phloem anatomy, they have been invaluable in increasing our knowledge on phloem function and potential limitations of the transport system [50–52]. Two persistent challenges include: (1) inserting tracers with minimal tissue damage, and (2) choosing

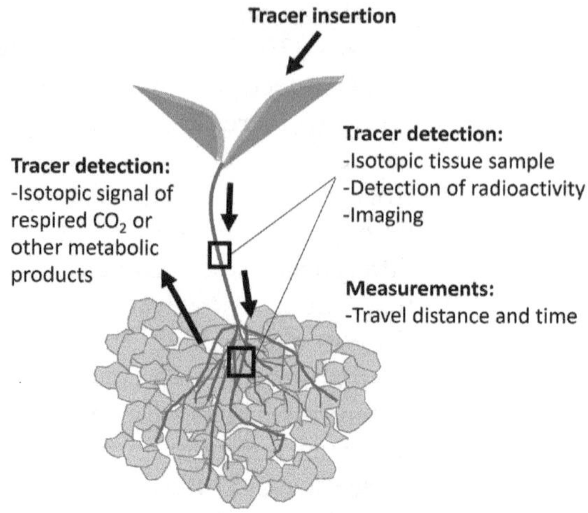

Fig. 1 A Schematic presentation of tracer methods for detecting phloem function. In these methods, a suitable tracer that tracks phloem flow, carbon movement, water movement, or metabolism is inserted to the plant, here into an upper leaf. Its propagation, indicated by black arrows, is followed by a suitable sensor at selected locations indicated here by black boxes one in the stem and one in the roots. The tracer can be tracked directly from different plant tissues by tissue sampling and chemical analysis, or by non-destructive imaging. Or it can be tracked from metabolic byproducts such as respired CO_2. Measurements focus on the travel distance and time of the tracer through the plant. This diagram explains how tracers track the movement and metabolism of substances within a plant, helping to study plant physiology and nutrient flow

tracers that faithfully follow bulk phloem flow. Environmental stress can exacerbate both challenges by altering loading rates and sink demand. For instance, drought often delays labeling of newly assimilated carbon or alters storage–remobilization dynamics [32, 53, 54], causing delays in phloem loading of newly assimilated carbohydrates and altering carbon storage and remobilization along the pathway [32, 53, 54]. This complicates the interpretation of isotope tracing results and has raised concerns about the reliability of carbon isotopic tracers as indicators of phloem transport velocity or carbon allocation speed under stress [55].

Commonly Used Tracers

Isotopic tracers, particularly ^{11}C, ^{13}C, and ^{14}C have long been used to map carbon allocation [56–65]. Stable isotopes permit field experiments but require destructive tissue sampling for mass-spectrometric analysis, limiting temporal resolution. Radioactive isotopes allow non-destructive, high-resolution tracking with radiation monitors but are generally restricted to controlled environments due to safety regulations [56, 64, 65]. Nevertheless, these approaches have revealed temperature and oxygen effects on phloem loading [56, 66, 67] and have been central to understanding drought impacts on carbon partitioning [32, 68, 69].

Heavy-water (D_2O) tracing provides complementary insights into xylem–phloem water exchange [70–72]. Using neutron or NMR imaging, D_2O movement can be visualized with high spatial and temporal resolution, yielding information on bulk flow and hydration status (see Chapt. 27). The caveat for phloem transport monitoring is that D_2O tracks the movement of water, not carbohydrates. To estimate carbohydrate movement based on water flow rates, an assumption that carbohydrates travel with water, or additional carbon tracking methods are needed. Because of current resolution limits in imaging systems, D_2O is most useful for assessing whole-tissue dynamics rather than single-conduit flow.

Fluorescent dyes and nanoparticles allow single-conduit visualization in herbaceous plants [40, 43, 73, 74]. Their advantages include real-time imaging and direct velocity measurement. These methods, however, have not been tested with plants under water stress, although phloem wounding responses have been addressed [75]. Their limitations include adsorption to cell walls, which can confound signal propagation, which can be exacerbated under drought or other stresses. Many dyes, including nanoparticles, tend to stick to plant cell walls or membranes between adjacent conduits [45]. This diversion from tracking the flow can be enhanced by lack of water. Careful planning, however, can make these label studies at least indicative of whether transport occurs or not under stress, even if the direct amount of transport or transport velocities could not be determined. Advances in dye chemistry now enable combined tracking of transport and unloading [40], improving data interpretation at whole plant level, which would be very helpful in drought response studies.

Medical Imaging Adaptations

An interesting development in studies of plant function is applying medical imaging techniques on plants. These techniques generally allow higher temporal and spatial resolution than traditional tracer detection methods, but are mostly limited to studies on potted, small size plants. Miniaturization of detectors, e.g., improved low-field NMR magnets [70] are making these approaches more scalable, although technical development in turning this to MRI imaging is still needed. Combining these methods with pressure or metabolite sensing would enhance mechanistic understanding of phloem stress responses.

The methods that have transformed phloem research [70, 71, 72, 73–92] include:

- **Magnetic Resonance Imaging (MRI/NMR)** that detects low-mass nuclei (H, C, and N) and provides non-destructive visualization of water content and movement. Flow-sensitive NMR sequences or tracer-based methods (using ^{13}C or D_2O) enable quantification of sap velocities and diurnal flow dynamics

[47, 81, 82]. Portable low-field NMR systems now permit long-term monitoring of tree stems in situ [77–80].

- **NMR spectroscopy** that identifies metabolite changes—sugars, amino acids, organic acids, and lipids—associated with stress or altered transport [49]. Several plant metabolomics databases exist to assist in the interpretation of NMR spectral data [49]. Related to plant stress, NMR spectroscopy techniques have mostly been applied to identify plant chemical responses to pathogenic microbes, UV-B light, or threats to reproductive success in crops. These techniques, however, can reveal interesting patterns of sugar allocation and provide insights at the scale of chemical process identification unavailable with other techniques [83]. Therefore, they would be specifically suitable for identifying changes in cellular-level biochemistry under stress.
- **Positron Emission Tomography (PET)** that uses short-lived positron-emitting isotopes (^{11}C, ^{13}N, ^{15}O, and ^{18}F) to map active transport and metabolic hotspots in three dimensions [85–90]. Combined PET–MRI systems [91] can simultaneously image structure and metabolism, offering unprecedented integration.

3.2 Carbon Balance Methods

Principle

Under stress, the possible occurrence of phloem transport or the importance of carbohydrate dynamics to plant stress responses and survival can be implied from studies of non-structural carbohydrate (NSC) pools of different tissues [39, 53, 93, 94]. The carbon balance method infers phloem activity by quantifying changes in NSC pools within tissues relative to photosynthetic input and respiratory output [39, 53, 93, 94, Chap. 26). By conservation of mass, an increase or depletion in NSCs indicates altered carbon flow (Fig. 2). Closing the carbon budget with knowledge on tissue-level NSC use rate, i.e., respiration, will help with data interpretation. When combined with gas-exchange measurements, this approach can be used to reveal whether stress-induced changes in carbon allocation reflect reduced transport or local metabolic demand. NSC accumulation combined with continued photosynthesis indicates sink limitation or restricted phloem function because of dehydration [13, 17, 53, 95]. NSC deficit would indicate source limitation and no or insignificant restriction of phloem transport. Overall, the carbon-balance method is inexpensive, widely applicable, and field-compatible, but indirect. It assumes that photosynthesis and respiration rates are known and cannot distinguish local from transported carbon without complementary measurements (e.g., tracers). Nonetheless, it remains an invaluable tool for scaling physiological measurements to ecosystem and modeling studies.

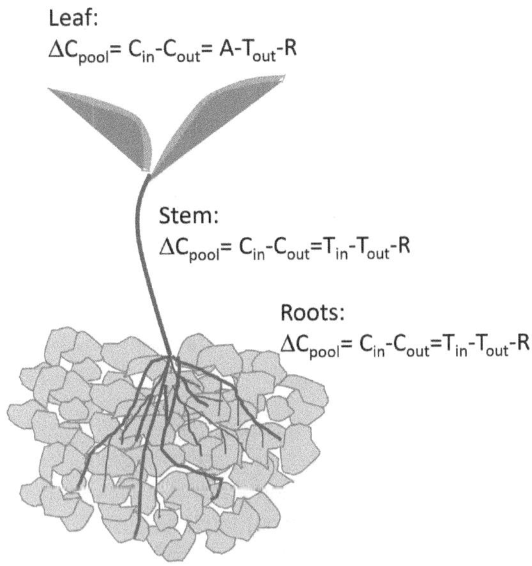

Fig. 2 A schematic presentation of the basis of carbon balance methods for detecting phloem function. In these methods, changes in carbon pools (ΔC_{pool}) of different tissues are measured and carbon allocation patterns and potential phloem transport are deducted from this information based on the mass balance equation and known carbon fluxes in and out of the tissue (C_{in} and C_{out}, respectively). These fluxes may consist of the carbon assimilation rate (A), the phloem transport rate (T), and the local carbon consumption rate approximated by respiration (R). The leaf formula states that the change in carbon pool equals carbon in minus carbon out, which also equals assimilation minus transport out minus respiration. The stem and roots formulas add that carbon in minus carbon out equals transpport in minus transport out minus respiration

Based on carbon balance studies, we know that the carbon pool responses of plants under drought follow a three-phase pattern where a carbon surplus in leaves due to sink limitation [13] is followed by a carbon deficit once stomata are closed [17]. This is then followed by a carbon limitation where access to carbohydrate reserves either via reduced phloem transport or general lack of water to facilitate use of reserves [53, 95, 96], finally results in tissue mortality [53] (Fig. 3). The sink limitation is due to the fact that growth decreases under drought before photosynthesis [17, 97]. The carbon deficit results from respiration continuing after photosynthesis stops [98]. After that, plants survive until something in the integrity of the water and carbon transport system gives in [53, 95]. How fast these phases follow each other depends on the plant and the experimental setup. A mild stress experiment may never proceed beyond the carbon surplus/sink limitation phase (*see* 13), while a severe drought might lead to plants going through these phases at different speeds depending on species [94] or even the individual [95]. Interestingly, plant mortality under drought usually occurs before carbohydrate

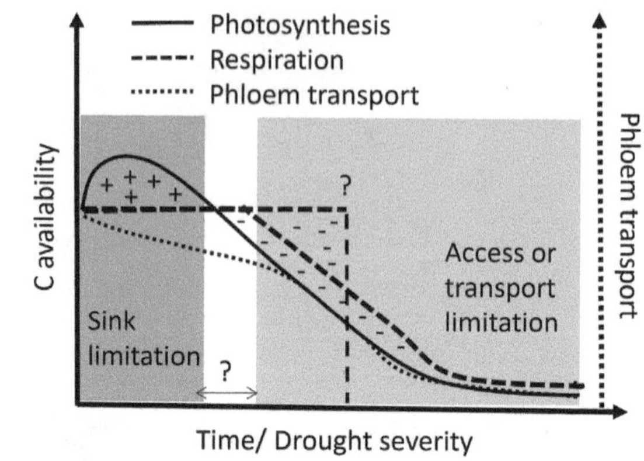

Fig. 3 Drought responses of carbon reserves in plant tissues depend on the severity and duration of drought. Two shaded areas mark phases: early drought causes sink limitation where carbon use is restricted despite availability, and later drought leads to access or transport limitation where carbon availability and transport are both low. Question marks indicate uncertainty in the transition between these phases. During mild drought, carbohydrate reserves may increase (+ signs indicate positive carbon balance) because growth ceases earlier than photosynthesis under stress resulting in a sink limitation for carbon transport. After stomatal closure, the carbohydrate reserves start to decline (– sign to indicate negative carbon balance), by the rate of declining depends on respiration rate, and the plant's ability to access carbohydrate reserves and transport carbon. The exact timing of stress severity at which plants go through these phases is unknown. Similarly, we don't know why some plants can access their carbohydrate reserves better than others, or why some plants can reduce their respiration rates under water stress. Most plants die before consuming all of their carbohydrate reserves

reserves are depleted [99], but well-watered, carbon-starved plants can use their NSC reserves almost entirely [53, 95]. This emphasizes the importance of water availability for access to carbohydrate reserves. Access to carbohydrate reserves, however, may not always increase plant survival time. Well-watered carbon-starved plants tend to consume their carbohydrate reserves faster than plants under drought [53, 95]. The longest survival time is observed for plants that can access their NSC reserves relatively well, but at the same time reduce their respiration rates (metabolic adjustment) with declining water availability [95].

Integration with Modeling

One of the advantages of the carbon balance method in evaluating the importance of phloem transport on plant stress responses and survival is its easy translation to dynamic modeling [100]. Mass balance-based models can simulate carbon allocation patterns and carbon pool dynamics at plant, canopy, and ecosystem scales [101–103]. They allow testing the effects of different parameterizations and controlling hypotheses on plant-scale carbon dynamics [20, 101, 104], as well as scaling the results up to canopy- and even ecosystem scale [103]. Some of these models explicitly describe phloem transport [100, 101, 104, 105] and allow testing hypotheses about phloem or plant structural parameters on plant function [106–109]. Others track changes in carbohydrate pools between tissues (103, *see* also 102), assuming no limitation to carbohydrate transport as long as the pools are not empty. Models in the latter class are currently the only ones computationally feasible for global-scale dynamic vegetation modeling. Even if lack of explicit description of phloem transport can be seen as a limitation, these models benefit from large amounts of calibration and benchmarking data from the eddy-covariance technique and a worldwide network of ecosystem measurement stations (e.g., Fluxnet). This data allows parameterizing the models to represent current conditions robustly [103]. Most of the measurement stations, however, are located in relatively moist environments, and the datasets contain few plant stress manipulation studies. The power of these models to predict plant fate under stress or the control points of plant carbon transport and use would thus benefit from data of ecosystem-scale plant stress studies [109].

3.3 Anatomical and Biophysical Assessments

Principle

Anatomical analyses can be used to link structural features to potential functional strategies under stress [22, 29, 32]. Typically, these analyses use fluid mechanical and biophysical principles to calculate phloem function-related parameters based on anatomical features such as conduit diameter, phloem thickness, or number of conductive cells and sieve plate structure [20, 110, 111]. Combined with phloem transport modeling, they can be used to indirectly evaluate the role of phloem transport under stress. For example, evidence of anatomical adaptations or acclimation that increase phloem transport capacity, such as increased phloem conduit diameter, or the amount of phloem tissue [15, 20] in plants grown under stress treatments, would indicate that continued phloem transport under stress is so important for plants that it has influenced anatomy [32]. Similarly, evidence of large carbohydrate storage tissues surrounding the phloem or individual phloem conduits could indicate adaptation or acclimation to serving osmoregulatory needs to avoid turgor loss and maintain hydraulic equilibrium between the xylem and the phloem under stress. Indeed, desiccation-tolerant (anisohydric) species often exhibit thicker phloem, greater inter-conduit connectivity, and more parenchyma buffering than their desiccation intolerant (isohydric) counterparts [29, 39].

Unfortunately, large surveys across plant types, species, and growth environments addressing stress-related hypotheses of phloem structure are currently lacking, and the only experimental study on the topic [29] gives somewhat obscure results. As to drought stress responses, the main unanswered questions that would determine what kind of anatomical features should be considered when evaluating phloem transport capacity are 1) to what degree the phloem conduits are directly hydraulically connected to the xylem, and/or the surrounding cells and 2) how much sieve plates increase flow resistance [110, 112]. Answering these questions would help to determine whether, for a certain species, the main limitation for phloem transport under drought is viscosity build-up or the potential of turgor loss, and how quickly transport limits might be crossed.

Techniques

Phloem anatomy is challenging to study because of the softness of the tissue makes sectioning and sample mounting challenging and the lack of contrast between living cells, cell walls, and lumen areas makes identifying functional conduits challenging (see Chaps. 1, 3, and 16 for methods). To obtain representative results of conduit dimensions, special care needs to be taken to retain tissue moisture and the original or close to original tissue dimensions for cell size measurements.

Light microscopy with staining to help differentiate between cell types has provided important information on the cellular composition, but identification of the cell types often requires sectioning both in horizontal and vertical directions to allow cell length guide conduit identification [74] and provide all the dimensional measurements needed for flow modeling (conduit diameter, length, tapering, and flow resistance of different elements). Light microscopy as well as electron microscopy have also been used to study structures of sieve plates that contribute to the resistance of phloem flow, and testing hypotheses about phloem transport mechanisms in plants [111, 113].

With the development of phase contrast imaging in X-ray tomography, 3D imaging of non-lignified phloem tissues has become available. Generally, X-ray tomography lends itself readily to studies of xylem anatomy because of the high contrast produced by lignified xylem conduit walls [114, 115]. In the soft phloem with low x-ray attenuation, obtaining enough contrast is challenging. In addition to traditional detection of x-ray transmission through the material, phase contrast imaging measures the phase shift in the radiation waves after interaction with the material. These phase shifts occur even if the material did not absorb significant amounts of x-ray energy and thus make the soft materials visible [116]. In the phloem, the cellular structure becomes clearly visible (Fig. 4), allowing 3D structural imaging, even if the low contrast does not allow automated sectioning and measurements of conduit or cell size.

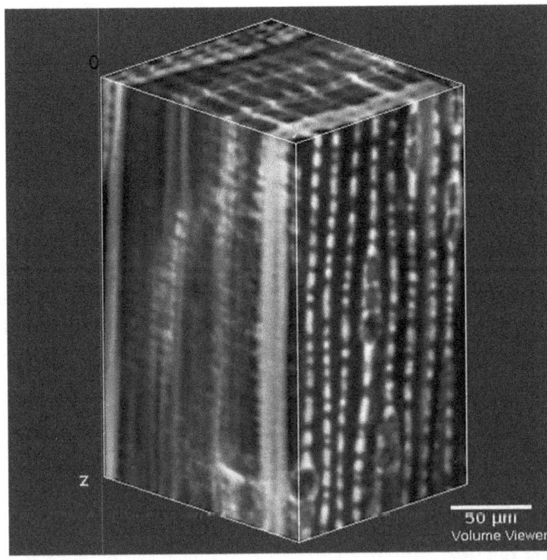

Fig. 4 A 3D x-ray tomography image of one-seed juniper (*Juniperus monosperma*) phloem from radial and tangential viewpoints. The radial surface (right) is characterized by conduit walls perforated by sieve areas and ray cells showing as clusters of 2–3 cell layers traveling through the tissue. The tangential and axial surfaces (left and top) show a structure where two conduits are separated by a layer of living cells, but separated from other such groups by thicker cell walls. The brightest areas indicate more lignified cells. Fibers in the phloem, and the xylem-phloem boundary at the top of the image

3.4 Stem Diameter Fluctuations

Principle

An interesting method for extracting information about phloem function and potential transport is detecting diurnal fluctuations in stem diameter. Tree stem diameter fluctuates diurnally with an amplitude of the order of magnitude of 10 μm [117, 118]. This fluctuation is due to the increasing tension in the tissue when water loss resulting from transpiration exceeds water input from the soil [119]. Initially, when the time resolution of dendrometric measurements for tree growth improved, these fluctuations were considered an artifact in the measurements that should be corrected for [120]. Soon, however, it was realized that they provide valuable information on the water dynamics of trees [121–123], and several methods for detrending the diameter fluctuation data from growth effects have been developed (124, 125, and *see* 126 for a review). While there still is some debate about what portion of the fluctuations are due to shrinking and swelling of the bark vs the xylem, and whether the shrinkage in the xylem is due to ray cell dehydration [127, 128], there is increasing evidence that simultaneous measurements of xylem and whole stem (measurement over bark) allow for calculating fluctuations in the size of the phloem tissue (here phloem includes the

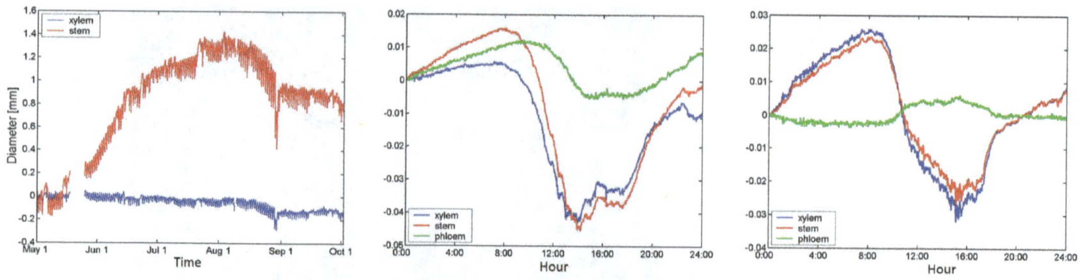

Fig. 5 An example of simultaneous xylem and whole stem measurements of tree stem diameter fluctuation on a Scots pine tree (left), and examples of phloem diameter fluctuations where phloem shrinks with the xylem (middle) and swells while the xylem shrinks (right) on a maple tree. Growth is seen as an overall increase in stem diameter (right), and the decline in tree stem diameter in September is due to late summer drought (*see* 133). The phloem diameter variation (middle and right) is calculated by subtracting the xylem diameter variation data from the data measured on the whole stem. Most of the time, the phloem shrinks with the xylem, but a small time lag in the shrinkage can indicate loading of osmolytes to the phloem tissue (a potential source area [118]. In the fall and early in the spring, or when hydraulically separated from the xylem, the phloem might swell when the xylem shrinks because of the accumulation of sugars or water in the phloem [134]

entire tissue between the xylem and the dead bark) [117, 118, 128, 129], and that these fluctuations are related to osmotic activity in the phloem [130].

Methods

Extracting information on phloem function from stem diameter fluctuation measurements requires simultaneous measurements of xylem and stem diameter fluctuations. Changes in the size of the phloem tissue are then calculated by subtracting the xylem diameter fluctuations from the measurements on the whole stem [117] (Fig. 5). There are several types of sensors that can be used for these measurements. These include dendrometer bands that measure changes in the stem circumference (e.g., 131), strain gauges installed on the bark [132], and linear displacement transducers attached to different kinds of holding frames [117, 124]. For xylem diameter variation measurements, the bark and phloem tissues are either removed completely underneath the sensor, or a screw is inserted through these tissues to the xylem. This avoids girdling the stem completely for xylem measurements, which could lead to xylem desiccation when large areas are exposed to air. The need to girdle the stem completely, and the tendency of the phloem to grow over any wound quickly makes use of dendrometer bands or strain gauges for the xylem measurement challenging. Generally, the resolution and measurement repeatability with linear displacement-type sensors is higher than with dendrometer bands, and therefore these types of sensors have grown in popularity.

Linear displacement transducers come in various sizes and shapes and can be attached to the plant with different methods. In trees, they can simply be attached to a metal frame screwed on the stem with an

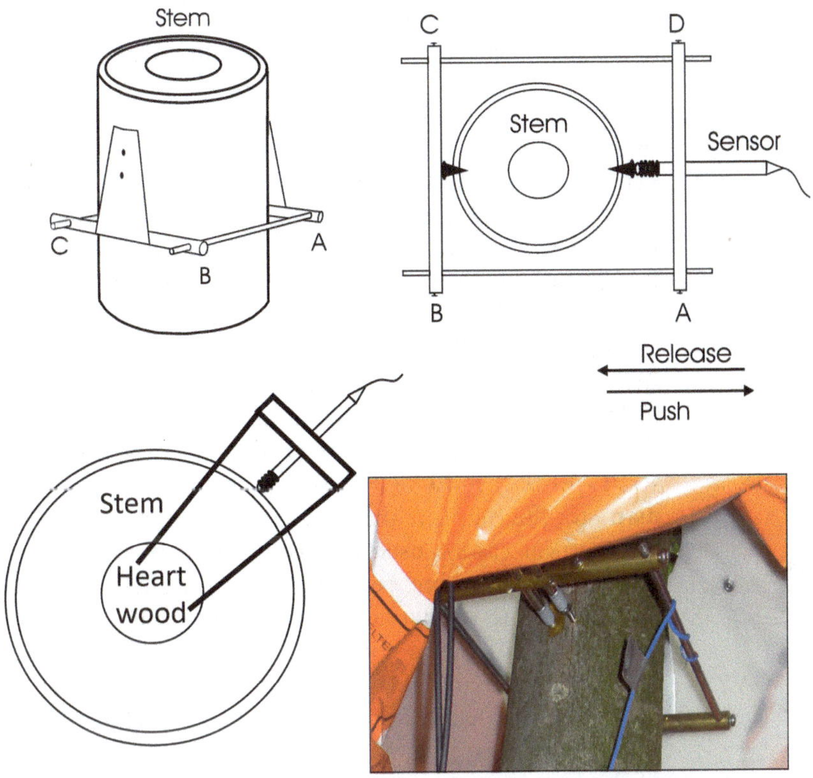

Fig. 6 A presentation of two different methods of mounting linear displacement transducers (LVDTs) on tree stems. The LVDT can be attached to the tree using a metal frame surrounding the stem (top), by anchoring the metal frame to the heartwood with long screws (bottom). The frame surrounding the stem is attached using metal plates screwed to the stem. In this setup, the stem diameter is measured directly with a reference point on the opposite side of the stem than the sensor. When anchoring the sensor to the heartwood, it is assumed that the heartwood does not change shape. In this setup, the diameter variations are measured from one side only. In both systems, xylem measurements can be obtained by inserting small screws into the xylem through the bark and the cambium (demonstrated at the top). For simultaneous xylem and stem diameter measurements, the LVDTs can be attached to the same metal bar (photo on bottom right). Here, we used a rescue blanket to shade the sensors from direct sunlight and measured frame temperature with a thermocouple (blue wire in the photo) for thermal expansion correction

attachment point either on opposite sides of the tree or anchored to the heartwood (Fig. 6). The first scenario allows direct measurement of diameter fluctuations. The second one measures the diameter fluctuations only on one side of the stem. This method generally works better on large stems where a frame surrounding the whole stem would become unstable. For soft or very small stems, different sensor stands can be developed. Generally, diameter fluctuation measurements give accurate and repeatable data, with high-resolution both at temporal and spatial scales. Separating the xylem and the phloem measurements is relatively easy. However, some caution and testing for homogeneity of the water tension field are required in plants showing high xylem sectoriality or secondary cambium.

Data Analysis

The principle behind extracting information on phloem function from diameter fluctuations is that the hydraulic connection between the xylem and the phloem ensures that the fluctuations in the phloem follow those of the xylem unless phloem activity makes them deviate from that. An increase in sugar concentration in the phloem would, for example, help retain water in the phloem or reverse the direction of the water flow between the xylem and the phloem to be towards the phloem [118]. This would be detected as "less than normal" shrinkage or even swelling of the phloem while the xylem shrinks [118, 130]. This activity would also cause a time lag between the shrinkage of the xylem and the phloem, or the xylem and the whole stem of the tree [118]. Based on this, when phloem and xylem diameter fluctuations are measured simultaneously, xylem diameter fluctuations can be used to model what the phloem would do if it acted as a passive water reservoir for the xylem. This prediction is then compared with the actual observed phloem fluctuations. Deviations from the prediction are interpreted as changes in osmotic potential inside the phloem [130]. Calculating phloem transport rates from these measurements requires theoretical modeling [101] and assumptions about pressure gradients and how changes in osmotic potential of the phloem tissue might influence transport inside the conduits [29]. The adequacy of the detected vertical pressure gradients for driving phloem transport has been a field of debate with some studies showing supporting evidence [50, 113, 135, 136], and others finding too small or no gradients (*for reviews, see* 52, 135). Generally, the Münch theory of phloem flow requires a pressure gradient between the sink and the source to drive the flow [23, 50, 113]. The lack of observed gradients has been explained by frequent loading and unloading zones along the transport pathway, which shorten the effective transport length of each conduit (the relay hypothesis) (*for review, see* 52). Similarly, if the phloem conduit walls are highly permeable to water but not the solutes, then an osmotic pressure gradient between the conduits and their surroundings is enough to drive the flow, and no long-distance vertical gradient is needed [137]. Hopefully, the recent advances in pressure sensor technology [73, 75] and new insights into sieve plate resistance [112] will help resolve the role of pressure gradients in phloem transport and shed light to the diversity of transport schemes and controls used by different plants.

The most direct evidence for possible phloem failure under stress with this method comes from excessive tissue shrinkage (phloem turgor collapse) [95] related to water loss from the phloem tissue. Even if the threshold for how large a shrinkage would lead to phloem transport failure is unknown, it can be argued that if the phloem loses most of its volume, transport is unlikely to remain intact. Interestingly, the timing of excessive phloem shrinkage might predict plant survival time quite accurately with 1 week delay in phloem collapse resulting in four weeks longer survival in piñon pine trees [95].

Stem diameter fluctuation measurements are an indirect method to measure phloem function, and the data interpretation is an active field of research [105, 124–126, 128–130]. These measurements, however, keep providing interesting information at a scale unavailable with most other methods. They offer a relatively robust, automated means of monitoring plant function in field conditions through seasonal changes and can be used in laboratory for highly targeted experiments as well.

4 Conclusions

Increasing experimental evidence suggests that environmental and biotic stress influence phloem transport. Changes and limitations in phloem function affect the productivity and survival of the entire plant. As a sensitive, soft, small, and diverse tissue, phloem structure, and function are challenging to study. Recent advances in chemical and metabolic analysis tools, and new flow tracers and imaging methods, have revealed a variety of principles covering phloem function in different plants. Extending the studies from phloem function in non-stress conditions to stress responses will allow understanding controls and limitations of phloem transport in more detail and provide a basis for formulating theories on plant stress responses, survival, and recovery. Theories that include both the water and carbon limitations of plant function and the interaction of water and carbon transport in plants are acutely needed for improving predictions of vegetation dynamics and their feedbacks to the climate under our changing environment. At the moment, we have some generalizable hypotheses about hydraulic, and structural limitations of phloem transport that influence phloem function and plant carbon dynamics under stress. The overall message from most studies, however, is that there is a high variability in responses depending on plant type, stress type, and severity. Therefore, applying available methods for studying both phloem function under stress and without stress is essential for building a more coherent picture that allows classification of phloem functional types and stress sensitivities to improve our conceptual models on phloem function and its influence on plant performance.

Conflict of Interest

The authors have no conflicts of interest to declare that are relevant to the content of this chapter.

References

1. Smith WK, Young DR, Carter GA, Hadley JL, McNaughton GM (1984) Autumn stomatal closure in six conifer species of central Rocky Mountains. Oecologia 63:237–242
2. Tyree MT, Sperry JS (1988) Do woody plants operate near the point of catastrophic xylem dysfunction caused by dynamic water stress? Answers from a model. Plant Physiol 88:574–580
3. Wilkinson S, Clephan AL, Davies WJ (2001) Rapid low temperature-induced stomatal closure occurs in cold-tolerant *Commelina communils* leaves but not in cold-sensitive tobacco leaves via a mechanism that involves apoplastic calcium but not abscisic acid. Plant Physiol 126:1566–1578
4. Kerchev PI, Fenton B, Foyer CH, Hancock RD (2012) Plant responses to insect herbivory: interactions between photosynthesis, reactive oxygen species and hormonal signaling pathways. Plant Cell Environ 35:441–453
5. Tombesi S, Nardini A, Frioni T, Soccolini M, Zadra C, Farinelli D, Poni S, Palliotti A (2015) Stomatal closure is induced by hydraulic signals and maintained by ABA in drought-stressed grapevine. Sci Rep 5: 12449
6. Martin-StPaul N, Delzon S, Cochard H (2017) Plant resistance to drought depends on timely stomatal closure. Ecol Lett 20:1437–1447
7. Hossain M, Poot P, Veneklaas E (2025) Interactive effects of xylem and phloem damage under drought on plant water and carbon dynamics: mimicking the impact of plant canker disease. bioRxiv. https://doi.org/10.1101/2025.03.03.641225
8. Roman DT, Novick KA, Brzostek ER, Dragoni D, Rahman F, Phillips RP (2015) The role of isohydric and anisohydric species in determining ecosystem-scale response to severe drought. Oecologia 179:641–654
9. McDowell N, Pockman WT, Allen CD, Breshears DD, Cobb N, Kolb T, Plaut J, Sperry J, West A, Williams DG, Yepez EA (2008) Mechanisms of plant survival and mortality during drought: why do some plants survive while others succumb to drought?. New Phytol 178:719–739
10. Domec JC, Warren JM, Meinzer FC, Brooks JR, Coulombe R (2004) Native root xylem embolism and stomatal closure in stands of Douglas-fir and ponderosa pien: mitigation by hydraulic redistribution. Oecologia 141:7–16
11. Meinzer FC, Johnson DM, Lachenbruch B, McCulloh KA, Woodruff DR (2009) Xylem hydraulic safety margins in woody plants; coordination of stomatal control of xylem tension with hydraulic capacitance. Funct Ecol 23:922–930
12. McCulloh KA, Woodruff DR (2012) Linking stomatal sensitivity and whole-tree hydraulic architecture. Tree Physiol 32:369–372
13. Sala A, Piper F, Hoch G (2010) Physiological mechanisms of drought-induced tree mortality are far from being resolved. New Phytol 186:274–281
14. Sala A, Woodruff DR, Meinzer FC (2012) Carbon dynamics in trees: feast or famine?. Tree Physiol 32:764–775
15. Savage JA, Clearwater MJ, Haines DF, Klein T, Mencuccini M, Sevanto S, Turgeon R, Zhang C (2016) Allocation, stress tolerance and carbon transport in plants: how does phloem physiology affect plant ecology?. Plant Cell Environ 39:709–725
16. Lemoine R, La Camera S, Atanassova R, Dedaldechamp F, Allario T, Pourtau N, Bonnemain JL, Laloi M, Coutos-Thevenot P, Maurousset L et al (2013) Source-to-sink transport of sugar and regulation by environmental factors. *Front Plant Sci* 2013, 4:272. https://doi.org/10.3389/fpls.2013.00272
17. McDowell NG, Sevanto S (2010) The mechanisms of carbon starvation: how, when, or does it even occur at all?. New Phytol 186:264–266
18. Stanfield RC, Bartlett MK (2022) Coordination between phloem loading and structure maintains carbon transport under drought. Front Plant Sci 13:787837. https://doi.org/10.3389/fpls.2022.787837
19. Xu Q, Chen S, Yunjuan R, Chen S, Liesche J (2017) Regulation of sucrose transporters and phloem loading in response to environmental cues. Plant Physiol 176:930–945. https://doi.org/10.1104/pp.17.01088
20. Hölttä T, Mencuccini M, Nikinmaa E (2009) Linking phloem function to structure: analysis with a coupled xylem – phloem transport model. J Theor Biol 259:325–337
21. Nikinmaa E, Hölttä T, Hari P, Kolari P, Mäkelä A, Sevanto S, Vesala T (2013) Assimilate transport in phloem sets conditions for leaf gas exchange. Plant Cell Environ 36:655–669
22. Sevanto S (2018) Drought impacts on phloem transport. Curr Opin Plant Biol 43:76–81
23. Sevanto S (2014) Phloem transport and drought. J Exp Bot 65:1751–1759
24. Thompson MW, Holbrook NM (2003) Scaling phloem transport: water potential equilibrium and osmoregulatory flow. Plant Cell Environ 26:1561–1577
25. Perri S, Katul GG, Molini A (2019) Xylem-phloem hydraulic coupling explains multiple osmoregulatory responses to salt stress.

New Phytol 224:644–662. https://doi.org/10.1111/nph.16072
26. Grange RI, Peel AJ (1978) Evidence for solution flow in the phloem of willow. Planta 138:15–23
27. Smith JAC, Milburn JA (1980) Osmoregulation and the control of phloem-sap composition in *Ricinus communis* L. Planta 148:28–34
28. Morison KR (2002) Viscosity equations for sucrose solutions: old and new. In: Gostomski PA, Morison KR (Eds.) Green processing and sustainability. Hill Publishing, Christchurch, New Zealand, Proceedings of APCChE Conference and CHEMECA 2002, 29 Sep–3 Oct 2002984
29. Sevanto S, Ryan MG, Dickman LT, Derome D, Patera A, Defraeye T, Pangle RE, Hudson PJ, Pockman WT (2018) Is desiccation tolerance and avoidance reflected in xylem and phloem anatomy of two co-existing arid-zone coniferous trees?. Plant Cell Environ 41:1551–1564
30. Lewis DH (1984) Occurrence and distribution of storage carbohydrates in vascular plants. In: Lewis DH (ed) Storage carbohydrates in vascular plants. Distribution, physiology and metabolism. Cambridge University Press, Cambridge UK, p 1–52
31. Woodruff DR (2014) The impacts of water stress on phloem transport in Douglas-fir trees. Tree Physiol 34:5–14
32. Dannoura M, Epron D, Desalme D, Massonnet C, Tsuji S, Plain C, Priault P, Gerant D (2018) The impact of prolonged drought on phloem anatomy and phloem transport in young beech. Tree Physiol 39:201–210. https://doi.org/10.1093/treephys/tpy070
33. Stanfield RC, Hacke UG, Laur J (2017) Are phloem sieve tubes leaky conduits supported by numerous aquaporins?. Am J Bot 104:719–732
34. Chaumont F, Tyerman SD (2014) Aquaporins: highly regulated channels controlling plant water relations. Plant Physiol 164:1600–1618
35. Maurel C, Verdoucq L, Rodrigues O (2016) Aquaporins and plant transpiration. Plant Cell Environ 39:2580–2587
36. Lang A (1979) A relay mechanism for phloem translocation. Ann Bot 44:141–145
37. Thompson MV, Zwieniecki MA (2005) The role of potassium in long distance transport in plants. In: Holbrook NM, Zwieniecki MA (eds), Vascular transport in plants. Elsevier Academic Press
38. Nikam PS, Ansari HR, Hasan M (2000) Density and viscosity studies of glucose and fructose solutions in aqueous and 0.5 mol dm3 aqueous NH4Cl. J Mol Liq 87:97–105
39. Dickman LT, McDowell NG, Sevanto S, Pangle RE, Pockman WT (2014) Carbohydrate dynamics and mortality in a pinon-juniper woodland under three future precipitation scenarios. Plant Cell Environ 38:729–739
40. Savage JA, Zwieniecki MA, Holbrook NM (2013) Phloem transport velocity varies over time and among vascular bundles during early cucumber seedling development. Plant Physiol 163:1409–1418
41. Roberts AG, Santa Cruz S, Roberts IM, Prior DAM, Turgeon R, Oparka KJ (1997) Phloem unloading in sink leaves of *Nicotiana benthamiana*: comparison of a fluorescent solute with a fluorescent virus. Plant Cell 9:1381–1396
42. Helfter C, Shephard JD, Martinez-Vilalta J, Mencuccini M, Hand DP (2007) A noninvasive optical system for measurements of xylem and phloem sapflow in woody plants of small stem size. Tree Physiol 27:169–179
43. Froelich DF, Mullendore DM, Jensen KH, Ross-Elliott TJ, Anstead JA, Thompson GA, Pelissier H, Knoblauch M (2011) Phloem ultrastructure and pressure flow: sieve-element-occlusion-related agglomerations do not affect translocation. Plant Cell 23:4428–4445
44. Wang Z, Xie X, Zhao J, Liu X, Feng W, White JC, Xing B (2012) Xylem- and phloem-based transport of CuO nanoparticles in maize (*Zea mays* L.). Environ Sci Technol 46:4434–4441
45. Zhai G, Walters KS, Peate DW, Alvarez PJJ, Schnoor JL (2014) Transport of gold nanoparticles through plasmodesmata and precipitation of gold ions in woody poplar. Environ Sci Technol Lett 1–2:146–151
46. Perez-de-Lugue A (2017) Interaction of nanomaterials with plants: what do we need for real applications in agriculture. Front Environ Sci 5:12. https://doi.org/10.3389/fenvs.2017.00012
47. Windt CW, Vergeldt FJ, de Jager PA, van As H (2006) MRI of long-distance water transport: a comparison of the phloem and xylem flow characteristics and dynamics in poplar, castor bean, tomato and tobacco. Plant Cell Environ 29:1715–1729
48. Hubeau M, Steppe K (2015) Plant-PET scans: *in vivo* mapping of xylem and phloem functioning. Trends Plant Sci 20:676–685
49. Deborde C, Moing A, Roch L, Jacob D, Rolin D, Giraudeau P (2017) Plant metabolism as studies by NMR spectroscopy. Prog Nucl Magn Reson Spectrosc 102–103:61–97
50. Knoblauch M, Knoblauch J, Mullendore DL, Savage JA, Babst BA, Beecher SD, Dodgen AC, Jensen KH, Holbrook NM (2016)

Testing the Münch hypothesis of long distance phloem transport in plants. eLife 5:e15341
51. Knoblauch M, Peters WS (2010) Münch, morphology, microfluidics – our structural problem with the phloem. Plant Cell Environ 33:1439–1452
52. De Schepper V, De Swaef T, Bauweraerts I, Steppe K (2013) Phloem transport: a review of mechanisms and controls. J Exp Bot 64:4839–4850
53. Hartmann H, Ziegler W, Kolle O, Trumbore S (2013) Thirst beast hunger –declining hydration during drought prevents carbon starvation in Norway spruce saplings. New Phytol 200:340–349
54. Gessler A, Treydte K (2016) The fate and age of carbon – insights into the storage and remobilization dynamics in trees. New Phytol 209:1338–1340
55. Mencuccini M, Hölttä T (2010) The significance of phloem transport for the speed with which canopy photosynthesis and belowground respiration are linked. New Phytol 185:189–203
56. Thorpe MR, Minchin PEH, Dye EA (1979) Oxygen effects on phloem loading. Plant Sci Lett 15:345–350
57. Kölling K, Müller A, Flütsch P, Zeeman SC (2013) A device for single leaf labelling with CO2 isotopes to study carbon allocation and partitioning in *Arabidopsis thaliana*. Plant Methods 9: 45
58. Werner C, Schnydeer H, Cuntz M, Keitel C, Zeeman MJ, Dawson TE, Badeck F-W, Brugnoli E, Ghashghaie J, Grams TEE, Kayler ZE, Lakatos M, Lee X, Maguas C, Ogee J, Rascher KG, Siegwolf RTW, Unger S, Welker J, Wingate L, Gessler A (2012) Progress and challenges in using stable isotopes to trace plant carbon and water relations across scales. Biogeosciences 9:3083–3111
59. Steinmann K, Siegwolf RTW, Saurer M, Körner C (2004) Carbon fluxes to the soil in a mature temperate forest assessed by 13C isotope tracing. Oecologia 141:489–501
60. Howarth WR, Pregitzer KS, Paul EA (1994) 14C allocation in tree soil systems. Tree Physiol 14:1163–1176
61. Mikan CJ, Zak DR, Kubiske ME, Pregitzer KS (2000) Combined effects of atmospheric CO2 and N availability on the belowground carbon and nitrogen dynamics of aspen mesocosms. Oecologia 124:432–445
62. Johnson D, Leake JR, Ostle N, Ineson P, Read DJ (2002) In situ 13CO2 pulse labelling of upland grassland demonstrates a rapid pathway of carbon flux from arbuscular mycorrhizal mycelia to soil. New Phytol 153:327–334
63. Högberg P, Högberg MN, Göttlicher SG, Betson NR, Keel SG, Metcalfe DB, Campbell C, Schindlbacher A, Hurry V, Lundmark T et al. (2008) High temporal resolution tracing of photosynthate carbon from the tree canopy to forest soil microorganisms. New Phytol 177:220–228
64. Babst BA, Karve AA, Judt T (2013) Radiometabolite analysis of carbon-11 biochemical partitioning to non-structural carbohydrates for integrated metabolism and transport studies. Plant Cell Physiol 54:1016–1025. https://doi.org/10.1093/pcp/pct045
65. Babst BA, Ferrieri R, Schueller M (2019) Detecting rapid changes in carbon transport and partitioning with carbon-11 (11C). Methods Mol Biol 2014:163–176. https://doi.org/10.1007/978-1-4939-9562-2_14
66. Minchin PEH, Thorpe MR (1983) A rate of cooling response in phloem translocation. J Exp Bot 34:529–536
67. Pickard WF, Minchin PEH (1990) The transient inhibition of phloem translocation in *Phaseolus vulgaris* by abrupt temperature drops, vibration and electric shock. J Exp Bot 41:1361–1369
68. Ruehr NK, Offermann CA, Gessler A, Winkler JB, Ferrio JP, Buchmann N, Barnard RL (2009) Drought effects on allocation of recent carbon: from beech leaves to soil CO2 efflux. New Phytol 184:950–961
69. Hommel R, Siegwolf R, Zavadlav S, Arend M, Schaub M, Galiano L, Haeni M, Kayler ZE, Gessler A (2016) Impact of interspecific competition and drought on the allocation of new assimilates in trees. Plant Biol 18:785–196
70. Malone MW, Yoder J, Hunter JF, Espy MA, Dickman LT, Nelson RO, Vogel SC, Sandin H, Sevanto S (2016) In vivo observation of tree drought response with low-field NMR and neutron imaging. Front Plant Sci 7:564. https://doi.org/10.3389/fpls.2016.00564
71. Zarebanadkouki M, Kim YX, Carminati A (2013) Where do roots take up water? Neutron radiography of water flow into the roots of transpiring plants growing in soil. New Phytol 199:1034–1044
72. Nelson RO, Vogel SC, Hunter J, Watkins EB, Losko AS, Tremsin AS, Borges NP, Cutler TE, Dickman LT, Espy M, Gautier C, Madden AC, Majewski J, Malone MW, Mayo DR, McClellan KJ, Montgomery D, Mosby S, Nelson AT, Ramos K, Schirato RC, Schroeder K, Sevanto S, Swift AL, Vo L, Williamson T, Winch N (2018) Neutron imaging at LANSCE – from cold to ultrafast. J Imaging 4:45
73. Knoblauch M, Vendrell M, de Leau E, Paterlini A, Knox K, Ross-Elliott TJ, Reinders

A, Brockman SA, Ward J, Oparka K (2015) Multispectral phloem-mobile probes – properties and applications. Plant Physiol 167:1211–1220
74. Truernit E (2014) Phloem imaging. J Exp Bot 65:1681–1688
75. Knoblauch J, Mullendore DL, Jensen KH, Knoblauch M (2014) Pico gauges for minimally invasive intracellular hydrostatic pressure measurements. Plant Physiol 166:1271–1279
76. Marion D (2013) An introduction to biological NMR spectroscopy. Mol Cell Proteom 12:3006–3025
77. Windt CW, Soltner H, van Dusschoten D, Blümler P (2011) A portable Halbach magnet that can be opened and closed without force: the nmr-cuff. J Magn Reson 208:27–33
78. Kimura T, Geya Y, Terada Y, Kose K, Haishi T, Gemma H et al. (2011) Development of a mobile magnetic resonance imaging system for outdoor tree measurements. Rev Sci Instrum 82:053704
79. Jones M, Aptaker PS, Cox J, Gardiner B, McDonald P (2012) A transportable magnetic resonance imaging system for *in situ* measurements of living trees: the tree hugger. J Magn Reson 218:133–140
80. Yoder J, Espy MA, Malone MW, Sevanto S (2014) Low-field NMR for the i*n* vivo study of water content in trees. Rev Sci Instrum 85:095110–095110-8
81. Homan NM, Windt CW, Vergeldt FJ, Gerkema E, van As H (2007) 0.7 and 3T MRI and sap flow in intact trees: xylem and phloem in action. Appl Magn Reson 32:157–170
82. Ilvonen K, Palva L, Perämäki M, Joensuu R, Sepponen R (2001) MRI-based D2O/ H2O –contrast methods to study water flow and distribution in heterogenous systems: demonstration in wood xylem. J Magn Reson 149:36–44
83. Bourgis F, Kilaru A, Cao X, Ngando-Ebongue G-F, Drira N, Ohlrogge JB, Arondel V (2011) Comparative transcriptome and metabolite analysis of oil palm and date palm mesocarp that differ dramatically in carbon partitioning. Proc Natl Acad Sci U S A 108:12527–12532
84. Brodribb TJ, Skelton RP, McAdam SA, Bienaime D, Lucani CJ, Marmottant P (2016) Visual quantification of embolism reveals leaf vulnerability to hydraulic failure. New Phytol 209:1403–1409
85. Ohya T, Tanoi K, Hamada Y, Okabe H, Rai H, Hojo J, Suzuki K, Nakanishi TM (2008) An analysis of lon-distance water transport in the soybean stem using H215O. Plant Cell Physiol 49:718–729
86. Mori S, Kiyomiya S, Nakanishi H, Ishioka NS, Watanabe S, Osa A et al. (2000) Visualization of 15 O-water flow in tomato and rice in the light and dark using a positron-emitting tracer imaging system (PETIS). Soil Sci Plant Nutr 46:975–979
87. Kiyomiya S, Nakanishi H, Uchida H, Nishiyama S, Tsukada H, Ishioka NS, Watanabe S, Osa A, Mizuniwa C, Ito T, Matsuhashi S, Hashimoto S, Sekine T, Tsuji A, Mori S (2001) Light activates H2 15O flow in rice: detailed monitoring using a positron-emmitting tracer imaging systems (PETIS). Physiol Plant 113:359–367
88. Nakanishi H, Bughio N, Matsuhashi S, Ishioka NS, Uchida H, Tsuji A, Osa A, Sekine T, Kume T, Mori S (1999) Visualizing real time [11c] methionine translocation in fe-sufficient and fe-deficient barley using a positron emitting tracer imaging system (PETIS). J Exp Bot 50:637–43
89. Meldau S, Woldemariam MG, Fatangare A, Svatos A, Galis I (2015) Using 2-deoxy-2-[18F]fluoro-D-glucose ([18F]FDG) to study carbon allocation in plants after herbivore attack. BMC Res Notes 8: 45
90. De Schepper V, Buhler J, Thorpe M, Roeb G, Huber G, van Dusschoten D, Jahnke S, Steppe K (2013) 11C-PET imaging reveals transport dynamics and sectorial plasticity of oak phloem after girdling. Front Plant Sci 4:200
91. Jahnke S, Menzel MI, Van Dusschoten D, Roeb GW, Buhler J, Minwuyelet S, Blumer P, Temperton VM, Hombach T, Streun M, Beer S, Khodaverdi M, Ziemons K, Coenen HH, Schurr U (2009) Combined MRI-PET dissects dynamic changes in plant structures and functions. Plant J 59:634–644
92. Mincke J, Courtyn J, Vanhove C, Vandenberghe S, Steppe K (2021) Guide to plant-PET imaging using 11CO2. Front Plant Sci 12, 602550. https://doi.org/10.3389/fpls.2021.602550
93. Hartmann H, Ziegler W, Trumbore S (2013) Lethal drought leads to reduction in nonstructural carbohydrates in Norway spruce tree roots but not in the canopy. Funct Ecol 27:413–427
94. Garcia-Forner N, Sala A, Biel C, Save R, Martinez-Vilalta J (2016) Individual traits as determinants of time to death under extreme drought in *Pinus sylvestris* L. Tree Physiol 36:1196–1209
95. Sevanto S, McDowell NG, Dickman LT, Pangle R, Pockman WT (2014) How do trees die? A test of the hydraulic failure and carbon starvation hypotheses. Plant Cell Environ 37:153–161
96. Hartmann H, Trumbore S (2016) Understanding the roles of nonstructural carbohydrates in forest trees – from what we

can measure to what we want to know. New Phytol 211:386–403
97. Pinheiro C, Chaves MM (2010) Photosynthesis and drought: can we make metabolic connections from available data?. J Exp Bot 62:869–882
98. Mitchell PJ, O'Grady AP, Tissue DT, White DA, Ottenschlaeger ML, Pinkard EA (2013) Drought response strategies define the relative contributions of hydraulic dysfunction and carbohydrate depletion during tree mortality. New Phytol 197:862–872
99. Dietze MC, Sala A, Carbone MS, Czimczik CI, Mantooth JA, Richardson AD, Vargas R (2014) Nonstructural carbon in woody plants. Annu Rev Plant Biol 65:667–687
100. Lacointe A (2000) Carbon allocation among tree organs: a review of basic processes and representation in functional-structural tree models. Ann For Sci 57:521–533
101. Hölttä T, Vesala T, Sevanto S, Perämäki M, Nikinmaa E (2006) Modeling xylem and phloem water flows in trees according to cohesion theory and Münch pressure flow hypothesis. Trees 20:67–78
102. McDowell NG, Fisher RA, Xu C, Domec JC, Hölttä T, Mackay DS, Sperry JS, Boutz A, Dickman LT, Gehres N, Limousin JM, Macalady A, Martinez-Vilalta J, Mencuccini M, Plaut JA, Ogee J, Pangle RE, Rasse DP, Ryan MG, Sevanto S, Waring RH, William AP, Yepez EA, Pockman WT (2013) Evaluating theories of drought-induced vegetation mortality using a multimodel-experiment framework. New Phytol 200:304–321
103. Klein T, Hoch G (2015) Tree carbon allocation dynamics determined using a carbon mass balance approach. New Phytol 205:147–159
104. Minchin PEH, Lacointe A (2005) New understanding on phloem physiology and possible consequences for modeling long-distance carbon transport. New Phytol 166:771–779
105. De Schepper V, Steppe K (2010) Development and verification of a water and sugar transport model using measured stem diameter variations. J Exp Bot 61:2083–2099
106. Nakad M, Domec JC, Sevanto S, Katul G (2023) Sucrose transport inside the phloem: bridging hydrodynamics and geometric characteristics. Phys Fluids 35, 061905. https://doi.org/10.1063/5.0151644
107. Nakad M, Domec JC, Sevanto S, Katul G (2023) Towards a realistic representation of sucrose transport in the phloem of plants. J Geophys Res Biogeosci 128:e2022JG007361. https://doi.org/10.1029/2022JG007361
108. Nakad M, Sevanto S, Domec J-C, Katul G (2023) Linking the water and carbon economies of plants in a drying and warming climate. Curr For Rep 9:383–400. https://doi.org/10.1007/s40725-023-00202-4
109. Youssef L, Nakad M, Domec JC, Sevanto S, Katul G (2025) Phloem catastrophe: Using bifurcation analysis to predict plant tipping points. ARC Geophys Res 1:14. https://doi.org/10.5149/ARC-GR.2526
110. Savage JA, Beecher SD, Clerx L, Gersony JT, Knoblauch J, Losada JM, Jensen KH, Knoblauch M, Holbrook NM (2017) Maintenance of carbohydrate transport in tall trees. Nat Plants 3:965–972
111. Jensen KH, Mullendore DL, Holbrook NM, Both T, Knoblauch M, Bruus H (2012) Modeling the hydrodynamics of phloem sieve plates. Front Plant Sci 3:151. https://doi.org/10.3389/fpls.2012.00151
112. Sevanto S (2024) Flow resistance of phloem sieve plates revisited using an experimental model. Phys Fluids 36, 031901. https://doi.org/10.1063/5.0188124
113. Jensen KH, Liesche J, Bohr T, Schulz A (2012) Universality of phloem transport in seed plants. Plant Cell Environ 35:1065–1076
114. Bouche PS, Delzon S, Choat B, Badel E, Brodribb TJ, Burlett R, Cochard H, Charra-Vaskou K, Lavigne B, Li S, Mayr S, Morris H, Torres-Ruiz JM, Zufferey V, Jansen S (2016) Are needles of *Pinus pinaster* more vulnerable to xylem embolism than branches? New insights from X-ray computed tomography. Plant Cell Environ 39:860–870
115. Patera A, Carl S, Stampanoni M, Derome D, Carmelier J (2018) A non-rigid registration methods for the analysis of local deformations in the wood cell wall. Adv Struct Chem Imaging 4: 1
116. Pfeiffer F, Weitkamp T, Bunk O, David C (2006) Phase retrieval and differential phase-contrast imaging with low-brilliance x-ray sources. Nat Phys. https://doi.org/10.1038/nphys26
117. Sevanto S, Vesala T, Perämäki M, Nikinmaa E (2002) Time lags for xylem and stem diameter variations in a Scots pine tree. Plant Cell Environ 25:1071–1077
118. Sevanto S, Vesala T, Perämäki M, Nikinmaa E (2003) Sugar transport together with environmental conditions controls time lags between xylem and stem diameter changes. Plant Cell Environ 26:1257–1265
119. Irvine J, Grace J (1997) Continuous measurements of water tensions in xylem of trees based on the elastic properties of wood. Planta 202:455–461
120. Leikola M (1969) Influence of environmental factors on the diameter growth of forest trees: axonometric study. Acta For Fenn 92:1–144. https://doi.org/10.14214/aff.7610

121. Lassoie JP (1973) Diurnal dimensional fluctuations in a Douglas-fir stem in response to tree water status. For Sci 19:251–255
122. Milne R, Ford ED, Deans JD (1983) Time lags in the water relations of Sitka spruce. For Ecol Manag 5:1–25
123. Neher HV (1993) Effects of pressures inside Monterey pine trees. Trees 8:9–17
124. Zweifel R, Haeni M, Buchmann N, Eugster W (2016) Are trees able to grow in periods of stem shrinkage?. New Phytol 211:839–849
125. Ziegler Y, Grote R, Alongi F, Knuver T, Ruehr NK (2024) Capturing drought stress signals: the potential of dendrometers for monitoring tree water status. Tree Physiol 44, tpae140. https://doi.org/10.1093/treephys/tpae140
126. Sevanto S (2025) Dendrometers – what are they good for? Tree Physiol 45, tpaf035. https://doi.org/10.1093/treephys/tpaf035
127. De Swaef T, De Schepper V, Vandergehuchte MW, Steppe K (2015) Stem diameter variations as a versatile research too in ecophysiology. Tree Physiol 35:1047–1061
128. Pfautsch S, Renard J, Tjoelker MG, Salih A (2015) Phloem as capacitor: radial transfer of water into xylem of tree stems occurs via symplastix transport in ray parenchyma. Plant Physiol 167:963–971
129. Pfautsch S, Hölttä T, Mencuccini M (2015) Hydraulic functioning of tree stems: fusing ray anatomy, radial transfer and capacitance. Tree Physiol 35:706–722
130. Mencuccini M, Hölttä T, Sevanto S, Nikinmaa E (2013) Concurrent measurements of change in the bark and xylem diameters of trees reveal a phloem-generated turgor signal. New Phytol 198:1143–1154
131. Pesonen E, Mielikäinen K, Mäkinen H (2004) A new girth band for measuring stem diameter changes. Forestry 77:431–438
132. Ueda M, Yoshikawa K, Okitu J (1996) Measurement of diurnal changes in stem and branch diameters using strain gauges. J For Res 1:139–142
133. Sevanto S, Hölttä T, Markkanen T, Perämäki M, Nikinmaa E, Vesala T (2005) Relationships between diurnal diameter variations and environmental factors in Scots pine. Boreal Environ Res 10:447–458
134. Sevanto S, Hölttä T, Holbrook NM (2011) Effects of the hydraulic coupling between xylem and phloem on diurnal phloem diameter variation. Plant Cell Environ 34:690–703
135. Fisher DB (1978) An evaluation of the Munch hypothesis for phloem transport in soybean. Planta 139:25–28
136. Gould N, Thorpe MR, Koroleva O, Minchin PEH (2005) Phloem hydrostatic pressure relates to solute loading rate: a direct test of the Munch hypothesis. Funct Plant Biol 32:1019–1026. https://doi.org/10.1071/FP05036
137. Phillips RJ, Dungan SR (1993) Asymptotic analysis of flow in sieve tubes with semi-permeable walls. J Theor Biol 162:465–485

Check for updates

Chapter 32

Isotope Measurements of Phloem-Transported Organic Matter: Tools to Obtain Information on Gas Exchange Physiology and Transport Processes, and to Better Understand Tree Ring Archives

Arthur Gessler

Abstract

Stable isotope analysis of phloem-transported organic matter provides valuable insights into plant physiological processes. This paper reviews methodologies for phloem sampling suitable for isotope measurements, including phloem bleeding from bark incisions, exudation into aqueous solutions, and centrifugation techniques. I highlight critical considerations for preventing sample contamination and degradation across these methods. I demonstrate how phloem isotope measurements offer crown-integrated information on gas exchange physiology, connecting stomatal conductance and photosynthetic assimilation. Isotopic modifications during phloem transport can be attributed to processes such as sugar pool mixing, metabolic conversions, and starch remobilization. The often observed ^{13}C enrichment during basipetal transport can be explained through a conceptual model involving unloading, metabolic conversion, and reloading processes. Particular attention is given to the application of phloem isotope analysis in improving interpretations of tree ring archives for paleoclimate reconstructions. We examine how oxygen atom exchange during cellulose synthesis and position-specific isotope signatures provide insights into source-sink relationships and plant metabolism. While acknowledging the complexities of isotope signal modifications during transport and metabolism, I emphasize the significant potential of phloem isotope analysis as a tool for advancing our understanding of plant gas exchange, phloem transport mechanisms, and the environmental information preserved in tree rings.

Keywords Carbon flux, Isotope analysis, Isotope modifications, Stable isotopes

1 Introduction

The elements carbon (C), hydrogen (H), nitrogen (N), oxygen (O), and sulfur (S) constitute the fundamental components of metabolites. For more than four decades, the field of plant ecology has used variations in the natural abundance of the stable isotopes of these elements to assess plant responses to environmental conditions.(e.g.,

Johannes Liesche and Sanna Sevanto (eds.), *Phloem*, Methods in Molecular Biology, vol. 3066,
https://doi.org/10.1007/978-1-0716-5488-0_32,

[1–5]). The main focus has been on carbon isotopes[1], since photosynthetic carbon isotope discrimination allows the reconstruction of gas exchange physiology from the carbohydrates assimilated [6]. Photosynthetic carbon isotope discrimination depends, in a first approximation, on the relationship between leaf intercellular (C_i) or chloroplastic (C_c) and ambient CO_2 concentration (C_a). Changes in the two main processes influencing C_i or C_c, i.e., the supply of CO_2 to the leaf through the stomata (and therefore directly related to stomatal conductance, g_s) on the one hand and the reduction of the CO_2 concentration in the leaf by RubisCO activity (and hence assimilation, A) on the other, influence the isotopic composition ($\delta^{13}C$) of the new assimilates. Therefore, photosynthetic carbon isotope discrimination is a proxy for intrinsic water use efficiency (iWUE = A/g_s). Carbon isotope measurements of organic matter can be combined with assessments of the oxygen isotope composition ($\delta^{18}O$). This dual-isotope approach can disentangle whether changes in water-use efficiency are driven by g_s or A, with $\delta^{13}C$ providing information about the relationship between photosynthesis and stomatal conductance while $\delta^{18}O$ helps differentiate the specific role of stomatal conductance. $\delta^{18}O$ of leaf water, which is imprinted on the new assimilates, is directly related to the vapor pressure deficit of the air, which in that approach is assumed to be the primary driver of g_s [7]. More recently, post-carboxylation isotope discrimination and atom exchange have been identified to modify $\delta^{13}C$ and $\delta^{18}O$ of organic compounds in processes downstream of photosynthesis as they are transported from the leaves via the phloem to the plant archives, such as tree rings [8]. These processes challenge the traditional interpretation of photosynthesis and gas exchange physiology from $\delta^{13}C$ and $\delta^{18}O$ but also open new perspectives to better understand phloem transport processes and heterotrophic metabolism.

Combining oxygen and hydrogen isotope analysis in organic compounds might be particularly helpful for the latter aspect. Plant water is the source of all organically bound H and most oxygen in organic molecules. Consequently, the relationship between $\delta^{18}O$ and $\delta^{2}H$ in organic compounds should show similar covariation as that in plant water. However, temporal variation in tree-ring chronologies indicates that both isotopes can be decoupled in cellulose. The reason for such decoupling might be different isotopic sensitivities to sink tissue metabolic processes, which can indicate metabolic dysregulation during stress [9].

[1] Clarivate Web of Science Core Collection searches (accessed December 29, 2024): carbon isotope* AND (plant* OR Tree*): 10694 publications; oxygen isotope* AND (plant* OR Tree): 2709; hydrogen isotope* AND plant*: 1149; nitrogen isotope* AND plants: 5811; (sulfur isotope* OR sulfur isotope*) AND plant*: 442.Web of Science categories: plant Science, geoscience multidisciplinary, ecology, and environmental sciences. It cannot be excluded that radioisotopes were sometimes included in the search, but an additional search for 14C OR C-14 within the carbon isotope results yielded only 562 hits. A considerable number of these publications deal with both stable isotopes and radioisotopes.

The nitrogen isotopic composition $\delta^{15}N$ of plant material is determined by the isotope ratio of the external nitrogen source and physiological mechanisms within the plant. Thus, $\delta^{15}N$ can be applied to characterize the root–shoot partitioning of nitrogen assimilation, loss of nitrogen, and resorption and reallocation of nitrogen [10–12]. While there are only a few papers on plant sulfur isotopes, there is an indication that they can be used to get insights into the assimilation and allocation of sulfur within different plant organs [13, 14].

While stable isotopes in plant material, such as leaves, have been determined for a long time, the natural carbon isotope composition in the organic matter transported in the phloem was only assessed in the late 1990s [15, 16]. This was mainly to obtain several day-averaged and whole-crown integrated iWUE information in trees [17]. A few years later, $\delta^{18}O$ analysis of phloem organic matter was established to enhance the understanding of leaf evaporative processes [18, 19].

Stable isotopes are often analyzed in "bulk" organic matter. This means that in the case of leaves or roots, dried plant material is homogenized and combusted in an elemental analyzer (C, N, and S) or pyrolyzed in high-temperature combustion processes (O, H). Thus, the isotope compositions of all carbon, nitrogen, oxygen, or hydrogen of the organic matter is determined. Phloem samples are initially obtained as liquid samples (see below) and are thus oven- or freeze-dried before combustion. The gases produced by combustion (e.g., CO_2 from organic C, N_2 from organic N) or pyrolysis (e.g., CO and H_2 from organic O and H, respectively) are transferred to an isotope ratio mass spectrometer where the isotope ratios of the elements are determined [20, 21]. The isotope ratios given as δ values (‰) are referenced against international standards. In addition, single chemical compounds can be separated via HPLC and then chemically oxidized [22] or via GC and then combusted or pyrolyzed [23] before IRMS analysis to obtain compound-specific isotope values. New high-resolution bioanalytical mass spectrometers (such as Orbitraps) coupled to ultra-high-performance liquid chromatography are now not only offering high-throughput metabolite identification but are also suitable for compound- and intramolecular position-specific isotope analysis in the natural isotope abundance range [24]. They might offer new insights into phloem transport by coupling metabolomics with stable isotope ecophysiology.

In this chapter, I will focus on carbon and oxygen isotopes and (1) the specific methodology for probing the phloem suitable for natural abundance stable isotope measurements. Moreover, (2) I will highlight the potential of the crown-integrating information on gas exchange physiology that can be obtained with the stable isotope analysis of phloem organic matter when (3) taking into account post-carboxylation modifications of the isotope composition, which is, however, helpful to describe processes related to phloem transport. I will (4) highlight how comparing phloem and tree ring isotopic signals can improve our understanding of the information stored in plant archives.

2 Collecting Phloem Organic Matter for Isotope Analyses

Mainly, four different techniques are used to obtain phloem organic matter for stable isotope analysis (Table 1).

Aphid-based techniques were applied in the first paper, which reported analyses of the natural abundance isotope composition of phloem-transported organic compounds [15]. In this study, aphids were caged onto specific parts of wheat plants at different growth stages. After 1 day, aphid stylets were severed using radio-frequency microcautery, allowing phloem exudates to accumulate on the plant surface. After three additional days, the exudates were brushed off. The exudates were diluted with demineralized water, transferred into tin capsules, and freeze-dried before analysis of nitrogen and carbon isotopes. The authors compared the $\delta^{15}N$ and $\delta^{13}C$ of the phloem exudates with that of the aphid biomass and the honeydew produced by the aphids. While $\delta^{13}C$ of honeydew and $\delta^{15}N$ of the aphids were comparable with the exudates, there was a ^{15}N and ^{13}C enrichment in the honeydew and aphids, indicating some trophic fractionation. The aphid stylet method is often assumed to produce relatively pure xylem sap. It is, however, complex to apply, produces only low amounts of exudate, and is restricted to particular species. In addition, contamination with aphid saliva and xylem sap can occur [29]. These issues and the problem of placing aphids under field conditions might be the main reasons for almost no follow-up studies. One remarkable exception is the study of Barbour, Hunt et al. [28]. In a field study, they took advantage of the fact that Nothofagus forests in New Zealand support large populations of the sooty beech scale insect (*Ultracoelostoma* spp.) that live within the bark,

Table 1
Methods for collecting phloem organic matter for isotope analyses

Method	Procedure	Isotopes measured	References
Phloem bleeding	Bleeding of phloem sap from cuts/ incisions due to the pressure in the sieve tubes	$\delta^{13}C$, $\delta^{18}O$, $\delta^{15}N$	[16, 17, 25]
Exudation	Collection of phloem sap in deionized water or in a solution containing chelating agents	$\delta^{13}C$, $\delta^{18}O$	[26, 25]
Centrifugation	Direct extraction from the bark through centrifugal force	$\delta^{13}C$	[27]
Aphids	Severing of the stylets of aphids feeding on stems and petioles and collecting the phloem sap from the cut stylets or collection of honey dew	$\delta^{13}C$, $\delta^{15}N$	[15, 28]

tap the phloem sap, and produce droplets of sugar-rich honeydew. This honeydew was regularly collected with 10 μL microcapillaries, allowing repeated non-destructive measurements and, thus, providing temporal courses of $\delta^{13}C$ in phloem organic matter. A comparison between phloem exudates from cut stylets and honeydew showed no significant differences.

Phloem bleeding from bark incisions with razor blades or scalpels can be applied to tree stems with soft bark, such as *Eucalyptus globulus* [16, 30] or *Fagus sylvatica* [17, 31], and to stems and petioles of cucurbits (cucumber, pumpkin) and *Ricinus communis* [12, 19, 32]. Droplets that bleed from the cuts are typically collected with microcapillaries to analyze the bulk isotope composition in the dried sap or for compound-specific isotope analysis. Although it is often assumed that the continued bleeding results from the pressure gradient in the sieve tubes between source and sink tissues and thus relatively pure phloem sap is collected [33], other results from cucurbits indicate that dilution with symplastic material from destroyed cells and contamination with apoplast and xylem constituents can occur [34].

The exudation of the phloem content from bark pieces into aqueous solutions is the most commonly used technique for collecting phloem organic matter for natural abundance isotope analysis in trees [26, 35–37]. For non-isotopic analysis of phloem loading and transport, exudation into EDTA-containing solutions is usually applied to prevent sealing of the severed sieve tubes through chelation of Ca^{2+} ions (see Chap. 13). Since EDTA contains C, N, and O atoms, the contents and the isotopic signatures of N, C, and O must be determined in the sampling solutions before phloem exudation and corrected for if the bulk isotope composition of the exudates is determined [25]. Alternatively, separating individual sugars with, e.g., HPLC and compound-specific isotope analysis can be done without such a correction. To avoid the potentially error-prone introduction of C, N, and O via the chelating agent in bulk isotope analyses, Gessler, Rennenberg et al. [25] compared EDTA-facilitated exudation of stem bark pieces with approaches applying polyphosphate as a chelator (without C and N) and with pure deionized water exudation. Since the experiments were conducted with *F. sylvatica*, where phloem bleeding from bark cuts can be applied, the exudation approaches were also evaluated against this technique. The authors found that the relative composition of the soluble amino N and sugar C pools, which are the most important compound classes for $\delta^{15}N$, $\delta^{13}C$, and $\delta^{18}O$, was comparable between exudates and phloem bleeding sap independent of the sample solution used for exudation. For the analysis of stable isotope composition in organic compounds transported in the phloem (which is a relative measure for which absolute concentration quantification is not necessary), the exudation with pure demineralized water seems to be appropriate since results comparable to the phloem bleeding technique were obtained. Even though lower absolute values of sugars were determined after an incubation time of 5 h in pure water compared to the EDTA- and polyphosphate-facilitated exudations, avoiding the loss of precision for the isotope determination by introducing error-prone isotopic mass corrections was assumed to justify the deionized water approach [25]. With a focus on $\delta^{18}O$, Lehmann, Egli et al. [26] assessed the effects of exudation time and EDTA concentration on the isotopic composition of phloem exudates. They purified the solutions from EDTA after exudation with an anion-exchange SPE cartridge and found no effect between 0, 5, and 20 mM EDTA on $\delta^{18}O$ of the exudate. However, with the increase in exudation time from 0.5 to 5 h (in the 0 mM EDTA approaches), the absolute sugar concentration and the relative abundance of sucrose decreased, leading to a time-dependent change in both bulk exudate and purified sugars.

These later findings indicate that sugars were degraded by microbial activity, pH changes, or invertase contamination from destroyed parenchymatic cells. In previous approaches, microbial and pH-related sugar degradation can and should be avoided by adding chloramphenicol and buffering the exudation solution at a pH of 7 [25, 38, 39]. Moreover, contamination of the exudate with invertase should also be monitored, which not only provides information on a potential reason for sucrose degradation but also indicates the contribution of organic compounds not originating from the sieve tubes. Schneider, Gessler et al. [38] showed for spruce and beech phloem exudates that when invertase activity was very low, and thus contamination with non-sieve tube contents <3%, no considerable sugar degradation occurred. The results of Lehmann, Goldsmith et al. [40] show that additional precautions to avoid sugar degradation and monitoring approaches for cellular contamination are required to correctly interpret the isotope composition of phloem organic matter. Measuring invertase activity, which should be negligible in the sieve tubes, could also be a way to assess contamination of the phloem sap obtained with the bleeding approach.

Centrifugation of phloem discs has also been tested as a method to obtain phloem-transported organic matter for isotope analysis. Devaux, Ghashghaie et al. [27] compared two methods for extracting phloem sap from pine trees: exudation in distilled water and centrifugation, in which phloem discs were placed in centrifugation vials above glass wool and centrifuged for 10 min at 14,000 × *g*. After centrifugation, the phloem sap at the bottom of the vial was collected. The authors observed that exudation extracted about 50% more compounds than centrifugation, but both methods yielded comparable carbon isotope composition.

Independent of the method applied, contamination and degradation need to be monitored and, if possible, avoided. Even though the analysis of the isotopic composition of phloem organic matter is now regularly applied, no comprehensive comparison of all available methods, as well as identification of potential sources of contamination and ways to avoid them, has been carried out. This calls for further research to define necessary quality control measures and best-practice approaches.

3 Crown-Integrating Information on Gas Exchange and Leaf-Level Processes

3.1 From the Assessment of Stomatal Conductance to Ecosystem Flux Partitioning

Sampling phloem-transported organic matter or purified sugars of the stem base of trees provides crown-integrated assimilation-weighted information on gas exchange physiology [41]. Such measurements have been applied to understand the impact of forest management and local to continental scale climate on stomatal conductance and photosynthetic assimilation [2, 42]. Keitel, Adams et al. [2] could show that $\delta^{13}C$ in phloem organic matter of *F. sylvatica* was significantly positively related to $\delta^{18}O$. According to the dual isotope model of Scheidegger, Saurer et al. [7], such a correlation indicates that $\delta^{13}C$ is affected by g_s and not by A. In fact, the authors could show that $\delta^{13}C$ of soluble carbon transported in the phloem was well correlated with canopy-integrated mean stomatal conductance in a 2-day integral before phloem sampling. Based on the canopy integrating information phloem transported organic matter provides, Rascher, Maguas et al. [41] proposed a simple model to derive forest overstory assimilation from phloem $\delta^{13}C$ and sap flux measurements. They derived C_i from phloem $\delta^{13}C$ and canopy g_s from sap flux measurements combined with an inverted Penman–Monteith equation. They then applied Fick's law to derive canopy-integrated A. Only recently, Stojanović, Jocher et al. [43] compared forest gross primary production from eddy covariance with the method based on sap flow and $\delta^{13}C$ of phloem contents. They observed strong agreement between the two methods for large parts of the growing season. As sap flow and phloem isotope measurements were conducted on individual trees throughout a growing season in a mixed broadleaf forest, this method allowed partitioning of primary productivity among the dominant overstory tree species (Fig. 1), which is impossible with the state-of-the-art eddy covariance technique. The study showed that among the three species that made up the forest, English oak showed the highest per-tree daily photosynthetic rates among tree species. Still, the smaller, more abundant common hornbeam contributed most to the stand-level summation of photosynthesis.

3.2 Temporal Dynamics of Phloem $\delta^{13}C$ Indicate Transitory Starch Build-Up and Release

In *R. communis*, phloem $\delta^{13}C$ showed substantial diel variations with relatively ^{13}C-depleted carbon exported from leaves during the day and enriched carbon at night [32]. Combined with isotopic modeling approaches accounting for isotope discrimination by aldolase in the Calvin cycle, these findings supported the idea that transitory starch

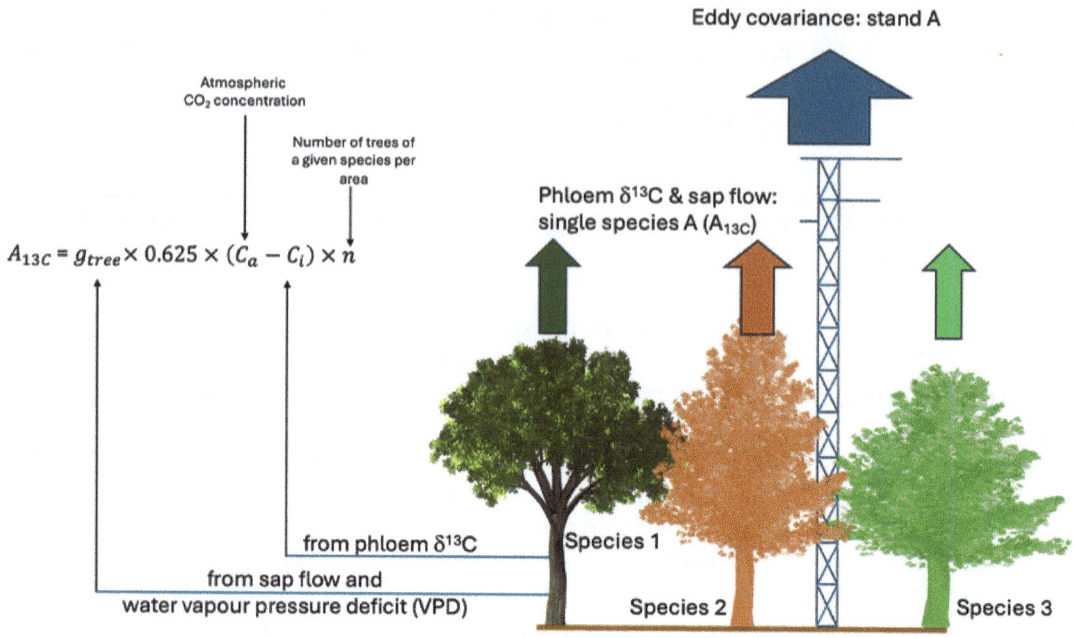

Fig. 1 The principle of disaggregation of canopy photosynthesis among different tree species in a mixed forest stand by applying phloem $\delta^{13}C$ measurements. State-of-the-art above-canopy covariance techniques assess stand-integrated gross primary productivity (stand A). Combined sap flux measurements and regular probing of phloem-transported organic matter for $\delta^{13}C$ allow calculating single tree photosynthesis and—by multiplying with the number of tree species per area—species-specific gross primary productivity (A_{13C}). In the equation, 0.625 is the ratio of the diffusivities of CO_2 to H_2O in air. For details, see [43]

accumulation and remobilization patterns governed the diel rhythm of $\delta^{13}C$ in organic matter exported by leaves.

4 Isotope Modifications and Processes Related to Phloem Transport

Changes in carbon or oxygen isotope composition of organic compounds have been observed during phloem transport along the plant axis. Since neither phloem loading nor the transport in the sieve tubes causes considerable isotope fractionation, other processes related to phloem transport must be responsible [8]. Assessing isotopic changes between leaf sugars and organic compounds in the phloem along the transport pathway can shed light on such processes.

4.1 Changes in $\delta^{13}C$ Related to Phloem Loading and Transport

There are two main changes in $\delta^{13}C$ directly related to phloem transport, one acting on the temporal (i.e., diel) scale and one on the spatial scale along the transport pathway. However, both changes are likely due to a common mechanism [44].

While in leaf sugars and the phloem of twigs of trees, there is a considerable diel variation of $\delta^{13}C$ related to transitory starch accumulation and remobilization (see above), an increasing dampening of these variations is observed along the tree axis in basipetal direction. This is assumed to result from the mixing of different sugar pools with various ages and metabolic histories during transport between the leaf and the tree trunk [45, 46]. Gessler, Ferrio et al. [8] proposed that the intrinsic properties of the dynamic Münch mass flow model can explain this sugar pool mixing during basipetal transport (see review by [47]). The model suggests that the sieve tubes act as a leaky pipe during transport: a proportion of the sucrose from sieve tubes is continually released to the surrounding parenchyma, and, to compensate for the loss, sugars from the surrounding tissues are reloaded back to the sieve tubes [48]. The loss and retrieval of sugars allow for continuous mixing of different sugar pools along the transport pathway towards the trunk base and, consequently, dampens the diel variations in $\delta^{13}C$. A combination of phloem mass flow assessments with the transport-distance-related dampening rate of diel $\delta^{13}C$ variations could be used to calculate rates of sugar loss from the phloem and retrieval.

The sugar loss and retrieval process can also explain the often observed ^{13}C enrichment of phloem sugars from the twig to the trunk base phloem [46]. The sugars leaking out of the phloem are used as substrates for different metabolic conversions in the parenchyma cells. If we assume that the compounds produced by these conversions are (1) deposited and thus not reloaded into the phloem (such as lignin) and (2) in their sum ^{13}C-depleted compared to the original sugar substrates originating from the phloem, the unreacted sugars, reloaded into the phloem, will be ^{13}C-enriched (Fig. 2). The continuous unloading, metabolic conversion and reloading of unreacted sugars will lead to the observed enrichment of phloem-transported sucrose in the basipetal direction.

4.2 Differences in $\delta^{18}O$ Between Leaf and Phloem Sugars

Gessler, Brandes et al. [36] observed that in three out of five tree species, sugars transported in the twig phloem were ^{18}O-depleted compared to sugars from adjacent leaves. Two of these species were conifers (*Pinus sylvestris* and *Larix decidua*). During phloem loading and transport of sucrose—the main constituent of phloem sugars in all species—such change in the isotope composition is not to be expected since sucrose contains no free carbonyl group. Only a carbonyl group allows the exchange between organic oxygen and oxygen from water, and thus could result in ^{18}O depletion, given that phloem water is less enriched than leaf water.

Loading of assimilates into the phloem in conifers is less understood than in angiosperm trees [49]. In contrast to angiosperm trees, the endodermis bundle sheath with cell walls locally suberized and lignified prevents any apoplastic radial transport in conifer needles, acting like a flow-controlling barrier similar to the root endodermis [50]. Consequently, the inward assimilate and outward water fluxes are constrained through a bottleneck formed by the plasmodesma sleeve within the cell wall of the endodermis. As the mass flow of

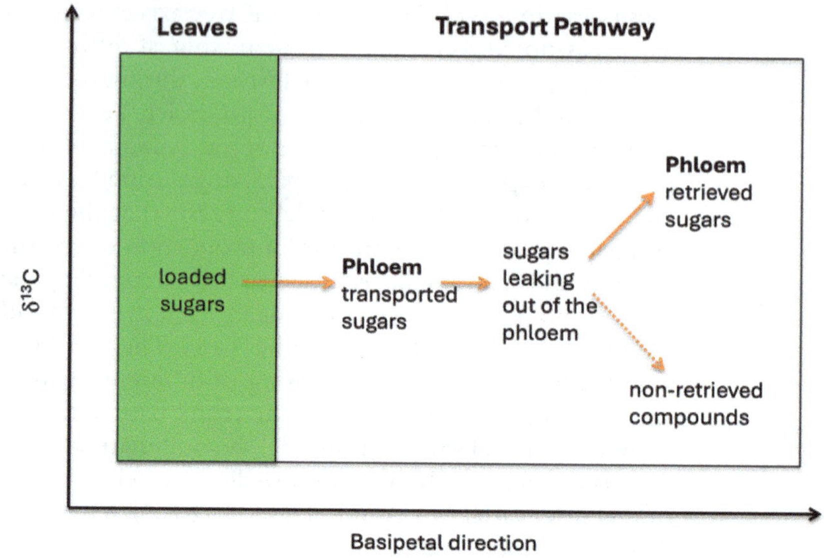

Fig. 2 Conceptual model of ^{13}C enrichment during phloem transport. The ^{13}C enrichment of carbohydrates transported in the phloem is due to the depletion of non-exported compounds generated from sugars leaking out of the phloem. Due to mass balance reasons, the remaining sugars retrieved are ^{13}C-enriched. For details, see [8]

water driven by transpiration is higher than the diffusive flow of assimilates in the opposite direction, the latter might become restricted if transpiration rates are high. It is also notable that in conifer needles, starch storage occurs not only in the mesophyll but also within the transfusion parenchyma [51]. If we assume that in phases of high transpiration during the daytime, no (or only few) new assimilates can be loaded into the phloem and—to maintain whole plant carbohydrate supply—phloem is fed by remobilized starch from the transfusion parenchyma, an ^{18}O depletion of the transported sugars becomes explainable. First, the water within the transfusion parenchyma is strongly influenced by non-evaporatively enriched xylem water and, thus, ^{18}O-depleted compared to the mesophyll water, which imprints its oxygen isotope signature on the new assimilates. During starch synthesis and breakdown in the transfer parenchyma, monosaccharides are produced as intermediate compounds, which can exchange carbonyl-oxygen with the non-enriched water. While only recently, Mai, Gao et al. [52] provided a model showing that bidirectional transport of an inbound diffusive sugar flux against an outbound advective water flux across the endodermis is, in principle, possible, these findings do not exclude the above-described phloem loading driven by transfusion parenchyma located starch. ^{13}C labeling experiments and temporally resolved in vivo localization of labeled starch with nano-SIMS could provide more information on phloem loading in conifers.

5 Improving our Understanding of the Tree Ring Archive

Stable isotopes in tree rings are used to reconstruct past climate variations and tree responses to these variations. Tree ring isotopes are, thus, the 'fingerprints' of past climate variations and plant performance [8, 53]. While it is often assumed that the information on iWUE and evaporative enrichment imprinted during photosynthesis in the $\delta^{13}C$ and $\delta^{18}O$, respectively, of new assimilates in the leaves is transferred into the tree ring archive without substantial alteration, there is now increasing evidence that processes downstream of the initial carboxylation in the Calvin cycle can alter the isotope signal [8, 44].

5.1 Oxygen Atom Exchange During Cellulose Synthesis Might Be Indicative of Source-Sink Relationships

Oxygen exchange occurs between sugars unloaded from the phloem and the non-evaporatively enriched reaction water during cellulose biosynthesis [54], leading to a depletion of cellulose in ^{18}O compared to the phloem transported sugars. The rate of oxygen atom exchange depends on the portion of hexose phosphates that cycle through triose phosphate before being incorporated into cellulose [55]. Exchange rates can be calculated by comparing the $\delta^{18}O$ of phloem-transported sugars during the growing season with $\delta^{18}O$ of the cellulose or wood produced simultaneously [56]. By doing so, it was observed that the exchange rates are variable and can range between close to 0% and 80% during one growing season. Song, Farquhar et al. [57] showed that the turnover time of carbohydrates and, thus, most likely, the relative availability of carbohydrates from source tissues (i.e., the carbon source strength) *vs.* the carbon demand in the sink tissues (i.e., the carbon sink strength) strongly affect the exchange rate. The lower the turnover time, the higher the sink demand compared to the source carbohydrate supply. A lower turnover time would give rise to a lower probability for hexose molecules to undergo triose cycling during the cellulose synthesis, and thus, less exchange would occur.

There is presently an intense controversy on the importance of source vs. sink control on vegetation growth and, hence, ecosystems' carbon balance [58]. Sink control thereby means that the demand for carbon by growth (but also respiration and other metabolic processes (together named sink activity) directly controls the plant carbon source activity, i.e., that there is a causal up-or down-regulation of the overall photosynthetic activity governed by the carbon demand. Source control, in contrast, indicates that the photosynthetic carbon availability directly controls growth and other sink activities. An accurate conceptual understanding of how the vegetation's carbon balance is determined by source or sink activity is essential to predict its carbon uptake and sequestration potential now and in the future. Still, there is no clear evidence whether one is dominating [58]. Combining assessments of tree ring growth with oxygen exchange during cellulose synthesis could, thus, help to better understand how source-sink relationships and their temporal dynamics are related to tree growth.

5.2 Position-Specific Isotope Signatures Can Provide Deeper Insights into Plant Metabolism

It is known that the hexoses (and hexose moieties of polysaccharides) show position-specific differences in the carbon isotope composition [59], which are assumed to contain information on specific metabolic processes not necessarily related to RuBisCO carboxylation. Wieloch, Ehlers et al. [59] postulated that carbon isotope effects in the glucose-6-phosphate isomerase-catalyzed reaction from fructose-6-phosphate to glucose-6-phosphate imprinted specific ^{13}C isotope signatures on the C-1 and C-2 positions of the tree ring glucose moieties. The isotope compositions of C-5 and C-6 are assumed to be influenced by leaf-level enolase, phosphoenolpyruvate carboxylase, pyruvate kinase, and 3-deoxy-D-*arabino*-heptulosonate 7-phosphate synthase [60]. In addition, chloroplastic aldolase and photorespiration, as well as transitory starch accumulation vs. export of triose-phosphate out of the chloroplast, affect the C-3 and C-4 positions of hexoses [61, 62]. Assessment of position-specific $\delta^{13}C$ in phloem sucrose might thus allow the integrated reconstruction of key metabolic processes in the leaves [24]. While still a lot of work is necessary to obtain quantitative information with such an approach, position-specific isotope analysis opens the path to label-free metabolic flux analysis, given that the relationship between enzyme-specific isotope effects, pathway commitments, and position-specific isotope compositions of metabolites is identified. The technological prerequisites are available as the new high-resolution bioanalytical mass spectrometers now offer high-throughput metabolite identification and are also suitable for compound- and intramolecular position-specific isotope analysis in the natural isotope abundance range [63]. Phloem sugars can, in the next step, link leaf metabolism to tree rings. Their position-specific isotope fingerprint can be used to quantify the transfer of isotopic information to the tree ring cellulose. The position-specific isotope composition in metabolites such as phloem sugars and tree ring cellulose can provide a linkage between metabolism and environment that is traceable with information from biological archives.

Conflict of Interest

The author has no conflicts of interest to declare that are relevant to the content of this chapter.

References

1. Helle G, Schleser G (2004) Beyond CO_2-fixation by Rubisco—an interpretation of C-13/C-12 variations in tree rings from novel intra-seasonal studies on broad-leaf trees. Plant Cell Environ 27(3):367–380
2. Keitel C et al (2003) Carbon and oxygen isotope composition of organic compounds in the phloem sap provides a short-term measure for stomatal conductance of European beech (*Fagus sylvatica L.*). Plant Cell Environ 26(7):1157–1168
3. Schulze E et al (1998) Carbon and nitrogen isotope discrimination and nitrogen nutrition of trees along a rainfall gradient in northern Australia. Aust J Plant Physiol 25(4):413–425
4. Trust BA, Fry B (1992) Stable sulphur isotopes in plants: a review. Plant. Cell Environ 15(9):1105–1110
5. Lehmann MM et al (2021) More than climate: hydrogen isotope ratios in tree rings as novel plant physiological indicator for stress conditions. Dendrochronologia 65:125788
6. Farquhar GD, O' Leary MH, Berry JA (1982) On the relationship between carbon isotope discrimination and the inter-cellular carbon-dioxide concentration in leaves. Aust J Plant Physiol 9(2):121–137
7. Scheidegger Y et al (2000) Linking stable oxygen and carbon isotopes with stomatal conductance and photosynthetic capacity: a conceptual model. Oecologia 125(3):350–357
8. Gessler A et al (2014) Stable isotopes in tree rings: towards a mechanistic understanding of isotope fractionation and mixing processes from the leaves to the wood. Tree Physiol 34(8):796–818
9. Vitali V et al (2024) Finding balance: tree-ring isotopes differentiate between acclimation and stress-induced imbalance in a long-term irrigation experiment. Glob Change Biol 30(3):e17237
10. Hogberg P (1997) Tansley review No 95 - N-15 natural abundance in soil-plant systems. New Phytol 137(2):179–203
11. Robinson D, Handley LL, Scrimgeour CM (1998) A theory for N-15/N-14 fractionation in nitrate-grown vascular plants. Planta 205(3):397–406
12. Peuke AD, Gessler A, Tcherkez G (2013) Experimental evidence for diel δ15N-patterns in different tissues, xylem and phloem saps of castor bean (*Ricinus communis* L.). Plant Cell Environ 36(12):2219–2228
13. Tcherkez G, Tea I (2013) 32S/34S isotope fractionation in plant sulphur metabolism. New Phytol 200(1):44–53
14. Cavallaro V et al (2022) Sulfur stable isotope discrimination in rice: a sulfur isotope mass balance study. Front Plant Sci 13:837517
15. Yoneyama T et al (1997) Variations of the natural abundances of nitrogen and carbon isotopes in Triticum aestivum, with special reference to phloem and xylem exudates. New Phytol 137(2):205–213
16. Pate J, Arthur D (1998) Delta C-13 analysis of phloem sap carbon: novel means of evaluating seasonal water stress and interpreting carbon isotope signatures of foliage and trunk wood of *Eucalyptus globulus*. Oecologia 117(3):301–311
17. Keitel C et al (2003) Carbon and oxygen isotope composition of organic compounds in the phloem sap provides a short-term measure for stomatal conductance of European beech (*Fagus sylvatica* L.). Plant Cell Environ 26(7):1157–1168
18. Barbour M et al (2000) Variation in the oxygen isotope ratio of phloem sap sucrose from castor bean. Evidence in support of the Peclet effect. Plant Physiol 123(2):671–679
19. Cernusak L, Wong S, Farquhar G (2003) Oxygen isotope composition of phloem sap in relation to leaf water in *Ricinus communis*. Funct Plant Biol 30(10):1059–1070
20. Werner RA, Bruch BA, Brand WA (1999) ConFlo III—an interface for high precision δ13C and δ15N analysis with an extended dynamic range. Rapid Commun Mass Spectrom 13(13):1237–1241
21. Werner RA (2003) The online 18O/16O analysis: development and application. Isot Environ Health Stud 39(2):85–104
22. Krummen M et al (2004) A new concept for isotope ratio monitoring liquid chromatography/mass spectrometry. Rapid Commun Mass Spectrom 18(19):2260–2266
23. Moerdijk-Poortvliet TCW et al (2015) Comparison of gas chromatography/isotope ratio mass spectrometry and liquid chromatography/isotope ratio mass spectrometry for carbon stable-isotope analysis of carbohydrates. Rapid Commun Mass Spectrom 29(13):1205–1214
24. Gessler A et al (2024) The marriage between stable isotope ecology and plant metabolomics—new perspectives for metabolic flux analysis and the interpretation of ecological archives. New Phytol 244(1):21–31
25. Gessler A, Rennenberg H, Keitel C (2004) Stable isotope composition of organic compounds transported in the phloem of European beech—Evaluation of different methods of phloem sap collection and assessment of gradients in carbon isotope composition during leaf-to-stem transport. Plant Biol 6(6):721–729

26. Lehmann MM et al (2020) Improving the extraction and purification of leaf and phloem sugars for oxygen isotope analyses. Rapid Commun Mass Spectrom 34(19):e8854
27. Devaux M et al (2009) Carbon stable isotope ratio of phloem sugars in mature pine trees throughout the growing season: comparison of two extraction methods. Rapid Commun Mass Spectrom 23(16):2511–2518
28. Barbour M et al (2005) Variation in the degree of coupling between delta C-13 of phloem sap and ecosystem respiration in two mature Nothofagus forests. New Phytol 166(2):497–512
29. Broussard L et al (2023) Phloem sap composition: what have we learnt from metabolomics?. Int J Mol Sci 24(8):6917
30. Cernusak LA et al (2003) Water relations link carbon and oxygen isotope discrimination to phloem sap sugar concentration in Eucalyptus globulus. Plant Physiol 131(4):1544–1554
31. Gessler A et al (2001) Radiation modifies the effect of water availability on the carbon isotope composition of beach (*Fagus sylvatica*). New Phytol 150(3):653–664
32. Gessler A et al (2008) Experimental evidence for diel variations of the carbon isotope composition in leaf, stem and phloem sap organic matter in *Ricinus communis*. Plant Cell Environ 31(7):941–953
33. Pate J et al (1998) Spatial and temporal variations in phloem sap composition of plantation-grown *Eucalyptus globulus*. Oecologia 117(3):312–322
34. Zhang C et al (2012) The origin and composition of cucurbit "phloem" exudate. Plant Physiol 158(4):1873–1882
35. Bogelein R, Thomas FM, Kahmen A (2017) Leaf water O-18 and H-2 enrichment along vertical canopy profiles in a broadleaved and a conifer forest tree. Plant Cell Environ 40(7):1086–1103
36. Gessler A et al (2013) The oxygen isotope enrichment of leaf-exported assimilates—does it always reflect lamina leaf water enrichment?. New Phytol 200:144–157
37. Gessler A et al (2007) Delta C-13 of organic matter transported from the leaves to the roots in Eucalyptus delegatensis: short-term variations and relation to respired CO_2. Funct Plant Biol 34(8):692–706
38. Schneider S et al (1996) Soluble N compounds in trees exposed to high loads of N: A comparison of spruce (*Picea abies*) and beech (*Fagus sylvatica*) grown under field conditions. New Phytol 134(1):103–114
39. Gessler A (1999) Untersuchungen zum Stickstoffhaushalt von Buchen (*Fagus sylvatica*) in einem stickstoffübersättigten Waldökosystem. 1. Aufl. ed. Frankfurt, M: Wiss.-Verl. Maraun
40. Lehmann MM et al (2020) The 18O-signal transfer from water vapour to leaf water and assimilates varies among plant species and growth forms. Plant Cell Environ 43(2):510–523
41. Rascher KG, Máguas C, Werner C (2010) On the use of phloem sap δ13C as an indicator of canopy carbon discrimination. Tree Physiol 30(12):1499–1514
42. Keitel C et al (2006) Carbon isotopic composition and oxygen isotopic enrichment in phloem and total leaf organic matter of European beech (*Fagus sylvatica* L.) along a climate gradient. Plant Cell Environ 29(8):1492–1507
43. Stojanović M et al (2024) Disaggregation of canopy photosynthesis among tree species in a mixed broadleaf forest. Tree Physiol 44(7):tpae064
44. Gessler A, Ferrio JP (2022) Postphotosynthetic fractionation in leaves, phloem and stem. In: Siegwolf RTW et al (ed) Stable isotopes in tree rings: inferring physiological, climatic and environmental responses. Springer International Publishing: Cham, pp 381–396
45. Kodama N et al (2008) Temporal dynamics of the carbon isotope composition in a *Pinus sylvestris* stand: from newly assimilated organic carbon to respired carbon dioxide. Oecologia 156:737–750
46. Brandes E et al (2006) Short-term variation in the isotopic composition of organic matter allocated from the leaves to the stem of *Pinus sylvestris*: effects of photosynthetic and post-photosynthetic carbon isotope fractionation. Glob Change Biol 12(10):1922–1939
47. Van Bel AJE (2003) The phloem, a miracle of ingenuity. Plant Cell Environ 26(1):125–149
48. Minchin P, Thorpe M (1987) Measurement of unloading and reloading of photo-assimilate within the stem of bean. J Exp Bot 38(187):211–220
49. Liesche J et al (2017) Height-related scaling of phloem anatomy and the evolution of sieve element end wall types in woody plants. New Phytol 214(1):245–256
50. Liesche J, Martens HJ, Schulz A (2011) Symplasmic transport and phloem loading in gymnosperm leaves. Protoplasma 248(1):181–190
51. Bäck J, Huttunen S, Kristen U (1993) Carbohydrate distribution and cellular injuries in acid rain and cold-treated spruce needles. Trees 8(2):75–84
52. Mai MH et al (2024) Relieving the transfusion tissue traffic jam: a network model of radial transport in conifer needles. New Phytol 244(6):2183–2196
53. Treydte K et al (2024) Recent human-induced atmospheric drying across Europe unprecedented in the last 400 years. Nat Geosci 17(1):58–65

54. Sternberg L et al (2006) Variation in oxygen isotope fractionation during cellulose synthesis: intramolecular and biosynthetic effects. Plant Cell Environ 29(10):1881–1889
55. Hill SA et al (1995) Rapid recycling of triose phosphates in oak stem tissue. Plant, Cell Environ 18:931–936
56. Offermann C et al (2011) The long way down-are carbon and oxygen isotope signals in the tree ring uncoupled from canopy physiological processes?. Tree Physiol 31(10):1088–1102
57. Song XIN et al (2014) Turnover time of the non-structural carbohydrate pool influences δ 18O of leaf cellulose. Plant Cell Environ 37:2500
58. Gessler A, Zweifel R (2024) Beyond source and sink control—toward an integrated approach to understand the carbon balance in plants. New Phytol 242:858
59. Wieloch T et al (2018) Intramolecular 13C analysis of tree rings provides multiple plant ecophysiology signals covering decades. Sci Rep 8(1):5048
60. Wieloch T et al (2022) Intramolecular carbon isotope signals reflect metabolite allocation in plants. J Exp Bot 73(8):2558–2575
61. Rossmann A, Butzenlechner M, Schmidt HL (1991) Evidence for a nonstatistical carbon isotope distribution in natural glucose. Plant Physiol 96(2):609–614
62. Tcherkez G et al (2004) Theoretical considerations about carbon isotope distribution in glucose of C-3 plants. Funct Plant Biol 31(9):857–877
63. Neubauer C et al (2023) Discovering nature's fingerprints: isotope ratio analysis on bioanalytical mass spectrometers. J Am Soc Mass Spectrom 34(4):525–537

Part VI

Grafting, Phloem Development, and Signaling

Check for updates

Chapter 33

High-Throughput Hypocotyl Grafting and Vascular Reconnection Assessment in Arabidopsis Seedlings

Kai Bartusch and Elisabeth Truernit

Abstract

Plant grafting is a powerful tool in research and agriculture, enabling the study of long-distance signaling and improving crop performance. Hypocotyl grafting in *Arabidopsis thaliana* has become a key method for investigating inter-tissue communication, root-to-shoot transport, and vascular regeneration. Here, we describe a high-throughput hypocotyl grafting protocol that consistently achieves success rates above 90% while minimizing complications, such as adventitious root formation, through an optimized hypocotyl cutting position. Our streamlined approach eliminates the need for silicone tubes and fine forceps, requiring only a standard scalpel for the cutting and joining procedure. With proper training, an experienced user can graft up to 1000 plants per day, facilitating large-scale experiments. Additionally, we present vascular connectivity assays using fluorescent dyes and the adjuvant Adigor to enable phloem and xylem transport analysis across the graft junction without the need for wounding. By simplifying the grafting process while maintaining high efficiency, our protocol makes Arabidopsis grafting more accessible and scalable, facilitating research on long-distance molecular transport, vascular development, and plant biotechnology applications.

Keywords *Arabidopsis thaliana*, Grafting, Chimeric plants, Hypocotyl, Long-distance transport, Shoot-root interaction, Vascular regeneration, Phloem transport, Fluorescent dyes

1 Introduction

Plant grafting is an ancient technique used to combine desirable traits from different plants. Traditionally, this involves cutting the shoot (scion) of one plant and attaching it to the root system (rootstock) of another [1, 2]. Grafting is widely employed in agriculture to increase crop yields, enhance resistance to biotic and abiotic stresses, and propagate plant material. To date, grafting remains a versatile tool, with innovations in techniques and automation expanding its application in both scientific research and agriculture [3–6].

More than 30 years ago, *Arabidopsis thaliana* (Arabidopsis) grafting was established, initially for inflorescence stems [7]. Introducing grafting of seedling tissues (micro-grafting) revolutionized

Johannes Liesche and Sanna Sevanto (eds.), *Phloem*, Methods in Molecular Biology, vol. 3066,
https://doi.org/10.1007/978-1-0716-5488-0_33,

Arabidopsis grafting due to its high success rates and efficiency [8]. This led to various techniques for grafting different tissues, including hypocotyls [8–22], epicotyls [16], roots [23], and cotyledons [24, 25]. Arabidopsis grafting has since become a key method for studying shoot-root communication and long-distance molecular transport, revealing the mobility of RNAs, peptides/proteins, metabolites, hormones, and nutrients [26, 27]. More recently, Arabidopsis rootstocks carrying a mobile genome-editing machinery have enabled transgene-free edits in grafted scions, significantly accelerating the development of new plant varieties and offering a powerful tool for modern breeding [28]. Furthermore, Arabidopsis grafting has been instrumental for investigating tissue communication, wound healing, and vascular differentiation during graft formation [1, 2]. Due to its short life cycle, minimal space requirements, and extensive genetic resources, Arabidopsis has become a widely used model for grafting in fundamental research and agricultural innovation [29].

Hypocotyl grafting has become the most commonly used grafting method in Arabidopsis [29]. While environmental factors affecting grafting, such as humidity and temperature, have been optimized [8, 11, 14, 16, 19], technical challenges remain, particularly in the grafting procedure itself. Different protocols suggest various materials, affecting success rates, cost, and efficiency [29]. For example, for setting clean and precise cuts during grafting, tools like injection needles [10, 20], razor blades [9, 11, 13, 16], scalpels [8, 9, 11, 13, 22], or specialized micro-knives [8, 9, 12–15, 19] have been used. Some methods employ silicone tubes to stabilize the graft junction, but this adds complexity and time, reducing throughput and limiting sample sizes [8, 9, 13, 16]. Additionally, innovations like grafting chips, designed to streamline the grafting procedure, have been associated with increased adventitious root formation and lower success rates compared to free-hand grafting [17, 18, 21]. Therefore, there remains a need for a simpler, more efficient hypocotyl grafting protocol that maintains high success rates while minimizing complications such as adventitious root formation. An optimal protocol should combine precision, cost-effectiveness, and high throughput, making it accessible for a wide range of experimental setups and research needs.

Here, we present a simplified, high-throughput hypocotyl grafting protocol for Arabidopsis that addresses the limitations of existing methods. Building on previously described hypocotyl grafting methods [8, 14, 15, 19], our protocol consistently achieves high success rates (>90%) at low cost, requiring only a standard scalpel and eliminating the need for silicone tubes or fine forceps during grafting. Our streamlined approach minimizes adventitious root formation through an optimized cutting procedure. With training, an experienced person can process up to 1000 plants per day, enabling large-scale experiments. Additionally, we demonstrate the versatility of our protocol in vascular regeneration studies. Using carboxyfluorescein diacetate (CFDA) and Rhodamine WT, we assess phloem reconnection by monitoring phloem transport. Incorporating the adjuvant Adigor into the dye solution eliminates the need for wounding leaves with fine forceps, a step required in other protocols [15, 30]. Overall, our hypocotyl grafting protocol is adaptable to various experimental setups and graft combinations, including interspecies grafting, and can be integrated with other methods such as two-shoot [8–10, 13, 14], petiole [25], or cotyledon grafting [30]. When planning experiments based on our protocol, researchers should consider that increasing procedural complexity extends the time required per plant, thereby reducing experimental throughput.

2 Materials

2.1 Preparation of Plants for Grafting

1. Arabidopsis seeds.
2. 1.5 mL microcentrifuge tubes.
3. Pipettes and tips.
4. 70% (v/v) ethanol.
5. 96% (v/v) ethanol.
6. Plates containing standard growth medium for Arabidopsis, such as ½ Murashige Skoog (MS) medium (pH 5.7) with 1%–1.2% (w/v) plant agar.
7. 3 M Micropore Surgical Tape.
8. Laminar flow hood.
9. Plant growth cabinet.

2.2 Hypocotyl Grafting Procedure

1. Arabidopsis plants (from Sect. 3.1).
2. Sterile deionized water.
3. 9 cm round Petri dishes.
4. Whatman 3 MM CHR cellulose chromatography paper, 46 × 57 cm; *see* **Note 1**.
5. Mixed cellulose ester (MCE) membrane, such as the MF-Millipore membrane filter with 0.45 μm pore size and 50 mm diameter (catalog number HAWP05000); *see* **Note 2**.
6. Scalpel blade No. 11 mounted on scalpel handle No. 3 (e.g., Swann-Morton or Aesculap); *see* **Note 3**.
7. Fine forceps.
8. Dissecting microscope.
9. Parafilm.
10. Laminar flow hood.
11. 70% (v/v) ethanol.

2.3 Post-Grafting Procedures

1. Grafted Arabidopsis plants, including controls (from Sect. 3.2).
2. Plant growth cabinet.
3. Fine forceps.
4. Dissecting microscope.
5. Laminar flow hood and 70% (v/v) ethanol (if sterile conditions are still needed).

2.4 Vascular Transport Assays

1. Grafted Arabidopsis plants, including controls (from Sect. 3.2).
2. 1.5 mL microcentrifuge tubes.
3. Adigor (Syngenta).
4. 5(6)-Carboxyfluorescein diacetate (CFDA); *see* **Note 4**.
5. Dimethyl sulfoxide (DMSO).
6. Rhodamine WT 20% in water (as an alternative to CFDA); *see* **Note 4**.
7. Deionized water (for diluting Rhodamine WT).
8. Aluminum foil.
9. Pipettes and tips.
10. Parafilm (for xylem assay).
11. Epifluorescence microscope with a suitable GFP/YFP filter to detect CFDA, or an RFP/mCherry filter for Rhodamine WT. We used a Leica M205 FCA with an ET GFP filter (excitation 470/40 nm, emission 525/50 nm) and an mCherry filter (excitation 560/40 nm, emission 630/75 nm).

3 Methods

In the following sections, we describe the preparation of plants for grafting (Sect. 3.1), the hypocotyl grafting procedure (Sect. 3.2), and post-grafting procedures (Sect. 3.3). These methods are adapted from previous protocols [8, 14, 15, 19], modified for higher throughput. Additionally, we detail optional vascular transport assays for assessing phloem and xylem reconnection in grafted plants (Sect. 3.4). A workflow overview of all described techniques is presented in Fig. 1.

3.1 Preparation of Plants for Grafting

1. Aliquot the seeds of the desired genotypes in 1.5 mL microcentrifuge tubes.
2. Sterilize seeds using a standard protocol. We recommend washing the seeds with 70% (v/v) ethanol for 1 min, followed by a second wash with 96% (v/v) ethanol for 1 min. Remove the remaining solution and air-dry the seeds in a laminar flow hood.

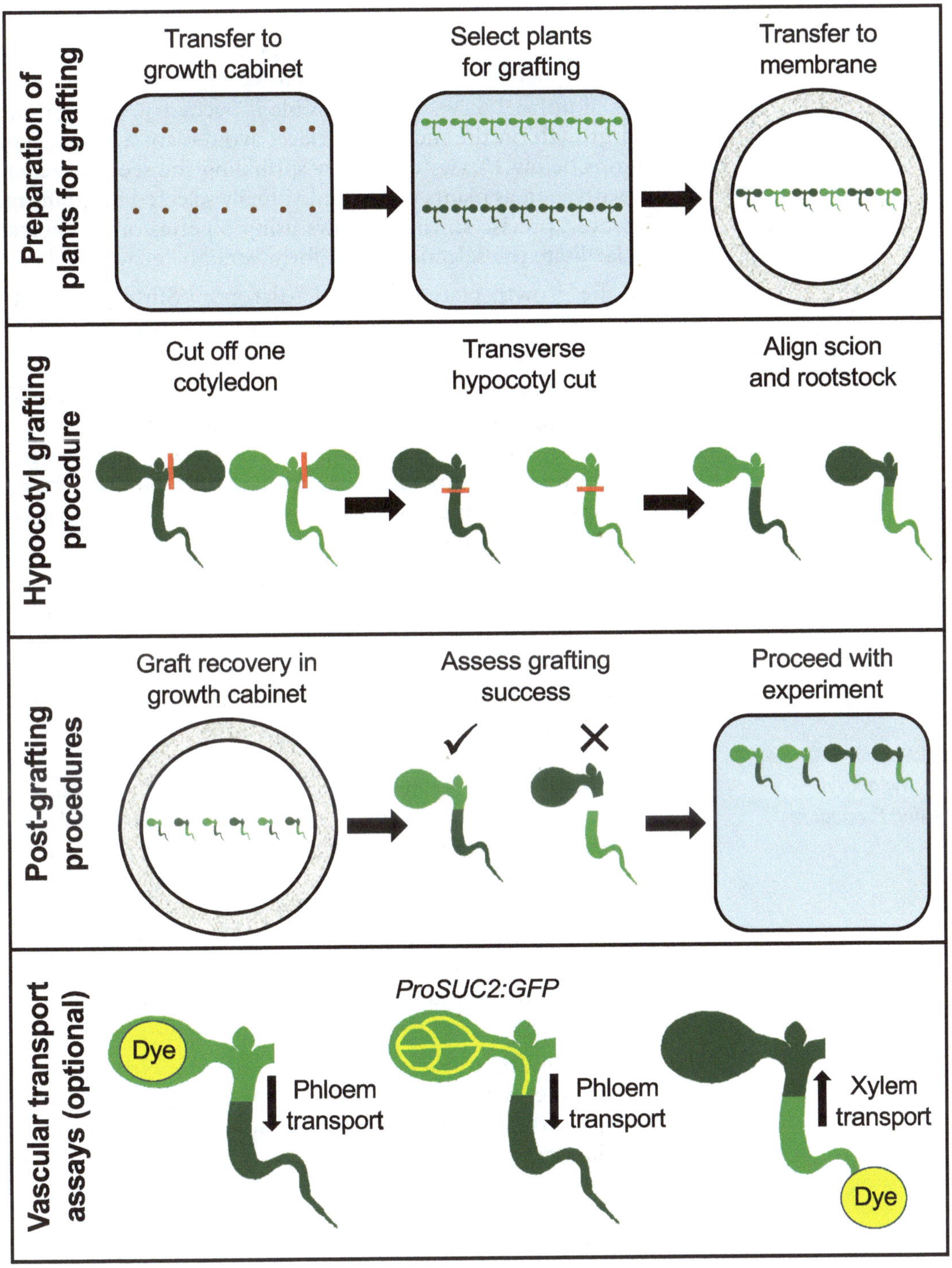

Fig. 1 Schematic overview of a typical hypocotyl grafting experiment. Vascular transport assays for assessing phloem and xylem reconnection are optional. Different shades of green (dark and light) represent distinct genotypes to be combined through grafting. Red lines indicate cut sites

3. Sow the seeds onto standard growth medium under sterile conditions in a laminar flow hood. Ideally, the medium should be free of sucrose and antibiotics. However, sucrose concentrations of up to 1% (w/v) are possible if necessary. To promote root growth on the medium surface, avoid plant agar concentrations below 1% (w/v). Simply sprinkling the seeds is generally sufficient, as plants will be individually selected for grafting. However, precise sowing in rows using pipettes or toothpicks can facilitate the selection of uniform, straight-growing plants.
4. Seal the growth plate with 3 M Micropore Surgical Tape to allow gas exchange. Stratify the seeds by incubating the plates in darkness at 4°C for at least 2 days.
5. After stratification, place the plates vertically in a growth cabinet under conditions tailored to your experiment. Temperature and light intensity influence plant development and can affect the grafting procedure (*see* **Note 5**). Based on our experience, optimal growth conditions range from 20 to 27°C with light intensities between 30 and 100 $\mu mol \cdot m^{-2} \cdot s^{-1}$.
6. Grow plants for 5 days under a 16 h light/8 h dark cycle. For shorter photoperiods (e.g., 12 h light/12 h dark or 8 h light/16 h dark), use 6- to 7-day-old plants for grafting. Ensure that the cotyledons are fully expanded while the first true leaves remain small. Although grafting older plants is possible, it requires additional time (*see* **Note 6**).

3.2 Hypocotyl Grafting Procedure

1. Prepare the following materials prior to grafting: Cut Whatman paper into circles using the bottom of a 9 cm Petri dish as a template to ensure a proper fit. Each Petri dish requires two circles. Additionally, cut Whatman paper into strips (~2 × 6 cm) for removing excess water during grafting. For alternative papers (*see* **Note 1**) or membrane materials (*see* **Note 2**), conduct a small test experiment first. Wrap the prepared Whatman papers and the membranes in aluminum foil and sterilize by autoclaving.
2. Perform grafting in a laminar flow hood (*see* **Note 7**) using a dissecting microscope. Clean the microscope, scalpels, and forceps with 70% (v/v) ethanol. Ensure scalpels and forceps remain sterile during grafting. Allow ethanol to fully evaporate before handling the plants, as residual ethanol impairs grafting success.
3. Fill a Petri dish with sterile deionized water (*see* **Note 8**).
4. Submerge two Whatman paper circles in the water, then carefully remove them, allowing excess water to drain off, and place them into a new Petri dish using forceps.

5. Submerge a membrane piece in water, drain excess water as above, and place it atop the Whatman papers (Fig. 2A).
6. Using fine forceps, carefully transfer six to ten plants onto the membrane in a row-like pattern (Fig. 2B). Select only straight-growing plants with fully expanded cotyledons positioned at right angles to the hypocotyl.

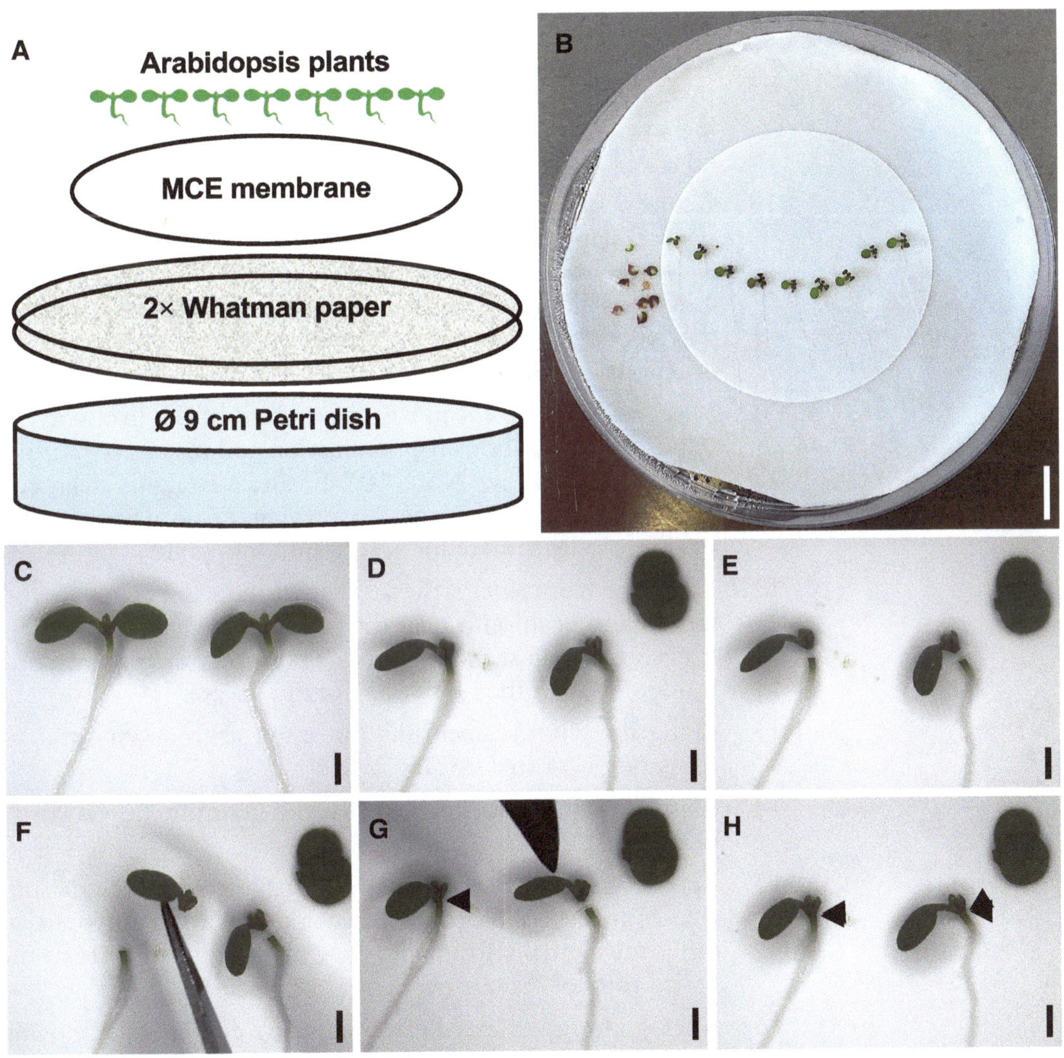

Fig. 2 Hypocotyl grafting methodology. (**A**) Grafting setup (adapted from [14, 19]): plants placed on a mixed cellulose ester (MCE) membrane, with two Whatman paper circles positioned underneath inside a Petri dish. (**B**) Setup with 14-day-old plants grafted at day 7. (**C–H**) Step-by-step hypocotyl grafting workflow for two plants. (**C**) Plants prior to grafting. (**D**) Excess water removed using a strip of Whatman paper, one cotyledon excised per plant. (**E**) Hypocotyls are transversely cut as close as possible to the shoot apical meristem. (**F**) Scions were carefully transferred near the recipient rootstocks using the blunt side of a scalpel. (**G**) Scions positioned against the recipient rootstocks. (**H**) Cut edges of scion and rootstock are properly aligned and barely visible. Black triangles indicate graft junctions. Scale bars: (**B**) 1 cm; (**C–H**) 1 mm

7. Arrange different genotypes in a logical order. For example, two genotypes can be placed alternately on a single membrane for pairwise grafting (Figs. 1 and 2C-H; *see* **Note 9**).
8. Remove excess water with a Whatman paper strip, ensuring only a thin film of water remains on the hypocotyl and root (Fig. 2C, D). Excess water causes scion cotyledons to adhere to the membrane, while insufficient moisture prevents proper adhesion at the graft junction.
9. The following cutting and joining procedure (Steps 10–14) should be performed using a scalpel (*see* **Note 3**). Throughout the procedure, careful handling is essential to avoid tissue damage. Work efficiently to prevent desiccation, and if necessary, add water to maintain humidity and prevent wilting.
10. Remove one cotyledon per plant by cutting through the petiole. This ensures that the hypocotyl lies flat (Fig. 2D), allowing precise scion-rootstock alignment. Retain the larger cotyledon if the other is smaller or bent. Place the excised cotyledons on one side of the Petri dish for better spatial organization (Fig. 2B).
11. Make a transverse cut in the upper third of the hypocotyl to minimize adventitious root formation and maximize grafting success (Fig. 2E; *see* **Note 10**). Ensure a straight, clean cut. Replace dull scalpel blades to prevent tissue damage. One blade typically suffices for several hundred grafts.
12. Using the blunt side of the scalpel, lift the scion below its cotyledon and position it near the recipient rootstock (Fig. 2F). If necessary, adjust the rootstock's position by gently rotating it sideways with the blunt edge of the scalpel.
13. Using the blunt scalpel side, move the scion down onto the rootstock's cut surface (Fig. 2G).
14. Align scion and rootstock so that the graft junction is barely visible (Fig. 2H).
15. Repeat Steps 10–14 for all plants (*see* **Note 11**). Include suitable controls (*see* **Note 12**). If working with genotypes that differ substantially from wild-type plants, adjustments to the basic protocol may be necessary (*see* **Note 13**).
16. After grafting, adjust the humidity by adding or removing water to maintain a minimal water film at the root. Excess moisture promotes adventitious root formation, while insufficient moisture causes wilting.
17. Seal the Petri dish with parafilm (*see* **Note 14**).

3.3 Post-Grafting Procedures

1. Position the plates vertically in a growth cabinet with overhead lighting. Adjust temperature (*see* **Note 15**) and light conditions (*see* **Note 16**) to optimize grafting success. In our experiments, temperatures of 20–27°C, white light at 90–100 μmol·m^{-2}·s^{-1}, and photoperiods ranging from 8 h light/16 h dark to 16 h light/8 h dark were suitable for graft healing.
2. Maintain grafted plants in the growth cabinet for at least 7 days. The rootstock generally resumes growth 4–5 days post-grafting, while the vasculature is typically fully connected within 6–7 days after grafting (Sect. 3.4).
3. After 7–10 days, examine grafted plants using a dissecting microscope. Successful grafts exhibit continued shoot and root growth, with a barely visible graft junction (Fig. 3A), absence of massive callus formation (Fig. 3B), and no adventitious roots above the graft junction (Fig. 3C).
4. If sterility is required for subsequent procedures, perform the following steps in sterile conditions. Remove the parafilm and open the grafting plates. Using fine forceps, gently pull the scion to assess tissue attachment without damaging any tissues. Proceed only with plants that have a stable graft connection.
5. Transfer grafted plants to fresh standard growth medium and seal the plates with 3 M Micropore Surgical Tape. Alternatively, transplant the plants directly into soil, ensuring that the graft junction remains above the soil surface to prevent adventitious root formation and allow for easy inspection (*see* **Note 17**). We recommend covering them with a lid for the first 6–7 days to maintain high humidity and promote acclimation.
6. After plant transfer, continue with the specific experimental procedure, adjusting growth conditions as necessary. At the experiment's conclusion, re-examine plants for growth arrest or adventitious roots (*see* **Note 17**).

Fig. 3 Assessment 7 days after grafting. (**A**) Successfully grafted plant. (**B**) Plant exhibiting excessive callus formation due to a misaligned graft junction. (**C**) Plant with an adventitious root emerging from the scion. Black triangles indicate graft junctions. Scale bars: 500 μm

3.4 Vascular Transport Assays

Vascular reconnection between scion and rootstock is essential, as insufficient reconnection, often indicated by adventitious root formation or stunted growth, leads to graft failure [1, 2]. To analyze vascular regeneration, physiological assessments of molecular transport across the graft junction provide high-throughput and technically simple approaches (Fig. 1). Phloem transport is typically restored within 4 days after grafting (Fig. 4), while xylem transport becomes functional after 6–7 days [31]. Here, we present an optimized phloem transport assay using fluorescent dye solutions, building on a previous protocol [30]. The addition of the adjuvant Adigor facilitates dye penetration into plant tissue [32]. For dye solution preparation and xylem reconnection assessment, we follow our previously published protocol [33].

1. Preparation of the CFDA solution: Weigh 46 mg CFDA into a 1.5 mL microcentrifuge tube and dissolve it in 1 mL DMSO to prepare a 100 mM stock solution. To prepare a 10 mM working solution with 0.5% (v/v) Adigor, mix 100 µL of the stock

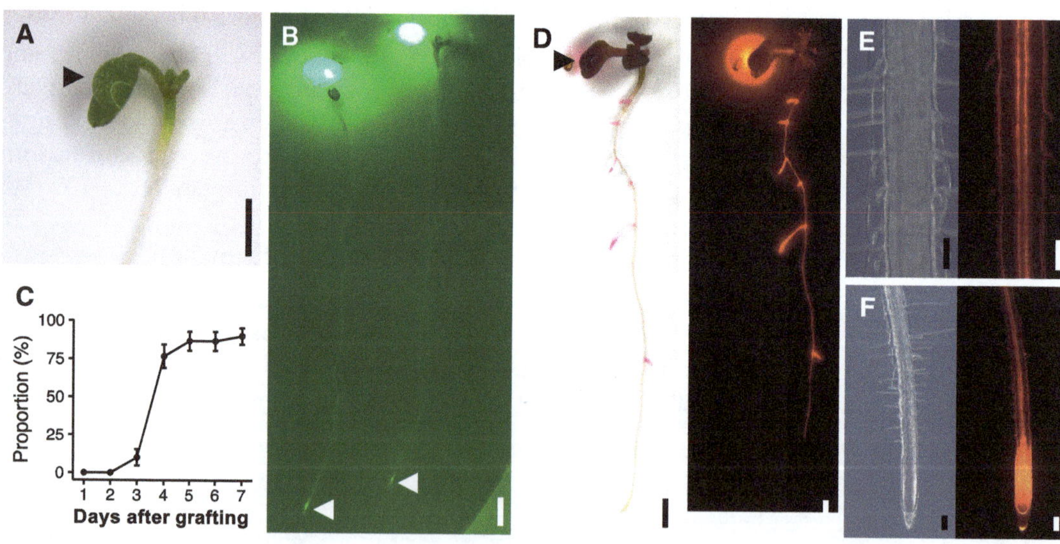

Fig. 4 Monitoring phloem transport restoration in grafted plants using fluorescent dyes. (**A**) A drop of CFDA solution pipetted onto the cotyledon, marked by a black triangle. (**B**) Detection of CFDA fluorescence in the rootstock using a GFP filter confirms successful phloem reconnection across the graft junction and restored phloem transport. White triangles indicate dye accumulation in the phloem unloading zone of the root tip. (**C**) Proportion of grafted Col-0 plants showing CFDA fluorescence in the rootstock. Phloem transport was assessed 1 h after dye application on consecutive days post-grafting, using a new set of plants for each time point, as plants cannot be reused in CFDA experiments. Error bars represent the standard error of the proportion ($n = 30$). Plants were 7 days old at grafting, and grown under a 12 h light/12 h dark cycle (100 µ mol·m^{-2}·s^{-1}) at 20°C. (**D–F**) Rhodamine WT transport visualized in brightfield mode (left) and fluorescence mode with an mCherry filter (right). (**D**) A drop of Rhodamine WT was pipetted onto the cotyledon, marked by a black triangle. (**E**) Differentiated root section showing Rhodamine WT fluorescence in the two phloem strands. (**F**) Root tip displaying dye unloading. Scale bars: (**A, B,** and **D**) 1 mm; (**E** and **F**) 100 µm

solution with 895 μL DMSO and 5 μL Adigor in a new 1.5 mL microcentrifuge tube (*see* **Note 18**).

2. Preparation of the Rhodamine WT solution (as an alternative to CFDA): Mix 283 μL Rhodamine WT (20% in water), 712 μL deionized water, and 5 μL Adigor. This 100 mM Rhodamine WT solution with 0.5% (v/v) Adigor can be used directly (*see* **Note 19**).
3. Wrap prepared dye solutions in aluminum foil and store at 4°C.
4. To analyze phloem reconnection, apply 1 μL of dye solution to the adaxial side of the cotyledon of a grafted plant using a pipette (Figs. 1 and 4A, D). If Adigor is not used, gently wound the cotyledon at the application site by pressing with fine forceps before applying the dye solution.
5. To analyze xylem reconnection, position grafted plants with their root tips on a parafilm strip placed on the grafting plate, ensuring that the shoot and most of the root system remain in contact with the moist membrane (*see* **Note 20**). Apply 1 μL of CFDA or Rhodamine WT solution (containing Adigor) to the root tip using a pipette (Fig. 1). If Adigor is not used, cut the root at the application site with a scalpel to facilitate dye uptake.
6. Cover the Petri dish with the lid to maintain humidity. Include non-grafted plants as controls (*see* **Note 12**).
7. After 1 h, inspect rootstocks (phloem assay) or scions (xylem assay) using an epifluorescence microscope. Use a GFP/YFP filter for CFDA detection or an RFP/mCherry filter for Rhodamine WT. At high Rhodamine WT concentrations, the dye can also be detected in bright-field mode (Fig. 4D). In grafted plants with reconnected phloem, the dye will be detected in the root and accumulate at the root tip (Fig. 4B, F). At higher magnification, the dye should be visible in the two phloem strands of the root (Fig. 4E). Xylem reconnection is confirmed if the dye reaches aerial tissues, such as the cotyledon. If the dye accumulates at the graft junction and is not detected in the rootstocks (phloem assay) or scions (xylem assay), the vasculature is disconnected.
8. To quantify the vascular reconnection rate for a single time point, record the number of plants exhibiting dye signal in the respective organ and calculate the proportion relative to the total number of analyzed plants. Time-course experiments investigating vascular regeneration can be designed (Fig. 4C; *see* **Note 21**).

4 Notes

1. To save preparation time, consider using precut circles (e.g., Whatman Grade 1 qualitative filter paper circles, 85 mm diameter). Thinner filter papers may require more than two

layers to retain sufficient moisture for grafting, whereas a single layer of thicker paper may suffice but should be tested.

2. We have found that mixed cellulose ester (MCE), nitrocellulose, and nylon membranes are effective. The membrane must provide an adequate water supply while maintaining a solid, flat surface. It should be hydrophilic, with an optimal pore size of ~45 μm. If not using precut circles, prepare 2×4 cm strips of MCE/nylon/nitrocellulose membrane. Grafting directly on Whatman paper without a membrane leads to fibers adhering to the plants and tools, interfering with the cutting and joining process.
3. Micro knives (e.g., Ultra Fine Micro Knives, catalog number 10315-12, Fine Science Tools) produce precise and clean cuts, but we found standard scalpels to be equally effective and a more cost-effective alternative. The tip of the scalpel blade No. 11 is comparable to specialized micro knives and offers the advantage of its blunt side for moving scions without puncturing leaves. When using micro knives, fine forceps are required for scion-rootstock alignment, which adds time to the grafting process. Razor blades were also tested but lacked the handling precision of scalpels and micro knives.
4. Here, we describe the use of the dyes CFDA and Rhodamine WT. However, alternative dyes can be considered for monitoring vascular transport. For instance, esculin has been suggested as a phloem-mobile probe, as it is specifically loaded into the phloem via the companion cell-specific sucrose transporter SUC2 in Arabidopsis. Moreover, esculin has the advantage of being degraded in plant tissue within a few hours, allowing for reloading at a later time point [32].
5. Low light levels and increased temperatures promote hypocotyl and petiole elongation [34], which facilitates the cutting and joining of scions during grafting. In contrast, plants with stunted growth or with cotyledons bent over their hypocotyls, whether due to genotype or environmental conditions, are technically more challenging to graft. Moreover, increased temperatures prior to grafting do not enhance graft healing [30], whereas higher temperatures post-grafting promote vascular regeneration (*see* **Note 15**).
6. When grafting older plants with true leaves, these leaves must be removed in addition to the cotyledons to ensure proper alignment of the scion and rootstock on the grafting plate.
7. Since our protocol does not require sucrose during grafting, it can be performed on a lab bench under non-sterile conditions, as previously reported [25]. However, this may occasionally lead to fungal contamination, reducing grafting success.

8. Adding sucrose or using liquid ½ MS medium should be avoided, as both reduce the success rate of our cutting and joining procedure. However, some studies report that sucrose supplementation in the Petri dish can enhance grafting success when both cotyledons are removed [11]. Additionally, hormone supplements such as indole-3-acetic acid (IAA) have been suggested to suppress adventitious root formation and improve grafting success [35]. In our experiments, however, IAA supplementation on the grafting plate decreased grafting success, whereas optimizing the cutting position was the key to achieving high success rates (*see* **Note 10**).
9. Alternatively, place the other genotype in a second row on an additional membrane and exchange scions between the lower and upper rows. To minimize desiccation, transfer plant organs only within the same Petri dish.
10. Previous protocols emphasized optimizing the transverse cut position, as grafting at a lower hypocotyl section increases adventitious root formation [8]. While some protocols recommend removing adventitious roots [17, 18, 21], we advise excluding plants with adventitious roots (*see* **Note 17**). To determine the optimal cut position, we systematically tested grafting at high, middle, and low hypocotyl positions (Fig. 5A–C). All plants grafted at the low position formed adventitious roots, whereas cutting at the high position largely suppressed adventitious root formation (Fig. 5D). This observation aligns with the previously reported increased competence for adventitious root formation in basal hypocotyl regions [36]. Additionally, plants with lower graft positions exhibited weaker graft junctions, with scions detaching when gently pulled with forceps. Thus, adventitious root formation correlates with overall grafting success (Fig. 5E). Our results confirm the importance of a high cut position to minimize adventitious roots and maximize grafting success.
11. To improve efficiency when grafting several plants at the same time, perform all cuts first, then remove the cotyledons and transfer the scions to the rootstocks at low magnification. Increase the magnification for precise graft alignment.
12. To identify general issues during grafting (e.g., wilting), include a set of non-grafted plants as controls and subject them to identical treatment, except for hypocotyl cutting and joining. Position these control plants on the membrane alongside the grafted plants and remove one cotyledon. Additionally, incorporate self-grafted control plants to confirm that any observed differences between combined genotypes in an experiment are not attributable to the grafting procedure itself. For this purpose, graft scions onto rootstocks

of the same genotype, either pairwise or within a single plant, by excising a small disc at the hypocotyl junction to prevent tissue self-attachment. In a typical reciprocal grafting experiment involving two genotypes, four combinations should be included: genotype A self-grafted, genotype B self-grafted, genotype A scion grafted onto a genotype B rootstock, and genotype B scion grafted onto a genotype A rootstock.

13. Our standard protocol is optimized for wild-type Arabidopsis but can be adapted to different genotypes. We recommend initially growing the genotypes of interest alongside wild-type controls under standard conditions. Examine the plants using a dissecting microscope to assess size differences. Place them on a grafting plate and test cutting and joining to ensure proper hypocotyl alignment. Ideally, both genotypes should have similar hypocotyl diameters. If they differ significantly, a section of the rootstock hypocotyl may be removed to accommodate smaller scions, while the scion should still be cut as close as possible to the apical meristem to minimize adventitious root formation (*see* **Note 10**). If size differences are substantial, consider adjusting growth conditions. Genotypes with stunted growth may require a longer growth period or altered light/temperature conditions to better align with controls. For instance, older *Eutrema salsugineum* plants have been shown to more closely match the development of faster-growing Arabidopsis plants [16]. Additionally, if anatomical features are affected (e.g., differently shaped cotyledons in some Arabidopsis mutants or other species, such as *Brassica napus* [37] and *Solanum lycopersicum* [11]), the cutting procedure must be adapted. In such cases, it may be necessary

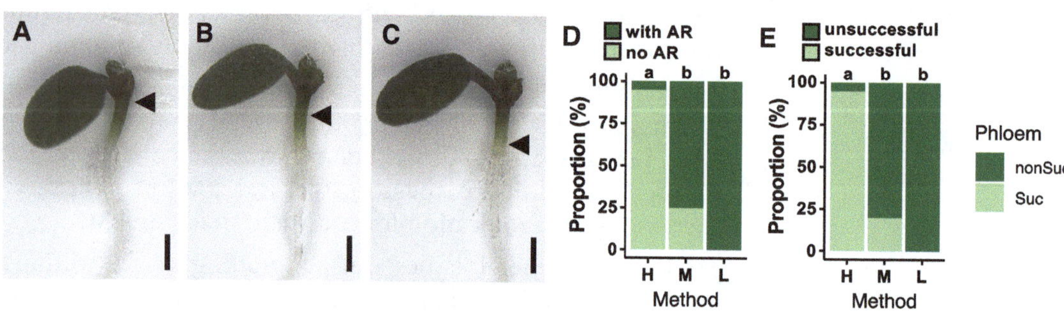

Fig. 5 Optimizing the cut position is crucial for grafting success. Examples of grafted plants with hypocotyls cut at a (**A**) high, (**B**) middle, or (**C**) low position. Black triangles indicate graft junctions. Grafted Col-0 plants were analyzed for (**D**) adventitious root formation and (**E**) grafting success ($n = 20$). Detached scions, excessive callus formation, or adventitious root formation were classified as unsuccessful (Fig. 3). Statistical significance was assessed using pairwise Fisher's exact tests, with p values adjusted by the Benjamini–Yekutieli method. Different letters indicate significant differences ($p < 0.05$). Plants were 7 days old at grafting, assessed 7 days later, and grown under a 12 h light/12 h dark cycle (100 μmol·m^{-2}·s^{-1}) at 20°C. Scale bars: 500 μm. H, high cut position; M, middle cut position; L, low cut position; and AR, adventitious root

to remove part or all of the second cotyledon to allow the shoot to lie flat on the grafting membrane. This additional cut has also been shown to improve cotyledon and petiole alignment [25]. Finally, our hypocotyl grafting protocol can be combined with other grafting methods to address specific research questions. For example, cotyledon grafting has been successfully integrated with hypocotyl grafting [30], highlighting the flexibility of the micro-grafting setup in enabling diverse cutting and joining patterns [29].

14. Do not use 3 M Micropore Surgical Tape for sealing grafting plates, as its gas exchange properties lead to plate desiccation and graft failure.

15. Elevated ambient temperatures accelerate vascular regeneration, with the maximum thermo-inductive effect observed at approximately 27°C [30]. This temperature effect may improve the success rate of delicate procedures, such as cotyledon grafting [25], but is generally unnecessary for hypocotyl grafting. In contrast, temperatures below 20°C delay graft formation. At 16°C, phloem reconnection is only completed by day 7 after grafting [30].

16. Light quality and quantity influence vascular regeneration during graft formation. Blue light was shown to promote vascular reconnection in grafted watermelon [38]. The molecular effects of different light conditions on graft healing in Arabidopsis hypocotyls remain unknown, but low-light conditions, such as short-day photoperiods (8 h light/16 h dark), have been shown to be beneficial [11, 16]. However, excessively low light intensities negatively affect cotyledon grafting [25] and, based on our observations, also impair hypocotyl graft formation.

17. Discard any plants that exhibit adventitious roots, as they indicate graft failure and can functionally replace the rootstock. Assessing adventitious root formation can be challenging in older plants. Therefore, consider genotyping grafted organs. Alternatively, use visual reporter lines (e.g., GFP, GUS, or RUBY [39]) to distinguish scion-derived adventitious roots from rootstock-derived roots.

18. Lower CFDA concentrations (100 μM–1 mM) can be used [25, 31], but higher concentrations enhance detectability in living tissues. When freshly prepared, CFDA stock and working solutions should not fluoresce. CFDA fluorescence is activated upon uptake into living cells, where cellular esterases cleave the diacetate group [40], or upon repeated use and light exposure [33].

19. Further dilutions are possible. However, highly concentrated Rhodamine WT exhibits a self-quenching effect and does not fluoresce. Fluorescence is activated upon dilution in plant tissues [33].

20. The parafilm strip prevents the dye drop from spreading into the membrane or Whatman paper, allowing for localized application. To avoid transferring individual grafted plants to the parafilm strip, position the plants before grafting at a lower position than in the standard hypocotyl grafting protocol (Sect. 3.2): the shoots and upper roots should contact the membrane, while the root tips remain in contact with Whatman paper. Just before applying the dye solution, use forceps to move the entire membrane onto the parafilm strip, enabling that all root tips are placed on the parafilm simultaneously. For further details, refer to our recently published protocol on monitoring xylem transport in Arabidopsis [33].
21. For time-course studies, graft sufficient plants for each time point of interest. Each plant can only be used once, as CFDA and Rhodamine WT persist in the tissues (*see* **Note 4**). For single time-point studies, test phloem and xylem reconnection on days 3–5 and 6–7 after grafting, respectively, to assess differences between mutants/treatments and wild-type/ mock controls. Deviations may suggest involvement of the tested gene/treatment in vascular regeneration. For large-scale phloem reconnection experiments, we adopted the phloem transport method by Yin et al. [41] as an alternative approach, using transgenic *ProSUC2:GFP* scions, which express mobile GFP in phloem companion cells [42]. For this, graft *ProSUC2:GFP* scions onto wild-type or mutant rootstocks (Fig. 1) and inspect the roots for GFP fluorescence at desired time points. Based on our experience, GFP fluorescence is more challenging to detect than CFDA or Rhodamine WT signals and is most readily detected in root tips due to accumulation in the phloem unloading zone. Researchers should consider that transgenic approaches are constrained by genetic background requirements, whereas dye-based methods offer greater flexibility and can be applied to both model and non-model species.

Acknowledgments

We are grateful to Thomas Assinger for his advice on adjuvants, to Orlando Maciel Rodrigues Junior for discussions, and to Syahfitri Retno Wulandari for help with Fig. 1. This work was supported by the Swiss National Science Foundation (Project 184762).

Conflict of Interest The authors have no conflicts of interest to declare that are relevant to the content of this chapter.

References

1. Loupit G, Brocard L, Ollat N, Cookson SJ (2023) Grafting in plants: recent discoveries and new applications. J Exp Bot 74:2433–2447. https://doi.org/10.1093/jxb/erad061
2. Feng M, Augstein F, Kareem A, Melnyk CW (2024) Plant grafting: molecular mechanisms and applications. Mol Plant 17:75–91. https://doi.org/10.1016/j.molp.2023.12.006
3. Mudge K, Janick J, Scofield S, Goldschmidt EE (2009) A history of grafting. In: Janick J (Ed.) Horticultural reviews. 1st edn. Wiley, New Jersey, 437–493
4. Lee J-M, Kubota C, Tsao SJ et al. (2010) Current status of vegetable grafting: diffusion, grafting techniques, automation. Sci Hortic 127:93–105. https://doi.org/10.1016/j.scienta.2010.08.003
5. Augstein F, Melnyk CW (2025) Modern and historical uses of plant grafting to engineer development, stress tolerance, chimeras, and hybrids. Plant J 121:e70057. https://doi.org/10.1111/tpj.70057
6. Kragler F, Bock R (2025) The biology of grafting and its applications in studying information exchange between plants. Nat Plants 11:955–966. https://doi.org/10.1038/s41477-025-01982-2
7. Tsukaya H, Naito S, Rédei GP, Komeda Y (1993) A new class of mutations in *Arabidopsis thaliana*, *acaulis1*, affecting the development of both inflorescences and leaves. Development 118:751–764. https://doi.org/10.1242/dev.118.3.751
8. Turnbull CGN, Booker JP, Leyser HMO (2002) Micrografting techniques for testing long-distance signalling in *Arabidopsis*. Plant J 32:255–262. https://doi.org/10.1046/j.1365-313X.2002.01419.x
9. Bainbridge K, Bennett T, Turnbull C, Leyser O (2006) Grafting. Methods Mol Biol Clifton NJ 323:39–44. https://doi.org/10.1385/1-59745-003-0:39
10. Notaguchi M, Daimon Y, Abe M, Araki T (2009) Adaptation of a seedling micro-grafting technique to the study of long-distance signaling in flowering of *Arabidopsis thaliana*. J Plant Res 122:201–214. https://doi.org/10.1007/s10265-008-0209-1
11. Marsch-Martínez N, Franken J, Gonzalez-Aguilera KL et al. (2013) An efficient flat-surface collar-free grafting method for *Arabidopsis thaliana* seedlings. Plant Methods 9: 14. https://doi.org/10.1186/1746-4811-9-14
12. Andersen T, Liang D, Halkier B, White R (2014) Grafting Arabidopsis. BIO-Protoc 4:e1164. https://doi.org/10.21769/BioProtoc.1164
13. Bainbridge K, Bennett T, Crisp P et al. (2014) Grafting in Arabidopsis. In: Sanchez-Serrano JJ, Salinas J (Eds.)Eds. Arabidopsis protocols. Humana Press, Totowa, NJ, 155–163
14. Melnyk CW (2017) Grafting with *Arabidopsis thaliana*. In: Kleine-Vehn J, Sauer M (Eds.) Eds. Plant hormones. Springer, New York, New York, NY, 9–18
15. Melnyk CW (2017) Monitoring vascular regeneration and xylem connectivity in *Arabidopsis thaliana*. In: De Lucas M, Etchhells JP (Eds.) Eds. Xylem. Springer, New York, New York, NY, 91–102
16. Li Y, Sun W, Liu F et al. (2019) Methods for grafting *Arabidopsis thaliana* and *Eutrema salsugineum*. Plant Methods 15: 93. https://doi.org/10.1186/s13007-019-0477-x
17. Tsutsui H, Yanagisawa N, Kawakatsu Y et al. (2020) Micrografting device for testing systemic signaling in Arabidopsis. Plant J 103:918–929. https://doi.org/10.1111/tpj.14768
18. Tsutsui H, Kawakatsu Y, Notaguchi M (2021) Micrografting in Arabidopsis using a silicone chip. BIO-Protoc 11:e4053. https://doi.org/10.21769/BioProtoc.4053
19. Vanderstraeten L, Sanchez-Muñoz R, Depaepe T et al. (2022) Mix-and-match: an improved, fast and accessible protocol for hypocotyl micrografting of Arabidopsis seedlings with systemic ACC responses as a case study. Plant Methods 18: 24. https://doi.org/10.1186/s13007-022-00859-1
20. Takahashi F (2022) Use of micrografting to study the role played by peptide signals in ABA biosynthesis. In: Yoshida T (Ed.)Eds. Abscisic acid. Springer, US, New York, NY, 101–109
21. Yell V, Li X (2023) Scalable, flexible, and cost-effective seedling grafting. J Vis Exp 191:64519. https://doi.org/10.3791/64519
22. Hsieh E-J, Grillet L (2023) Arabidopsis micro-grafting to study the systemic signaling of nutrient status. In: Jeong J (Ed.)Eds. Plant iron homeostasis. Springer, US, New York, NY, 113–120
23. Wang T, Li F, Xu S et al. (2011) The time course of long-distance signaling in radiation-induced bystander effect in vivo in *Arabidopsis thaliana* demonstrated using root micro-grafting. Radiat Res 176:234–243. https://doi.org/10.1667/RR2486.1
24. Yoo SJ, Hong SM, Jung HS, Ahn JH (2013) The cotyledons produce sufficient FT protein to induce flowering: evidence from cotyledon micrografting in Arabidopsis. Plant Cell Physiol 54:119–128. https://doi.org/10.1093/pcp/pcs158

25. Bartusch K, Trenner J, Melnyk CW, Quint M (2020) Cut and paste: temperature-enhanced cotyledon micrografting for *Arabidopsis thaliana* seedlings. Plant Methods 16: 12. https://doi.org/10.1186/s13007-020-0562-1
26. Tsutsui H, Notaguchi M (2017) The use of grafting to study systemic signaling in plants. Plant Cell Physiol 58:1291–1301. https://doi.org/10.1093/pcp/pcx098
27. Thomas HR, Frank MH (2019) Connecting the pieces: uncovering the molecular basis for long-distance communication through plant grafting. New Phytol 223:582–589. https://doi.org/10.1111/nph.15772
28. Yang L, Machin F, Wang S et al. (2023) Heritable transgene-free genome editing in plants by grafting of wild-type shoots to transgenic donor rootstocks. Nat Biotechnol 41:958–967. https://doi.org/10.1038/s41587-022-01585-8
29. Bartusch K, Melnyk CW (2020) Insights into plant surgery: an overview of the multiple grafting techniques for *Arabidopsis thaliana*. Front Plant Sci 11:613442. https://doi.org/10.3389/fpls.2020.613442
30. Serivichyaswat PT, Bartusch K, Leso M et al. (2022) High temperature perception in leaves promotes vascular regeneration and graft formation in distant tissues. Development 149:dev200079. https://doi.org/10.1242/dev.200079
31. Melnyk CW, Schuster C, Leyser O, Meyerowitz EM (2015) A developmental framework for graft formation and vascular reconnection in *Arabidopsis thaliana*. Curr Biol 25:1306–1318. https://doi.org/10.1016/j.cub.2015.03.032
32. Knoblauch M, Vendrell M, De Leau E et al. (2015) Multispectral phloem-mobile probes: properties and applications. Plant Physiol 167:1211–1220. https://doi.org/10.1104/pp.114.255414
33. Bartusch K, Blanco-Touriñán N, Rodriguez-Villalón A, Truernit E (2024) Monitoring xylem transport in *Arabidopsis thaliana* seedlings using fluorescent dyes. In: Agusti J (Ed.) Eds. Xylem. Springer, US, New York, NY, 3–15
34. Quint M, Delker C, Franklin KA et al. (2016) Molecular and genetic control of plant thermomorphogenesis. Nat Plants 2:15190. https://doi.org/10.1038/nplants.2015.190
35. Rus A, Baxter I, Muthukumar B et al. (2006) Natural variants of AtHKT1 enhance Na+ accumulation in two wild populations of Arabidopsis. PLoS Genet 2:e210. https://doi.org/10.1371/journal.pgen.0020210
36. Damodaran S, Strader LC (2024) Factors governing cellular reprogramming competence in Arabidopsis adventitious root formation. Dev Cell 59:2745–2758.e3. https://doi.org/10.1016/j.devcel.2024.06.019
37. Ostendorp A, Pahlow S, Deke J et al. (2016) Protocol: optimisation of a grafting protocol for oilseed rape (*Brassica napus*) for studying long-distance signalling. Plant Methods 12: 22. https://doi.org/10.1186/s13007-016-0122-x
38. Bantis F, Panteris E, Dangitsis C et al. (2021) Blue light promotes vascular reconnection, while red light boosts the physiological response and quality of grafted watermelon seedlings. Sci Rep 11: 21754. https://doi.org/10.1038/s41598-021-01158-w
39. He Y, Zhang T, Sun H et al. (2020) A reporter for noninvasively monitoring gene expression and plant transformation. Hortic Res 7:152. https://doi.org/10.1038/s41438-020-00390-1
40. Breeuwer P, Drocourt JL, Bunschoten N et al. (1995) Characterization of uptake and hydrolysis of fluorescein diacetate and carboxyfluorescein diacetate by intracellular esterases in *Saccharomyces cerevisiae*, which result in accumulation of fluorescent product. Appl Environ Microbiol 61:1614–1619. https://doi.org/10.1128/aem.61.4.1614-1619.1995
41. Yin H, Yan B, Sun J et al. (2012) Graft-union development: a delicate process that involves cell–cell communication between scion and stock for local auxin accumulation. J Exp Bot 63:4219–4232. https://doi.org/10.1093/jxb/ers109
42. Imlau A, Truernit E, Sauer N (1999) Cell-to-cell and long-distance trafficking of the green fluorescent protein in the phloem and symplastic unloading of the protein into sink tissues. Plant Cell 11:309–322. https://doi.org/10.1105/tpc.11.3.309

Check for updates

Chapter 34

Efficient Establishment of Interfamily Heterograft of *Nicotiana benthamiana* and *Arabidopsis thaliana*

Kentaro Okada, Koji Okayasu and Michitaka Notaguchi

Abstract

The grafting technique has been applied to study systemic signaling in plants, to investigate whether gene action is graft-transmissible and/or gene products such as RNAs and proteins are transported systemically. Here, we describe two protocols, stem grafting and micrografting, for an interfamily heterograft system between *Nicotiana benthamiana* scion plants and Arabidopsis stock plants for the identification of systemic mobile signals. Since these plants belong to evolutionally distant families and both genome databases are available, this system can be used to identify mobile substances transported from one plant to the other without doubt.

Keywords Grafting, Interfamily grafting, Heterograft, Wedge-graft, *Nicotiana*, Arabidopsis, Mobile signals, Phloem transport

1 Introduction

Grafting has been an important technique in agriculture to propagate trees by explanting stems onto rootstocks [1]. In scientific research, grafting techniques have been applied to test the transmissibility of physiological factors, such as phytohormones, RNA, and proteins, from one to the other grafted plant, the so-called long-distance systemic signaling [2–4]. In recent years, grafting experiments using model and non-model plants, including *Arabidopsis thaliana* (Arabidopsis), lotus, tomato, and grape, have revealed that numerous small RNAs, mRNAs, peptides, and proteins are systemically transported organ-to-organ via the phloem and the xylem and exert their roles in coordinating the development and growth in relation to the environmental conditions. Various grafting experiments conducted in Arabidopsis, vegetables, and trees have shown that many physiological processes in plants, including flowering, branching, tuberization, nutrient absorption, symbiosis, and defense responses, are controlled systemically [2, 5, 6]. Facilitated by the significant progress on the

Johannes Liesche and Sanna Sevanto (eds.), *Phloem*, Methods in Molecular Biology, vol. 3066,
https://doi.org/10.1007/978-1-0716-5488-0_34,

analysis of phloem function (see other sections), plant systemic signaling mechanisms and the role of the phloem in signal transport are now well recognized in plant biology.

There are various methods of grafting. Grafting of one shoot and one rootstock has been a frequently used method to test the shoot-root signaling/transport. Two-shoot, two-root, interstock, bud, and leaf grafting methods have been applied depending on the research purpose [2]. Most of these grafting experiments were performed using the same species or between closely related species. Therefore, to investigate molecular transport, knockout mutants or transgenic lines were required as graft partners. Alternatively, grafting could be performed using different genotypes to distinguish the transported substances from the two plants. In the latter case, nucleotide sequencing analysis has been employed to recognize the origin of RNA species [7–9], and mass spectrometry analysis has been used to detect the transported proteins [10]. Thus, the difference in the nucleotide or amino acid sequence has facilitated the identification of mobile signals by overcoming the challenge of limited molecular differences among closely related plant varieties/species. Here, we demonstrate that two types of heterograft systems between a *Solanaceae* species, *Nicotiana benthamiana*, used as scion, and a *Brassicaceae* species, Arabidopsis, used as stock, can be established with a high success rate. Since these plants belong to distant botanical clades, molecular identification is easier than in traditional homograft experiments. One of the methods, stem grafting, described here has been used to identify mobile transcripts derived from Arabidopsis by RT-PCR [11] and RNA-Seq analysis [12]. In these interfamily heterograft experiments, up to 100 transcripts were identified from the sink leaves and the stems of the *N. benthamiana* recipient scion as mobile transcripts in Arabidopsis. The other method, hypocotyl micrografting, described here, has been used to observe autophagosome formation for wound repair using a confocal laser scanning microscope [13]. The micrografting allows grafting to be performed between 4-day-old seedlings of Arabidopsis and *N. benthamiana*. This method shortens the grafting preparation period and is conducted under sterile conditions, thereby allowing the assessment of graft interactions without the microbial effects. It also minimizes both the duration and the spatial requirements of the experiment compared to the stem grafting.

2 Interfamily Stem Heterografting

2.1 Materials

Experiments require well-grown, healthy plant material. Thicker Arabidopsis stems make grafting easier and increase the success rate compared to thin stems.

Preparation of Plant Material

1. Sterile seeds.
2. 10% bleach solution: Add 5 mL conventional household bleach to 45 mL distilled water and mix in a 50 mL conical tube.
3. High Speed Mixer (EYELA CM-1000).
4. Clean bench.
5. Autoclaved distilled water.
6. Agar medium: 0.5× Murashige and Skoog mineral salts, 1% sucrose, 0.8% phytoagar. Add 2.16 g of Murashige and Skoog mineral salts to 0.9 L of distilled water. Mix and adjust pH to 5.7 with 1 M KOH. Make up to the final volume of 995 mL and add 8 g of phytoagar. Autoclave 20 min at 121°C. After the media cools, add 5 mL of 20% sucrose solution.
7. Round pots of 6 cm diameter and square pots of 7.5 cm side length.
8. Trays (20 cm × 30 cm).
9. Peat moss (Supermix A; Sakata Seed Corporation).
10. 5 mL disposable plastic droppers.
11. Plastic bags (20 cm height × 14 cm width).
12. Soil (compost mix: vermiculite = 1:1).
13. Nutrient-enriched water: 1/1000× Hyponex solution. Add 1 mL Hyponex to 1 L of tap water and mix.
14. Tweezers.
15. Square petri dishes (9 cm × 14 cm).
16. 20 μL pipettes and pipette tips.

Grafting

1. Parafilm (ca 2–2.5 cm height × 4 cm width).
2. Plastic sticks (ca 25 cm length).
3. Zipper-locked plastic bags (17 cm height × 12 cm width).
4. Stainless razor blades (2 cm height × 4 cm width).
5. Petri dishes (circle type, 9 cm diameter).
6. A water spray bottle.

2.2 Methods

Preparation of Arabidopsis Plants

1. Soak Arabidopsis seeds in 10% bleach solution in a tube and surface-sterilize by shaking for about 5 min (*see* Sect. 2.3, **Note 1**). Hereafter, proceed all manipulations under a clean bench.
2. Wash seeds with autoclaved distilled water three times.

Fig. 1 Stock and scion plants. (**A**) A 5-week-old Arabidopsis plant for the stock. (**B**) An Arabidopsis stock prepared by trimming the upper region of the primary stem and cauline leaves and cutting a slit (inset). (**C** A 4-week-old *N. benthamiana* plant for the scion. (**D**) *N. benthamiana* scion prepared by cutting the inflorescence stem and trimming the stem to a wedge (insets). Bars: 5 cm (**A–C**), 2 cm (**D**)

3. Remove water and add 0.1% agar solution into the tube.
4. Place the tube at 4°C and incubate for 2–3 days to break dormancy.
5. Sow seeds into pots containing peat moss that absorbed water beforehand using a dropper (around 100 seeds/pot) (*see* Sect. 2.3, **Note 2**).
6. Cover the pot with a plastic bag to maintain moisture and place the pot at 22°C for 10 days under continuous light conditions (around 60 μmol m^{-2} s^{-1}) (*see* Sect. 2.3, **Note 3**).
7. Transplanted seedlings: Prepare new pots (6 cm diameter) filled with the soil (compost mix: vermiculite = 1:1).
8. Place the pots into a tray (*see* Sect. 2.3, **Note 4**) and add nutrient water (1000-fold diluted solution of Hyponex) to the tray until the soil is thoroughly wetted (*see* Sect. 2.3, **Note 5**).
9. Transplant germinated seedlings prepared by the previous step (*see* Sect. 2.3, **Note 6**) with tweezers from the peat moss to the new pots, taking care not to wound roots (*see* Sect. 2.3, **Note 7**).
10. Cover the entire tray with a plastic film and incubate for 1 day to maintain moisture until the roots adapt to the soil.
11. Partially remove the film (only two sides out of four sides of the tray) and acclimate for another 2 days.
12. At the end of acclimation, remove the film completely (*see* Sect. 2.3, **Note 8**).
13. Add nutrient water to the tray once or twice a week until use (*see* Sect. 2.3, **Note 9**).

14. Use bolting-stage Arabidopsis plants, around 5 weeks old, for grafting. At this time, the Arabidopsis stems are 15–20 cm in length and 1–1.5 mm in width (Fig. 1A) (*see* Sect. 2.3, **Note 10**).

Preparation of N. benthamiana *Plants*

1. Mix and pour agar solution into plastic petri dishes.
2. Sterilize *N. benthamiana* seeds in the same way as for Arabidopsis seeds (see Sect. 2.2.1).
3. Sow seeds on agar media by using a pipette. Sow 50 seeds on a square dish.
4. Place the plates at 4°C and stand for 2–3 days to stimulate seed germination.
5. Transfer the plates to a growth chamber at 27°C and grow plants for 7 days.
6. Prepare soil pots, transplant germinated seedlings prepared by the previous step to the pots (*see* Sect. 2.3, **Note 11**) and acclimate them in a similar fashion as *Arabidopsis* seedlings (see Sect. 2.2.1).
7. Add nutrient solution to the tray once, twice, or three times a week until use (*see* Sect. 2.3, **Note 12**).
8. Use *N. benthamiana* plants around 4 weeks old for grafting. At this time, the *N. benthamiana* stems are 8–12 cm in length and 4–5 mm in width (Fig. 1C) (*see* Sect. 2.3, **Note 10**).

Preparing Scion and Stock

Prepare plant materials and all equipment on the working table. Wear gloves or wash hands before carrying out the grafting procedure. To avoid the scion wilting after removal of the roots, prepare the stock plant first and then cut the scion plant immediately before assembling the stock and scion. Making one graft takes around 3 min.

1. Cut the primary bolting stem of 5-week-old Arabidopsis horizontally (Fig. 1A) with a razor blade around 10 cm above the ground (Fig. 1B).
2. Cut and remove the secondary inflorescence stems if present.
3. Cut and remove the cauline leaves on the primary stem if they are close to the grafting site. If not, removal of the cauline leaves is not necessary.
4. Use a petri dish as a cutting board to make a 1.5–2 cm slit at the center of the cut primary stem (*see* Sect. 2.3, **Note 13**). When making a slit, hold the stem just below the part to be cut and stabilize with a finger on the side (Fig. 1B inset) (*see* Sect. 2.3, **Note 14**).
5. Cut the primary stem of a 4-week-old *N. benthamiana* plant horizontally (Fig. 1C) to take 7–10 cm of the top of the plant (Fig. 1D).

Fig. 2 Assembling of the stock and the scion plants by wrapping the cut stems with parafilm. (**A, B**) A piece of parafilm for grafting before use (**A**) and after stretching and making a loop (**B**). (**C**) A stem of an Arabidopsis stock plant was put through the loop of a stretched parafilm. (**D**) Sequential processes of graft assembly. First, a side of parafilm (*) was folded to the other side, and then the grafting point was wrapped up while stretching the parafilm. Bar: 5 cm (**D**)

6. Remove the expanded leaves and only retain one to three small young leaves (~5 cm).
7. Place the cut stem on a petri dish used as a cutting board.
8. Hold the stem just at the upper part (*see* Sect. 2.3, **Note 14**) and trim the cut edge into a 1.5–2 cm wedge by making two cuts (Fig. 1D insets) (*see* Sect. 2.3, **Note 15**).

Grafting

1. Pull a parafilm sheet once (Fig. 2A) to make it stretched (Fig. 2B) and make a loop (Fig. 2C).
2. Place the looped parafilm to surround the slit of the stock stem.
3. While holding the stock stem with parafilm and fingers, gently insert the scion stem into the slit of the stock (Fig. 2D) (*see* Sect. 2.3, **Note 16**).
4. Fix the grafting point by wrapping with parafilm (*see* Sect. 2.3, **Note 17**). Fold a side of the stretched parafilm to the other side across the stem, pinch the stretched parafilm with your fingers, leaving a small gap for the plant tissue. In this way, wrap the graft union with several layers of parafilm (Fig. 2D). Wrapping should be done cautiously but quickly without crushing the tissues.
5. Stick the plastic bar to the potted soil along the stems of the scion and stock and wrap the rest of parafilm at the grafting point around the bar and the grafted plants to avoid bending of the scion.
6. Spray water mist to the inside of a zipper-locked plastic bag (Fig. 3A).

Fig. 3 Curing and establishment of graft union. (**A**) A zip-lock plastic bag sprayed with water mist inside. (**B**) Grafted plants were tied with a plastic stick as a support and covered with a water-sprayed plastic bag. (**C**) An *N. benthamiana*/Arabidopsis graft at 3 weeks after grafting. The *N. benthamiana* scion formed inflorescences. Bars: 5 cm (**B**,**C**)

7. Cover the entire scion with the water-sprayed plastic bag and zip the bag (Fig. 3B). The grafting point can be either inside or outside of the plastic bag (*see* Sect. 2.3, **Note 18**).

Curing of the Graft

1. Grow the grafted plants in a growth chamber at 27°C under relatively weak light conditions (around 35 μmol m^{-2} s^{-1}) for 7 days (*see* Sect. 2.3, **Note 19**).
2. Transfer the grafted plants to the plant growth room at 22°C under continuous light conditions (around 60 μmol m^{-2} s^{-1}).
3. Release the moisture of the plastic bag by making a 3 cm cut at an edge of the plastic bag and opening the zip lock for acclimation (*see* Sect. 2.3, **Note 20**).
4. The next day, remove the plastic bag completely (*see* Sect. 2.3, **Note 21**).
5. Keep the grafted plants growing until further experiments are conducted (Fig. 3C) (*see* Sect. 2.3, **Note 22**).

Grafting Success and Further Experiments

1. Evaluate the success of grafting by judging whether scion plants are alive at 4 weeks post grafting (*see* Sect. 2.3, **Note 23**). To confirm the effect of grafting, perform a control experiment where *N. benthamiana* stem scion is prepared but not grafted onto Arabidopsis. The cut *N. benthamiana* stem is placed in a plastic bag for a week, as for grafts, and then taken out from the bag. Without roots, the cut *N. benthamiana* stem will be

completely dried within another week. Using the proper grafting procedure, a grafting success rate of at least 90% can typically be reached.

2. Using this heterograft system, the long-distance transport of the transcripts from the Arabidopsis stock to the *N. benthamiana* scion across the graft junction can be tested by analyzing the *N. benthamiana* scions (*see* Sect. 2.3, **Note 24**). RNA-Sequencing can be applied to identify mobile transcripts in an untargeted, genome-wide manner [12]. Moreover, RT-PCR can also be applied for investigating the presence and/or abundance of specific RNAs [11].

2.3 Notes

1. Shake the seeds gently on a tabletop mixer. To further reduce the contamination rate, pre-wash with 70% ethanol for about 1 min. We have often used an Arabidopsis ecotype, Col-0, as plant materials [11, 12].
2. Sowing seeds on solid media in sterile conditions is another choice, but this requires more care for transplanting and acclimating. Further information on handling Arabidopsis seeds can be found in the literature [14].
3. During this period, do not add excess water to the pots. Keeping the water level to "just moist," not soaked, will promote root growth and make plants vigorous and healthy. If the growth conditions are good, the seeds will start to germinate within 2 days and expansion of cotyledons will be observed within 3 days.
4. We often use 20 × 30 cm trays and place 15 pots per tray.
5. After adding nutrient water, discard excess water that remains on the tray.
6. At this time, the first true leaves will have emerged, the size of which is around 1/3 of a cotyledon. The growth of them is slightly slower than that of leaves on plants growing on solid media.
7. Make a hole in the soil of the new pot before transplanting the seedling. Take a seedling out of the old pot by taking hold of the peat moss and roots together. Bring the roots to the hole using tweezers and gently cover the entire roots with soil using fingers. At this moment, the leaves should not touch the soil.
8. Using a plastic cover to maintain humidity is another choice.
9. Avoid excess watering. To check the watering level, weigh the pots or touch the soil surface. Once the soil is almost dry, pour water into the tray until the soil is wet but not soaked. Do not pour water over the pots. The fFrequency of watering will need to be increased depending on the plant size.

10. To grow plants uniformly, shuffle the positions of the pots sometimes to reduce position effects caused by light and temperature gradients of growth chambers and growth rooms.
11. During this period, the *N. benthamiana* seedlings develop their first true leaves. The procedure to transplant *N. benthamiana* seedlings from the agar medium to the soil is the same as for Arabidopsis. If the agar surface is dry and rigid, pouring water on the media will help to take the roots out smoothly. We place 12 pots per tray.
12. During the small seedling stage, watering should be less to maintain soil aeration and enhance root growth. After approximately 3 weeks of growth, pour a larger amount of water to enlarge the size of *N. benthamiana* plants.
13. If cut off-center, the tissues of the thin side will not unite with the scion tissue.
14. To make an accurate slit or a V-shaped cut, hold the stem by pressing the stem to the petri dish using fingers at a region just beside the cutting point, adjust the position of the razor blade, and cut.
15. To avoid the *N. benthamiana* tissue breaking at the tip of the cut edge of the scion when inserting the scion to the slit of the stock, cutting the tip region of the V-shaped trimmed stem of the scion around 1 mm is optional. Adjust the length of the slit made on the stock and of the V-shaped trimmed region of the scion to fit each other.
16. To avoid the damage of the scion during the fixation of the graft, insert the trimmed region of the scion into the slit of the stock at an angle. Touch the side of the trimmed region of the scion to the slit of the stock first, not the tip of the trimmed region, and then slide the scion deeply into the slit of the stock, adjusting the angle of the scion stem to become straight.
17. We prefer using parafilm over fixtures for grafting sites, such as clips, tubes, and pins [15], because it fits well even though the stem diameters are different for *N. benthamiana* and Arabidopsis.
18. The grafted region should not be drowned by water droplets out of the sprayed plastic bag. Spraying once is enough to make the inside of the plastic bag moist. Leaving the grafting point outside of the plastic bag is not a problem, since that region is wrapped with parafilm and the water loss from the stem surface is limited.
19. Transferring to weaker light conditions can reduce photo-damage for which plants are more susceptible after wounding. During this period, tissues of grafting point are united and start to translocate water, carbohydrates, and

other substances between the stock and the scion plants. In general, growth temperature between 22–30°C are adequate to establish this grafting.

20. As the establishment of the graft union, such as xylem connection, is first limited and gradually develops over several weeks, the watering level should be increased at this stage to reduce the risk of water loss stress of the scion.
21. Failure of grafts can be recognized by wilting of the scion plants.
22. The *N. benthamiana* scion growth can be observed by 3 weeks after grafting (see Fig. 3C).
23. Additional tests can be carried out to evaluate the success of the grafting, e.g., microscopic analysis of a transverse cut across the graft union, measurement of electrical resistance across the graft interface, thermal imaging of leaf temperature, and assessment of vascular connections by tracer dye transport [15–18].
24. To avoid Arabidopsis tissues, such as pollen grains, contaminating the tissues of *N. benthamiana* scions, cut newly grown Arabidopsis axillary inflorescence stems and keep covering the entire *N. benthamiana* scion with a paper or a plastic bag until sampling [12].

3 Interfamily Hypocotyl Micro-Heterografting

3.1 Materials

A hypocotyl micro-heterografting procedure can be performed using a micrografting device originally designed for Arabidopsis homografting [19, 20].

Preparation of Plant Material

1. Sterile *N. benthamiana* and Arabidopsis seeds.
2. 6% Bleach solution: Add 3 mL of sodium hypochlorite to 50 mL of distilled water and mix in a 50 mL conical tube.
3. High-speed orbital shaker capable of rotating at 1500 rpm (EYELA CM-1000).
4. Autoclaved distilled water.
5. Square Petri dishes with grid (100 × 15 mm; Simport Scientific, Beloeil, Canada).
6. Agar medium: 0.5× Murashige and Skoog (MS) mineral salts, 0.05% MES, 1% sucrose, and 1% phytoagar. Add 2.3 g of MS and 10 g of sucrose to 0.9 L of distilled water. Mix and adjust pH to 5.7 with 1 M KOH. Add distilled water to make the final volume 1000 mL, then add 10 g of agar. Autoclave for 20 min at 121°C.

7. Autoclaved nylon membrane, Hybond-N+ (GE Healthcare, Chicago, USA): 1 × 2 cm.
8. Micrografting devices (BMS, Tokyo, Japan).
9. Surgical tape.
10. Forceps.
11. 20 μL Pipettes and pipette tips.
12. Stereotype microscope.
13. Clean bench or open-type clean bench (KOACH T-500F; KOKEN Ltd., Tokyo, Japan).

Grafting

1. 70% ethanol.
2. Agar medium: 0.5× Murashige and Skoog (MS) mineral salts, 0.05% MES, 1% sucrose, and 2% phytoagar.
3. Autoclaved distilled water.
4. Autoclaved filter paper (70 mm diameter).
5. Petri dishes (circle type, 90 mm diameter).
6. Disposable dropper.
7. Forceps.
8. Micro scissor.
9. Surgical tape.
10. Disposable surgical scalpel No.11 (KAI Corporation and KAI Industries Co., Ltd., Tokyo, Japan)
11. Stereotype microscope.
12. Clean bench or open-type clean bench.

3.2 Methods

Preparation of N. benthamiana *Seedlings*

1. Sterilize *N. benthamiana* seeds with 6% bleach solution in a 1.5 mL microtube by shaking at 1500 rpm for approximately 10 min. Hereafter, proceed with all manipulations with a stereomicroscope in a clean bench.
2. Wash the seeds with autoclaved distilled water five times.
3. Remove water and add 0.1% agar solution into the tube.
4. Sow *N. benthamiana* seeds onto autoclaved membranes (four seeds per row, two rows per membrane sheet) on 1% agar medium using a 20 μL pipette (*see* Sect. 3.3, **Note 1**).
5. Seal the plate with surgical tape.
6. Place the plates horizontally at 4°C and incubate them in darkness for 3 days to break dormancy.

Fig. 4 Stock and scion plants for hypocotyl micro-heterografting. (**A**) 4-day-old *N. benthamiana* seedlings for the scion. (**B**) An open-type clean bench for sowing Arabidopsis seeds onto chips and performing grafting. Clean zone creators generated a filtered air flow and established a sterilized zone. All procedures were performed under a stereomicroscope within the zone. (**C**) For stock preparation, Arabidopsis seeds were sown in pockets of a micrografting chip. The upper and lower panels show the chip before and after covering with the chip cover, respectively. (**D**) The chip was placed upright on the membrane and the seedlings were grown for 4 days. This panel shows a top-view image of 4-day-old Arabidopsis seedlings used as the stock plants. Bars: 5 mm (**A,C**), 2 mm (**D**)

7. Place the plates vertically in a growth chamber at 27°C and grow the seedlings under continuous light conditions for 4 days (25 µmol m^{-2} s^{-1}; Fig. 4A; *see* Sect. 3.3, **Note 2**).

Preparation of Arabidopsis Seedlings

1. Sterilize Arabidopsis seeds in the same way as for *N. benthamiana* seeds.
2. Wash the seeds with autoclaved distilled water five times.
3. Remove water and add 0.4% agar solution to the tube (*see* Sect. 3.3, **Note 3**).
4. Place the tubes at 4°C and incubate them in darkness for 3 days to break dormancy.
5. Place autoclaved membranes onto 1% agar medium, and put sterilized micrografting chips [19] onto the membranes with the seed pocket facing upward (*see* Sect. 3.3, **Note 4**).

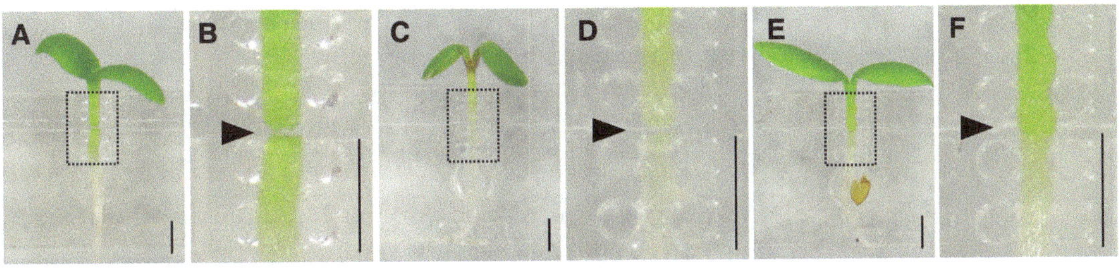

Fig. 5 Hypocotyl micro-heterografting in the micrografting chip. (**A**) A 4-day-old *N. benthamiana* seedling was transferred to the micro-path of the micrografting chip. A dashed rectangle indicates the area shown in (**B**). (**B**) The *N. benthamiana* hypocotyl was cut horizontally along the arrowhead. The upper part was used as the scion. (**C**) A 4-day-old Arabidopsis seedling grown in a micrografting chip. A dashed rectangle indicates the area shown in (**D**). (**D**) The Arabidopsis hypocotyl was cut horizontally along the arrowhead. The bottom part was used as the stock. (**E**) A hypocotyl micro-heterograft of the *N. benthamiana* scion and the Arabidopsis rootstock. A dashed rectangle indicates the area shown in (**F**), the grafting site. (**F**) Assembly of the *N. benthamiana* scion and the Arabidopsis stock, with the two cut surfaces in contact. An arrowhead indicates the grafting site. Bars: 500 μm

6. Sow an Arabidopsis seed into the seed pocket of the micrografting chip and fill the seed pocket with 0.4% agar solution to keep it wet, using a stereomicroscope under sterile conditions made by a clean bench or open-type clean bench (Fig. 4B, C; *see* Sect. 3.3, **Note 5**).
7. Cover the chip with the chip cover and place the chip upright on the membrane.
8. Seal the plate with surgical tape.
9. Wrap the plates with aluminum foil, place them horizontally in a growth chamber at 22°C, and grow seedlings in darkness for 2 days.
10. Remove the aluminum foil and grow the seedlings under continuous light conditions for another 2 days (60 μmol m^{-2} s^{-1}; Fig. 4D; *see* Sect. 3.3, **Note 6**).

Grafting

Prepare *N. benthamiana* and Arabidopsis seedlings, 2% agar medium, and all equipment on the working clean bench (Fig. 4B; *see* Sect. 3.3, **Note 7**). Wear gloves or wash hands before carrying out the grafting procedure. Wipe the working table and all equipment with 70% ethanol before performing grafting. Wash the inside of a dropper with autoclaved distilled water twice. Hereafter, proceed with all manipulations with a stereomicroscope under a clean bench or open-type clean bench.

1. Place an autoclaved filter paper in a petri dish and moisten it with autoclaved distilled water using a dropper.
2. Place a new autoclaved micrografting chip with the seed pocket facing upward on the filter paper.

3. Transfer 4-day-old *N. benthamiana* seedlings to the micropath of the micrografting chip (Fig. 5A).
4. Transfer a micrografting chip, in which 4-day-old Arabidopsis seedlings are grown, onto the filter paper where another micrografting chip with *N. benthamiana* seedlings is placed.
5. Transfer the membranes originally placed on the 1% agar medium to the new 2% agar medium.
6. Conduct micrografting. Firstly, cut the *N. benthamiana* hypocotyl horizontally along the slit made in the micrografting chip with a disposable surgical scalpel (Fig. 5B; *see* Sect. 3.3, **Note 8**).
7. Change the angle of the micrografting chip of Arabidopsis with the seed pocket facing upward. Remove the chip cover and remove any seedlings that have unopened cotyledons (Fig. 5C; *see* Sect. 3.3, **Note 9**).
8. Cut the Arabidopsis hypocotyl horizontally along the slit made in the micrografting chip with a disposable surgical scalpel (Fig. 5D).
9. Remove the shoot part of Arabidopsis hypocotyl and retain the rootstock part.
10. Transfer the shoot part of *N. benthamiana* hypocotyl onto the rootstock part of Arabidopsis hypocotyl (Fig. 5E).
11. Align the cut surfaces of the *N. benthamiana* scion and the Arabidopsis stock using forceps (Fig. 5F; *see* Sect. 3.3, **Note 10**).
12. Transfer the micrografting chip containing the grafted seedlings vertically onto the membrane placed on the 2% agar medium.
13. Seal the plates with surgical tape.

Curing of the Graft

1. Grow the grafted plants by placing the agar plates containing the grafted seedlings horizontally in a growth chamber at 27°C under continuous low light conditions for 6 days (25 μmol m^{-2} s^{-1}).
2. Cut and remove adventitious roots that emerged from the *N. benthamiana* scion side with micro scissors and forceps at 4 days after grafting (Fig. 6A, B; *see* Sect. 3.3, **Note 11**).
3. Seal the plates with surgical tape.
4. Place the plates horizontally in the same chamber at 27°C.
5. Transfer the grafted seedlings to a new 1% agar medium at 6 days after grafting (*see* Sect. 3.3, **Note 12**). Cut and remove adventitious roots if emerged from the *N. benthamiana* scion. Ensure that the entire Arabidopsis roots are in contact with the agar medium (*see* Sect. 3.3, **Note 13**).

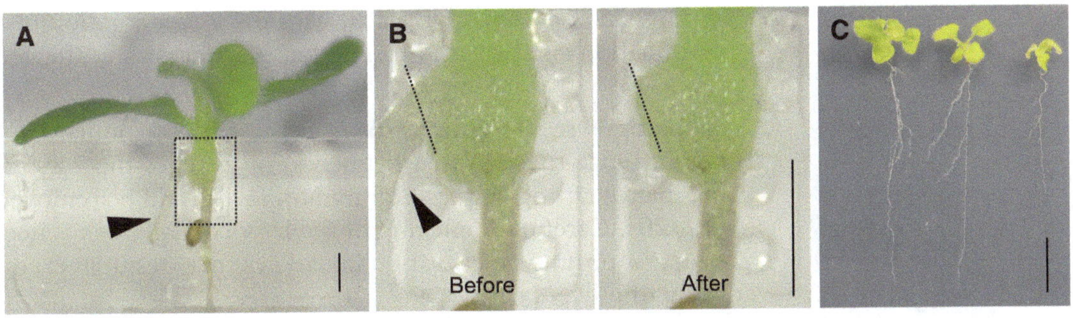

Fig. 6 Curing and establishment of micro-heterograft. (**A**) An *N. betnthamiana*/Arabidopsis micro-heterograft at 6 days after grafting. A dashed box indicates the area shown in (**B**). (**B**) Cutting and removal of an adventitious root emerged from the *N. betnthamiana* scion. The left and right panels show the graft before and after the removal, respectively. An arrowhead indicates the adventitious root. Dashed lines indicate the cutting site. The heterograft will be transferred to a new medium, and adventitious roots will be removed every 2 days thereafter. (**C**) *N. betnthamiana*/Arabidopsis micro-heterografts at 14 days after grafting. The two grafts on the left were successful, and the one on the right failed. Bars: 500 μm (**A,B**), 1 cm (**C**)

6. Seal the plates with surgical tape.
7. Place the plates vertically in a growth chamber at 22°C.
8. Cut and remove adventitious roots of the scion part every 2 days thereafter.

Grafting Success and Further Experiments

1. Evaluate the graft success by assessing whether roots at 10–14 days after grafting are elongated compared with 6 days after grafting or the scion part is alive (Fig. 6C; *see* Sect. 3.3, **Note 14**). Typically, a success rate of at least 50% can be reached.
2. Using the heterograft system with young seedling, fluorescent signals during heterografting can be observed with a confocal laser scanning microscope [13]. This supports the visualization of systemic mobile signals from shoots to roots compared with stem grafting.

3.3 Notes

1. *N. benthamiana* seeds in the second row are sown between those in the first row. An autoclaved membrane can be used to prevent root penetration into the agar. If the roots penetrate the agar media, the seedlings should be excluded from grafting.
2. *N. benthamiana* seedlings with opened cotyledons are used for grafting (Fig. 4A). 4-day-old seedlings correspond to this stage under continuous light conditions.
3. 0.4% agar solution is thoroughly mixed with a stir bar until the temperature cools to room temperature after autoclaving.
4. Used micrografting chips are wrapped with aluminum foil, autoclaved once, and dried in an oven before use.

5. If the seed pockets are not filled with 0.4% agar solution, the germination and growth of Arabidopsis are adversely affected (Fig. 4C).
6. Arabidopsis seedlings are grown in the dark to induce etiolation. Once the etiolated hypocotyl tip reaches the edge of the micrografting chip, the aluminum foil should be removed [19, 20]. Careful monitoring of the hypocotyl growth is necessary, as prolonged etiolation in darkness may result in excessively thin hypocotyls.
7. The humidity in the laboratory should be maintained at 50%–60%. Low humidity causes seedlings to dry out during grafting.
8. We place the scalpel blade perpendicular to the hypocotyl surface and make a horizontal cut.
9. Seedlings with unopened cotyledons should be removed, as they interfere with grafting and transplantation.
10. The cut surface of the hypocotyl dries quickly after cutting. Therefore, alignment should be performed as soon as possible. The cut surfaces of the scion and the stock should be aligned as closely as possible to ensure contact.
11. The adventitious roots tend to develop more readily in heterografts than in homografts and elongate along the hypocotyl. Therefore, close monitoring and removal of the emerged adventitious roots are necessary.
12. If the primary root is curved or misaligned, gently straighten it using forceps. Be careful not to damage the primary root tip. The damage in this region suppresses primary root elongation.
13. The roots of grafted seedlings tend to detach from the agar medium shortly after being transferred to a new medium. We often confirm that the entire root is properly in contact with the medium.
14. At 14 days after grafting, the scion of *N. benthamiana* exhibited slight chlorosis in the leaves. While some individuals showed elongation of the primary root, others displayed limited root growth but continued shoot development compared to 6 days after grafting (Fig. 6C).

Acknowledgments

We thank Drs. Johannes Liesche and Sanna Sevanto for careful reading and editing of the manuscript. This work was supported by JSPS grant 23H04196, 24K02043, and 25H01341 to M.N.

Conflict of Interest The authors have no conflicts of interest to declare that are relevant to the content of this chapter.

References

1. Hartmann HT, Kester DE (1975) Plant propagation: principles and practices, 3rd edn. Prentice Hall, NJ
2. Mudge K, Janick J, Scofield S, Goldschmidt EE (2009) A history of grafting. In: Janick J (ed) Horticultural reviews, vol 35. John Wiley & Sons Inc, Hoboken, pp 437–493
3. Goldschmidt EE (2014) Plant grafting: new mechanisms, evolutionary implications. Front Plant Sci 5:727
4. Tsutsui H, Notaguchi M (2017) The use of grafting to study systemic signaling in plants. Plant Cell Physiol 58:1291–1301
5. Notaguchi M, Okamoto S (2015) Dynamics of long-distance signaling via plant vascular tissues. Front Plant Sci 6:161
6. Spiegelman Z, Golan G, Wolf S (2013) Don't kill the messenger: long-distance trafficking of mRNA molecules. Plant Sci 213:1–8
7. Thieme CJ, Rojas-Triana M, Stecyk E, Schudoma C, Zhang W, Yang L et al (2015) Endogenous Arabidopsis messenger RNAs transported to distant tissues. Nat Plants 1:15025
8. Yang Y, Mao L, Jittayasothorn Y, Kang Y, Jiao C, Fei Z et al (2015) Messenger RNA exchange between scions and rootstocks in grafted grapevines. BMC Plant Biol 15: 251
9. Zhang Z, Zheng Y, Ham BK, Chen J, Yoshida A, Kochian LV et al (2016) Vascular-mediated signalling involved in early phosphate stress response in plants. Nat Plants 2:16033
10. Lin MK, Belanger H, Lee YJ, Varkonyi-Gasic E, Taoka K, Miura E et al (2007) FLOWERING LOCUS T protein may act as the long-distance florigenic signal in the cucurbits. Plant Cell 19:1488–1506
11. Notaguchi M, Wolf S, Lucas WJ (2012) Phloem-mobile Aux/IAA transcripts target to the root tip and modify root architecture. J Integr Plant Biol 54:760–772
12. Notaguchi M, Higashiyama T, Suzuki T (2015) Identification of mRNAs that move over long distances using an RNA-Seq analysis of Arabidopsis/*Nicotiana benthamiana* heterografts. Plant Cell Physiol 56:311–321
13. Kurotani KI, Shinozaki D, Okada K, Tabata R, Kawakatsu Y, Sugita R et al (2025) Autophagy is induced during plant grafting to promote wound healing. Nat Commun 16:3483
14. Arabidopsis Biological Resource Center (2013) The ABRC register. https://abrc.osu.edu/seed-handling. Accessed 13 May 2018
15. Lee JM, Oda M (2003) Grafting of herbaceous vegetables and ornamental crops. In: Janick J (ed) Horticultural reviews, vol 28. John Wiley & Sons Inc, Hoboken, pp 61–124
16. Notaguchi M, Daimon Y, Abe M, Araki T (2009) Adaptation of a seedling micro-grafting technique to the study of long-distance signaling in flowering of Arabidopsis thaliana. J Plant Res 122:201–214
17. Melnyk CW, Schuster C, Leyser O, Meyerowitz EM (2015) A developmental framework for graft formation and vascular reconnection in Arabidopsis thaliana. Curr Biol 25:1306–1318
18. Matsuoka K, Sugawara E, Aoki R, Takuma K, Terao-Morita M, Satoh S et al (2016) Differential cellular control by cotyledon-derived phytohormones involved in graft reunion of Arabidopsis hypocotyls. Plant Cell Physiol 57:2620–2631
19. Tsutsui H, Yanagisawa N, Kawakatsu Y, Ikematsu S, Sawai Y, Tabata R et al (2020) Micrografting device for testing systemic signaling in Arabidopsis. Plant J 103:918–929
20. Tsutsui H, Kawakatsu Y, Notaguchi M (2021) Micrografting in *Arabidopsis* using a silicone chip. Bio Protoc 11:e4053

Check for updates

Chapter 35

VISUAL-PH and VISUAL-CC: Modified Culture Systems for Phloem Differentiation in Arabidopsis

Shunji Shimadzu and Yuki Kondo

Abstract

Vascular cell culture systems have served as powerful tools for identifying key regulators of cell differentiation and function. Vascular cell Induction culture System Using Arabidopsis Leaves (VISUAL) is a tissue culture system that can induce not only xylem vessel-like cells but also sieve element-like cells, thus providing a culture-based system for studying phloem cell differentiation. Presented here are two improved versions of the VISUAL assay: VISUAL-PH, which predominantly induces sieve element-like cells with high specificity, and VISUAL-CC, which enables the induction of companion cell-like cells.

Keywords Culture system, VISUAL, Companion cell, Sieve element, Cell differentiation

1 Introduction

Because vascular tissues are deeply embedded within the plant body, tracing their developmental processes in vivo has been technically challenging. However, the advent of vascular cell culture systems has greatly advanced research on vascular development, especially in xylem development [1–6]. Efficient induction systems for xylem vessel-like cells have been developed since 1980, but efficient induction of sieve elements (SE)-like cells was not achieved until the development of a tissue culture system named VISUAL in 2016 [7]. In VISUAL, Arabidopsis cotyledons are cultured in liquid medium supplemented with auxin, cytokinin, and bikinin, a chemical inhibitor of Arabidopsis GSK3 proteins, including BRASSINOSTEROID INSENSITIVE 2 (BIN2) [8]. VISUAL takes advantage of the abundant genetic resources available in Arabidopsis, including mutants and fluorescent markers, thereby accelerating molecular genetic studies of vascular development [9–16].

Single-cell transcriptome analysis revealed the detailed differentiation process in VISUAL. Mesophyll cells in cotyledons synchronously undergo dedifferentiation, acquire a procambial-like identity,

Johannes Liesche and Sanna Sevanto (eds.), *Phloem*, Methods in Molecular Biology, vol. 3066,
https://doi.org/10.1007/978-1-0716-5488-0_35,

transition to a cambium stem cell-like state, and then differentiate into SE-like and vessel-like cells [17]. This analysis also showed that cytokinin plays an important role in the transition from procambial-like cells into cambium stem cell-like cells. Because cytokinin is required for xylem differentiation competence in cambium stem cells, removing cytokinin from the conventional VISUAL medium predominantly induces phloem SE-like cells with high specificity [17]. Therefore, we named this culture system "VISUAL-PH."

However, VISUAL and VISUAL-PH cannot induce phloem companion cells (CCs), which maintain SEs and ensure phloem transport. Through a screen for modified culture media, we developed a new culture system, VISUAL-CC, that enables the induction of CC-like cells [18]. Genetic analysis using VISUAL-CC revealed that GSK3s play a role in fate determination between SEs and CCs, highlighting the utility of tissue culture systems as powerful tools for uncovering hidden regulatory mechanisms in plant development. Altering the composition of plant hormones or other compounds in the culture medium enables manipulation of the ratio of differentiated cell types. In this chapter, we describe the detailed methods for VISUAL and its modified versions developed for phloem differentiation.

2 Materials

2.1 Materials for Growth of Plant Samples Prior to VISUAL Assay

1. MS growth medium: prepare medium containing 2.2 g/L MS basal medium (Sigma), 10 g/L sucrose, and 0.5 g/L 2-morpholinoethanesulfonic acid monohydrate (MES) in deionized water. Adjust pH to 5.7 using KOH. Autoclave at 120°C for 20 min. The medium can be stored at room temperature for several weeks (*see* **Note 1**).
2. Sterilizing solution: dilute sodium hypochlorite solution in deionized water at a 1:9 (v/v) ratio, and add 0.1% Triton X-100. Prepare freshly before use.
3. Sterilized 6-well plates.
4. Purified autoclaved deionized water.
5. Surgical tape.
6. Growth chamber: continuous light at 22°C and 45–55 μmol m^{-2} s^{-1}.
7. Orbital shaker: suitable for 6- or 12-well plates and capable of shaking at around 110 rpm. Typically, three to five plates are placed simultaneously.

2.2 Materials for VISUAL, VISUAL-PH, and VISUAL-CC Assay

1. VISUAL base medium: dissolve 2.2 g/L MS basal medium and 50 g/L D(+)-glucose (*see* **Note 2**) in deionized water. Adjust pH to 5.7 with KOH (*see* **Note 3**). Dispense into glass bottles

and autoclave at 120°C for 20 min. Store at room temperature for up to several weeks.

2. VISUAL-CC base medium: as above, but with 10 g/L D(+)-glucose instead of 50 g/L. Note the difference in glucose concentration from the standard VISUAL base medium.
3. 2,4-D stock solution: dissolve in ethanol and dilute in sterilized deionized water to a final concentration of 2.5 g/L. Sterilize using a 0.22 μm filter, aliquot 200 μL into sterile tubes, and store at −20°C.
4. Kinetin stock solution: dissolve and dilute in 0.1 M KOH to 0.5 g/L. Sterilize through a 0.22 μm filter, aliquot 200 μL into sterile tubes, and store at −20°C.
5. Bikinin stock solution: dissolve and dilute in DMSO to a final concentration of 10 mM. Sterilize using a 0.22 μm filter (*see* **Note 4**), aliquot 50 μL into tubes, and store at −80°C.
6. Sterile 12-well plates.
7. Growth chamber: continuous light at 22°C and 60–70 μmol m^{-2} s^{-1}.
8. Orbital shaker.

2.3 Materials for Observation of Induced SE-Like and CC-Like Cells

1. ClearSee solution [19]: a clearing reagent obtained by dissolving 25% (w/v) urea, 15% (w/v) xylitol, and 10% (w/v) sodium deoxycholate in deionized water with gentle stirring until fully dissolved. Store at 4°C.
2. Fixative solution: 4% (w/v) paraformaldehyde and 0.01% Triton X-100 in 1× phosphate-buffered saline (PBS). Prepare fresh before use.
3. Phosphate-buffered saline (PBS): produce a 1× solution by mixing one part of 10× PBS with nine parts of deionized water.
4. Vacuum chamber: any standard laboratory vacuum chamber capable of achieving around −50 kPa is sufficient for infiltration during fixation.
5. Fluorescence microscope: a confocal laser scanning microscope or a wide-field fluorescence microscope equipped with appropriate illumination and filters. A 4× objective is commonly used for capturing the overall view of induced cotyledons, whereas 10× –40× objectives are typically used for observing individual cells.

3 Methods

The VISUAL and VISUAL-PH assays share the same base medium (VISUAL base medium) throughout the 4-day culture period, but *differ in hormone composition*—kinetin is omitted in VISUAL-PH. In contrast, the VISUAL-CC assay involves a two-step culture: the first 2 days use the VISUAL base medium with standard hormone supplementation, while the subsequent 4 days *use a distinct medium* (VISUAL-CC base medium) with modified concentrations of auxin and bikinin (Fig. 1).

3.1 Growth of Arabidopsis Seedlings Prior to VISUAL Assays

1. Add sterilizing solution to Arabidopsis seeds in 1.5-mL tubes and rotate gently for 5 min. Transfer the tubes to a laminar flow hood and let them stand for another 5 min.
2. Remove the sterilizing solution with a pipette and wash the seeds at least three times with autoclaved deionized water.
3. Dispense 10 mL of MS growth medium into each well of a sterile 6-well plate.
4. Sow eight to ten seeds per well (*see* **Note 5**).
5. Seal the plate with surgical tape and incubate for 6–7 days under continuous light (22°C, 45–55 μmol m^{-2} s^{-1}) on an orbital shaker at 110 rpm (Fig. 2).

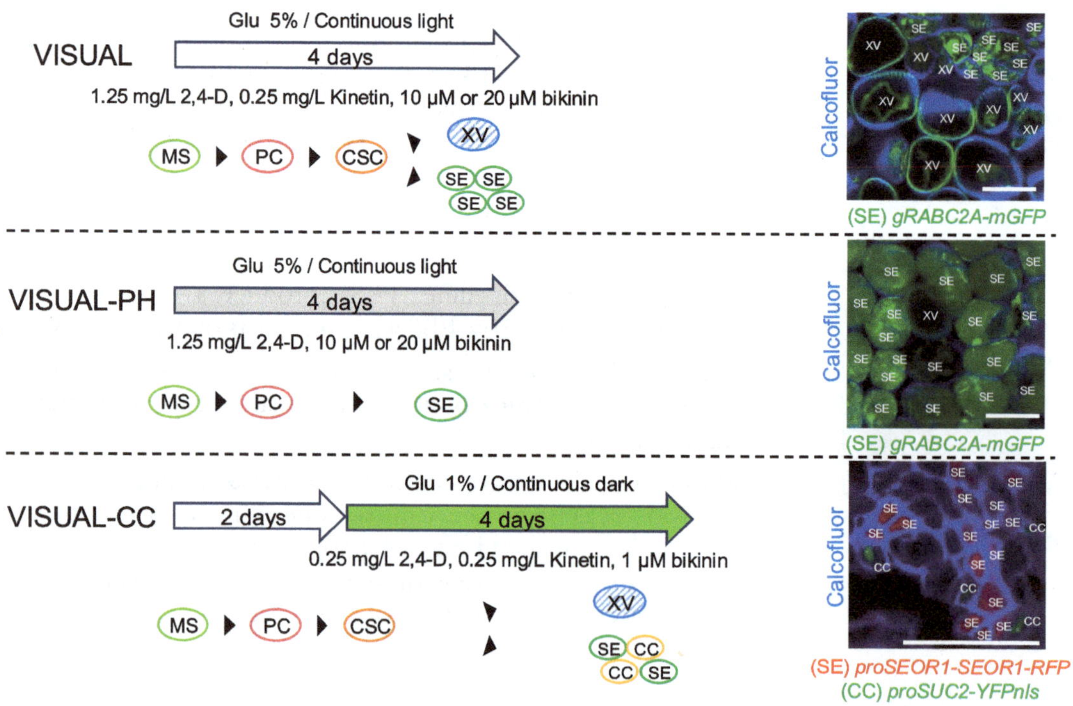

Fig. 1 Overview of experimental procedures for VISUAL-based assays. Summary of hormone and culture conditions in VISUAL-based assays for inducing vascular cell types. Differences in light exposure, hormone composition, and duration distinguish the three methods. MS: mesophyll cells, PC: procambium-like cells, CSC: cambium stem cell-like cells, XV: xylem vessel-like cells, SE: sieve element-like cells, and CC: companion cell-like cells. Bars: 50 μm

3.2 Conventional VISUAL Assay

1. Thaw 2,4-D, kinetin, and bikinin stock solutions at room temperature and bring them into the laminar flow hood (*see* **Note 6**).
2. Prepare VISUAL induction medium by adding 1.25 mg/L 2,4-D, 0.25 mg/L kinetin, and either 10 μM or 20 μM bikinin to the VISUAL base medium (final concentrations) (*see* **Note 7**).
3. Dispense 2.5 mL of the medium into each well of a sterile 12-well plate.
4. Cut 6–7-day-old seedlings (*see* **Note 8**) at the middle of the hypocotyl to remove the roots using a sterilized scalpel or dissecting scissors (*see* **Note 9**). Use the aerial part (cotyledons and upper hypocotyl) as explants (*see* **Notes 10 and 11**) (Fig. 3A, B).
5. Transfer three to four explants into each well and seal the plate with surgical tape (*see* **Note 12**) (Fig. 3C, D).
6. Incubate for 4 days under continuous light (22°C, 60–70 μmol m^{-2} s^{-1}) on an orbital shaker at 110 rpm (*see* **Note 13**).

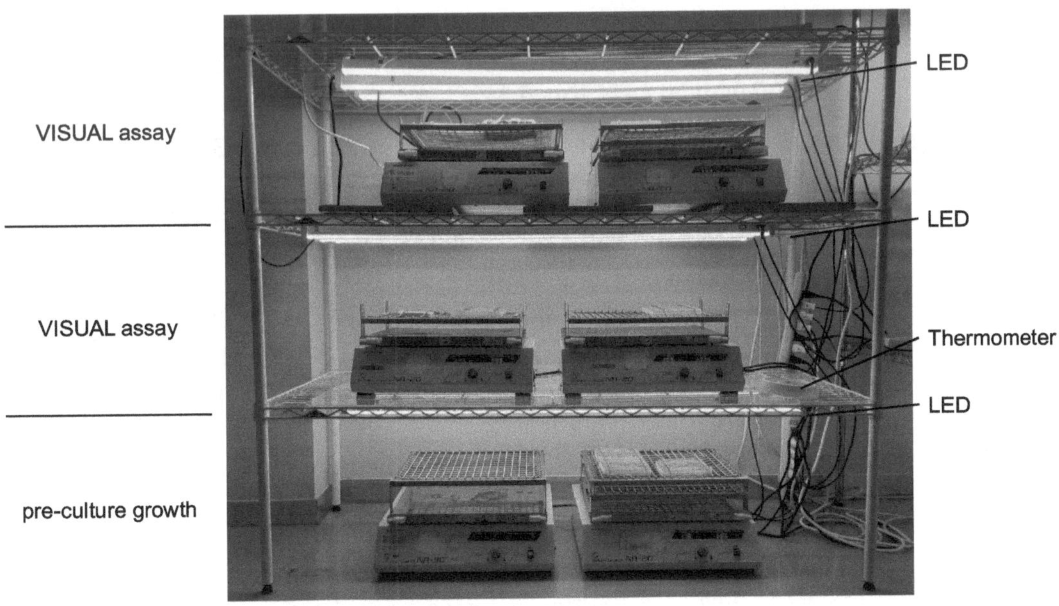

Fig. 2 Culture chamber setup for the VISUAL assay. Photograph of a three-tier steel rack equipped with six orbital shakers (two per level), used for Arabidopsis seedling cultivation and vascular cell induction in the VISUAL assay. The top two tiers are dedicated to the 4-day VISUAL-based induction under continuous light, while the bottom tier is used for pre-culture growth of seedlings in liquid MS medium. Each level is fitted with LED lighting, and thermometers are placed to ensure uniform environmental control throughout the culture process

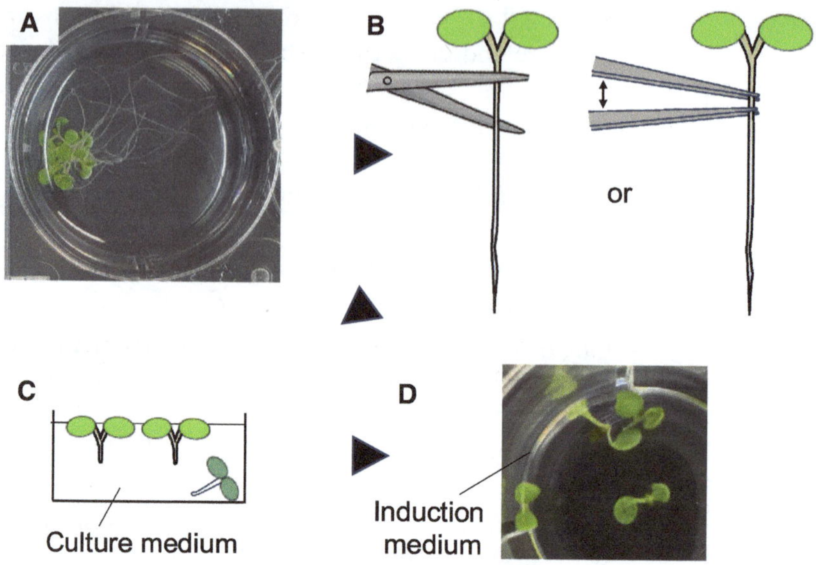

Fig. 3 Experimental workflow of VISUAL-based assay. Representative steps in the VISUAL-based vascular cell induction protocol. (**A**) Arabidopsis seedlings were grown in liquid MS medium for 6–7 days prior to induction. (**B**) Schematic illustration of how the hypocotyl is cut to isolate the aerial part (cotyledons and upper hypocotyl). (**C**) Selection of suitable explants by floating them in MS growth medium; damaged or unhealthy explants typically sink. (**D**) Explants were placed in VISUAL induction medium for culture

3.3 VISUAL-PH Assay

Follow the same procedure as described in Sect. 3.2, except that kinetin is not added to the induction medium. All other steps remain unchanged (*see* **Note 14**).

3.4 VISUAL-CC Assay

1. Prepare the induction medium as in Sect. 3.2 using VISUAL base medium with 1.25 mg/L 2,4-D, 0.25 mg/L kinetin, and 10 μM bikinin (final concentrations). Incubate explants for2 days under continuous light (22°C, 60–70 μmol m^{-2} s^{-1}) on an orbital shaker at 110 rpm.
2. Thaw stock solutions and add 0.25 mg/L 2,4-D, 0.25 mg/L kinetin, and 1 μM bikinin to VISUAL-CC base medium (final concentrations). Note that 2,4-D and bikinin concentrations are lower than in VISUAL medium.
3. Dispense 2.5 mL of the new medium into each well of a sterile 12-well plate.
4. Transfer the pre-treated explants from step 1 into the VISUAL-CC medium and culture under dark conditions at 22°C for 4 additional days, with shaking at 110 rpm (*see* **Note 14**).

3.5 Observation of Induced SE-Like Cells and CC-Like Cells

Fluorescent reporter lines are required to visualize the induced cell types. The following are representative examples commonly used in previous studies:

SE-like cells: *pSEOR1-SEOR1-YFP or RFP* [18, 20], *gRAB-C2A-mGFP* [11].

CC-like cells: *pSUC2-YFPnls* [18].

Xylem vessel-like cells can be distinguished by their secondary cell walls exhibiting autofluorescence.

1. Isolate cotyledons and fix them in fixative solution (4% paraformaldehyde and 0.01% Triton X-100 in 1× PBS) under vacuum for 3 h using a vacuum chamber.
2. Wash the fixed samples twice with 1× PBS, then transfer them to ClearSee solution.
3. Replace the ClearSee solution with fresh solution every 2 days for 3–4 weeks to achieve sufficient tissue clearing.
4. For calcofluor staining, which should be performed approximately 1 week prior to observation, add 0.1% (w/v) calcofluor white to the ClearSee solution. Stain overnight, then wash the samples with ClearSee solution without dye.
5. Continue washing with ClearSee until the time of imaging.
6. Observe the samples using a wide-field fluorescence microscope or a confocal laser scanning microscope.

4 Notes

1. Seedlings grown on solid medium can also be used for the VISUAL assay, for example, in time-lapse imaging under a microscope [21]. However, compared to seedlings cultured in liquid medium, their cotyledon mesophyll cells may turn purplish after induction starts, likely due to stress, which can reduce differentiation efficiency.
2. Glucose is essential for the induction of vessel-like and SE-like cells, and its concentration must be precisely adjusted. Using sucrose instead of glucose leads to the accumulation of cambium stem cell-like cells and inhibits their differentiation [22].
3. Accurate pH adjustment is critical when preparing the induction medium. The acceptable range is narrow (5.65–5.75). Although the absence of MES buffer makes pH adjustment more difficult, MES negatively affects vascular induction and should therefore be excluded.
4. Wear rubber gloves when handling bikinin to protect the skin.

5. Avoid sowing more than 12 seeds per well, as overcrowding can prevent individual seedlings from receiving sufficient light and space for healthy growth.
6. Repeated freeze–thaw cycles of bikinin stock solution are not recommended, as they may reduce its activity. Reuse after thawing should be done one time at most. This limitation does not apply to 2,4-D or kinetin.
7. 20 μM bikinin improves induction efficiency, but 10 μM should be used in VISUAL-CC and control conditions, as the initial 2-day induction period under VISUAL induction medium was specifically optimized for this concentration. Using 20 μM may accelerate differentiation and disrupt the intended timing of the protocol.
8. Seedlings should be excluded if the two true leaves have become clearly visible and distinct, as this indicates an advanced developmental stage that negatively affects differentiation efficiency (Fig. 4).
9. Alternatively, the hypocotyl can be severed by pinching the hypocotyl at two closely spaced points using a pair of forceps and gently pulling them apart (Fig. 3B). All tools must be sterilized with ethanol prior to use.
10. Avoid wounding the leaves or exposing explants to air for prolonged periods, as this may lead to dehydration. Briefly placing explants in MS growth medium after cutting can help identify unsuitable ones, as wounded or unhealthy explants tend to sink (Fig. 3C).

Fig. 4 Suitable growth stages for VISUAL assay. Representative Arabidopsis seedlings at different developmental stages in relation to suitability for the VISUAL assay. Symbols indicate suitability: double circle (◎), optimal stage; triangle (△), suitable but less optimal stage; and cross (╳), too mature for reliable induction

11. Although the hypocotyl is typically cut to keep the cotyledons intact, it may be preferable to cut at the petiole and use cotyledons alone when signals from vascular tissue in the hypocotyl are undesirable—for example, during time-lapse measurement of luminescence signals throughout the differentiation process in the VISUAL assay [17].
12. When using 12-well plates, up to six explants can be cultured effectively in a single well. In contrast, 48-well plates are suitable for culturing only one explant per well due to limited space and medium volume.
13. High temperatures or insufficient light intensity can markedly reduce the efficiency of vascular cell differentiation. Therefore, environmental conditions must be strictly regulated throughout the culture period.
14. Attempts to combine VISUAL-PH and VISUAL-CC (i.e., a cytokinin-free VISUAL-CC condition) to enhance CC-like cell induction have not yet been successful. Without cytokinin, cell divisions that typically precede SE differentiation are reduced, likely hindering the asymmetric divisions required for CC specification.

Acknowledgments

This work was supported by the Japan Society for the Promotion of Science (JSPS) Grants-in-Aid for Scientific Research (KAKENHI) (21J20775 to S.S.; 17H05008, 20K15815, 22H02647, and 25K02295 to Y.K.), the Ministry of Education, Culture, Sports, Science, and Technology of Japan (MEXT) Grants-in-Aids for Scientific Research on Innovative Areas (17H06476, 20H05407, and 22H04720 to Y.K.), the Japan Science and Technology Agency (JST) FOREST Program (JPMJFR224Q to Y.K.), and Sasakawa Scientific Research Grant from The Japan Science Society (2019-4018 to S.S.).

Conflict of Interest The authors have no conflicts of interest to declare that are relevant to the content of this chapter.

References

1. Fukuda H, Komamine A (1980) Establishment of an experimental system for the study of tracheary element differentiation from single cells isolated from the mesophyll of *Zinnia elegans*. Plant Physiol 65(1):57–60. https://doi.org/10.1104/pp.65.1.57
2. Kubo M, Udagawa M, Nishikubo N, Horiguchi G, Yamaguchi M, Ito J, Mimura T, Fukuda H, Demura T (2005) Transcription switches for protoxylem and metaxylem vessel formation. Genes Dev 19(16):1855–1860. https://doi.org/10.1101/gad.1331305

3. Oda Y, Fukuda H (2012) Initiation of cell wall pattern by a Rho- and microtubule-driven symmetry breaking. Science 337(6100):1333–1336. https://doi.org/10.1126/science.1222597
4. Derbyshire P, Ménard D, Green P, Saalbach G, Buschmann H, Lloyd CW, Pesquet E (2015) Proteomic analysis of microtubule interacting proteins over the course of xylem tracheary element formation in Arabidopsis. Plant Cell 27(10):2709–2726. https://doi.org/10.1105/tpc.15.00314
5. Kondo Y, Ito T, Nakagami H, Hirakawa Y, Saito M, Tamaki T, Shirasu K, Fukuda H (2014) Plant GSK3 proteins regulate xylem cell differentiation downstream of TDIF-TDR signalling. Nat Commun 5: 3504. https://doi.org/10.1038/ncomms4504
6. Kondo Y, Fujita T, Sugiyama M, Fukuda H (2015) A novel system for xylem cell differentiation in *Arabidopsis thaliana*. Mol Plant 8(4):612–621. https://doi.org/10.1016/j.molp.2014.10.008
7. Kondo Y, Nurani AM, Saito C, Ichihashi Y, Saito M, Yamazaki K, Mitsuda N, Ohme-Takagi M, Fukuda H (2016) Vascular cell induction culture system using Arabidopsis leaves (VISUAL) reveals the sequential differentiation of sieve element-like cells. Plant Cell 28(6):1250–1262. https://doi.org/10.1105/tpc.16.00027
8. De Rybel B, Audenaert D, Vert G, Rozhon W, Mayerhofer J, Peelman F, Coutuer S, Denayer T, Jansen L, Nguyen L, Vanhoutte I, Beemster GT, Vleminckx K, Jonak C, Chory J, Inzé D, Russinova E, Beeckman T (2009) Chemical inhibition of a subset of *Arabidopsis thaliana* GSK3-like kinases activates brassinosteroid signaling. Chem Biol 16(6):594–604. https://doi.org/10.1016/j.chembiol.2009.04.008
9. Wallner ES, López-Salmerón V, Belevich I, Poschet G, Jung I, Grünwald K, Sevilem I, Jokitalo E, Hell R, Helariutta Y, Agustí J, Lebovka I, Greb T (2017) Strigolactone- and karrikin-independent SMXL proteins are central regulators of phloem formation. Curr Biol 27(8):1241–1247. https://doi.org/10.1016/j.cub.2017.03.014
10. Saito M, Kondo Y, Fukuda H (2018) BES1 and BZR1 redundantly promote phloem and xylem differentiation. Plant Cell Physiol 59(3):590–600. https://doi.org/10.1093/pcp/pcy012
11. Nurani AM, Ozawa Y, Furuya T, Sakamoto Y, Ebine K, Matsunaga S, Ueda T, Fukuda H, Kondo Y (2020) Deep imaging analysis in VISUAL reveals the role of YABBY genes in vascular stem cell fate determination. Plant Cell Physiol 61(2):255–264. https://doi.org/10.1093/pcp/pcaa002
12. Furuya T, Saito M, Uchimura H, Satake A, Nosaki S, Miyakawa T, Shimadzu S, Yamori W, Tanokura M, Fukuda H, Kondo Y (2021) Gene co-expression network analysis identifies BEH3 as a stabilizer of secondary vascular development in Arabidopsis. Plant Cell 33(8):2618–2636. https://doi.org/10.1093/plcell/koab151
13. Matsuoka K, Sato R, Matsukura Y, Kawajiri Y, Iino H, Nozawa N, Shibata K, Kondo Y, Satoh S, Asahina M (2021) Wound-inducible ANAC071 and ANAC096 transcription factors promote cambial cell formation in incised Arabidopsis flowering stems. Commun Biol 4(1):369. https://doi.org/10.1038/s42003-021-01895-8
14. Iwase A, Kondo Y, Laohavisit A, Takebayashi A, Ikeuchi M, Matsuoka K, Asahina M, Mitsuda N, Shirasu K, Fukuda H, Sugimoto K (2021) WIND transcription factors orchestrate wound-induced callus formation, vascular reconnection and defense response in Arabidopsis. New Phytol 232(2):734–752. https://doi.org/10.1111/nph.17594
15. Liu C, Li A, Guo Z, Chen N, Wang Y, Tang W, Wu Y, Liu J, Wang Z, Li L, He XQ (2024) MicroRNA analysis reveals 2 modules that antagonistically regulate xylem tracheary element development in Arabidopsis. Plant Cell 37(1):koaf011. https://doi.org/10.1093/plcell/koaf011
16. Torii K, Inoue K, Bekki K, Haraguchi K, Kubo M, Kondo Y, Suzuki T, Kubota A, Uemoto K, Shimizu H, Saito M, Fukuda H, Araki T, Endo M (2022) A guiding role of the Arabidopsis circadian clock in cell differentiation revealed by time-series single-cell RNA sequencing. Cell Rep 40(2):111059. https://doi.org/10.1016/j.celrep.2022.111059
17. Shimadzu S, Yonekura T, Furuya T, Kojima M, Ishizaki K, Asahina M, Ohashi-Ito K, Sakakibara H, Fukaki H, Fukuda H, Kondo Y (2025) A cytokinin response maximum induces and activates bifacial stem cells for radial growth. Nat Plants 11(8):1558–1571. https://doi.org/10.1038/s41477-025-02051-4
18. Tamaki T, Oya S, Naito M, Ozawa Y, Furuya T, Saito M, Sato M, Wakazaki M, Toyooka K, Fukuda H, Helariutta Y, Kondo Y (2020) VISUAL-CC system uncovers the role of GSK3 as an orchestrator of vascular cell type ratio in plants. Commun Biol 3(1):184. https://doi.org/10.1038/s42003-020-0907-3
19. Kurihara D, Mizuta Y, Sato Y, Higashiyama T (2015) ClearSee: a rapid optical clearing reagent for whole-plant fluorescence imaging. Development 142(23):4168–4179. https://doi.org/10.1242/dev.127613

20. Froelich DR, Mullendore DL, Jensen KH, Ross-Elliott TJ, Anstead JA, Thompson GA, Pélissier HC, Knoblauch M (2011) Phloem ultrastructure and pressure flow: sieve-element-occlusion-related agglomerations do not affect translocation. Plant Cell 23(12):4428–4445. https://doi.org/10.1105/tpc.111.093179
21. Shimadzu S, Furuya T, Ozawa Y, Fukuda H, Kondo Y (2022) Spatio-temporal imaging of cell fate dynamics in single plant cells using luminescence microscope. Quant Plant Biol 3:e15. https://doi.org/10.1017/qpb.2022.12
22. Narutaki A, Kahar P, Shimadzu S, Maeda S, Furuya T, Ishizaki K, Fukaki H, Ogino C, Kondo Y (2023) Sucrose signaling contributes to the maintenance of vascular cambium by inhibiting cell differentiation. Plant Cell Physiol 64(12):1511–1522. https://doi.org/10.1093/pcp/pcad039

Chapter 36

Measurement of Phloem-Based Electrical Signals in Plants

Alexandra C. U. Furch, Matthias R. Zimmermann, Henk Oppermann and Torsten Will

Abstract

Electrical signals are vital components of long-distance signaling in plants, enabling rapid systemic responses to environmental stimuli. These signals, which include action potentials (APs), variation potentials (VPs), and system potentials (SPs), that can also combine into electropotential waves (EPWs), are primarily propagated through the vascular system, with the phloem playing a central role. EPWs, encompassing APs, VPs, and SPs, are integral to systemic signaling and inter-organ communication via the phloem, and are closely associated with plant responses to pests and pathogens. The complex architecture of the phloem, embedded deep within plant tissues and surrounded by multiple cell layers, poses significant challenges for direct electrophysiological measurements. Traditional electrophysiological methods for studying electrical long-distance signals often require wounding, which can alter the physiological conditions being measured. The use of aphids as bio-electrodes offers a promising alternative, as it avoids mechanical wounding and allows for flexible selection of measurement points. This chapter highlights the advantages and limitations of various techniques available for studying electrical signals in intact plants, which is essential for understanding how plants respond to environmental stimuli.

Keywords Electrical signals, Membrane potential, Microelectrodes, Sieve elements, Systemic signaling, Aphid, Bioelectrode

Abbreviations

AP	Action potential
EPG	Elctrical penetration graph
EPW	Electropotential wave
PM	Plasma membrane
SE	Sieve element
SP	System potential
VP	Variation potential

Johannes Liesche and Sanna Sevanto (eds.), *Phloem*, Methods in Molecular Biology, vol. 3066,
https://doi.org/10.1007/978-1-0716-5488-0_36,

1 Introduction

A plant cell is surrounded by a plasma membrane (PM), which separates the cell from its external environment and creates an internal space that facilitates chemical reactions in a controlled environment. The PM is composed of a lipid bilayer and contains a diverse array of proteins. These proteins may be located on either side of the bilayer or span both layers, such as ion channels. Ion channels play a key role in generating the electrical transmembrane potential.

In plant cells, the membrane potential refers to the electrical potential difference across the PM, which arises from the unequal distribution of ions such as Ca^{2+}, Cl^-, K^+, and H^+. This distribution is maintained by ion pumps and carriers. Typically, the membrane potential is negative inside the cell relative to the outside and serves as a crucial secondary energy source for various cellular processes [1]. Additionally, changes in the membrane potential can act as signals, enabling the plant to respond to environmental stimuli [2]. These signals include action potentials (APs), variation potentials (VPs), and system potentials (SPs), which can combine into electropotential waves (EPWs). These signals are briefly described below.

2 Electrical Signals in the Phloem

An action potential (AP) is a brief depolarization of the plasma membrane (PM) characterized by a specific voltage pattern. Interestingly, this phenomenon occurs in both animals and plants. Despite their distinct origins, plant and animal APs share several core features: (1) a characteristic voltage profile, (2) adherence to the all-or-nothing principle, (3) the ability to self-propagate, and (4) the presence of a refractory period. However, they differ in two key aspects: (1) the duration of the signal and (2) the molecular mechanisms involved [2]. In plant APs, the initial depolarization is thought to involve PM Ca^{2+} channels (cf. [3, 4], leading to Ca^{2+} influx and the activation of Ca^{2+}-dependent anion channels [5, 6], which results in anion efflux. The recovery phase involves passively driven K^+ influx and active symport of H^+/Cl^- [5].

Variation potentials (VPs) are transient depolarizations of the SE PM Hier sollte eingefügt werden in welchen Zellen sich das MP verändert, da der Text hier suggeriert, dass es sich auch um das Xylem handeln könnte, welches aber natürlich kein MP und auch kein PM hat., characterized by variable shape, amplitude, and duration (e.g., [7–9]). In VPs, the activation of mechanosensitive Ca^{2+} channels in response to hydraulic pressure waves propagated via the xylem [10], along with the additional shutdown of the PM H^+-pump [11], leads to prolonged PM depolarization [12].

In contrast to APs and VPs, system potentials (SPs) are systemically transmitted hyperpolarization event that propagate along the apoplast. These are driven by the relayed activation of proton pumps, enabling self-propagation from cell to cell [13].

Electropotential waves (EPWs) integrate features of distinct classes of electrical signals, including APs and VPs. The molecular identity of the ion channels underlying EPWs remains incompletely understood, partly because EPWs combine characteristics of both rapid, transient APs and slower, wave-like VPs. It has been hypothesized that Ca^{2+} influx associated with EPWs results from the coordinated activity of multiple Ca^{2+}-permeable channels. The repertoire of Ca^{2+}-permeable channels involved in electrical signal propagation likely varies depending on the nature of the eliciting stimulus [8, 14, 15].

The phloem's extensive symplastic connectivity through sieve pores and plasmodesmata makes it an ideal conduit for electropotential wave (EPW) transmission. Its systemic reach enables whole-plant coordination, influencing processes such as leaf movement and systemic plant defense, including phloem occlusion. Accurate electrophysiological recordings from phloem sieve elements (SEs) are therefore critical for understanding plant signaling networks.

3 Measuring Electrical Signals in the Phloem

Long-distance electrical signals can be recorded using microelectrodes inserted symplastically into phloem SEs, microelectrodes blindly pierced into either symplastic or apoplastic regions of the tissue surrounding SEs, or by employing aphids as bio-electrodes, with their stylets penetrating SEs (Fig. 1).

To measure electrical signals *symplastically within SEs using microelectrodes*, a so called phloem observation window must be

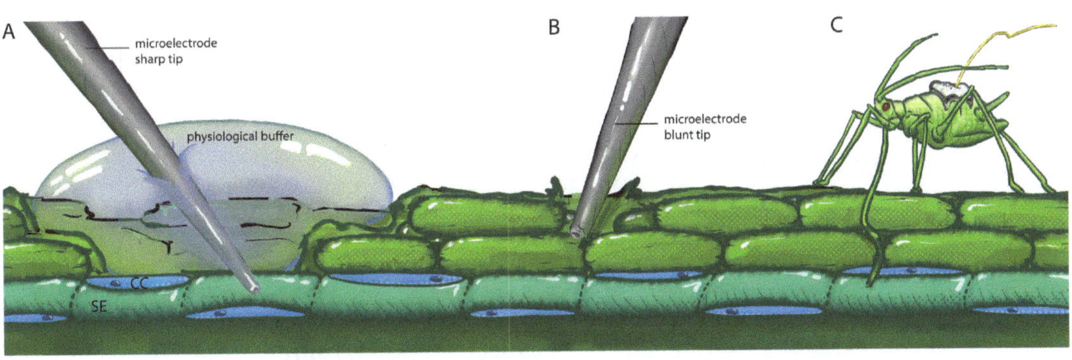

Fig. 1 Schematic drawing of the measurement methods. **(A)** Intracellular measurement, a sharp-tipped microelectrode is shown penetrating cells covered by a physiological buffer. **(B)** Blind piercing, a blunt-tipped microelectrode is inserted blindly. **(C)** Aphid as bio-electrode.

prepared in a way that maintains phloem tissue integrity while ensuring optimal visibility and accessibility of the SEs. The cortical cell layers, typically consisting of epidermal, mesophyll, and parenchyma tissue, are removed from the abaxial side of the main vein of a mature leaf still attached to an intact plant, exposing the phloem. Some plants have a perixylematic vascular bundle structure, such as *Convallaria majalis* and *Iris germanica*, where the xylem completely surrounds the phloem. In these cases, the xylem must also be partially removed, which may negatively impact phloem function due to the interaction between neighboring xylem and phloem tissues [16]. The cell layers are removed manually via paradermal cutting with a razor blade, taking care to avoid damage to the phloem [15]. The integrity of the phloem tissue must be visually verified under a microscope using a water immersion lens and a physiological buffer solution [17]. Deposits on sieve plates, for instance, indicate that an SE has been damaged during preparation and is no longer intact.

Intact SEs are punctured using sharp, laterally broken microelectrodes with an outer diameter of approx. 1 mm, made from aluminum silicate microcapillaries (Fig. 1A). This is performed using a micromanipulator under constant microscopic observation. Particular attention is given to experimental conditions, such as stable lighting and electrical shielding using a Faraday cage, which are essential for accurate measurements. However, accessing phloem cells for electrophysiological measurements is challenging and only possible through wounding. It is important to note that wounding can lead to the release of damage-associated molecular patterns, which may trigger so-called slow wave potentials [7].

For extracellular electrical recordings using the *blind piercing technique*, microelectrodes must be specifically fabricated with a mechanically blunted tip to ensure stable tissue penetration without causing excessive damage. These electrodes are typically pulled from borosilicate glass capillaries and are subsequently heat-polished or mechanically broken to achieve the desired blunt profile. The insertion of microelectrodes is performed without visual guidance, hence the term "blind piercing." It targets vascular tissues within the primary vein of fully expanded mature leaves or the stem of physiologically intact, undisturbed plants (Fig. 1B). Electrical signal acquisition begins only after the extracellular potential stabilizes, a process that can take between 5 and 24 h depending on the species, tissue condition, and electrode placement. A key limitation of the extracellular approach is its inherently low spatial resolution due to the lack of effective electrical shielding. The recorded potential reflects the composite activity within an undefined tissue volume, encompassing an unknown number of adjacent vascular bundles. Consequently, the extracellular signal often represents a superimposed summation of several localized electrophysiological events, with the corresponding voltage traces displaying diverse kinetic profiles and time courses. Such variability complicates the unambiguous identification of individual electrical events and their specific cellular or tissue origins [14, 18].

An alternative approach that avoids injury-induced limitations in both experimental design and data interpretation is the use of *aphids as bio-electrodes* as part of the electrical penetration graph setup. Aphids possess flexible mouthparts (stylets) that can penetrate SEs with minimal damage to the tissue (Fig. 1C, cf. [19]). When SEs and other cells are punctured, aphid saliva—specifically gel saliva—seals the micro-injury [20, 21]. Additionally, the release of watery saliva suppresses plant defense responses [22].

The electrical penetration graph (EPG) technique was originally developed to observe the feeding behavior of aphids. The technical setup integrates aphids and plants into an electrical circuit [23]. A low, constant electrical voltage is applied and modified by the aphid and plant, which together act as a variable resistor, generating specific waveforms as a result of different behavioral patterns. Additionally, the method allows for the measurement of membrane potentials in punctured plant cells, represented as so-called "potential drops," which correspond to one of the EPG waveforms [24]. Studies such as those by Furch et al. [18] and Salvador-Recatalà et al. [25] have demonstrated the utility of EPG recordings with aphids for monitoring EPWs over long distances within the plant body. Furthermore, the technique enables simultaneous recordings at multiple locations, facilitating comprehensive spatial analysis of signal propagation. For example, Paulmann et al. [26] showed that infection with the *barley yellow dwarf virus* negatively affects the transmission of APs, reducing both their velocity and propagation distance along a barley leaf.

Unlike the piercing process required for microelectrode techniques, which necessitates tissue preparation and microscopic observation, the EPG method allows easy access to multiple measurement sites without extensive preparation. These sites can be located on a single leaf, along a stem, or even on leaves at different developmental stages, demonstrating the high flexibility of this approach.

Despite its advantages, the EPG technique also has limitations. Aphid saliva contains various effector proteins with largely unknown functions [23], which may influence the activity of membrane channels, pumps, and transporters within SEs. Additionally, aphid infestation can activate plant defense mechanisms, such as the production of hydrogen peroxide, as observed in *Myzus persicae* feeding on *Arabidopsis thaliana* [27]. Hydrogen peroxide has been shown to inhibit ATP hydrolysis and proton pump activity of the PM H^+-ATPase [28], which may not only affect the membrane potential but also influence aphid behavior. To minimize the effects of plant defense responses on signal generation and propagation during aphid-host plant interactions, it is advisable to select plant genotypes with high susceptibility to aphids.

Another limitation of the EPG technique is that EPWs recorded using aphids differ in amplitude, form, and intensity compared to those obtained with classical microelectrode techniques. Although the reasons for these differences remain unclear, it is likely that the measured voltage does not correspond to the actual voltage due to factors such as voltage drops. When measured with aphids, the voltage is typically lower than that recorded with traditional electrodes, possibly due to the specific electrical resistances of the biological components in the system. Additionally, electrical artifacts such as flow potentials can overlap with the actual EPW signals, further complicating datainterpretation.

4 Conclusion

The diversity of electrical signals, APs, VPs, SPs, and EPWs, reflects the complex nature of plant electrophysiological signaling. APs are rapid, all-or-nothing responses characterized by a distinct biphasic depolarization of the plasma membrane. In contrast, VPs are slower, variable in amplitude, and are thought to primarily result from the inhibition of H^+-ATPases. SPs represent unique hyperpolarization events linked to the activation of these same ATPases. EPWs, on the other hand, are composite signals that merge features of APs and VPs. The distinct temporal dynamics and signal morphologies underscore the nuanced roles these signals play in plant responses to environmental stress.

The techniques described above enable the measurement of these electrical signals, each with its own advantages and disadvantages. These range from high-accuracy methods that require tissue wounding to flexible approaches that allow for multiple measurement sites on a single plant but may introduce signal modifications. While the use of aphids and the EPG technique offer new opportunities for minimally invasive in vivo electrophysiological measurements in plants without the need for extensive tissue preparation, the application of electrophysiological techniques in other contexts, such as plant phenotyping for breeding, appears limited. This is due to the low number of plants that can be tested and the physiological variables affected by the different methods.

Nevertheless, beyond basic physiological observations, such as the activity of ion channels and transporters, especially the use of aphid bio-electrodes has significantly expanded the application of electrophysiological techniques. This is particularly due to the method's low technical requirements and its ability to provide valuable insights into plant signaling processes.

Conflict of Interest

The authors have no conflicts of interest to declare that are relevant to the content of this chapter.

References

1. Dainty (1962) Ion transport and electrical potentials in plant cells. Ann Rev Plant Biol 13:379–402
2. Zimmermann MR, Mithöfer A (2013) Electrical long-distance signaling in plants. In: Baluska F (Ed.)Eds. Long-distance systemic signaling and communication in plants. Springer, Heidelberg, 291–308pp
3. White PJ (2000) Calcium channels in higher plants. Biochim Biophys Acta 1465:171–189
4. Demidchik V, Maathuis FJM (2007) Physiological roles of nonselective cation channels in plants: from salt stress to signalling and development. New Phytol 175:387–404
5. Felle HH, Zimmermann MR (2007) Systemic signalling in barley through action potentials. Planta 226:203–214
6. van Bel AJE, Furch ACU, Will T, Buxa SV, Musetti R, Hafke JB (2014) Spread the news: systemic dissemination and local impact of Ca^{2+} signals along the phloem pathway. J Exp Bot 65:1761–1787
7. Farmer EE, Gao Y-Q, Lenzoni G, Wolfender J-L, Wu Q (2020) Wound-and mechanostimulated electrical signals control hormone responses. New Phytol 227:1037–1050
8. Paulmann MK, Zimmermann MR, Wegner L, van Bel AJE, Kunert G, Furch ACU (2021) Species-specific and distance-dependent dispersive behaviour of forisomes in different legume species. Int J Mol Sci 22:492
9. Furch ACU, Zimmermann MR, Noll GA, Wrobel LS, Scholz SS, Kleeberg SV, Hafke JB, Fliegmann J, Mithöfer A, Ehlers K, Haufschild T, Nötzold J, Koch AM, Grabe V, Teutemacher F, Maaß J-P, Prüfer D, Oelmüller R, Peiter E, Kogel K-H, van Bel AJE (2025) Transformation of flg22 perception into electrical signals decoded in vasculature leads to sieve tube blockage and pathogen resistance. Sci Adv 11:eads6417
10. Stahlberg R, Cosgrove DJ (1997) The propagation of slow wave potentials in pea epicotyls. Plant Physiol 113:209–217
11. Stahlberg R, Stephens NR, Cleland RE, Van Volkenburgh E (2006) Shade-induced action potentials in *Helianthus annuus* L. originate primarily from the epicotyl. Plant Signal Behav 1:15–22
12. Fromm J, Lautner S (2007) Electrical signals and their physiological significance in plants. Plant Cell Environ 30:249–257
13. Zimmermann MR, Maischak H, Mithöfer A, Boland W, Felle HH (2009) System potentials, a novel electrical long-distance apoplastic signal in plants, induced by wounding. Plant Physiol 149:1593–1600
14. Zimmermann MR, Mithöfer A, Will T, Felle HH, Furch ACU (2016) Herbivore-triggered electrophysiological reactions: candidates for systemic signals in higher plants and the challenge of their identification. Plant Physiol 170:2407–2419
15. Furch ACU, van Bel AJE, Fricker MD, Felle HH, Fuchs M, Hafke JB (2009) Sieve element Ca^{2+} channels as relay stations between remote stimuli and sieve tube occlusion in *Vicia faba*. Plant Cell 21:2118–2132
16. Furch ACU, Zimmermann MR, Kogel K-H, Reichelt M, Mithöfer A (2014) Direct and individual analysis of stress-related phytohormone dispersion in the vascular system of *Cucurbita maxima* after flagellin 22 treatment. New Phytol. https://doi.org/10.1111/nph.12661
17. Zimmermann MR, Hafke JB, Paulmann MK, Furch ACU (2014) Measurement of electropotential waves in intact phloem sieve elements using microelectrodes. In: Liesche J (ed) Phloem methods and protocols. Springer, Heidelberg, pp 439–447
18. Furch ACU, Zimmermann MR, Will T, Hafke JB, van Bel AJE (2010) Remote-controlled stop of phloem mass flow by biphasic occlusion in *Cucurbita maxima*. J Exp Bot 61:3697–3708
19. Tjallingii WF, Hogen Esch TH (1993) Fine-structure of aphid stylet routes in plant-tissues in correlation with EPG signals. Physiol. Entomol. 18:317328
20. Will T, van Bel AJE (2006) Physical and chemical interactions between aphids and plants. J Exp Bot 57:729–737
21. Will T, Vilcinskas A (2015) The structural sheath protein of aphids is required for phloem feeding. Insect Biochem Mol Biol 57:34–40
22. Will T, Furch ACU, Zimmermann MR (2013) How phloem-feeding insects face the challenge of phloem-located defenses. Front Plant Sci 4:336

23. Tjallingii WF (1988) Electrical recording of stylet penetration activities. In: Minks AK, Harrewijn P (eds) Aphids, their biology, natural enemies and control, vol. 2B. Elsevier, Amsterdam, pp 95–108
24. Prado E, Tjallingii WF (1994) Aphid activities during sieve element punctures. Entomol Exp Appl 72:157–165
25. Salvador-Recatalà V, Tjallingii WF, Farmer EE (2014) Real-time, in vivo intracellular recordings of caterpillar-induced depolarization waves in sieve elements using aphid electrodes. New Phytol 203:674–684
26. Paulmann MK, Kunert G, Zimmermann MR, Theis N, Ledwig A, Meichsner D, Oelmüller R, Gershenzon J, Hebekuss A, Ordon F, Furch ACU, Will T (2018) *Barley yellow dwarf virus* infection leads to higher chemical defense signals and lower electrophysiological reactions in susceptible compared to tolerant Barley genotypes. Front Plant Sci 9:145
27. Xu J, Padilla CS, Li J, Wickramanayake J, Fischer HD, Goggin FL (2021) Redox responses of *Arabidopsis thaliana* to the green peach aphid, *Myzus persicae*. Mol Plant Pathol 22:727–736
28. Cosse M, Seidel T (2021) Plant proton pumps and cytosolic pH-homeostasis. Front Plant Sci 12:672873

Part VII

Theoretical Models of Phloem Function

Check for updates

Chapter 37

Modeling the Hydraulic Conductivity of Phloem Sieve Elements

Sabrina Gennis and Kaare H. Jensen

Abstract

Phloem transport of photoassimilates affects nearly all aspects of plant life, from growth to reproduction. This chapter summarizes mathematical techniques to quantify the impact of sieve element anatomy on phloem transport processes.

Keywords Hydraulic resistance, Hydraulic conductivity, Fluid mechanics, Sieve plate

1 Introduction

Phloem sugar transport is key to plant life; hence, it is essential to understand the basic constraints that limit transport through this system. The phloem enables the distribution of photoassimilates (sugars) in vascular plants through a network of cellular conduits to sustain metabolism, growth, and storage. The flow is driven by pressure differences between sources and sinks. Phloem sap contains a plethora of osmolytes, but the primary determinant of phloem pressure is believed to be the concentration of sugars and small ions [1, 2]. This chapter provides a concise overview of techniques used to quantify phloem hydraulics, including parameters essential in studies of long-distance sugar transport [3], and the hydraulic architecture of leaves [4–6].

The phloem provides a microfluidic pipe network that links distal parts of the organism (Fig. 1). Flow phenomena in pipes and channels have been widely studied; in fact, this is one of the most important topics in fluid dynamics [7–9]. The relation between applied force and flow depends on the structure of the phloem sieve elements, i.e., on the detailed anatomy of the cell lumen and sieve plates (Fig. 1), and on the number of phloem channels running in parallel. The structure of the conduits and the network they form is strikingly complex, yet for liquid flows dominated by viscous friction in small

Johannes Liesche and Sanna Sevanto (eds.), *Phloem*, Methods in Molecular Biology, vol. 3066,
https://doi.org/10.1007/978-1-0716-5488-0_37,

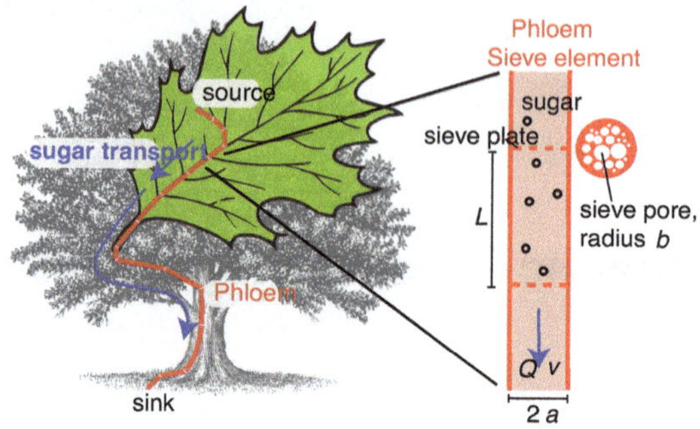

Fig. 1 Phloem sieve elements form a fluid channel network responsible for sugar transport in vascular plants. The conductive cells are approximately cylindrical of radius *a* and length *L*. Perforated sieve plates of thickness *t* connect adjacent sieve elements. Flow from source to sink (speed *v* and flow rate *Q*) is illustrated by the blue arrows

channels, evidence has accrued for a (nearly) universal law: there is a linear relation between the applied pressure Δp (Pa) and the flow rate Q (m^3). This is often expressed as,

$$Q = \frac{\Delta p}{R}, \tag{1}$$

where R (Pa/m^3) is the hydraulic resistance (*see* Table 1 for a list of parameters; *see* **Notes 2 and 3**). A familiar example is the pressure-drop/flow rate relation for a cylindrical pipe $Q = \pi a^4/(8\eta L)\Delta p$ (due to Hagen and Poiseuille), and for a circular hole in a thin membrane $Q = b^3/(3\eta)\Delta p$ (Sampson flow). Here, L (m) is the channel length, η (Pa s) is the liquid viscosity, and a (m) and b (m) are the channel and pore radius, respectively (*see* **Notes 1 and 2**). Most of these classical results were obtained in the mid-to-late nineteenth century, *see* e.g., [10–12]. We note a strikingly strong dependence on the transverse conduit and pore dimensions, which enter in the equations raised to the third and fourth power, indicating a strong geometric dependence.

1.1 Hydraulic resistance of phloem sieve elements

To elucidate how changes in sieve element geometry affect phloem flow, we proceed to consider the hydraulic properties of the combined lumen-sieve plate system. Our analysis follows that of [13] and [14]. Because liquid flows sequentially through the lumen and across the sieve plate, we can consider the system as a serial coupling with two components. Serial couplings of two hydraulic elements follow the well-known result from electrical circuit theory: $R_{tot} = R_1 + R_2$ [7]. Therefore, we can write the hydraulic resistance of a sieve element as,

$$R_{SE} = R_{lumen} + R_{plate}. \tag{2}$$

Table 1
List of parameters

Parameter	Symbol (unit)
Sieve element radius	a (m)
Sieve element length	L (m)
Sieve pore radius	b (m)
Sieve plate thickness	t (m)
Sieve pore radius, mean	$\langle b \rangle$ (m)
Sieve pore radius, std	σ (m)
Sieve pore number	N
Liquid viscosity	η (Pa s)
Conductivity/permeability	k (m^2)
Statistical moments	M_3, M_4
Hydraulic resistance	R (Pa/m^3)
Liquid density	ρ (kg/m^3)
Reynolds number	Re

We assume that the lumen is well-approximated by a cylindrical pipe, such that,

$$R_{\text{lumen}} = \frac{8\eta L}{\pi a^4}. \tag{3}$$

As liquid approaches the sieve plate, the current is split into smaller fragments that flow through each of the pores. Therefore, we can consider a sieve plate with N pores as a parallel coupling. Applying the rules of circuit theory ($R_{\text{tot}}^{-1} = R_1^{-1} + R_2^{-1}$) leads to,

$$R_{\text{Plate}}^{-1} = \sum_{n=1}^{N} \left[\frac{3\eta}{b_i^3} \left(1 + \frac{8}{3\pi} \frac{t}{b_i} \right) \right]^{-1}, \tag{4}$$

where we have explicitly accounted for the variance in pore radius b_i by taking the sum over all pores. Resistance and flow scales with the 3rd and 4th power of the pore radius, hence a few large pores could strongly reduce the plate resistance. Combining the lumen and plate resistance, the total SE resistance (Eq. (2)) becomes,

$$R_{\text{SE}} = \frac{8\eta L}{\pi a^4} + \left(\sum_{n=1}^{N} \left[\frac{3\eta}{b_i^3} \left(1 + \frac{8}{3\pi} \frac{t}{b_i} \right) \right]^{-1} \right)^{-1}. \tag{5}$$

This expression for the SE resistance has the distinct disadvantage that it requires knowledge of each pore radius b_i. If all pores are equal, Eq. (5) reduces to,

$$R_{SE} = \frac{8\eta L}{\pi a^4} + \frac{1}{N}\frac{3\eta}{\langle b\rangle^3}\left(1 + \frac{8}{3\pi}\frac{t}{\langle b\rangle}\right). \tag{6}$$

However, it is well established that pore sizes are not uniform [13]. For normally distributed pores with mean radius $\langle b\rangle$ (m) and standard deviation σ (m), Eq. (5) is well approximated by,

$$R_{SE} \simeq \frac{8\eta L}{\pi a^4} + \frac{1}{N}\frac{3\eta}{\langle b\rangle^3}\left(\frac{1}{M_3} + \frac{8}{3\pi}\frac{t}{\langle b\rangle}\frac{1}{M_4}\right), \tag{7}$$

where $M_3 = 1 + 3\left(\frac{\sigma}{\langle b\rangle}\right)^2$ and $M_4 = 1 + 6\left(\frac{\sigma}{\langle b\rangle}\right)^2 + 3\left(\frac{\sigma}{\langle b\rangle}\right)^4$ are moments of the distribution (*see* detailed discussion, including results for other distribution functions in [14]).

To illustrate the effect of pore size distribution on the sieve element resistance, we consider the typical case $\sigma/\langle b\rangle = 0.25$ [13]. For these parameters $1/M_3 \simeq 0.8$ and $1/M_4 \simeq 0.6$ and the sieve plate resistance contribution to (Eq. (7)) is reduced by approximately 20%–40% when compared to the uniform pore case (Eq. (6)).

1.2 Conductivity of phloem sieve elements

The preceding analysis has focused on the relationship between volumetric flow rate Q and the applied pressure drop Δp. Experiments on phloem flow using radioactive or fluorescent tracers (*see* Part III), however, typically quantify the flow speed $v = Q/(\pi a^2)$ (m/s) averaged over the channel cross-section. A formalism similar to Eq. (1) exists for the velocity/pressure drop relation,

$$v = \frac{k}{\eta}\frac{\Delta p}{L}, \tag{8}$$

where $v = Q/(\pi a^2)$ (m/s) is the velocity averaged over the channel cross-section. Equation (8) is known as Darcy's law, an equation that is widely used to describe fluid flow in porous media [15]. The parameter k (m^2) is the channel conductivity (or permeability), related to resistance by

$$k = \frac{\eta L}{\pi a^2}\frac{1}{R_{SE}}. \tag{9}$$

Using the conductivity k to quantify the hydraulic properties of phloem sieve elements has the advantage that it is scaled by length (L) and area (πa^2), *see* Eq. (9). This means that we can directly com-

Table 2

A selection of plants spanning a wide range of conductivity values (*k*). All relevant parameters needed to calculate *k* are provided. These include the sieve element length (*L*) and radius (*a*), the number of pores (*N*), their mean diameter (⟨*b*⟩) and standard deviation (*σ*), as well as the sieve plate thickness (*t*). The parameters are obtained from [16, 17]. Changing the sieve element radius (*a*) by +/− 20% leads to different values in the conductivity. Sometimes, an increase in conductivity can be observed with either increasing or decreasing radius

Plant	*L* (μm)	*a* (μm)	*N*	⟨*b*⟩ (μm)	*σ* (μm)	*t* (μm)	*k* (μm²)	*k* for *a* +20% (μm²)	*k* for *a* −20% (μm²)
Arabidopsis thaliana [16]	107	2	52	0.26	0.01	1.0	0.28	0.27	0.24
Pinus pinea [18]	2800	10	420	0.2	0.037	2.2	0.62	0.44	0.90
Gossypium barbadense [19]	210	11	161	0.5	0.1	1.0	1.60	1.18	2.17
Solanum lycopersicum [17]	130	10.7	121.3	0.61	0.15	0.52	2.07	1.55	2.67
Metasequoia sempervirens [18]	2850	6	468	0.25	0.05	1.3	3.40	3.87	2.54
Cupania furfuracea [18]	380	8.5	65	0.85	0.17	0.04	6.57	7.32	5.01
Tectona grandis [18]	280	19.5	97	1.25	0.3145	0.8	9.60	7.44	11.62
Juglans cinerea [18]	300	12	300	1.275	0.29	0.9	15.35	19.08	10.76
Sabal palmetto [19]	700	18	287	0.95	0.2375	0.5	22.06	21.34	19.31
Curcubita maxima [17]	341	25.65	54.8	2.54	0.86	1.27	31.05	26.81	31.42

pare tubing systems with different geometries that occupy the same tissue volume.

In summary, we have determined two metrics to quantify the hydraulic properties of sieve elements: the hydraulic resistance (Eq. (2)) and conductivity/permeability (Eq. (9)).

1.3 Examples

Having established the theoretical framework, we conclude this chapter by considering a few representative examples. Specifically, we will compute and compare values of conductivity (k) in the ten species listed in Table 2.

We first note that the magnitude of the parameters affecting the conductivity (e.g., sieve tube radius a, *see* Eqs. (9) and (7)) varies substantially between species, tissues, and tissue age. For instance, the sieve element radius a spans roughly one order of magnitude. As a result, conductivity values also differ. For this data, the conductivity k ranges from about 0.3 μm² to 30 μm², covering roughly two orders of magnitude.

Naturally, the conductivity value varies when one or more of the geometric parameters are changed. An example of this variation is adjusting the sieve element radius (a) by, for instance, plus or minus 20 % (*see* Table 2). According to the Hagen-Poiseuille law (*see* Eq. (3)), one would expect an increase in radius a to lead to a rise in conductivity k. However, the change in conductivity does not follow this pattern, possibly due to the growing influence of the sieve plate (*see* Eq. (7)).

2 Materials

Knowledge of the following anatomical parameters is required to determine the hydraulic resistance and conductivity of phloem sieve elements. The relevant experimental methods are outlined in Chaps. 1–3 and 5.

2.1 Mean sieve tube length

L.

2.2 Mean sieve tube radius

a.

2.3 Mean sieve pore number

N.

2.4 Mean sieve pore radius

$\langle b \rangle$.

2.5 Standard deviation of the sieve pore radius

σ.

2.6 Mean sieve plate thickness

t.

3 Methods

To compute the sieve element resistance and conductivity, we proceed in the following steps (*see* **Note 3**).

1. Determine the statistical moments

$$M_3 = 1 + 3\left(\frac{\sigma}{\langle b \rangle}\right)^2,$$

and

$$M_4 = 1 + 6\left(\frac{\sigma}{\langle b\rangle}\right)^2 + 3\left(\frac{\sigma}{\langle b\rangle}\right)^4.$$

If σ is unknown, use $M_3 = M_4 = 1$ in the following

2. Compute the hydraulic resistance

$$R_{SE} = \frac{8\eta L}{\pi a^4} + \frac{1}{N}\frac{3\eta}{\langle b\rangle^3}\left(\frac{1}{M_3} + \frac{8}{3\pi}\frac{t}{\langle b\rangle}\frac{1}{M_4}\right).$$

3. Compute the conductivity/permeability

$$k = \frac{\eta L}{\pi a^2}\frac{1}{R_{SE}}.$$

4 Notes

1. Equation (1) and the subsequent analysis are valid for flows dominated by viscous friction. Specifically, we require that the Reynolds number $\mathrm{Re} = \rho v a/\eta \ll 1$, where ρ (kg/m^3) is the liquid density.
2. Equation (1) and the subsequent analysis are valid for flows in rigid pipes. Specifically, we require that stress-induced elastic deformations of the channel are small relative to the initial state [20].
3. It is recommended to use a computer for the calculations outlined in steps 1–3.

Acknowledgments

This work was supported by research grants (13166 and 17587) from VILLUM FONDEN.

Conflict of Interest Statements The authors have no conflicts of interest to declare that are relevant to the content of this chapter.

References

1. Stroock AD, Pagay VV, Zwieniecki MA, Michele Holbrook N (2014) The physicochemical hydrodynamics of vascular plants. Ann Rev Fluid Mech 46:615–642
2. Jensen KH, Berg-Sørensen K, Bruus H, Holbrook NM, Liesche J, Schulz A, Bohr T (2016) Sap flow and sugar transport in plants. Rev Mod Phys 88(3):035007

3. Thompson MV, Michele Holbrook N (2003) Application of a single-solute non-steady-state phloem model to the study of long-distance assimilate transport. J Theor Biol 220(4):419–455
4. Ronellenfitsch H (2015) Scaling of phloem structure and optimality of photoassimilate transport in conifer needles. Proc R Soc Lond B Biol Sci 282(1801):20141863
5. Carvalho MR (2017) The hydraulic architecture of ginkgo leaves. Am J Bot 104(9):1285–1298
6. Carvalho MR (2017) The scaling of the hydraulic architecture in poplar leaves. New Phytol 214(1):145–157
7. Bruus H (2008) Theoretical microfluidics. Oxford University Press, Oxford
8. Tabeling P (2005) Introduction to microfluidics. Oxford University Press, Oxford
9. Kirby BJ (2010) Micro-and nanoscale fluid mechanics: transport in microfluidic devices. Cambridge University Press, Cambridge
10. Happel J, Brenner H (2012) Low Reynolds number hydrodynamics: with special applications to particulate media. Springer
11. Prandtl L, Tietjens OG (1934) Applied hydro- and aerodynamics. Dover, New York
12. Darrigol O (2005) Worlds of flow: a history of hydrodynamics from the Bernoullis to Prandtl. Oxford University Press, Oxford.
13. Jensen KH, Mullendore DL, Holbrook NM, Bohr T, Knoblauch M, Bruus H (2012) Modeling the hydrodynamics of phloem sieve plates. Front Plant Sci 3:151
14. Jensen KH, Valente AX, Stone HA (2014) Flow rate through microfilters: influence of the pore size distribution, hydrodynamic interactions, wall slip, and inertia. Phys Fluids 26(5):052004
15. Bird RB, Stewart WE, Lightfoot EN (2004) Transport phenomena. 2002. John Wiley & Sons, New York
16. Thompson MV, Wolniak SM (2008) A plasma membrane-anchored fluorescent protein fusion illuminates sieve element plasma membranes in Arabidopsis and tobacco. Plant Physiol 146:1599–161
17. Mullendore DL, Windt CW, Van As H, Knoblauch M (2010) Sieve tube geometry in relation to phloem flow. Plant Cell 22(3):579–593
18. Liesche J (2017) Height-related scaling of phloem anatomy and the evolution of sieve element end wall types in woody plants. New Phytol 214:245–256
19. Tyree MT, Christy A, Lawrence, Ferrier JM (1974) A simpler iterative steady state solution of münch pressure-flow systems applied to long and short translocation paths. Plant Physiol 54:589–600
20. Duprat C, Stone HA (eds) (2015) Fluid-structure interactions in low-Reynolds-number flows. Royal Society of Chemistry

Chapter 38

Phloem Catastrophe: Identifying Plant Tipping Points Through Bifurcation Analysis

Mazen Nakad, Louis Youssef, Jean-Christophe Domec, Sanna Sevanto and Gabriel Katul

Abstract

This chapter presents a mathematical framework to identify conditions leading to phloem failure using bifurcation analysis and catastrophe theory. The approach assumes that at equilibrium, sucrose production through leaf photosynthesis is balanced by sucrose transport within the phloem. Leaf photosynthesis is modeled according to stomatal optimality theories, which propose that stomatal opening adjusts to maximize carbon uptake while being constrained by resource availability (usually water per unit leaf area in the rooting zone). Sucrose transport in the phloem vasculature is described by the Münch mechanism, which is the basis for the pressure flow hypothesis (PFH). The PFH argues that both water and sucrose transport are due to a pressure-driven flow based on osmotic potential differences between source and sink tissues. The proposed framework models this transport process within a leaf-xylem–phloem network, allowing for an examination of how key factors, such as xylem water potential and finite sucrose removal rate along the entire flow path, and anatomical features of the phloem (tube length and radius) influence the rate, efficiency, and failure tipping points of sucrose transport. By systematically varying these control parameters, the number and stability of equilibrium points necessary to maintain a balance between sucrose production and transport are identified. The analysis reveals that phloem failure emerges as an imperfect pitchfork bifurcation, where two stable equilibria at low and high sucrose loading concentrations are separated by an unstable equilibrium at intermediate sucrose loading. The lower stable equilibrium sucrose concentration is shown to be within the range of reported sucrose loading concentration values across many species. The unstable equilibrium branch operates at loading sucrose concentration values commensurate with those responsible for the negative feedback between elevated sucrose concentration levels and reduced leaf-level photosynthesis. This behavior characterizes a phloem catastrophe, which shares mathematical analogies with other ecological catastrophes. The framework thus provides a theoretical basis for understanding the necessary conditions under which phloem transport may collapse, offering insights into the robustness and vulnerability of plant vascular systems and how anatomical traits shape this safety-efficiency trade-off. It also foreshadows a necessary coordination between plant photosynthetic and phloem hydraulic properties, such as plant height, and suggests testable hypotheses that can be evaluated experimentally.

Keywords Bifurcation, Carbohydrates, Münch mechanism, Optimal stomatal control, Osmotically driven flow, Pressure flow hypothesis, Xylem water potential

Johannes Liesche and Sanna Sevanto (eds.), *Phloem*, Methods in Molecular Biology, vol. 3066,
https://doi.org/10.1007/978-1-0716-5488-0_38,

1 Introduction

The pressure-flow hypothesis (PFH), first formulated by E. Münch in 1930, remains the foundational theory describing long-distance assimilate transport in plants [1]. Arguably, its significance to plant hydraulics rivals that of cohesion-tension theory (CTT) describing water transport in the xylem [2–7] but with some notable exceptions—"*It is virtually out of experimental reach—it is deeply embedded into other tissues—and is very 'touchy'—it shows a range of immediate defense reactions following any kind of manipulation*" [8]. Thus, mathematical models are indispensable for testing competing hypotheses about phloem function—and more critically—phloem dysfunction, the topic of this chapter.

According to the PFH, osmotic gradients drive bulk flow under positive water pressure. Sucrose loading into the phloem (leaves) and sucrose consumption at the final sink sites (e.g., trunk or roots) generate a positive pressure potential that moves water and solutes toward sink tissues [8, 9]. While widely accepted, the PFH application to tall plants and trees still presents some challenges. The main one is that sucrose loading concentrations in trees and shrubs are comparable, but the hydraulic resistance in tall trees is much larger [10], suggesting that the osmotic gradient may not be sufficient to drive flow in tall trees [11, 12]. Others include phloem elasticity, the presence of densely packed sieve plates along the flow path, and sucrose unloading, which complicate some aspects of the PFH [10, 13]. Modern refinements of the PFH emphasize the role of hydraulic coupling between xylem and phloem, particularly in response to environmental stresses [14–16]. Experimental studies confirm that lateral water exchange between these two vascular systems regulates flow efficiency, allowing plants to adapt to fluctuating conditions, such as drought, freezing, and changes in stomatal conductance [15–19]. Additionally, factors such as sieve plate resistance, sieve tube geometry, and elastic properties influence phloem transport [11, 20–23]. These empirical and modeling studies have expanded the understanding and utility of the PFH to describe how plants regulate phloem flow under diverse physiological and environmental constraints. While much research has focused on optimizing phloem transport [24, 25], less is known about the mechanisms leading to phloem failure. The collapse of PFH can occur due to hydraulic imbalances, osmotic stress, or excessive viscosity, disrupting carbon transport and ultimately affecting plant survival [26]. To address this gap, a framework based on bifurcation analysis and catastrophe theory is proposed to investigate phloem failure in the most generic terms. This approach enables the exploration of how key control parameters such as xylem water potential arising from CTT and sucrose removal rate along the flow path, when gradually varied, influence transport stability. By identifying critical thresholds or tipping points at which the flow can collapse, precursors to phloem failure and potential recovery pathways can be elucidated. Catastrophe theory is used here as a mathematical tool to locate such thresholds in phloem transport described by the PFH, rather than to imply an irreversible "catastrophe" in the physiological sense.

Initiated by H. Poincare in the late 1800s [27] and further formalized and popularized by R. Thom in the 1960s [28–30] and others in the 1970s within ecology [31–33], bifurcation analysis provides a mathematical framework for understanding how gradual changes in key parameters (referred to as control variables) induce sudden and irreversible shifts in system behavior [34, 35]. These abrupt transitions, referred to as catastrophes, arise when the system (described by a set of interacting non-linear equations) reaches a threshold beyond which it can no longer maintain prior equilibrium states. Some of the more studied local bifurcations include saddle-node (or fold), transcritical, pitchfork, period-doubling, and the so-called Poincaré—Andronov–Hopf bifurcation associated with the emergence or disappearance of periodic or limit cycles [35]. To be clear, catastrophe theory does have its own critics, especially in the context of social and biological sciences [36–38], but it also has numerous ardent enthusiasts and supporters who have shown its diagnostic and prognostic utility in ecology [7, 39–44], especially in food webs and population dynamics [32, 33, 45–51].

In this chapter, it will be shown that in the context of phloem dysfunction, bifurcation analysis does allow for a systematic study of how variations in xylem water potential, phloem geometric characteristics (length and diameter of tubes), and sucrose removal efficiency affect the stability of sucrose flow when described by the PFH. At its core, bifurcation analysis examines the relation between state variables such as sucrose concentration and functional and structural control parameters, including background xylem water potential, sucrose unloading rate, phloem radius, and tree height. Under certain stable conditions, the phloem operates efficiently, maintaining a steady flow of carbohydrates from leaves to sink tissues along the phloem path all the way to the roots. However, environmental stressors such as prolonged drought (reducing xylem water potential) or increased sucrose removal (e.g., herbivory) can push the phloem system beyond a tipping point, leading to irreversible phloem failure. By treating these two control parameters as slowly adjustable variables, bifurcation analysis, in conjunction with the Münch mechanism, enables mapping system features that ensure equilibrium stability in phloem transport. In the next section, key aspects of catastrophe theory and the physics behind PFH are reviewed and then combined to examine how exogenous and endogenous factors impact phloem failure. Ecological consequences connecting control parameters to catastrophes are then discussed, paving the way for new hypotheses about phloem resilience to be explored. Because catastrophe theory relies on perturbations from equilibrium states (i.e., sucrose loading balancing transport capacity described by the PFH), it cannot predict the entire trajectory of phloem failure—much like stability analysis cannot predict the entire pathway by which unstable modes in a laminar flow excite other modes that then grow (nonlinearly) and lead to a state of turbulence.

2 Review of Key Concepts

An overview of bifurcation analysis and the key components of the Münch mechanism are covered here for completeness and are aimed to supplement material in other chapters.

2.1 Equilibrium, Bifurcation Analysis, and Catastrophes

Catastrophe theory has a long tradition in the ecological sciences that dates back to the 1970s [31, 33, 40, 52–56]. Within the phloem, the buildup of sucrose concentration c_0^* at the loading site is an outcome of sucrose supply from leaf photosynthesis and the transport capacity of the phloem as described by the Münch mechanism. Both supply and transport capacity depend on sucrose concentration c_0^*, the state variable here, as well as many other environmental, physiological (e.g., potassium ions), geometric, and hydraulic variables labeled as control variables. In the most "generic" form, this balance may be expressed as

$$\frac{dc_0}{dt} = S(c_0) - T(c_0) = f(c_0), \tag{1}$$

where c_0^* is time, c_0^* is a simplified sucrose loading rate (usually in mass of sucrose per mass of water per time) that depends on leaf photosynthesis and its conversion to c_0^* (usually in mass of sucrose per mass of water), and c_0^* is a simplified transport capacity (usually in mass of sucrose per mass of water per time) of the phloem that reflects the outcome of the competing effects of driving and resistive forces. For c_0^*, both drivers (osmotic potential) and resistance to flow (viscosity) are assumed to only vary with c_0^* in a phloem with predetermined geometry. The control variables that determine c_0^* are numerous and include photosynthetically active radiation, air temperature, air relative humidity, atmospheric carbon dioxide concentration, xylem water potential, among others (e.g., biochemical limitation of photosynthesis). The control variables impacting c_0^* include tube diameter, tree height, background xylem potential variations, temperature, the distribution of sucrose sinks along the phloem, among others. The connection between $S(c_0)$ and c_0 will be made explicit assuming that leaves obey stomatal optimality theories with a priori imposed negative feedback at high c_0, whereas the relation between $T(c_0)$ and c_0 will be formulated using the PFH without the details of the movement of other ions such as potassium. Before doing so, a number of features about the system given by Eq. (1) are now pointed out to introduce key concepts about the use of catastrophe theory here.

Equilibrium Analysis

While Eq. (1) can be solved (analytically) for externally supplied initial conditions at $t = 0$ and for all times t, equilibrium analysis seeks to establish the "end" points of this solution if one waits long enough (i.e., $t \rightarrow \infty$). This is the foresight obtained from equilibrium analysis: a connection between any starter concentration (at $t = 0$) and all possible concentrations obtained by waiting sufficiently long without needing all the details between them provided by the general solution for Eq. (1). In dynamical systems, the "sufficiently long wait" implies an approach to steady-state or the attainment of an equilibrium state. These two terms, steady-state and equilibrium, are used interchangeably in the ecological science literature, though thermodynamically, they are different. In keeping with the dynamical systems literature [35], the term equilibrium is employed throughout. At equilibrium, $dc_0/dt = f(c_0) = 0$and the solution to this algebraic equation (instead of an ordinary differential equation) defines sucrose concentration values c_0^* whereby $S(c_0^*) = T(c_0^*)$. The number of equilibrium points depends on the nonlinearity of $f(c_0)$, which encodes the physiological processes impacting $S(c_0)$ and the transport/physical processes in $T(c_0)$. For one-dimensional systems or a system with a single state variable c_0, such as the one in Eq. (1), each equilibrium point is classified as stable or unstable. To determine the stability of an equilibrium concentration point, a small perturbation analysis can be conducted [35]. A perturbation δc_0 is first introduced to unbalance c_0^* from its equilibrium value by a small amount. Stability of c_0^* is then determined as to whether the added δc_0 grows in time and the system moves away from c_0^* (unstable) or decays in time and the system returns to the equilibrium c_0^* after some relaxation period (stable). Mathematically, this analysis may be formulated by decomposing c_0 into an equilibrium c_0^* and a perturbation δc_0. Inserting this decomposition into Eq. (1) yields

$$\frac{d}{dt}(c_0^* + \delta c_0) = \frac{dc_0^*}{dt} + \frac{d\delta c_0^*}{dt} = f(c_0^* + \delta c_0). \tag{2}$$

Since δc_0 is assumed to be small, $f(c_0^* + \delta c_0)$ may be expanded using a Taylor series around c_0^* to yield

$$\begin{aligned} f(c_0^* + \delta c) &= f(c_0^*) + f'(c_0^*)\delta c + \frac{1}{2!}f''(c_0^*)(\delta c_0)^2 \\ &+ \frac{1}{3!}f'''(c_0^*)(\delta c_0)^3 + \cdots \end{aligned} \tag{3}$$

where $f'(c_0^*)$, $f''(c_0^*)$, $f'''(c_0^*)$, etc... are the first, second, and third derivative of $f(c_0)$ with respect to c_0. All the derivatives are evaluated at the equilibrium point $c_0 = c_0^*$ and are thus known if $f(c_0)$ is known. Ignoring momentarily the higher-order terms in the small δc_0 limit, noting that $dc_0^*/dt = 0$ and $f(c_0^*) = 0$ at equilibrium by definition yields

$$\frac{d(\delta c_0)}{dt} = f'(c_0^*)(\delta c_0). \tag{4}$$

This equation can be rearranged to solve for the growth or decay of δc_0 in time using variable separation

$$\frac{d(\delta c_0)}{(\delta c_0)} = f'(c_0^*)dt. \tag{5}$$

While $f(c_0^*) = 0$, $f'(c_0^*)$ need not be. The solution of this first-order ordinary differential equation can be obtained for $\delta c_0(t)$ by integrating both sides to yield

$$\delta c_0 = \exp\left[m\frac{t}{\tau_0}\right],\ \tau_o = \frac{1}{|f'(c_0^*)|},\ m = sgn\,[f'(c_0^*)], \tag{6}$$

where sgn(.) is the sign of $f'(c_0^*)$ (=+1 or −1). Thus, δc_0 exponentially grows in time when $m > 0$ and exponentially decays in time when $m < 0$. In practice, this means that an equilibrium concentration c_0^* is unstable to a small concentration perturbation δc_0 when $f'(c_0^*) > 0$ and is stable when $f'(c_0^*) < 0$. The term $|1/f'(c_0^*)|$ can also be interpreted as a characteristic or e− folding time to return to the equilibrium state c_0^* after a perturbation δc_0^* is introduced. This linear analysis fails when $f'(c_0^*) = 0$, thereby requiring higher-order terms in δc_0 to be retained but this special case is not treated here. It suffices to note that the occurrence of $f'(c_0^*) = 0$ is often associated with what is termed critical slowdown [35] as $|1/f'(c_0^*)| \to \infty$, and this approach has been popularized in ecological studies to fingerprint regime change in stability [42].

Bifurcation Analysis

Having determined the equilibrium states whereby $c_0 = c_0^*$ satisfies $dc_0/dt = 0$, bifurcation analysis proceeds to study how the stability of c_0^* shifts when certain control variables are gradually altered. These are precisely the control variables to be studied in this chapter. For illustration, a simplified case where a constant sucrose loading rate into the phloem $S(c_0) = a$ independent of c_0 is applied and treated as one of the main control variables to be varied. The ability of the phloem to translocate c_0 at this loading rate A_0 is described by $T(c_0)$. In choosing a "generic" model for $T(c_0)$, a competition between the linear driver anticipated from an osmotic potential and a nonlinear resistance term in c_0 expected by viscous forces must also be considered. Because the nonlinearity of viscosity to concentration changes is, to a leading order, a fluid property, not a control variable, it is imposed here with no external control variables. For this reason, we consider a $T(c_0) = bc_0 - c_0^3$. The parameters $a \geq 0$ and $b > 0$ are assumed for simplicity, and their dimensions or units are not relevant for this illustration. The coefficient b is controlled by geometric factors such as phloem tube diameter and length and will be later elaborated upon using the PFH. With this formulation,

$$f(c_0) = S(c_0) - T(c_0) = (a) - (bc_0 - c_0^3). \tag{7}$$

In reality, $T(c_0)$ need not be given by such a polynomial but, to a leading order, the nonlinearity linking $T(c_0)$ to c_0 may have this mathematical form. The term "normal form" [35] is used to describe the leading-order nature of the nonlinearity (cubic here). Bifurcation analysis seeks to answer the question: What happens to the stability of the many equilibrium points c_0^* when the control parameters A and b gradually vary? Gradually here means that variations in a and b occur on time scales much longer than the dynamics governing dc_0/dt (or its e-folding time). As a check on logical expectations, the case where the loading $S(c_0) = 0$ (or $a = 0$) is first considered. In this case, the equilibrium points c_0^* are given by the solution to the cubic equation

$$f(c_0) = -c_0(b - c_0^2) = 0. \tag{8}$$

Mathematically, three equilibrium points emerge: $c_0^* = 0$ and $c_0^* = \pm\sqrt{b}$. Since $c_0^* \geq 0$ are sought, only the $c_0^* = +\sqrt{b}$ is considered, and the $c_0^* = -\sqrt{b}$ is momentarily discarded. The equilibrium status (stable versus unstable) of these two physically plausible equilibrium points can be readily assessed by evaluating $f'(c_0)$ at these two equilibrium points: $c_0^* = 0$ and $c_0^* = +\sqrt{b}$ values. Noting that

$$f'(c_0) = 3c_0^2 - b, \tag{9}$$

and $b > 0$ leads to $f'(0) = -b < 0$(stable) and $f'(\sqrt{b}) = 3b - b = 2b > 0$ (unstable). Thus, in the absence of any loading (i.e., $a = 0$), the phloem vasculature has a stable background sucrose concentration $c_0^* = 0$ as intuitively expected and as shown in Fig. 1. Moving toward a finite loading ($a > 0$), the stability of the phloem vasculature now becomes more dynamically interesting. For a graphical illustration, let $a \geq 0$ be the control variable and set $b = 9$ as momentarily fixed. That is, the bifurcation analysis is arresting the $T(c_0)$ but varying the $S(c_0)$. The aforementioned case where $a = 0$ (horizontal line) is also shown with $c_0^* = 0$ (stable) and $c_0^* = +\sqrt{b} = 3$ (unstable). When a

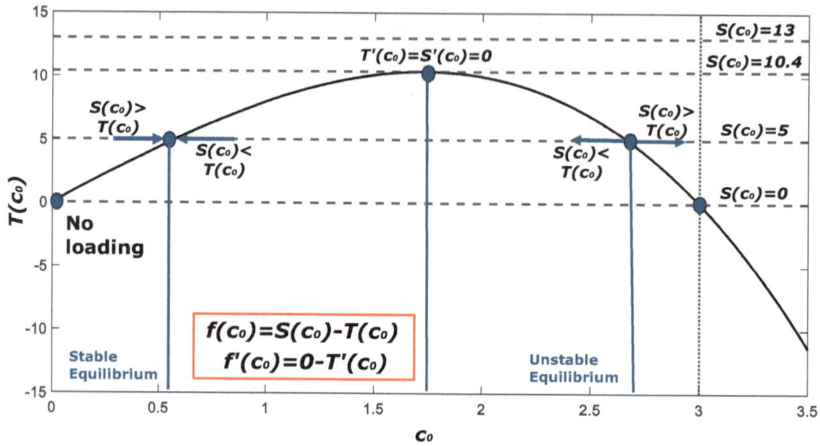

Fig. 1 Illustration of the bifurcation analysis for $f(c_0) = S(c_0) - T(c_0)$ where the loading $S(c_0) = a = 0, 5, 10.4, 13$ is a control variable (arbitrary units) and $b = 9$ (arbitrary units) is momentarily fixed. The intersections $S(c_0) = T(c_0)$ identify the equilibrium points c_0^*. When $a < 10.4$, two equilibrium c_0^* values emerge. At $a = 10.4$, only one c_0^* emerges, and for $a > 10.4$, no equilibrium points can be found signaling a catastrophe. The graphical case where $T'(c_0) > 0$ on the diagram correspond to $f'(c_0) < 0$ and all intersections between $S(c_0) = a$ and $T(c_0)$ are stable. However, for $T'(c0)$ 0, the intersections, when they exist, lead to an unstable equilibrium c_0^*. Stability can also be visualized by noting that two convergent arrows indicate a stable c_0^* (i.e. when perturbed, $\delta c_0 \rightarrow 0$) whereas two divergent arrows indicate an unstable c_0^*. The direction of the arrow is positive when $S(c_0) > T(c_0)$ and negative when $S(c0)$ $T(c_0)$ in the vicinity of $c_0 = c_0^*$. A negative $T(c_0)$ is retained only for illustrating the stability of equilibrium points

becomes finite but smaller than some threshold to be later determined, the two equilibrium points (stable and unstable) begin to approach each other as graphically presented for the case where $S(c_0) = 5$. Here, the two resulting c_0^* values are $c_0^* \approx 0.58$ (stable) and $c_0^* \approx 2.67$ (unstable). Their stability can be graphically determined in several ways. One common way is to note that when $dc_0/dt > 0$, c increases relative to c_0^* and must move to the right of c_0^* (shown by an arrow), whereas when $dc_0/dt < 0$, c_0 must move to the left of c_0^* (shown by an arrow as well). In the vicinity of the target equilibrium point c_0^*, $dc_0/dt > 0$ occurs when $S(c_0) > T(c_0)$ and $dc_0/dt < 0$ when $S(c_0) < T(c_0)$as demonstrated in Fig. 1. Two convergent arrows indicate a stable equilibrium c_0^* (i.e., when perturbed, $\delta c_0 \rightarrow 0$), whereas two divergent arrows indicate an unstable c_0^*. In the unstable case, δc_0 grows in time, and the phloem vasculature cannot operate at this unstable equilibrium point. So, what happens to this phloem system when $S(c_0) = 5$? If the initial or antecedent concentration is $c_0 < 2.67$, then the phloem vasculature will head toward $c_0^* \approx 0.58$ or the stable equilibrium. If the starting concentration $c_0 > 2.67$ with a set $S(c_0) = 5$, then the phloem will experience a "runaway" concentration buildup—mathematically leading to what is termed a finite-time singularity or blowup [35, 56, 57]. In fact, when $c_0^* > 3$, the $T(c_0)$ unrealistically reverses sign and begins to contribute to sucrose pileup above and beyond the loading term—signaling a catastrophe. In reality, other finite-size effects will be activated, preventing c_0^* from increasing indefinitely, and those effects are not captured by the current dynamics governing $dc_0/dt = f(c_0)$ However, they will naturally emerge from the PFH governing $T(c_0)$ described later.

Returning to the determination of the stability of equilibrium points c_0^*, an alternative graphical method is to directly inspect $f'(c_0^*)$. In this particular case, $f'(c_0^*) = 0 - T'(c_0^*)$ provided that the constant sucrose loading rate is independent of c_0. As before, a stable equilibrium is anticipated when $f'(c_0^*) < 0$ or when $T'(c_0^*) > 0$. Conversely, an unstable equilibrium is expected when $f'(c_0^*) > 0$ and $T'(c_0^*) < 0$. Fig. 1 demonstrates that $T'(c_0^*) > 0$ for $c_0^* \approx 0.58$ (stable) and $T'(c_0^*) < 0$ for $c_0^* \approx 2.67$ (unstable). As the loading rate a continues to increase due to enhanced photosynthesis, the stable c_0^* increases whereas the unstable c_0^* decreases relative to their values evaluated at the "no-loading" case $a = 0$. At some loading rate $S(c_0) = a$ (near 10), the two equilibrium points "collide." The collision of these two equilibrium c_0^* values (one stable and one unstable) occurs when the tangent of the loading curve $S'(c_0)$ intersects the tangent of the transport capacity $T'(c_0)$. Once again, because of the constant loading imposed here, $S'(c_0) = 0$. Thus, the single equilibrium c_0^* occurs when $T'(c_0) = S'(c_0^*) = 0$ (i.e., $T(c_0)$ is at its maximum transport capacity in this case). Solving for this single equilibrium whereby the loading and transport tangents are zero yields a $c_0^* = +\sqrt{b/3} = \sqrt{3}$ for $b = 9$. The loading a at this zero-tangent intersection point is the same as $T(c_0^*)$ evaluated at $c_0^* = \sqrt{3}$ and is given by $a = 6\sqrt{3} \approx 10.4$ (see Fig. 1). For $a > 10.4$, the phloem fails and no c_0^* that balances loading and transport—at least for $b = 9$—can be found. Beyond this c_0^*, photosynthesis must be operationally reduced to accommodate the maximum transport capacity of the phloem, and a negative feedback between c_0^* and a must be activated. The foresight about this negative feedback will be implemented when representing photosynthesis using a realistic formulation based on optimality theory.

The illustration with a treated as a single control parameter and b as a constant illustrates what is termed as "Saddle Node" bifurcation. It occurs when the loading parameter a is gradually varied to maintain steady-state concentration, but b is fixed. In general, saddle node bifurcations occur when two equilibrium points of opposite stability (as shown here) collide, annihilate, and vanish with further increases in the control parameter a [35].

Thus far, the bifurcation analysis focused on fixing b as prescribed by the Münch mechanism and varying the loading rate a. In reality, both a and b may vary and the bifurcation analysis must be conducted in two control variables instead of a only. Before doing so, it is constructive to ask how changes in b impact c_0^* when the loading a is held constant. Fig. 2 offers a preview of how changes in b impact the

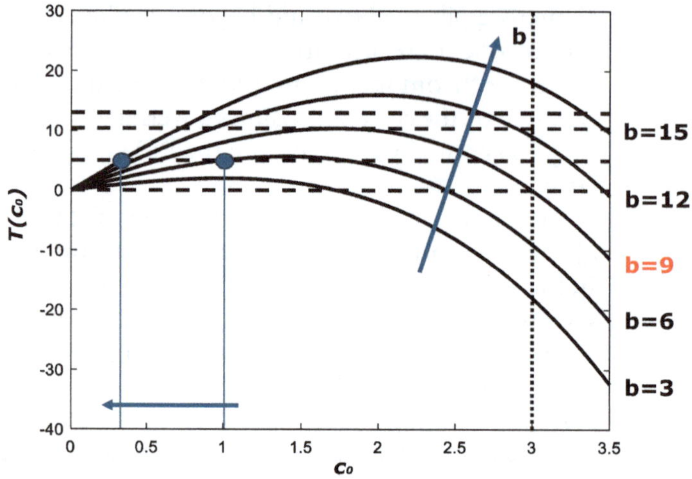

Fig. 2 Illustration of the bifurcation analysis for $f(c_0) = S(c_0) - T(c_0)$ where the loading $S(c_0) = a = 0, 5, 10.4, 13$ are shown as horizontal lines and $b = 3, 6, 9$ (as before), 12, and 15 are presented (units are arbitrary). Intersections $S(c_0) = T(c_0)$ identify the equilibrium points c_0^*. Note that when setting $a = 5$, increasing b results in a reduced stable equilibrium c_0^*. Moreover, increasing b shifts the occurrence of a catastrophe (i.e. the a at which $T'(c_0) = 0$) to higher a

phloem vasculature for the a values flagged in Fig. 1. Increasing b leads to a reduced stable c_0^* for a fixed $a = 5$. Moreover, increasing b enables the phloem to accommodate higher loading a before the catastrophe or phloem failure occurs. The c_0^* associated with the incipient point of catastrophe also increases with increasing b, as anticipated—meaning higher b, higher transport capacity, higher loading, and higher concentrations can be accommodated before failure.

The joint effects of the two control parameters a and b on the equilibrium points c_0^* and their stability are shown in Fig. 3. As anticipated, the "collision" between the two equilibrium points as a function of a and b is the hallmark of the so-called "fold" catastrophe. For certain combinations of a and b, no equilibrium points exist, signaling phloem failure. The unstable "manifold" (in red) separates phloem function from phloem failure. Depending on the choices of a and b, any initial c at $t = 0$ or $c_0(0)$ below this manifold of c_0^* will push the phloem vasculature to operate on the stable branch. Conversely, any initial $c_0(0)$ above the unstable manifold will lead to runaway sucrose accumulation and incipient phloem dysfunction in this reduced model. This runaway is a theoretical construct aimed at showing the limits of the modeling framework for phloem transport, rather than a literal physiological prediction, since in reality feedback mechanisms and concentration gradients would act to prevent such unbounded accumulation.

Before completing this section, we wish to draw further attention to the mathematical form of $f(c_0)$ when all the roots of the algebraic

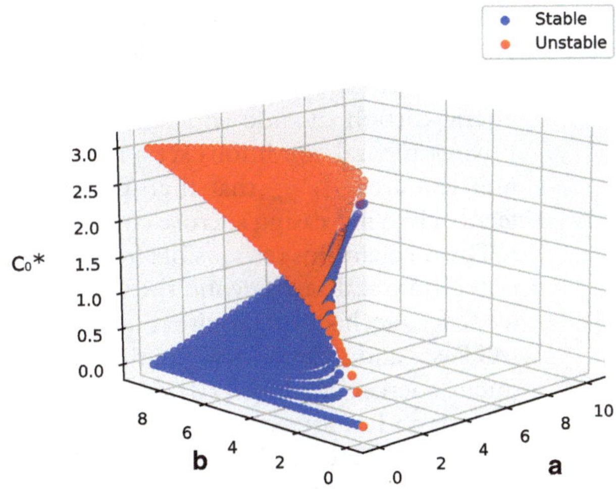

Fig. 3 Equilibrium sucrose concentration (c_0^*) as a function of source strength (*a*) and sink strength (*b*), illustrating the fold catastrophe structure (units are arbitrary). The red surface denotes the unstable manifold separating functional phloem transport from failure, and the blue surface denotes the stable one

expression $f(c_0) = 0$ are admissible, not just the positive roots. Equation 7 leads to the much-studied "cusp catastrophe" reported for the spruce bud-worm outbreak [33] and bubble expansion in xylem vessels [7, 56]. In this case, the $T(c_0)$ experiences a super-critical pitchfork bifurcation (due to the cubic nature and the three solutions) and a acts as a symmetry-breaking term (i.e., $f(-c_0) = f(c_0)$ no longer holds for finite a) discussed elsewhere [35]. For this reason, symmetry breaking is necessary for catastrophes to occur.

2.2 The Münch Mechanism

The prior analysis did not make explicit contact between $T(c_0)$ and the Münch Mechanism describing the PFH. In this section, approximations common to many mathematical models describing the Münch mechanism in all its stages are covered. While an approximate analytical function for $S(c_0)$ may be possible, an explicit mathematical function for $T(c_0)$ is not. Instead, numerical analysis that balances loading and transport capacity is better suited for achieving mathematical tractability. The four stages approximating the PFH are now reviewed with different levels of resolution and complexity depending on the problem and science questions [6, 9, 15, 19, 58–62].

- *Stage 1—Sucrose loading*: The products of photosynthesis are first loaded into the phloem, increasing the sucrose concentration at these loading sites to create an osmotic potential gradient. This stage requires representing photosynthesis and its conversion to sucrose loading concentration for the entire plant. Photosynthesis per unit leaf area can be modeled using various formulations for stomatal conductance, including empirical models and those based on optimality theory [63–69]. Such models link carbon dioxide uptake by leaves from the atmosphere to environmental conditions, such as photosynthetically active radiation, air temperature, leaf water potential, atmospheric carbon dioxide concentration, vapor pressure deficit. Environmental conditions set a maximum photosynthesis per unit leaf area $A_{0,\max}$ that then gets reduced instantly once a feedback to increasing sucrose concentration is initiated at high c. The mechanisms responsible for this feedback and their associated time scale are beyond the scope of the analysis here. The feedback mechanisms require resolving processes such as the H^+ -ATPases and the role of K^+ in the electrical balance and subsequent sucrose loading [70]. For the purposes of bifurcation analysis and catastrophe, $A_{0,\max}$ itself can be treated as a bifurcation parameter in lieu of the variables used to determine its magnitude. For completeness, its estimate from environmental parameters and physiological factors will be elaborated upon later. Once the rate of CO_2 uptake from the atmosphere is determined at the leaf scale, a carbon balance equation is required to allocate carbon supply to various sinks, including phloem loading. For simplicity, it is assumed that all assimilated carbon, denoted by the production rate "PR," is directly loaded into the phloem; i.e., $S(c_0)$=PR. Additional terms can be incorporated into this balance as discussed in **Note 1**. In reality, phloem loading can also be regulated independently of photosynthesis and is influenced by transporter activity and concentration gradients [71] at loading concentration time scales.

- *Stage 2—Osmotic potential buildup*: The sucrose concentration loaded into the phloem increases the osmotic potential at the source end. This higher sucrose concentration draws water at a rate J_o into the phloem from the surrounding xylem that is already operating at negative background water potential Ψ_x. Because water is incompressible, its influx into the phloem leads to a rapid buildup of turgor or positive pressure P_p locally at the source end. The Münch mechanism is unique in that it enables the passive transport of sucrose from sources to sinks without requiring additional metabolic energy to generate a pressure gradient. Instead, the pressure gradient emerges naturally from osmotic differences between the phloem and the xylem. This should not be mistaken for the fact that the mechanism of loading and unloading of sucrose along the phloem pathway can be active or passive, but the force exerted on the sap to allow movement along the phloem is driven solely by the passive chemical (osmosis) potential difference of water molecules between the xylem and the phloem. It should be noted that ions, particularly K^+, can also contribute substantially to phloem turgor and partly decouple osmotic pressure from sucrose concentration [72], but this effect is ignored here for simplicity. The flux of water due to osmosis from xylem to phloem through the membrane separating them can be written as

$$J_o = k(\Psi_x - \Psi_p), \tag{10}$$

where J_o denotes the osmotic flux of water molecules, k is the membrane permeability, Ψ_x is the xylem water potential determined from the water balance along the xylem pathway from the root to the leaf using CTT, and Ψ_p is the phloem water potential, which is the sum of phloem turgor pressure P_p and osmotic potential of sucrose $\Pi(c)$. For dilute solutions, the osmotic potential of the phloem can be approximated using the Van't Hoff relation $\Pi = -R_g Tc$, where R_g is the universal gas constant, and T is the sap temperature. At higher concentrations, deviations from linearity between Π and c can occur due to non-ideal solution behavior, as further discussed **Note 2**.

- *Stage 3—Sucrose transport*: The osmotic pressure difference between the source and sink regions creates a pressure gradient along the phloem, which drives sap flow. This pressure gradient is the primary force behind phloem transport and can be determined using conservation laws for water mass and momentum. As discussed in earlier chapters, when viscous resistive forces dominate inertia, the volumetric sap flow rate can be expressed as:

$$Q = \frac{\Delta P_p(c)}{R(c)} \tag{11}$$

where $R(c)$ is the hydraulic resistance and $\Delta P_p(c)$ is the pressure gradient along the phloem, both of which depend on sucrose concentration. The resistance r increases with sucrose concentration because the sap dynamic viscosity $\mu(c)$ increases as the sap becomes more sucrose concentrated, especially at high sucrose levels. The turgor pressure gradient ΔP_p depends on sucrose concentration because osmotic potential influences turgor pressure, which in turn drives bulk flow. The total transport of sucrose within the phloem is given by the mass flux equation

$$J_c(c) = Q(c)\,\rho(c)c, \tag{12}$$

where $\rho(c)$ is the sap density, which varies with sucrose concentration but to a much lesser extent than $\mu(c)$. Because the molecular diffusion coefficient of sucrose in water is sufficiently small, it is momentarily ignored relative to its advective counterpart in this simplistic model. Despite its apparent simplicity, this formulation has been shown to capture a number of complexities inherent to phloem hydrodynamics [24, 73]. First, because phloem cells are living, their thickness can expand or contract with turgor pressure, meaning local hydraulic resistance depends not only on sap viscosity but also on phloem geometry and wall elasticity. Second, local pressure must adjust to maintain mass conservation, and since water movement is driven by osmosis, these pressure adjustments are influenced by local sucrose concentration. This local sucrose concentration follows another conservation equation where both diffusion and dispersion as well as local source-sink dynamics play a role. Sieve plates that are distributed along the phloem length introduce additional energy losses [6, 21]. However, such losses appear minor when compared to the overall wall friction [9]. It was recently conjectured that the presence of sieve plates adds rigidity (i.e., acting like structural dampers), thereby allowing much of the osmotic potential at the loading site to pressurize water instead of expanding the phloem volume [73]. For this reason, the role of sieve plates and membrane elasticity is not considered further.

- *Stage 4—Sucrose removal and sinks*: The sucrose concentration has to be conserved in this formulation and whatever is loaded at the leaf level must either be removed at the root level or along the phloem or stored within the phloem cells over some long averaging time. When this conservation law is written for a control volume (such as a group of phloem cells), the local variation in sucrose concentration can be expressed as:

$$\frac{dc}{dt} = \Delta J_c + S_0(c) \tag{13}$$

where $\Delta J_c = J_{c,in} - J_{c,out}$ represents the net sucrose mass flux (the difference between inflow and outflow), and $S_0(c)$ is the local source/sink term governing sucrose loading and unloading. The transport of sucrose is influenced by two primary mechanisms: bulk flow (movement with the sap driven by pressure gradients) and molecular diffusion or spreading due to concentration differences as described by Fick's law. This transport can be approximated by solving flow dynamics along the phloem axis, though additional effects, such as Taylor dispersion (which accounts for radial variations affecting the axial sucrose distribution), can be included for refinement [62].

3 A Reduced Mathematical Version of the PFH

The number of parameters describing the four stages of the PFH is large and unwieldy for bifurcation analysis. Hence, the focus here is on a select few parameters that deal with key controls, such as background xylem water potential and the less studied sink strengths, as well as phloem geometric properties, such as radius and length. The reduced model is based on a single tree of height L and total leaf area L_a. The phloem geometry is represented by a fixed number of tubes per tree collectively occupying a total cross-sectional area P_a. Each tube is characterized by L and diameter $2r$. The sap density ρ is assumed to be weakly dependent on c, and this dependence is hereafter ignored. The dependence of dynamic viscosity μ on the loading concentration is retained as it is the most relevant factor influencing stability and transition dynamics.

3.1 Stage 1: Sucrose Loading Through Photosynthesis

As discussed earlier, the total assimilated carbon PR is directly loaded into the phloem at steady state. This production rate depends on the photosynthetic rate per unit leaf area A_0 and the total leaf area L_a. The A_0 can be determined from optimality theories that assume that stomatal conductance g to carbon dioxide adjusts so as to maximize carbon gain for a given amount of water available in the soil per unit leaf area. The A_0 is down-regulated with increasing leaf sucrose accumulation c_0 [mol/m^3] to reflect the negative feedback discussed earlier in the bifurcation analysis of Sect. 2.1.2. That is,

$$\mathrm{PR}=L_a\times\begin{cases} A_0, & c_0<1200\\ A_0[1-(c_0-\alpha)/\delta], & 1200\leq c_0<2800\\ 0, & c_0\geq 2800,\end{cases} \tag{14}$$

where coefficients α and δ describe the strength of the interaction between c_0 and A_0, and A_0 is the photosynthetic rate for the set of environmental conditions. One approach to determining A_0 is to use stomatal optimality theories assuming that no down-regulation due to increased c_0 occurs [65, 66, 74]. These optimality theories couple the Farquhar photosynthesis model for biochemical demand [75] with diffusional transport of carbon dioxide molecules from the atmosphere into the mesophyll cells, complemented by the assumption that g operates so as to maximize carbon dioxide gain for a given amount of water available in the rooting system per unit leaf area [54, 67, 68]. The general solution to this set of equations has already been derived for static [65] and dynamic marginal water use efficiencies [67, 68], and a linearized version of the static solution is given here only for illustration. The full version that accommodates all the limitations on photosynthesis in the static version is used for the actual model calculations reported here. Those formulations are elaborated upon elsewhere [65] and not repeated here. The linearized version, which shares similar parameters to the full version employed here, is given by [65, 69, 76]

$$A_0 = A_{0,\max}\left(1 - \sqrt{\frac{1.6\text{VPD}}{c_a}\lambda_o}\right),\ A_{0,\max} = \frac{\beta_1 c_a}{\beta_2 + \beta_3 c_a}, \tag{15}$$

where β_1 [umol · m^{-2}s^{-1}]] and β_2 [umol · mol^{-1}] are determined from whether RuBP (Ribulose-1,5-bisphosphate) or RuBisCO (Ribulose-1,5-bisphosphate carboxylase/oxygenase) limit $A_{o,\max}$ [77, 78], $\beta_3 = c_i/c_a \in [0.7, 0.9]$ is a long-term internal (c_i) to ambient atmospheric dioxide concentration ratio for C3 plants (setting β_3 to a constant is the linearization), c_a [umol · mol^{-1}] is the atmospheric carbon dioxide concentration, VPD [kPa] is the vapor pressure deficit determined from air temperature and relative humidity, and λ_o [umol · mol^{-1}kPa^{-1}] is the Lagrange multiplier, or marginal water use efficiency, which increases with the magnitude of leaf water tension (p_m), or equivalently with reduced water availability per unit leaf area in the rooting zone. The maximum λ_o that can be attained while keeping photosynthesis finite is when

$$1 - \sqrt{\frac{1.6\text{VPD}}{c_a}\lambda_o} = 0, \text{ or } \lambda_{o,\max} = \frac{c_a}{1.6\text{VPD}}. \tag{16}$$

For simplicity, it is also assumed that as p_m increases, λ_o linearly increases from a minimum value $\lambda_{o,\min}$ at p_m=0 MPa until a maximum value is reached at p_m=3 MPa. That is

$$\lambda_o = \lambda_{o,\min} + \frac{p_m}{3}(\lambda_{o,\max} - \lambda_{o,\min}). \tag{17}$$

Other functions, including those predicted from dynamic optimality theories [68], can also be used instead. Non-stomatal controls on $A_{0,\max}$, such as mesophyll conductance, can be included, but for the purposes of the bifurcation analysis here, those controls are not explicitly treated as they impact $A_{0,\max}$ predictably [79–84]. The expressions for the photosynthetic parameters β_1 and β_2 are further explored in **Notes 3 and 4** under RuBisCO and RuBP limitations. In this linear solution for A_0, as c_a increases, A_0 increases but begins to saturate at very high $c_a \gg \beta_2/\beta_3$. Similarly, increases in atmospheric aridity increase VPD and reduce A_0. Also, increases in p_m increase λ_o and reduce A_0 as already predicted from stomatal optimality theories and confirmed empirically [85]—whether they be static or dynamic [67, 68].

For the purposes of bifurcation analysis, photosynthetic parameters β_1 and β_2 remain constant since VPD, air temperature, photosynthetically active radiation, and c_a are set to typical daytime values, leaving λ_o (through variations in p_m) as a bifurcation parameter. We note that the formulation for A_0 presented here for illustration purposes assumes a linear biochemical demand function, which is reasonable when RuBisCO limitations govern A, not RuBP. In this case, β_1 is determined from the maximum carboxylation capacity $V_{c,\max}$ that varies across species types (high for shrubs and grasses, low for conifers), and leaf nitrogen levels among other factors [86]. Roughly, RuBP and RuBisCO limit photosynthesis about 50% of the time in forested canopies. In summary, the PR determined in stage 1 follows an inverse logistic profile with sucrose loading concentration, c_0. At low c_0, not much sucrose is transported inside the plant, which prompts it to perform the highest photosynthesis per unit leaf area set by A_0. At moderate c_0, plant demand for sucrose is met, and any excess starts accumulating. To prevent sucrose build-up and potential phloem clogging, the plant reduces its photosynthesis. At high c_0, where the phloem and sink tissues are fully saturated with sucrose, photosynthesis completely shuts down as a drastic measure against definite failure. Fig.4 shows these PR curves as a function of maximum background pressure p_m and loading concentration c_0.

3.2 Stage 2: Phloem–Xylem Hydraulic Coupling

When coupling xylem and phloem transport, the xylem tension Ψ_x acts as a boundary condition on the phloem. To solve for Ψ_x, the head losses encountered as water moves through the xylem must be determined [16] using CTT. Those head losses depend on the vulnerability curve of the xylem, which requires their specification for each plant organ [56]. In lieu of such detailed analysis, a simpler representation is adopted for specifying Ψ_x. The water in the xylem is assumed to be quasi-hydrostatic so that

$$\Psi_x(x) = -p_m\left(1 - \frac{x}{L}\right), \qquad (18)$$

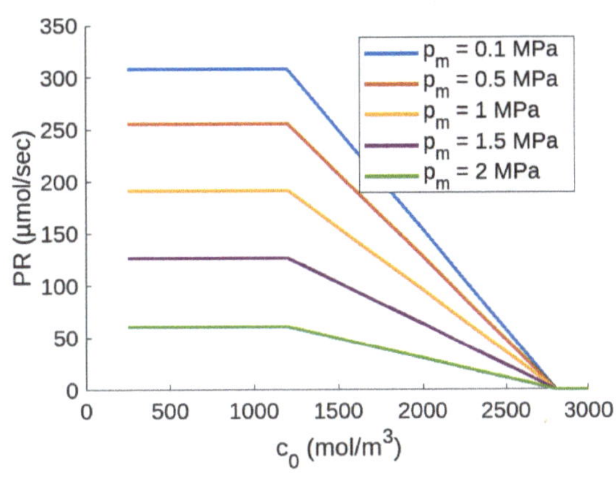

Fig. 4 Sucrose production rate (PR, μmol/s) from photosynthesis as a function of leaf sucrose concentration (c_0, mol/m^3). Different colors denote different leaf water tensions (p_m, MPa)

where p_m is impacted by leaf water potential, which in turn, reflects soil water stress in the plant. Because water is moving in the xylem, Ψ_x must deviate from linearity. Nevertheless, a linear profile captures the leading-order effect that Ψ_x is at its maximum tension, $-p_m$ near the top of the tube ($x=0$) and drops to 0 MPa (near atmospheric) at the base of the tube ($x=L$). The p_m is impacted by soil–root water availability and atmospheric aridity [87]. It is expected that p_m is large (in magnitude) during droughts and is smaller for well-watered soil conditions.

3.3 Stage 3: Pressure-Driven Sap Flow

Applying the conservation of water mass and momentum to a small segment of the phloem, the difference in water potential drives both osmotic flux and advection. This leads to the following equation for sap velocity u (area-averaged) as a function of location along the flow pathway [59]

$$\frac{r}{2}\frac{\partial u}{\partial x} = k\left[R_g Tc - P_p - p_m\left(1 - \frac{x}{L}\right)\right], \tag{19}$$

Even when p_m is specified using external factors (e.g., drought), the equation contains three unknowns: c, u, and P_p requiring two more equations to complete the mathematical description. One of the two needed equations can be obtained from the Hagen–Poiseuille approximation. This equation, derived for tubes under the assumption of low Reynolds number flow, links the two unknowns u and P_p by $u = -\left[r^2/8\mu\left(c\right)\right]\left(\partial P_p/\partial x\right)$. When combined with Eq. (19), u can be eliminated, and an expression for P_p can be derived

$$\frac{r^3}{16\mu(c)k}\partial^2 P_p \partial x^2 - P_p = -R_g Tc + p_m\left(1 - \frac{x}{L}\right). \tag{20}$$

Still, the problem remains mathematically not closed (one equation in two unknowns), necessitating one additional equation. This equation can be obtained from considerations of the solute mass balance in Stage 4.

3.4 Stage 4: Conservation of Solute Mass

The standard equation governing sucrose transport for the Münch mechanism is derived elsewhere [59]. This equation is extended to accommodate additional terms that incorporate quantitative representation of environmental conditions and mixing. These extensions are now highlighted:

1. a Taylor dispersion term that is derived elsewhere [62] and arises due to area-averaging of the conservation of sucrose mass and fluid momentum. This term can be locally positive or negative at position x.

2. a term $(c - c_b)/t_r$ that represents a sucrose sink along L where c_b is the background concentration around the phloem and t_r is a removal time treated as a control variable. A higher rate of sucrose unloading—such as during fruit development or plant growth in general—occurs when t_r is small or the concentration gradient, $c - c_b$, is large. Other formulations for this sink term can be used, as further discussed in **Note 5**.

When including these two revisions, the steady-state conservation of sucrose mass is now given as

$$\frac{\partial}{\partial x}[(1+\underbrace{\frac{r^2}{24D}\frac{\partial u}{\partial x}}_{TD_{\text{advection}}})cu]=\frac{\partial}{\partial x}[(\underbrace{D\frac{r^2u^2}{48D^2}}_{TD_{\text{dispersion}}}+D)\frac{\partial c}{\partial x}]-\left(\frac{c-c_b}{t_r}\right) \quad (21)$$

where D is the molecular diffusivity of sucrose in water. The terms labeled as $TD_{advection}$ and $TD_{dispersion}$ arise due to the effects of Taylor dispersion on advection and diffusive sucrose transport in pipes with membrane walls instead of solid walls. For the conventional Taylor dispersion in closed tubes, $TD_{advection}=0$. In the absence of any Taylor dispersion, both $TD_{advection}$ and $TD_{dispersion}$ can be ignored and the point-equation is assumed to apply to cross-sectional area-averaged quantities. Equations (20) and (21) can be simultaneously solved to obtain P_p, c, and u from the Hagen–Poiseuille approximation. These equations must be solved numerically once r, L, and external factors setting p_m, c_b, and t_r are supplied. The TR is then determined from the sucrose mass flux. Moreover, the intersections with PR (solved numerically) enable the determination of the equilibrium sucrose concentration points that can be analyzed via bifurcation diagrams.

4 Materials

The following parameters were used in all model calculations unless otherwise stated. The A_0 was computed from the optimal photosynthesis model that includes all nonlinearities in the biochemical demand described elsewhere [65]. Equation (15) was presented here simply for illustration purposes. For clarity, all symbols and parameters used in the calculations are summarized with their descriptions and units in Table 1. Some of the terms below are also defined in Sect. 7.

- *Environmental variables*: Atmospheric carbon dioxide concentration was set to $c_a = 400$ ppm, leaves experience a mean photosynthetically active radiation $I = 800\ \mu\text{mol}\cdot\text{m}^{-2}\cdot\text{s}^{-1}$, a mean air temperature of 25°C, and a relative humidity of 40%.

Table 1

Summary of the symbols used in the biophysical model, including physical variables, parameters, and constants, along with their definitions and units

Symbol	Description	Unit
c_0	sucrose loading concentration	mol/m^3
Jo	osmotic flux of water molecules	m/s
k	membrane permeability	m/(Pa·s)
ϕ_x	xylem water potential	Pa
ϕ_p	phloem water potential	Pa
P_p	phloem turgor pressure	Pa
Π	osmotic potential of sucrose	Pa
R_g	ideal gas universal constant	J/(mol·K)
T	temperature	K
Q	volumetric sap flow rate	m^3/s
R	hydraulic resistance	Pa·s/m^3
J_c	mass flux	mol/(m^3·s)
ρ	sap density	kg/m^3
S_o	local source/sink term	mol/(m^3·s)
L	plant height	m
L_a	total leaf area	m^2
P_a	total phloem cross-sectional area	m^2
r	phloem tube radius	m
μ	viscosity	Pa·s
P_R	total assimilated carbon	mol/s
A_o	photosynthetic rate per leaf area	mol/(m^2·s)
α	interaction strength between c_0 and A_o	mol/m^3
δ	interaction strength between c_0 and A_o	mol/m^3
β_1	photosynthetic parameter	µmol/(m^2·s)
β_2	photosynthetic parameter	µmol/mol
β_3	photosynthetic parameter	unitless
c_a	atmospheric carbon dioxide concentration	ppm
VPD	vapor pressure deficit	kPa
λ_p	Lagrange multiplier/marginal water use efficiency	µmol/(mol·kPa)
$V_{o,\max}$	max carboxylation capacity = β_1 if RuBisCO limited	µmol/(m^2·s)
p_m	leaf water tension	MPa
x	location along phloem tube	m

Table 1 (continued)

Symbol	Description	Unit
u	sap flow axial velocity	m/s
c	sap flow concentration	mol/m^3
c_b	background concentration	mol/m^3
t_r	removal time	s
D	molecular diffusivity of sucrose in water	m^2/s
T_R	phloem transport capacity	mol/s
L_{th}	max transport length	m
R_e	leaf respiration	mol/s
S_s	starch accumulation during growth	mol/s
L_o	sucrose export into the phloem	mol/s
$K_{c,o}$	Michaelis constant for CO_2 fixation	μmol/mol
$K_{o,a}$	Michaelis constant for O_2 inhibition	mmol/mol
$C_{o,a}$	oxygen cocentration in air	mmol/mol
I	photosynthetically active radiation	μmol/(m^2·s)
α_p	apparent quantum yield coeff.	(μmol electron)/(μmol photon)
$J_{e,\max}$	max rate of electron transport	μmol/(m^2·s)

- *Physiological variables*: The high I here means that photosynthesis is limited by RuBisCO. Thus, the photosynthetic parameters for A_0 are: $V_{c,\max} = 59 \mu\text{mol} \cdot \text{m}^{-2} \cdot \text{s}^{-1}$, the minimum marginal water use efficiency is $\lambda_{0,\min} = 50$ mmol·mol^{-1}· kPa^{-1}, compensation point $c_p = 40\text{ppm}$, the Michaelis–Menten constants for oxygen fixation $K_{c,o} = 300\text{ppm}$, and for oxygen inhibition $K_{o,a} = 300\text{mmol/mol}$, the oxygen concentration in the atmosphere is assumed $C_{o,a} = 210\text{mmol/mol}$, apparent quantum yield for electron transport coefficient, $\alpha_p = 0.064\text{molelectronmol}^{-1}\text{photon}$, and in the description of the down-regulation of photosynthesis $\alpha = 1200$ mol/m^3 and $\delta = 1600$ mol/m^3

- *Phloem variables*: The cross-sectional area of all phloem tubes within a stem/trunk of any length was set to $P_a = 0.000825\, m^2$. The membrane permeability k was set to $10^{-14}\, m \cdot \text{Pa}^{-1} \cdot s^{-1}$. The dynamic viscosity at steady-state was assumed to vary with the loading concentration following a predefined equation arising from neural network approximation [88]. This implies that μ is locally independent of c at steady-state and is set to its highest value (loading concentration). The goal of catastrophe theory is to identify the limiting resistance to the flow, which is guaranteed when selecting the most limiting viscosity at c_0. The molecular diffusivity D weakly depends on c. For simplicity, it was set to a constant with $D=4\times10^{-10}\ m^2/s$. The temperature of the sap was assumed constant with $T=298$K, and the ideal gas constant is $R_g=9.314\ \text{J}\cdot\text{mol}^{-1}\cdot\text{K}^{-1}$.

- *Boundary conditions*: At $x=L$, roots are assumed to be an infinite sink so that $c(L)=0$. At $x=0$, the loading concentration matches the transport capacity. The phloem and xylem are assumed to be in mechanical equilibrium at the leaf and root levels. This assumption sets the boundary conditions for the turgor pressure to be $p\,|_{x=0}=R_gTc(0)-p_m$ and $p\,|_{x=L}=0$. The background concentration c_b in the sink term can be varied along the phloem to model different unloading intensity. For this case, it was assumed to be constant and set to 10 mol/m^3.
- *Solving the equations*: The one-dimensional phloem model is discretized using a central finite difference approximation with 50 grid points. The loading concentration c_0 was varied from 250 mol/m^3 to 3000 mol/m^3. At $c_0>2000$ mol/m^3, the Van't Hoff assumption of dilute solutions will break and the sap fluid will behave like a non-Newtonian fluid. However, in this regime, the flow stagnates and the accuracy of such a flow lies outside the scope of this analysis. Due to the nonlinear nature of the problem, an iterative method was used for each c_0 where the osmotic balance was solved first, followed by the conservation of sucrose mass. The process was repeated until convergence.

5 Critical Transitions in Phloem Transport

To demonstrate the power of bifurcation analysis in evaluating critical features for sustainable phloem transport, the bifurcation analysis was conducted for steady-state conditions by identifying the sucrose concentrations where supply from photosynthesis PR and phloem transport capacity TR intersect. In these model calculations, the environmental conditions that were varied were related to p_m, while the structural properties that were varied included L, r, and the removal time t_r. The list of parameters used to perform the bifurcation analysis is presented in Sect. 4. To solve for equilibrium (at the intersection), the following procedure was used:

- The leaf water potential p_m was first imposed, representing values that range from well-watered soils to drought. The variations in

p_m were then used to estimate the marginal water use efficiency λ_o so that A_0 can be calculated for each imposed p_m.

- The loading concentration c_0 was set, enabling the estimation of PR for each choice of c_0. The c_0 was systematically varied so as to allow PR and TR curves to be determined.
- The phloem geometry was prescribed, and a xylem water potential profile along the phloem was set. To solve for TR, the phloem length was discretized for the numerical scheme.
- With the sink intensity known (c_b was fixed and t_r varied for the bifurcation analysis), the transport capacity was solved using numerical methods.
- TR was obtained by multiplying the mass flux (numerically obtained as described in **Note 6**) and the total cross-sectional area of phloem tubes in a stem/trunk, P_a.
- Now that photosynthesis and transport flow rates have been established and solved for as a function of c_0, the c points at which they intersect reflect the phloem's equilibrium sucrose loading concentrations or concentrations at which the supply equals the transport capacity, an ideal condition for plants.
- The equilibrium points for sucrose concentration were then categorized into stable and unstable based on the stability analysis discussed in Sect. 2.1.1. The number of equilibrium points can vary from 0 (phloem dysfunction) to 3 (usually two stable at the extrema and one unstable intermediate value).

5.1 Impact of Control Variables on Transport Capacity and Production Rate

A set of TR-c_0 curves for variables t_r, p_m, L, and r are shown in Fig. 5 for illustration. All of these curves resembled those featured in Fig. 1 and 2 qualitatively in that they all exhibited a rise, a peak, and a drop in transport capacity with increasing c_0. While the peak value of TR was affected by the control variables, the loading concentration at which this peak occurred remained relatively consistent, between 1000 and 1250 mol/m^3. For $c_0 < 500$mol/m^3, TR increased approximately linearly with loading concentration as expected from the Van't Hoff formulation when viscosity does not vary appreciably with c_0. However, when $c_0 > 1200$mol/m^3, the nonlinear viscosity effects become limiting and TR drops. At very high c_0, TR was appreciably reduced, leading to phloem dysfunction. At these high c_0, corrections to the Van't Hoff equation must be included. However, phloem failure would have likely occurred well before the attainment of such a high operating c_0. For this reason, those corrections were not implemented.

Fig. 6 illustrates the impact of the control parameters on the equilibrium points. Since PR is only affected by environmental conditions, increasing p_m decreased PR through its effect on the marginal water use efficiency, ultimately leading to a decrease in A_0. An inter-

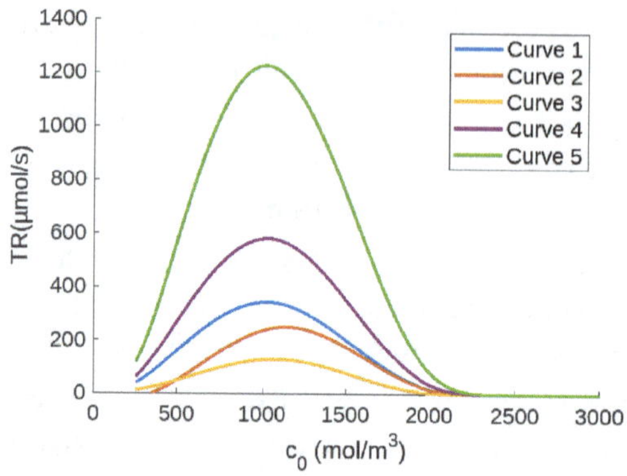

	t_r (s)	P_m (MPa)	L (m)	r (μm)
Curve 1	750	0.5	10	10
Curve 2	750	2	10	10
Curve 3	60	0.5	10	10
Curve 4	750	0.5	6	10
Curve 5	750	0.5	10	20

Fig. 5 Phloem transport capacity (TR) as a function of loading concentration (c_0) for different combinations of control variables (curves 1 to 5), where the sucrose sink time t_r, background xylem tension at the canopy top p_m, tube length L and tube radius r are varied. Note the similarity in shapes between these modeled TR using the Münch physics and the analytical shape for $T(c_0)$ in Figs. 1 and 2

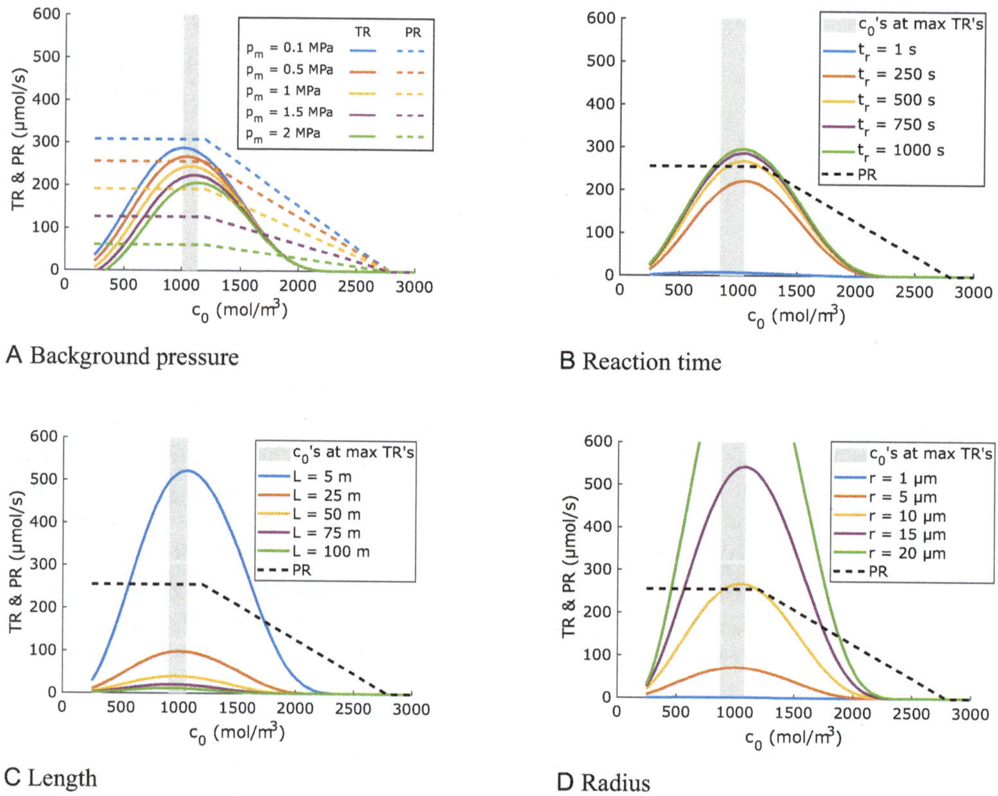

Fig. 6 Effect of sucrose loading concentrations and geometrical/environmental parameters on sucrose supply and removal rates in phloem tubes

esting observation from this figure is that a c_0 around 1250 mol m^{-3} is near the maximum TR but is also close to the limit at which elevated c_0 reduces $A_{0,\max}$.

For changes in r and t_r, the trends were similar: as the magnitude of the control variable increased, the maximum TR increased, and the corresponding equilibrium points (excluding the one at high loading concentration) became more separated. This behavior was consistent with the shape of the TR curve—when the peak value of TR increased at approximately the same c_0, the rate of its rise and fall became steeper, and vice versa. In contrast, increases in p_m and L resulted in a reduction in the peak TR value. In the case of p_m, this occurred due to feedback on PR, leading to a decline in both PR and TR. However, because PR decreased more sharply, the equilibrium points remained well separated. For L, the drop in TR was more pronounced, and equilibrium points were only feasible for short vegetation heights ($L < 5$m). At first glance, this finding links back to the main critique for the Münch mechanism that suggests its failure to explain long-distance transport at large L while fixing A_0.

As shown in Fig. 6, a reduction in PR due to lower A_0 values restored intersections and equilibrium between PR and TR. This point is discussed further in a later section. At this stage, it suffices to note that leaf age is expected to impact A_0 in several ways. Younger leaves typically have higher photosynthetic capacities when compared to their older leaf counterpart [89]—meaning β_1 is reduced with age as documented by several studies reviewed elsewhere [90–92]. However, the effect of age on photosynthetic capacity or A_0 is by no means a settled issue, with some studies reporting weak reductions [93], no significant changes [94], or confounding effects from nutritional status and tree height [86, 95, 96] with aging.

5.2 Stability Analysis

The question to be addressed next is whether those intersections between TR and PR lead to stable equilibrium values—and whether those c_0 intersections are within the observed range for a functioning phloem. To address these questions, c_0 associated with the intersections between TR and PR curves are shown in Fig. 6 (featured as an ordinate) and their stability discussed in Fig. 7 using variations in one control parameter at a time (presented as an abscissa). Equilibrium c_0 was classified as stable (red or blue) or unstable (green). For reference, the measured sucrose loading concentration ranges across a variety of species [24] were shown as a violet shaded region in Fig. 7. This region served as a "plausibility" check on the c_0 outcomes from the model. For this reason, we labeled the violet region as acceptable operating ranges in c_0 for the phloem vasculature. The approximate region where loading concentration begins to influence PR was marked in light-shaded orange for reference.

Equilibrium points corresponding to the highest concentration values (red dots), while stable, were not realized in practice (well

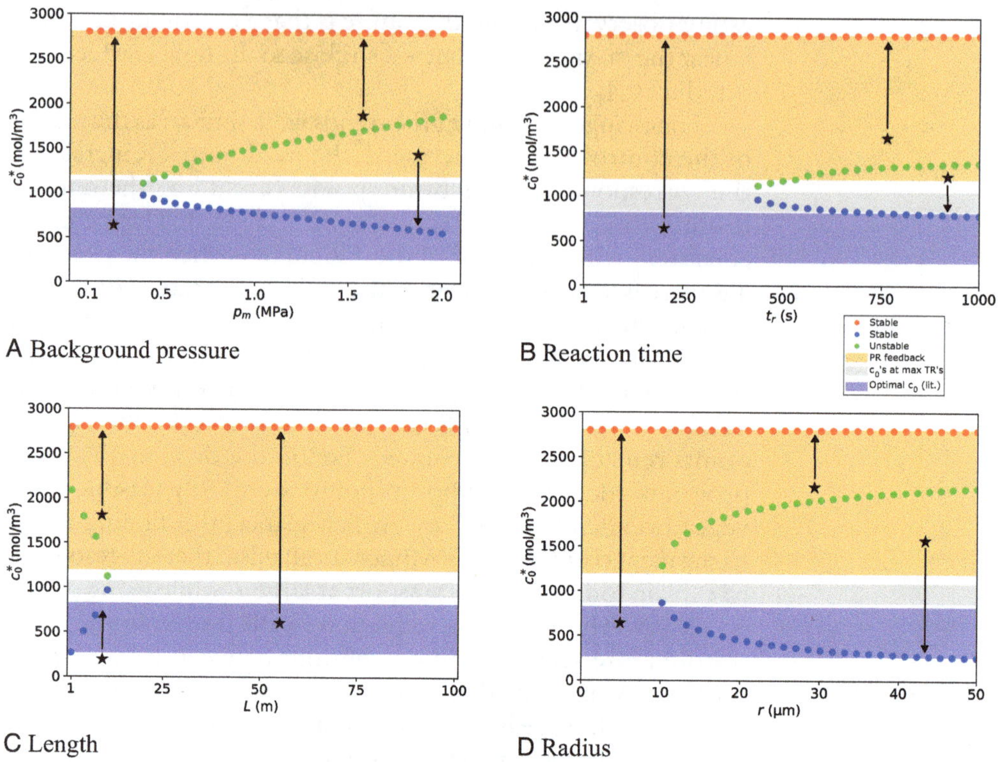

Fig. 7 Bifurcation curves of different control variables: a) of radius at $L = 10$ m, $t_r = 500$ sec, $p_m = 0.5$ MPa, $c_a = 400$ ppm, b) of length at $r = 10$ μm, $t_r = 500$ sec, $p_m = 0.5$ MPa, $c_a = 400$ ppm, c) of reaction time at $r = 10$ μm, $L = 10$ m, $p_m = 0.5$ MPa, $c_a = 400$ ppm, d) of background pressure at $r = 10$ μm, $L = 10$ m, $t_r = 500$ sec, $c_a = 400$ ppm. The violet shaded area indicates the reported c_0 in the literature whereas the shaded orange area indicates the c_0 range whereby photosynthesis is impaired by the loading concentration. Vertical arrows indicate the expected transition from an initial loading concentration (star) to the equilibrium loading concentration (red or blue)

above the violet shaded area). For these stable but high-loading concentrations, the entire phloem becomes dysfunctional as both TR and PR drop to levels that prevent the plant from supporting its own construction and maintenance respiration costs. The intermediate sucrose loading concentrations (green dots) were deemed unstable. Any slight deviation in sucrose concentration around the concentration associated with green dots led to a self-reinforcing cycle instead of a return to equilibrium. Referring to Fig. 6, if the system began slightly to the right of the intermediate equilibrium concentration point (sucrose loading concentration increases), photosynthesis exceeded the removal rate, causing a further increase in loading concentration and, thereby, pushing the system away from equilibrium. If the system started to the left (loading concentration decreases), the removal rate exceeded photosynthesis, leading to sucrose depletion and reinforcing the decline in concentration.

A similar analysis on the lowest loading concentration (blue dot) revealed that this concentration is locally stable.

5.3 Biophysical Limitations on Phloem Function

Returning to the bifurcation analysis on the control variables, the following results can be drawn from Figs. 7 and 8:

- *Tube length:* For longer tubes (i.e., $L > 5$ m), when the loading concentration was below approximately 2750 mol/m^3, PR appeared to consistently exceed TR. As a result, the change in loading concentration over time remained positive, causing c_0 to continuously increase. This accumulation led to a system clogging, ultimately causing failure. Therefore, for large L, no equilibrium points existed below a concentration of 2750 mol/m^3 and categorically led to phloem failure. The only viable operating condition for these systems was when the concentration exceeded 2750 mol/m^3, but this was accompanied by trivial TR and phloem dysfunction. As earlier noted, this finding hints that PR is unrealistically high for plants with high L operating at this TR. Physiologically, this implies that a high $V_{c,\max}$ required in the determination of $A_{0,\max}$ cannot be sustained for such high L values at this given leaf area. The value of $V_{c,\max}$ used in A_0 here is comparable to values reported for shrubs and grasses instead of tall trees. Fig. 8 repeated these calculations by varying $V_{c,\max}$ and finding the maximum acceptable length for safe transport, denoted as L_{th}. It is clear that as $V_{c,\max}$ decreased, L_{th} increased, all other conditions held constant. Indirect evidence points to the fact that $V_{c,\max}$ at a reference temperature of 25°C is on the order of 60 $\mu\text{mol}\cdot\text{m}^{-2}\cdot\text{s}^{-1}$ for crops and grasses (the value assumed here) whereas it is on the order of 40 $\mu\text{mol}\cdot\text{m}^{-2}\cdot\text{s}^{-1}$ for deciduous forests and 35 $\mu\text{mol}\cdot\text{m}^{-2}\cdot\text{s}^{-1}$ for tall coniferous

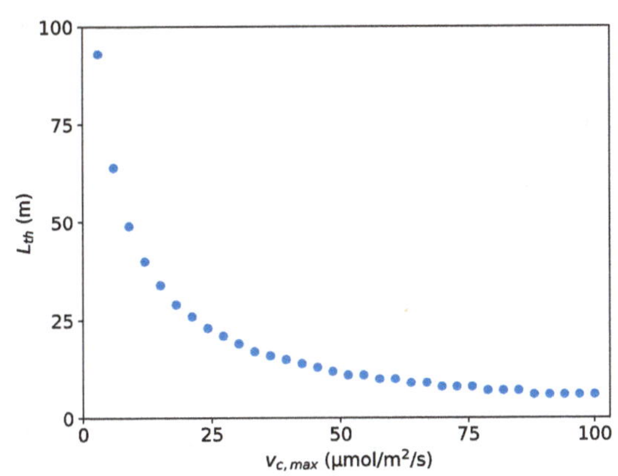

Fig. 8 The maximum phloem tube length before phloem transport stops or transitions to the red stable branch at different $v_{c,\max}$ values, which are associated with different plant heights, ranging from trees (low $v_{c,\max}$) to shrubs and grasses (high $v_{c,\max}$)

forests [97]. Paradoxically, a drop of a factor of 2 in $V_{c,\max}$, $A_{0,\max}$ (and PR) would have avoided the phloem catastrophe for tall plants, not caused the catastrophe. This drop in $A_{0,\max}$ restored the balance between reduced transport capacity due to longer L and photosynthesis. Notably, the computed equilibrium c_0 remained consistent with the range of values reported in the literature (violet region, Fig. 7).

It must be emphasized in Fig. 8 that path-length or tree height (or L) does not control $V_{c,\max}$ per se. Larger trees might have more leaves and greater overall photosynthetic capacity—all leading to higher sucrose loading. What is of significance here is the "critical" $V_{c,\max}$ of individual leaves at a preset leaf area, which is mostly influenced by leaf nitrogen, chlorophyll content, and leaf age, may drop with higher L if phloem dysfunction is the "bottleneck" to the overall plant hydraulics. Another interesting finding from Fig. 8 is that when $V_{c,\max} > 60 \mu\text{mol} \cdot \text{m}^{-2} \cdot \text{s}^{-1}$, the sensitivity to L diminished and any $L < 5\text{m}$ can accommodate such a high $V_{c,\max}$. The inverse relation that emerged between $V_{c,\max}$ and L was apparent for $L > 25\text{m}$ at a preset leaf area. For tall trees, chlorophyll display that enhances sucrose loading along a distributed crown (in conifers about $2/3L$) makes effective L shorter.

- *Sieve tube radius:* For $r < 0.1 \times 10^{-4}\text{m}$, the TR was too small and no intersections between PR and TR were found except for the highest loading concentrations (red). As noted earlier, this stable equilibrium is indicative of a collapse of the phloem. However, as r increased, the stable loading concentration decreased as shown in Fig. 7. This decrease began gradually to "flatten out" while the unstable equilibrium branch increased into the zone where photosynthesis experienced its negative feedback and was reduced by elevated c_0. Thus, increasing r beyond 0.4×10^{-4} m has no clear hydraulic advantages to phloem efficiency (lower stable branch) but may confer safety advantages by elevating the unstable equilibrium c_0 and its associated bifurcation branch.
- Reaction *time:* For the reaction time, once it exceeded a threshold value (high t_r implies less sink strength), a saddle-node bifurcation occurred. The sucrose consumption along the entire length was sufficiently reduced, and a stable equilibrium (blue) emerged independently of the reaction time. The unstable branch (green) also emerged from the saddle-node bifurcation. This unstable branch operated at a loading concentration higher than those associated with a maximum TR discussed in Fig. 6 but still within the range where c_0 was found to reduce photosynthesis (orange shaded area). The lower stable branch operated at a loading concentration below that associated with a

maximum TR but within the observed c_0 range in the literature.

- *Leaf water potential:* Increasing p_m affected both modeled TR and PR through its influence on λ_0. Beyond a threshold (around 0.5 MPa), the stable equilibrium concentration shifted well within the observed operating c_0 (violet region). Instead of a reduction in $V_{c,\max}$, the increase in λ_0 induced by elevated p_m was sufficient to lower PR and generate intersections with TR that were both stable and within the physiologically acceptable sucrose concentration range. Interestingly, the unstable equilibrium concentration branch increased rapidly with increasing p_m and entered a regime where PR was suppressed due to the negative feedback at high c_0. Unlike increases in L, which drove the system toward phloem dysfunction, elevating p_m separated stable and unstable branches away from the dysfunctional regime. This finding implies that while increasing L (not sensed directly by PR) and p_m reduced the maximum TR, they were fundamentally different in how the saddle node bifurcation emerged.

These findings can be used to define a new safety-efficiency tradeoff in the phloem transport system. The phloem will rarely operate at a loading concentration larger than the one associated with a maximum TR (maximum efficiency)—as any positive increases in loading concentration away from the one associated with maximum TR can trigger severe dysfunction, potentially leading to a catastrophic breakdown of phloem transport. Instead, the operating loading concentration is likely to be lower (increased safety) despite some loss in TR (reduced efficiency). The unsafe c_0 is set by the unstable branch, which appears to coincide with the concentration range at which the negative feedback on leaf photosynthesis occurs. Apparent "sub-optimal" loading strategies in some species may therefore reflect an adaptive shift toward safety, reducing the risk of phloem transport failure under fluctuating or stressful conditions rather than a true inefficiency.

Future work can focus on testing these ideas using experimental data from species with variable loading strategies (e.g., facultative apoplastic/symplastic loaders) across contrasting environments, to see whether systems exposed to greater hydraulic or osmotic variability tend to favor the "safer" configurations highlighted by this model. In addition, the bifurcation analysis can be extended to a wider range of environmental drivers (e.g., atmospheric carbon dioxide, temperature, humidity, photosynthetically active radiation, soil moisture), physiological traits ($V_{c,\max}$, λ_0), and phloem structural properties using more complex flow hydrodynamics (e.g., explicitly including sieve plates, elasticity of the phloem, radial-axial flow dynamics, imperfect mechanical equilibrium between the xylem and phloem, etc....). The outcome of such analysis is better kept as a topic for the future.

Perhaps the most important "take-home" message from Figs. 6 and 7 is the use of catastrophe theory to expand understanding of phloem dysfunction. The full "armory" of mathematical techniques in dynamical systems has been available for quite some time now within the ecological science literature (at least since the 1970s). However, the application of this approach to plant hydraulics for interpreting failure studies in the xylem and phloem is just beginning to mature. This is especially critical given the pervasive uncertainty surrounding the key physiological and environmental parameters that govern phloem and xylem function, such as pressure gradients, conductivity, and vulnerability to embolism or blockage. In this context, advancing such analysis is becoming increasingly essential for evaluating the structural and functional resilience of plants under changing environmental conditions. A deeper understanding of hydraulic failure modes will not only inform predictions of plant survival and productivity but also support the development of more robust models of ecosystem response to climate variability and stress.

6 Summary of Key Findings

The phloem dysfunction was analyzed using catastrophe theory and bifurcation analysis. The bifurcation analysis was applied to an equilibrium sucrose loading concentration c_0 that equates sucrose production from photosynthesis (PR) to transport capacity (TR) of the phloem vasculature. The control variables considered include tube length L, tube radius r, a time scale associated with sucrose sinks t_r along the flow path, and the leaf water tension p_m. Leaf photosynthesis was assumed to obey a stomatal optimality principle whereby stomata maximize carbon gain for a given amount of water in the rooting zone per unit leaf area. The PR computed from such optimality hypothesis was then reduced to account for a negative feedback on leaf photosynthesis when loading concentrations are high. The TR was assumed to follow a steady-state pressure flow hypothesis modified to include Taylor dispersion in this one-dimensional (axial) representation. The Taylor dispersion terms arise due to area-averaging of the point equations for momentum and sucrose mass balance.

Among all the control parameters considered, only increases in p_m (i.e., less water available to the plant) simultaneously reduced PR and TR. When the feedback between photosynthesis and c_0 was accounted for, the balance between PR and TR can lead to 3, 2, or 1 c_0 equilibrium points. The single equilibrium c_0 was shown to be a stable equilibrium. However, its value was much larger than the operational c_0 values reported for loading concentrations in the literature. That is, this single stable equilibrium at very high c_0 reflects a state of phloem dysfunction despite equality of PR and TR, as both PR and TR are so small that they cannot cover the plant construction and maintenance respiration costs. For certain combinations of control variables, systematic variations in one of these control variables result in a catastrophe through a saddle-node bifurcation whereby one stable and one unstable c_0 emerge above and beyond the aforementioned large and stable c_0. The unstable equilibrium c_0 was shown to be operating within the limits where leaf photosynthesis is reduced by elevated c_0. The other stable equilibrium c_0 from this saddle-node bifurcation was shown to be within the operating range reported in the literature. The unstable c_0 was shown to be higher than the c_0 at which TR is maximum, whereas the stable c_0 is lower (less efficient but safe). Increasing the radius and reaction times generate stable equilibrium c_0 values that appear plausible and commensurate with literature c_0 values. Moreover, these increases in radius and reaction time tend to reduce the stable c_0 value and increase the unstable c_0 value.

The work here also showed that the phloem does not operate at a c_0 that maximizes TR for almost all combinations. Instead, the phloem operates on a c_0 value always below the c_0 associated with maximum TR. The quasi-quadratic shape of TR with c_0 means that plants can afford to sacrifice some efficiency (reduced TR) to gain safety and operate below the unstable equilibrium c_0 branch. The model results also shed light on the role of increased L and the conventional critique of the Münch mechanism. For tall trees, the attainment of a stable yet operational c_0 that is safe and efficient requires a reduction in PR to match the reduced TR arising from the increased L when all other control variables are held constant. Comparing the maximum carboxylation capacity $V_{c,\max}$ at a reference temperature for trees versus grasses and shrubs seems to suggest that trees operate with lower $V_{c,\max}$ compared to grasses and shrubs. Conifers, for example, experience a $V_{c,\max}$ that is almost a factor of 2 smaller than the $V_{c,\max}$ for grasses and shrubs (used here). This finding demonstrates the power of bifurcation analysis to reveal new insights and points to a previously under-studied coordination between the photosynthetic machinery and phloem geometry independent of leaf water potential. With regard to increasing the magnitude of leaf water potential, the stable and operational equilibrium c_0 balancing TR and PR seems to hold across a wide range of leaf water potential values ($\in[0.5, 2\text{MPa}]$). Decreasing leaf water potential also increases the unstable c_0 branch, thereby allowing the phloem to operate over a wider range of c_0 values (and leaf water potential values). This finding may explain why the phloem is relatively robust to droughts.

Several simplifying assumptions underlie this framework and delimit how the results should be interpreted. First, we assume steady-state phloem transport and a one-dimensional pipe geometry, so we do not resolve transient recovery processes or anatomical heterogeneity along the transport pathway. Second, phloem loading is tied directly to leaf photosynthesis and sucrose is treated as the dominant osmolyte, whereas in real plants loading, storage, and ionic composition (e.g., K^+), can be regulated independently of instantaneous assimilation. Third, sink demand is represented by a single reaction time scale t_r, without explicit distinction among growth, respiration, and storage. These choices make the bifurcation structure analytically tractable and help identify where multiple equilibria and tipping points could occur, but they should not be read as quantitative predictions for any given species. Within these constraints, bifurcation analysis provides an organizing framework for describing theoretical tipping points and possible safety mechanisms for phloem transport.

7 Notes

The notes here elaborate on aspects of the PFH referenced in the main model development.

1. *Carbon balance equation:* The carbon balance equation partitions gross carbon input from photosynthesis PR and outputs through leaf respiration R_e, transient storage such as starch accumulation S_s that is used during the growth period, and sucrose export into the phloem—labeled as L_0 [related to $S(c)$ earlier described]. The L_0 into the phloem is influenced by both passive diffusion and active transport using proton-sucrose symporters. The overall balance for the partition of PR can be expressed as:

$$\mathrm{PR} = R_e + S_s + L_0. \tag{22}$$

   In general, L_0 is linked to PR mediated by feedback when c is large. Mathematically, L_0 serves as a boundary condition for the conservation of sucrose mass and momentum within the phloem vasculature in Stage 2.

2. *Osmotic potential for non-ideal solution:* The Van't Hoff prediction is accurate for dilute solutions where the activity of sucrose in water is approximately equal to its concentration. However, this equivalence breaks down at higher concentrations, leading to non-ideal behavior that necessitates corrections to Van't Hoff's law. In such cases, the osmotic potential must be re-derived from thermodynamic principles if the activity of sucrose is known. Since this is not always practical, these deviations may be approximated by introducing a nonlinear correction

factor $\Phi(c) < 1$, which reduces the Van't Hoff prediction. That is, the non-ideal osmotic potential resulting from c is given by

$$\Pi = -\Phi(c) R_g T c. \tag{23}$$

This $\Phi(c) \leq 1$ correction accounts for "molecular crowding," which reduces the effective osmotic pressure and hydration effects, where sucrose–water interactions limit free water availability, deviating from the ideal Van't Hoff behavior.

3. *Photosynthetic parameters under RuBisCO limitations:* In the case of RuBisCO limitations on photosynthesis (as is the case here for high I), β_1 is the maximum carboxylation capacity ($V_{c,\max}$) as earlier noted and β_2 is related to CO_2 fixation and oxygen inhibition and is given by

$$\beta_2 = K_{c,o}\left(1 + \frac{C_{o,a}}{K_{o,a}}\right), \tag{24}$$

where $K_{c,o}$ and $K_{o,a}$ are the Michaelis constants for CO_2 fixation and oxygen inhibition, respectively, and $C_{o,a}$ is the oxygen concentration in air. The $K_{c,o}$, $K_{o,a}$ (or β_2), and $V_{c,\max}$ (or β_1) are all temperature dependent—meaning that for a given species and an externally specified temperature, these coefficients can be treated as constants.

4. *Photosynthetic parameters under RuBP limitations:* When RuBP limitations govern photosynthesis, β_1 is linearly related to the photosynthetically active radiation (I) through the apparent quantum yield of electron transport coefficient (α_p) and saturates at the maximum rate of electron transport ($J_{e,\max}$) at high photosynthetically active radiation. This form leads to an approximate expression

$$\beta_1 = \frac{\alpha_p I}{\sqrt{1 + \left(\alpha_p I / J_{e,\max}\right)^2}}. \tag{25}$$

The $\beta_2 = 2c_p$, where c_p is the CO_2 compensation point that varies with temperature. The $J_{e,\max}$ is linearly related to $V_{c,\max}$ with many studies showing $J_{e,\max}/V_{c,\max} \approx 2$ at a reference temperature of 25°C [98].

5. *Local sink/source within the phloem:* The local source/sink term $S_o(c)$ depends on biological mechanisms governing sucrose exchange with surrounding cells. This process can be approximated using different formulations:

-A reaction-like equation following first-order kinetics $S_o \sim k_r c$, where k_r represents the rate of sucrose loading/unloading.
-A diffusion-like equation $S_o \sim D_{eff}(c - c^o)$, where D_{eff} is an effective diffusion coefficient and c^o is a target concentration maintained by nearby living cells.
-A Michaelis–Menten relation $S_o \sim V_1 c/(V_2 + c)$, where V_1 is the maximum loading/unloading rate and V_2 is a Michaelis constant, capturing saturation effects in transporter-mediated sucrose exchange.
These different source/sink formulations provide flexibility in modeling how sucrose is dynamically regulated within the phloem.

6. *Numerical estimation of sucrose mass flux:* After numerically solving for the flow velocity u and sucrose concentration c, the mass flux can be approximated from

$$J = J_{\text{advection}} + J_{\text{dispersion}} = \left(1 + \frac{r^2}{24D}\frac{\partial u}{\partial x}\right)cu - \left(\frac{r^2 u^2}{48D^2} + 1\right)D\frac{\partial c}{\partial x}, \quad (26)$$

which includes advection, molecular diffusion, and corrections due to Taylor dispersion.

Acknowledgments

G. Katul acknowledges support from the U.S. National Science Foundation (NSF-AGS-2028633) and the U.S. Department of Energy (DE-SC0022072). S. Sevanto acknowledges support from U.S. Department of Defense, Strategic Environmental Research and Development Program (SERDP) grant RC25-0189. M. Nakad acknowledges support from the startup support from the Lebanese American University, and the NASA Modeling, Analysis, and Prediction (MAP) Program (Grant No. 80NSSC21K1496)

Conflict of Interest The authors have no conflicts of interest to declare that are relevant to the content of this chapter.

References

1. Münch E (1930) Stoffbewegungen in der Pflanze. G. Fischer, Jena, Germany
2. Dixon H, Joly J (1895) On the ascent of sap. Philos Trans R Soc Lond B 186(0):563–576
3. Van den Honert TH (1948) Water transport in plants as a catenary process. Discuss Faraday Soc 3(0):146–153
4. Tyree MT (1997) The cohesion-tension theory of sap ascent: current controversies. J Exp Bot 480(10):1753–1765
5. Angeles G, Bond B, Boyer JS, Brodribb T, Brooks J, Burns MJ, Cavender-Bares J, Clearwater M, Cochard H, Comstock J et al (2004) The cohesion-tension theory. New Phytol 1630(3):451–452
6. Jensen KH, Berg-Sørensen K, Bruus H, Holbrook NM, Liesche J, Schulz A, Zwieniecki MA, Bohr T (2016) Sap flow and sugar transport in plants. Rev Mod Phys 880(3): 035007

7. Konrad W, Katul G, Roth-Nebelsick A, Jensen KH (2018) Xylem functioning, dysfunction and repair: a physical perspective and implications for phloem transport. Tree Physiol 390(2):243–261
8. van Bel AJE (2003) The phloem, a miracle of ingenuity. Plant Cell Environ 260(1):125–149
9. Thompson MV, Holbrook NM (2003) . Application of a single-solute non-steady-state phloem model to the study of long-distance assimilate transport. J Theor Biol 2200(4):419–455
10. Fensom DS (1981) Problems arising from a Münch-type pressure flow mechanism of sugar transport in phloem. Can J Bot 590(4):425–432
11. Savage JA, Beecher SD, Clerx L, Gersony JT, Knoblauch J, Losada JM, Jensen KH, Knoblauch M, Holbrook NM (2017) Maintenance of carbohydrate transport in tall trees. Nat Plants 30(12):965
12. Knoblauch M, Knoblauch J, Mullendore DL, Savage JA, Babst BA, Beecher SD, Dodgen AC, Jensen KH, Holbrook NM (2016) Testing the Münch hypothesis of long distance phloem transport in plants. Elife 5(0):e15341
13. Turgeon R (2010) The puzzle of phloem pressure. Plant Physiol 1540(2):578–581
14. Sevanto S, Mcdowell NG, Dickman LT, Pangle R, Pockman WT (2014) How do trees die? A test of the hydraulic failure and carbon starvation hypotheses. Plant, Cell Environ 370(1):153–161
15. Hölttä T, Vesala T, Sevanto S, Perämäki M, Nikinmaa E (2006) Modeling xylem and phloem water flows in trees according to cohesion theory and Münch hypothesis. Trees 20:67–78
16. Huang CW, Domec JC, Palmroth S, Pockman WT, Litvak ME, Katul GG (2018) Transport in a coordinated soil-root-xylem-phloem leaf system. Adv Water Resour 119:1–16
17. Hölttä T, Mencuccini M, Nikinmaa E (2009) Linking phloem function to structure: analysis with a coupled xylem—phloem transport model. J Theor Biol 2590(2):325–337
18. Mencuccini M, Hölttä T, Sevanto S, Nikinmaa E (2013) Concurrent measurements of change in the bark and xylem diameters of trees reveal a phloem-generated turgor signal. New Phytol 1980(4):1143–1154
19. Nikinmaa E, Hölttä T, Hari P, Kolari P, Mäkelä A, Sevanto S, Vesala T (2013) Assimilate transport in phloem sets conditions for leaf gas exchange. Plant Cell Environ 360(3):655–669
20. Jensen KH, Mullendore DL, Holbrook NM, Bohr T, Knoblauch M, Bruus H (2012) Modeling the hydrodynamics of phloem sieve plates. Front Plant Sci 3:151
21. Sevanto S (2024) Flow resistance of phloem sieve plates revisited using an experimental model. Phys Fluids 360(3):031901
22. Stanfield RC, Schulte PJ, Randolph KE, Hacke UG (2019) Computational models evaluating the impact of sieve plates and radial water exchange on phloem pressure gradients. Plant Cell Environ 420(2):466–479
23. Nakad M, Domec J-C, Sevanto S, Katul G (2023a) Toward a realistic representation of sucrose transport in the phloem of plants. J Geophys Res Biogeosci 1280(3):e2022JG007361
24. Jensen KH, Lee J, Bohr T, Bruus H, Holbrook NM, Zwieniecki MA (2011) Optimality of the Münch mechanism for translocation of sugars in plants. J R Soc Interface 80(61):1155–1165
25. Ronellenfitsch H, Liesche J, Jensen KH, Holbrook NM, Schulz A, Katifori E (2015) Scaling of phloem structure and optimality of photoassimilate transport in conifer needles. Proc R So B Biol Sci 2820(1801):20141863
26. Sevanto S (2018) Drought impacts on phloem transport. Curr Opin Plant Biol 43:76–81
27. Poincare H (1885) L'Équilibre d'une masse fluide animée d'un mouvement de rotation. Acta Math 7:259–380
28. Thom R (1969) Topological models in biology. Topology 80(3):313–335
29. Thom R (1977) Structural stability, catastrophe theory, and applied mathematics. SIAM Rev 190(2):189–201
30. Thom R (2018) Structural stability and morphogenesis. CRC press
31. Zeeman EC (1976) Catastrophe theory. Sci Am 2340(4):65–83
32. May RM, Oster GF (1976) Bifurcations and dynamic complexity in simple ecological models. Am Nat 1100(974):573–599
33. Ludwig D, Jones DD, Holling CS et al (1978) Qualitative analysis of insect outbreak systems: the spruce budworm and forest. J Animal Ecol 470(1):315–332
34. Saunders PT (1980) An introduction to catastrophe theory. Cambridge University Press
35. Strogatz SH (2001) Nonlinear dynamics and chaos: with applications to physics, biology, chemistry, and engineering (studies in nonlinearity), vol 1. Westview press
36. Zahler RS, Sussmann HJ (1977) Claims and accomplishments of applied catastrophe theory. Nature 2690(5631):759–763
37. Sussmann HJ (1970) Catastrophe theory: a skeptic. Science 4306(1977):820–821
38. Boutot A (1993) Catastrophe theory and its critics. Synthese 96:167–200
39. Craig Loehle (1989) Catastrophe theory in ecology: a critical review and an

example of the butterfly catastrophe. Ecol Model 490(1-2):125–152
40. Scheffer M, Carpenter S, Foley JA, Folke C, Walker B (2001) Catastrophic shifts in ecosystems. Nature 4130(6856):591–596
41. Ives AR, Carpenter SR (2007) Stability and diversity of ecosystems. Science 3170(5834): 58–62
42. Scheffer MA, Bascompte J, Brock WA, Brovkin V, Carpenter SR, Dakos V, Held H, Van Nes EH, Rietkerk M, Sugihara G (2009) Early-warning signals for critical transitions. Nature 4610(7260):53–59
43. Meron E (2015) Nonlinear physics of ecosystems. CRC Press, Boca Raton, FL
44. Liu Y, Kumar M, Katul GG, Porporato A (2019) Reduced resilience as an early warning signal of forest mortality. Nat Clim Change 90(11):880–885
45. May RM (1977) Thresholds and breakpoints in ecosystems with a multiplicity of stable states. Nature 2690(5628):471–477
46. Duckstein L, Casti J, Kempf J (1979) Modeling phytoplankton dynamics using catastrophe theory. Water Resour Res 150(5):1189–1194
47. Hastings A (1988) Food web theory and stability. Ecology 690(6):1665–1668
48. Turchin P (2013) Complex population dynamics: a theoretical/empirical synthesis (mpb-35). In: Complex population dynamics. Princeton University Press
49. Hastings A, Powell T (1991) Chaos in a three-species food chain. Ecology 720(3):896–903
50. Hastings A, Hom CL, Ellner S, Turchin P, Godfray HCJ (1993) Chaos in ecology: is mother nature a strange attractor? Annual Review of Ecology and Systematics, 24:1–33
51. Turchin P (2001) Does population ecology have general laws?. Oikos 940(1):17–26
52. Scheffer M, Carpenter SR, Dakos V, van Nes EH (2015) Generic indicators of ecological resilience: inferring the chance of a critical transition. Annu RevEcol Evol Syst 460(1):145–167
53. Konrad W, Roth-Nebelsick A (2003) The dynamics of gas bubbles in conduits of vascular plants and implications for embolism repair. J Theor Biol 2240(1):43–61
54. Konrad W, Roth-Nebelsick A, Grein M (2008) Modelling of stomatal density response to atmospheric CO_2. J Theor Biol 2530(4):638–658
55. Parolari AJ, Katul GG, Porporato A (2014) An ecohydrological perspective on drought-induced forest mortality. J Geophys Res Biogeosci 1190(5):965–981
56. Johnson DM, Katul G, Domec JC (2022) Catastrophic hydraulic failure and tipping points in plants. Plant Cell Environ 450(8):2231–2266
57. Von Foerster H, Mora PM, Amiot LW (1960) Doomsday: Friday, 13 November, AD 2026: at this date human population will approach infinity if it grows as it has grown in the last two millenia. Science 1320(3436):1291–1295
58. Nikinmaa E, Sievänen R, Hölttä T (2014) Dynamics of leaf gas exchange, xylem and phloem transport, water potential and carbohydrate concentration in a realistic 3-d model tree crown. Ann Bot 1140(4):653–666
59. Jensen KH, Rio E, Hansen R, Clanet C, Bohr T (2009) Osmotically driven pipe flows and their relation to sugar transport in plants. J Fluid Mech 636:371–396
60. Jensen KH, Savage JA, Holbrook NM (2013) Optimal concentration for sugar transport in plants. J R Soc Interface 100(83):20130055
61. Sevanto S (2014) Phloem transport and drought. J Exp Bot 650(7):1751–1759
62. Nakad M, Witelski T, Domec J-C, Sevanto S, Katul G (2021) Taylor dispersion in osmotically driven laminar flows in phloem. JFluid Mech 913:A44001–A44026. https://doi.org/10.1017/jfm.2021.56
63. Cowan IR, Troughton JH (1971) The relative role of stomata in transpiration and assimilation. Planta 970(4):325–336
64. Hari P, Mäkelä A, Korpilahti E, Holmberg M (1986) Optimal control of gas exchange. Tree Physiol 20(1-2-3):169–175
65. Katul G, Manzoni S, Palmroth S, Oren R (2010) A stomatal optimization theory to describe the effects of atmospheric CO_2 on leaf photosynthesis and transpiration. Ann Bot 1050(3):431–442
66. Medlyn BE, Duursma RA, Eamus D, Ellsworth DS, Prentice I, Barton CVM, Crous KY, De Angelis P, Freeman M, Wingate L (2011) Reconciling the optimal and empirical approaches to modelling stomatal conductance. Glob Change Biol 170(6):2134–2144
67. Manzoni S, Vico G, Palmroth S, Porporato A, Katul G (2013) Optimization of stomatal conductance for maximum carbon gain under dynamic soil moisture. Adv Water Resour 62:90–105
68. Mrad A, Sevanto S, Domec J-C, Liu Y, Nakad M, Katul G (2019) A dynamic optimality principle for water use strategies explains isohydric to anisohydric plant responses to drought. Front For Glob Change 2:49
69. Nakad M, Sevanto S, Domec J-C, Katul G (2023b) Linking the water and carbon economies of plants in a drying and warming climate. Curr For Rep 90(6):383–400
70. Daniel RB (1993) Proton-coupled sugar and amino acid transporters in plants. Annu Rev Plant Biol 440(1):513–542
71. David MB (2022) Phloem loading and unloading of sucrose: what a long, strange

trip from source to sink. Annu Rev Plant Biol 730(1):553–584
72. Babst BA, Braun DM, Karve AA, Baker RF, Tran TM, Kenny DJ, Rohlhill J, Knoblauch J, Knoblauch M, Lohaus G et al (2022) Sugar loading is not required for phloem sap flow in maize plants. Nat Plants 80(2):171–180
73. Nakad M, Domec J-C, Sevanto S, Katul G (2023) Sucrose transport inside the phloem: bridging hydrodynamics and geometric characteristics. Phys Fluids 35(6):061905
74. Katul GG, Palmroth S, Oren R (2009) Leaf stomatal responses to vapour pressure deficit under current and CO_2-enriched atmosphere explained by the economics of gas exchange. Plant Cell Environ 320(8):968–979
75. Farquhar GD, von Caemmerer S, Berry JA (1980) A biochemical model of photosynthetic CO_2 assimilation in leaves of C3 species. Planta 1490(1):78–90
76. Palmroth S, Berninger F, Nikinmaa E, Lloyd J, Pulkkinen P, Hari P (1999) Structural adaptation rather than water conservation was observed in scots pine over a range of wet to dry climates. Oecologia 1210(3):302–309
77. Collatz GJ, Ball JT, Grivet C, Berry JA (1991) Physiological and environmental regulation of stomatal conductance, photosynthesis and transpiration: a model that includes a laminar boundary layer. Agric For Meteorol 540(2):107–136. http://www.sciencedirect.com/science/article/pii/0168192391900028, ISSN 0168-1923. https://doi.org/10.1016/0168-1923(91)90002-8
78. Leuning R (1995) A critical appraisal of a combined stomatal-photosynthesis model for c3 plants. Plant Cell Environ 180(4):339–355
79. Dewar R, Mauranen A, Mäkelä A, Hölttä T, Medlyn B, Vesala T (2018) New insights into the covariation of stomatal, mesophyll and hydraulic conductances from optimization models incorporating nonstomatal limitations to photosynthesis. New Phytol 2170(2):571–585
80. Salmon Y, Lintunen A, Dayet A, Chan T, Dewar R, Vesala T, Hölttä T (2020) Leaf carbon and water status control stomatal and nonstomatal limitations of photosynthesis in trees. New phytol 2260(3):690–703
81. Qiu R, Katul GG (2020) Maximizing leaf carbon gain in varying saline conditions: an optimization model with dynamic mesophyll conductance. Plant J 1010(3):543–554
82. Dewar R, Hölttä T, Salmon Y (2022) Exploring optimal stomatal control under alternative hypotheses for the regulation of plant sources and sinks. New Phytol 2330(2):639–654
83. Matthews A, Katul G, Porporato A (2024) Multiple time scale optimization explains functional trait responses to leaf water potential. New Phytol 2440(2):426–435
84. Potkay A, Cabon A, Peters RL, Fonti P, Sapes G, Sala A, Stefanski A, Butler E, Bermudez R, Montgomery R et al (2025) Generalized stomatal optimization of evolutionary fitness proxies for predicting plant gas exchange under drought, heatwaves, and elevated CO_2. Glob Change Biol 310(1):e70049
85. Manzoni S, Vico G, Katul G, Fay PA, Polley W, Palmroth S, Porporato A (2011) Optimizing stomatal conductance for maximum carbon gain under water stress: a meta-analysis across plant functional types and climates. Funct Ecol 250(3):456–467
86. Palmroth S, Katul GG, Maier CA, Ward E, Manzoni S, Vico G (2013) On the complementary relationship between marginal nitrogen and water-use efficiencies among pinus taeda leaves grown under ambient and CO_2-enriched environments. Ann Bot 1110(3):467–477
87. Manzoni S, Katul G, Porporato A (2014) A dynamical system perspective on plant hydraulic failure. Water Ress Res 500(6):5170–5183
88. Bouchard C, Granjean BPA (1995) A neural network correlation for the variation of viscosity of sucrose aqueous solutions with temperature and concentration. LWT-Food Sci Technol 280(1):157–159
89. Op de Beeck M, Gielen B, Jonckheere I, Samson R, Janssens IA, Ceulemans R (2010) Needle age-related and seasonal photosynthetic capacity variation is negligible for modelling yearly gas exchange of a sparse temperate Scots pine forest. Biogeosciences 70(1):199–215
90. Field C, Mooney HA (1983) Leaf age and seasonal effects on light, water, and nitrogen use efficiency in a California shrub. Oecologia 56:348–355
91. Han Q, Kawasaki T, Nakano T, Chiba Y (2008) Leaf-age effects on seasonal variability in photosynthetic parameters and its relationships with leaf mass per area and leaf nitrogen concentration within a *Pinus densiflora* crown. Tree Physiol 280(4):551–558
92. Tejera M, Boersma NN, Archontoulis SV, Miguez FE, VanLoocke A, Heaton EA (2022) Photosynthetic decline in aging perennial grass is not fully explained by leaf nitrogen. J Exp Bot 730(22):7582–7595
93. Niinemets U, Tenhunen JD, Beyschlag W (2004) Spatial and age-dependent modifications of photosynthetic capacity in four Mediterranean oak species. Funct Plant Biol 310(12):1179–1193
94. Delzon S, Bosc A, Cantet L, Loustau D (2005) Variation of the photosynthetic capacity across a chronosequence of maritime pine correlates with needle phosphorus concentration. Ann ForSci620(6):537–543
95. Han Q (2011) Height-related decreases in mesophyll conductance, leaf photosynthesis

and compensating adjustments associated with leaf nitrogen concentrations in Pinus densiflora. Tree Physiol 310(9):976–984
96. Rijkers T, Pons TL, Bongers F (2000) The effect of tree height and light availability on photosynthetic leaf traits of four neotropical species differing in shade tolerance. Func Ecol 140(1):77–86
97. Wang X, Chen JM, Ju W, Zhang Y (2022) Seasonal variations in leaf maximum photosynthetic capacity and its dependence on climate factors across global FLUXNET sites. J Geophys Res Biogeosci 1270(5):e2021JG006709
98. Walker AP, Beckerman AP, Gu L, Kattge J, Cernusak LA, Domingues TF, Scales JC, Wohlfahrt G, Wullschleger SD, Woodward FI (2014) The relationship of leaf photosynthetic traits—Vc_{max} and j_{max}—to leaf nitrogen, leaf phosphorus, and specific leaf area: a meta-analysis and modeling study. Ecol Evol 40(16):3218–3235

Chapter 39

Predicting Phloem Transport Rate in Crops

Guillaume Tcherkez, R. Anjal P. Nugroho, Louis Broussard, Rudy Lukman, Jean-Pierre Caliman and Marilyn Ball

Abstract

Accessing phloem sap movement properties, such as transport rate and front velocity, is difficult and involves technologies that are not routinely implementable at high throughput. This can be problematic if quantitative information on phloem transport is to be used for comparing varieties for breeding. We present here how to calculate an average phloem transport rate from agronomic data, using two crops, wheat and oil palm, as examples. When physiological variables are also known for the same agronomic trial, phloem transport rate can be related to other processes such as water movement in xylem sap or nutrient utilization.

Keywords Yield, Gross photosynthesis, Respiration, Phloem surface area, Front sap velocity

1 Introduction

Phloem sap physical properties, such as sap composition and velocity, loading and unloading capacity, are probably important crop biomass determinants and thus limiting factors in productivity. In fact, it has been suggested that improving phloem loading or unloading could be an avenue to improve yield [1, 2]. In several species, including rice and wheat, phloem transport capacity has been suggested to play a role in grain, seed, or fruit production [3–7]. For example, in wheat, it has been proposed that grain development is limited by phloem characteristics (sieve element or sieve pore radii, sieve element cross-sectional area, phloem loading activity, or pressure gradient) as much as by source leaf size and photosynthetic assimilation [8]. However, a strong limitation of yield by anatomical phloem properties does not appear very likely because the size and/or the number of sieve tubes scale to sink (such as grain) size and number [9, 10], and a positive relationship has been found between minor vein number or phloem cross-sectional area and photosynthesis [11–13]. Rather,

Johannes Liesche and Sanna Sevanto (eds.), *Phloem*, Methods in Molecular Biology, vol. 3066,
https://doi.org/10.1007/978-1-0716-5488-0_39,

yield should be related to phloem functional properties, such as loading and unloading capacity, and transport rate by phloem sap.

Measuring phloem transport rate is not easy, because it not only implies in vivo experiments but also requires non-invasive methods such as resonance magnetic imaging (MRI) [14, 15], fluorescent labeling [16], or isotopic tracing (see, e.g., [17–19] for ^{14}C utilization) that are all technologically challenging. Consequently, these methods cannot be implemented routinely at high throughput in crops to compare cultivars, germplasms, or accessions. Furthermore, they would be limited to specific measurement conditions and are thus not fully representative of phloem transport rate over the cultivation life cycle of the crop of interest. Finally, these methods are not easily applicable to trees simply because following the fluorescent dye would be virtually impossible in woody organs/tissues, isotopic labeling would be expensive, and trees may be too big for MRI. Taken as a whole, these limitations are problematic if one desires to use phloem transport rate as a quantitative variable for breeding, an index of crop performance, or simply to explore yield limitations.

We have recently proposed an exploitation of agronomic data that includes yield and other variables giving access to gross carbon fixation, to calculate an estimate of phloem transport rate and applied this method to wheat cultivars [20]. In this chapter, we explain in more detail how to perform these calculations and show numerical results obtained in two crops, wheat and oil palm. These two species have been selected because they represent major crops at the global scale (first cereal grain production, first plant oil production worldwide, respectively), have two very different plant architectures (annual herbaceous and perennial tree, respectively) and are associated with accessible agronomic datasets allowing cultivar or germplasm comparison. Oil palm was also included in this chapter because carbon allocation is believed to have some importance for production in this species. For example, the persistence of non-matured fruits or empty fruits on bunches (which is detrimental to yield) has been suggested to relate to inefficient phloem redistribution of assimilates within bunches [21]. In addition, an important aspect of the evolution of palm species is the reconfiguration of vascular tissues, with an increase in phloem cross-sectional surface area and the appearance of two phloem fascicles (strands) within a vascular bundle in modern palm tribes such as *Cocoideae* [22, 23].

2 Theory and Method

The overall process to perform calculations is shown in Fig. 1. To carry out calculations, the two essential pieces of information are (1) the total amount of transported carbon, and (2) the phloem cross-sectional area across which this carbon is transported. Both are potentially difficult to access when anatomical and agronomical

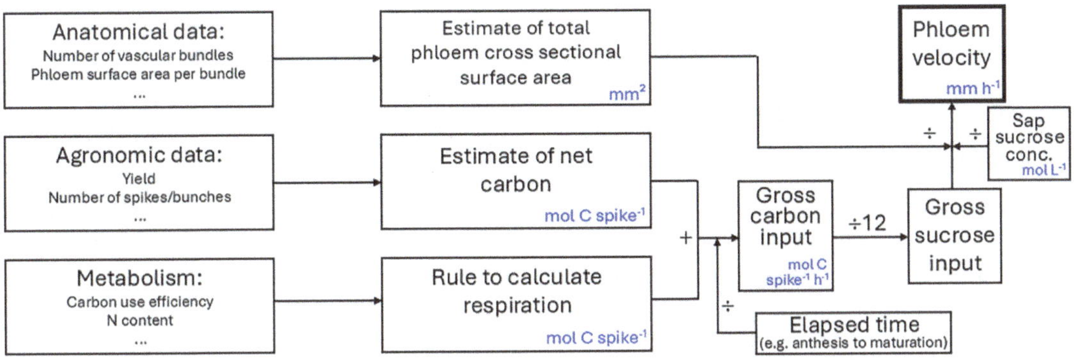

Fig. 1 Overview of the calculation method to compute an estimate of average phloem velocity. Examples of units (typically in wheat) are shown in blue. Note the use of phloem sap sucrose concentration, which can be a critical parameter (*see* **Note 1**)

knowledge is limited. This is why this type of calculation is best adapted to crops, which are associated with considerable documentation, such as yield, biomass production, respiration loss, elemental contents, or vascular histology. The parameterization of calculations in wheat ears and oil palm bunches is summarized in Table 1. We do not describe here methods to compute respiratory carbon loss and gross carbon assimilation and simply provide adapted references.

2.1 Phloem Cross-Sectional Surface Area

Calculating phloem sap transport rate expressed by cross-sectional surface area of phloem tissue, or phloem velocity from the total amount of imported sucrose requires the knowledge of phloem surface area (σ_{phloem}) in the stem or peduncle that feeds the sink organs of interest. In wheat, σ_{phloem} can be estimated using the relationship between spike size (grain no. per spike, g) and stem phloem surface area; that is, the transverse surface area of phloem at the position where all phloem must pass en route to the sink, that is, the ear (for an illustration of phloem location and cross section surface area in wheat stems, see [20]). Using a regression analysis, it has been found that σ_{phloem} (mm^2) $= 7.93 \times 10^{-3} + g \times 7.49 \times 10^{-4}$ [24, 25] and thus phloem surface area per stem falls within the range 0.04–0.05 mm^2, which agrees with values found elsewhere [10, 28]. In oil palm, advantage can be taken from bunch peduncle cross- section (which is measured in agronomic palm assessment of germplasms) and average vascular bundle density (in the inner part of peduncle) of 1 mm^{-2} [31, 37], leading to an average value of 2.4 vascular bundles per fruit. Of course, this value (2.4) is less than the reported total number of original bundles in bunch rachis at early stages of bunch development (6 to 10 bundles per female flower, plus 6 to 10 central vascular bundles) [38]. This is because many bundles develop into fibrous, non-vascular bundles (with little or no functional xylem and phloem). Whenever sucrose export rate from leaves is calculated (estimates in oil palm provided in Fig. 2), phloem cross-sectional

Table 1
Parameterization of calculations of phloem transport rate in wheat ears and oil palm bunches

Variable	Value or range	References	Comments
Wheat (*Triticum aestivum*)			
Phloem cross-section	0.033–0.057 mm^2 $spike^{-1}$	[24, 25]	Calculated via the relationship with grain number
Respiratory loss	0.032–0.068 mol C $spike^{-1}$	[26, 27]	Calculated using either %N increment during ear development or coefficients relating to growth rate, and scaled with harvest index; alternatively, could be calculated with grams of grain per spike and carbon use efficiency (0.5–0.6)
Phloem sap sucrose concentration	0.40 mol L^{-1}	[28–30]	Average of values found in these references
Time	41–63 days	–	Source agronomic dataset (recorded for each cultivar)
Oil palm (*Elaeis guineensis*)			
Phloem cross- section	46–83 mm^2 $bunch^{-1}$	[31–33]	Calculated via either the number of normal fruits per bunch, or functional vascular (inner) bundle density, and phloem cross- section per bundle
Respiratory loss	239–350 mol C $bunch^{-1}$	[34]	Calculated via bunch carbon use efficiency
Phloem sap sucrose concentration	0.42 mol L^{-1}	[35]	Average found in many palm bunches via exudation
Time	165 days	[36]	Average time from anthesis to mature fruits, with very small variations (up to 5 days)

[1]Variables that are not mentioned here (yield, ear, or bunch number, %N, harvest index, etc.) are part of source datasets and thus come from experiments and agronomic trials. When a range of values is given, it means that a specific value was obtained for each cultivar or germplasm, and the amplitude (min–max) is reported

surface area can be estimated from vascular bundle density in petiole (also 1 mm^{-2}) and phloem surface area per vascular bundle measured in rachis (0.035 mm^2) [37, 32, 33, 39, 40]. Taken as a whole, collecting anatomical information (including dimensions) is essential to get proper estimates of phloem surface area. Potential limitations associated with this process are provided in **Note 1**.

2.2 Phloem Transport Rate

The average phloem transport rate from vegetative material to sink organ (ear, bunch) can be first expressed in moles of sucrose per cultivation surface area in m^2 or ha (or alternatively, per organ or whenever ear, tree, bunch, or stem density is known) and per time unit. It is denoted as S because it is expressed in sucrose equivalents.

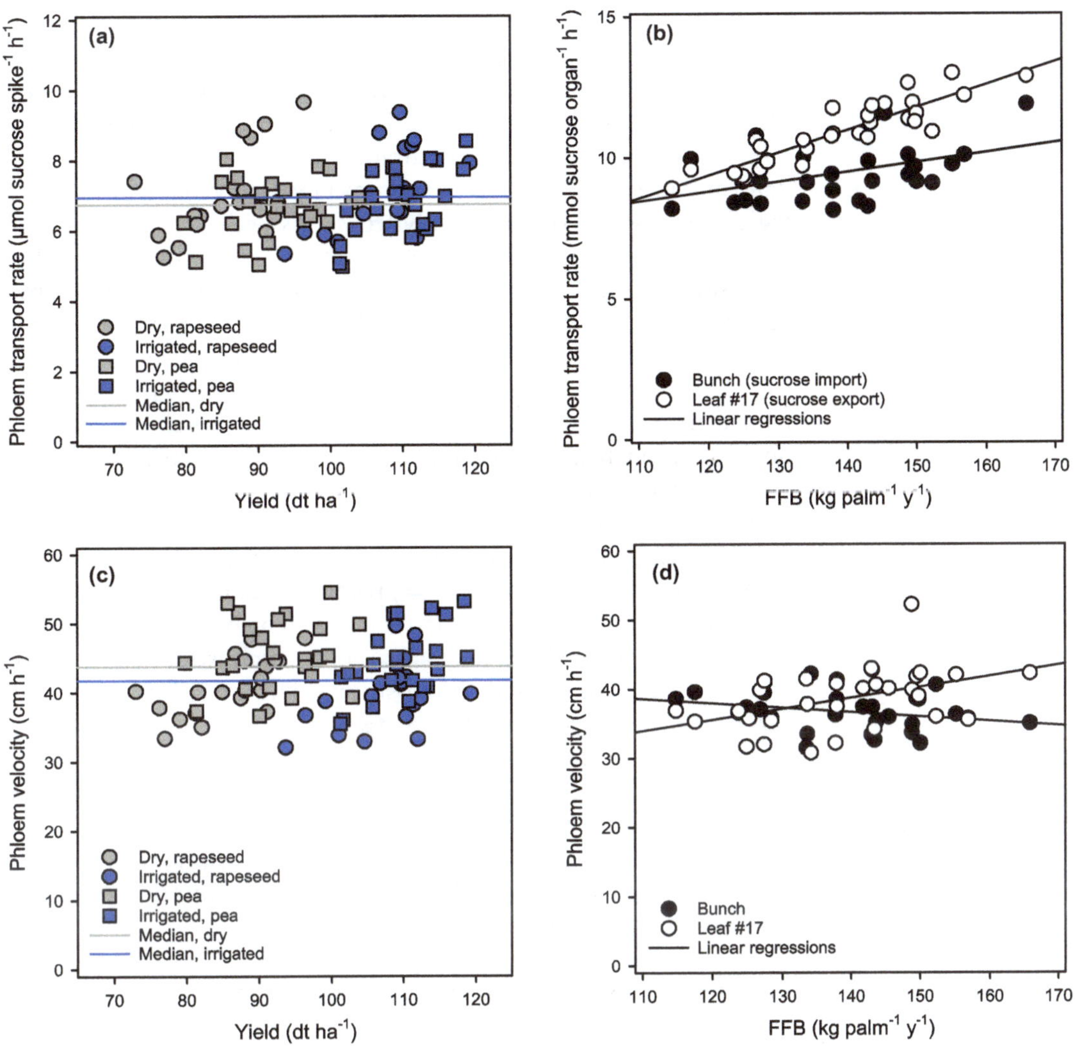

Fig. 2 Predicted phloem transport rate in wheat (**a, c**) and oil palm (**b, d**). Phloem transport is expressed in μ mol sucrose organ^{-1} h^{-1} (**a, b**) or in front sap velocity (**c, d**). In wheat, four growth conditions are shown (dry or irrigated, cultivation on plot previously used for rapeseed or pea). In oil palm, leaf sucrose export (in leaf no. 17, which is the most photosynthetically active) is also shown to compare with bunch sucrose import. In (**b**), linear regressions are significant ($P < 0.01$) and explain 80% (leaf) and 18% (bunch) of total variance (R^2 values). The biological significance of these regressions is provided in the main text. Regressions in (**d**) are not significant ($P > 0.01$). In (**a**) and (**c**), there is no significant relationship with yield. Production is shown on *x* axes using the commonly used variables: annual yield in dt grain (15% moisture) per ha in wheat, and fresh fruit bunches in kg fruits per palm per year. In this figure, each individual point is an average value for a 6-m² cultivation plot (wheat) or 18 trees of the same germplasm (oil palm)

$$S = \frac{\Upsilon \cdot x_c + R_g - \Lambda}{\upsilon \cdot M_c \cdot T_{\mathrm{am}}}, \tag{1}$$

where Y is yield (g dry weight m^{-2}), R_g is respiratory loss (g C m^{-2}), x_c is average %C in the product of interest (40% in wheat grains, 49% in oil palm bunches), M_c is the molecular weight of carbon (12 g mol^{-1}), υ is the number of carbon atoms in sucrose ($\upsilon = 12$), and T_{am} is time from anthesis to maturation. The components of Y in oil palm and its relationship to the commonly used variable, FFB (fresh fruit bunches), are provided in **Note 2**. Λ is a correction accounting for the contribution of non-sucrose phloem constituents to carbon provision. If one desires to express total carbon provision in sucrose equivalents, then $\Lambda = 0$. It is how the calculation has been carried out in oil palm in Fig. 2 since protein content in bunches is relatively low. In fact, fruit mesocarp contains less than 0.1% protein [41]. Thus, the contribution of amino acids to phloem carbon flux was neglected. In wheat, however, protein content is about 13% in grains, and thus the contribution of carbon flux that goes along N input in developing grains cannot be neglected. In that case, one may use the following relationship: $\Lambda = \lambda \times Y \times x_n$, where x_n is %N determined from grain protein content (x_n = protein content in %/5.62), and λ is the average mass stoichiometric coefficient between carbon and nitrogen in sap amino acids (set to 3.5, which is representative of wheat phloem amino acid composition). The contribution of other carbon sources is addressed in **Note 3**. The use of an equation like Eq. 1 applied to nitrogen is discussed in **Note 4**. Note that Eq. 1 can be easily adapted to sucrose export rate by leaves. In that case, $x_c = 1$, Y is cumulated net photosynthetic carbon input (A, in g C m^{-2}) corrected for allocation to export (denoted as γ; thus $Y = \gamma A$), and R_g is cumulated respiratory loss by night respiration (g C m^{-2}) during the time window of interest (here, T_{am}).

2.3 Phloem Sap Velocity

Phloem velocity, F_{phloem}, in international units, m s^{-1}, is calculated as:

$$F_{phloem} = \frac{S}{10^3 \cdot C_{suc} \cdot \sigma_{phloem}}, \tag{2}$$

where 1000 is the conversion factor from mol L^{-1} to mol m^{-3}, and S is defined by Eq. 1. This represents phloem sap front velocity, that is, the speed of liquid movement in sieve tubes. It is common practice, however, to express phloem velocity with more convenient units, such as cm h^{-1}. This can be obtained from F_{phloem} by multiplying by 3.6×10^5. Of course, a critical parameter in Eq. 2 is sucrose sap concentration. In wheat and palms, measured values are available (Table 1). However, measurements of sucrose sap concentration in pure sap samples are relatively scarce, and thus, it might be difficult to use Eq. 2 in some species. Also, sucrose concentration may vary with conditions. This issue is discussed in detail in [20] for wheat. Parenthetically, Eq. 2 can be used in a slightly modified version to compute xylem sap velocity as explained in **Note 5**.

3 Implementation

3.1 Numerical Application: Transport Rate in Sucrose Equivalents

Figure 2 provides examples of calculations with Eqs. 1 and 2, in wheat ears and oil palm bunches, using (1) 20 plots representing height cultivars grown under irrigated and non-irrigated conditions after pea or rapeseed have been grown the previous year (wheat, Fig. 2a) and (2) 4-year palms representing thirty *Dura* × *Dura* germplasms (genotypes) cultivated in the same plantation in Sumatra (oil palm, Fig. 2b). Sucrose transport rate is plotted against production: grain yield in wheat, and FFB in oil palm. Quite understandably, the scale is completely different between the two species, due to the substantial difference in weight: an ear represents a few grams, while a bunch weighs several kilograms and thus, we do expect three orders of magnitude between their sucrose transport rate, as observed here. Also, in oil palm, the sucrose export rate by leaf no. 17 is similar to the bunch sucrose import rate. Although oil palm has about 48 leaves (fronds), leaf no. 17 is conservatively the most photosynthetically active in this species and is responsible for a substantial proportion of photosynthetic input. In addition, one palm tree carries about 20 bunches. Therefore, about the same order of magnitude is expected between leaf and bunch sucrose transport rate, as observed here. In terms of absolute numbers, wheat phloem transport is on average 7 μmol sucrose spike^{-1} h^{-1}. It is easy to see that this number translates to 7 μmol C m^{-2} s^{-1} (× 1/12 × 1/3600 × 300 spikes m^{-2}), which is consistent with the order of magnitude of net photosynthesis. Similarly, average leaf export rate in oil palm is 11 mmol sucrose leaf^{-1} h^{-1}, which converts to about 9 μmol C m^{-2} s^{-1} (× 1000 × 1/12 × 1/3600 × 4 m^2 leaf^{-1}), which is also consistent with the order of magnitude of net photosynthesis. Taken as a whole, expressing sucrose transport rate in μmol sucrose per organ per hour is the first important step to check consistency across numbers obtained by calculations.

3.2 Numerical Application: Phloem Sap Velocity

Phloem sap velocity, calculated using sucrose concentration from the literature for these species, is shown in Fig. 2c, d. Despite some scattering, it is worth noting that the order of magnitude is the same, regardless of cultivars and species, that is, 30–50 cm h^{-1}. In effect, calculated phloem front velocity should be within 10 and 100 cm h^{-1}, which is the common range of values (low values being more typical of trees and high values being observed in herbaceous crops with rapid growth). Also, in C_3 crops where water use efficiency is about 2% (thus a water-to-sucrose mol:mol ratio of about 50 × 12), the scaling factor with xylem sap velocity must be lower than one order of magnitude (since 50 × 12 × 1/55 (where 1/55 is the sucrose-to-water concentration ratio) = 10), thus < 1000 cm h^{-1}. We used the term "lower than" because the cross-sectional area of metaxylem is usually larger than that of phloem sieve tubes. Collectively, with wheat and oil palm data, the xylem-to-phloem sap velocity ratio is 2 to 8 (*see* **Note 1**), which is satisfactory in terms of order of magnitude (<10).

3.3 Important Considerations to Interpret Results

Representing phloem transport rate as a function of production, such as yield, as in Fig. 2, is very useful for understanding underlying relationships with parameters involved in calculations. In wheat, there is no relationship at all of phloem transport rate or sap velocity with yield (Fig. 2a, c). At first glance, it might appear surprising because yield is used to compute S (Eq. 1). Nevertheless, amongst wheat cultivars, yield is not related to the number of grains and thus vascular bundles per spike but rather relates to the number of ears. Therefore, there is no link with phloem transport rate, which is expressed per spike, nor with phloem velocity, which involves phloem cross-sectional area, which is in turn unrelated to yield.

Conversely, in oil palm, the impact of the use of production (FFB) to calculate S is readily visible in Fig. 2b for leaf export rate. FFB is indeed a proportional component of total carbon input, with, here, an allocation of photosynthetically fixed C to fruit of 27%. From a biological point of view, it also reflects the fact that in this trial with young palms, FFB was mostly driven by photosynthesis via intercepted light (energy-limited situation), with very little differences in nutrient content between germplasms (likely due to the close genetic distance between them, all germplasms being the result of advanced breeding cycle). In particular, the content of potassium, which is critical for oil palm yield, photosynthetic activity, and sap circulation, was nearly constant, at 1.14 ± 0.05 % K in leaflet material (not shown). However, it is interesting to note the weak relationship between bunch sucrose import rate and FFB, simply reflecting the poor link between yield and bunch number or weight in this trial. Also, there is no significant relationship between FFB and phloem velocity (slope not significantly different from zero; Fig. 2d), simply because FFB itself relates to fruit number in bunches, scales to bundle number and therefore phloem surface area. Similarly, in leaves, photosynthesis and FFB are linked to leaf size (limiting overlapping and increasing intercepted light) and thus vascular bundle number. As a result, phloem sap velocity, which is normalized to phloem surface area, appears mostly independent of FFB. Taken as a whole, hidden relationships must be kept in mind to avoid misinterpretation of how calculated phloem transport rate links to production and yield.

3.4 Utilization

Keeping in mind precautions described above in Sect. 3.3, it is remarkable that there is some scattering in Fig. 2, showing that there is some inter-cultivar or inter-germplasm variance in phloem properties. Whenever phloem velocity is high, it may suggest that phloem properties (either allocation, sap sucrose concentration, pressure, or phloem surface area) did not align with other cultivars/germplasms. For example, the way phloem cross-sectional surface area was calculated could be non-applicable to a specific germplasm because of peculiar vascular geometry or atypical bundle density. In that sense, the average phloem transport rate and velocity calculated here can be useful to characterize crop lines, revealing phloem properties that would not have been apparent at first glance. Also, it might be useful to (1) compare with xylem sap transport rate and velocity (*see* **Note 5**), (2) explore relationships with available agronomic parameters that are not involved in calculations, such as nutrient content or (3) look at the effect of environmental stress that can cause changes in vascular anatomy (see, e.g., [42–44] in wheat). In wheat, it has been suggested that phloem sap flow scales to xylem sap flow, likely reflecting the concurrent change in water potential in xylem vessels and phloem tissue [20]. In oil palm, leaf export rate correlates with the product of elemental contents that are the most reflective of the photosynthetic machinery, N·Mg·K (14% variance explained; not shown here). This perhaps reflects the adjustment of phloem properties to photosynthetic capacity, as hypothesized elsewhere in herbaceous plants [11, 13, 45, 46]. Also, this might offer a new perspective for breeding in this species since phloem transport rate could be related to abiotic stress resistance, such as tolerance to drought [47]. In addition, genetic analyses (such as GWAS) are currently underway to identify genetic markers associated with leaf petiole properties to improve palm compactness and increase yield expressed per ha (e.g., [48]). However, in principle, manipulating leaf size and petiole anatomy to increase compactness should impact vascular bundle density and thus phloem function. Correlation analyses amongst different oil palm genotypes have effectively shown a link between petiole cross-section surface area and bunch index (bunch-to-vegetative dry matter ratio) [49]. In bunches, a link has been found between peduncle weight (and thus probably bundle number and/or density) and oil production per palm [50]. As such, phloem-related properties could be interesting features for future genetic association studies or breeding initiatives. That is, calculating phloem transport rate could be a useful quantitative trait to appreciate the trade-off between production per ha and carbon allocation imperatives in this species.

4 Notes

1. It should be noted that a limitation in the use of phloem surface area reported in the literature is that it generally represents not only sieve elements but also companion cells and phloem parenchyma cells. As such, the phloem cross-section is probably slightly overestimated and thus the phloem velocity is probably slightly underestimated. That said, the order of magnitude is probably not far from the actual phloem transport rate because the ratio found with the xylem sap transport rate falls in an acceptable range. For example, in wheat, the xylem-to-phloem sap velocity ratio has been

found to be within 5.7 and 7.7, which falls within the ratio found in vivo with MRI [15]. In oil palm, a very detailed examination of metaxylem in bundles shows vessel cross-sectional surface area is about 0.044 mm^2 per bundle [51]. This value, combined with fruit transpiration rate (up to 10 mg water g^{-1} fruit h^{-1}; [52]), leads to an average xylem sap velocity of 77 cm h^{-1} and thus a xylem-to-phloem sap ratio of 2.1. This value is not high, suggesting that phloem surface area was not dramatically overestimated.

2. It is not common practice to express annual production in oil palm using yield Υ in g bunch DW m^{-2}. Rather, oil palm production is measured in kg fresh fruit bunches per ha or per palm, denoted as FFB. In order to calculate total organ weight, it may be useful to add 8% of FFB to account for non-harvested material, such as stalk (peduncle). The precise value can be calculated for each palm tree or germplasm, the average value being 8.1%. Also, the total fresh weight value must be converted to dry weight (bunch dry-to-fresh mass ratio is near 0.55). In Eq. 1, the conversion factor to carbon (%C) used is 49%, which is a good approximation. A more rigorous way to obtain yield in terms of carbon content is to account for bunch oil (75% C) along with bunch oil fraction, and non-oil bunch matter (40% C). This can be calculated for each palm tree or germplasm, and values fall within a narrow range, 48.9 ± 1.1%.
3. A potential limitation is that Eq. 1 neglects the contribution of sink organ photosynthesis prior to maturity. For example, ear (glume) photosynthesis could be involved in wheat ear carbon mass balance. However, it is relatively safe to assume that all carbon atoms found in wheat grains had to be fixed by leaf photosynthesis and thus transported via the phloem. The net contribution of ear photosynthesis is small due to substantial refixation of respired CO_2 [53], and in terms of mass balance, the contribution of glume carbohydrates and proteins to grain filling has been estimated near 5% only [54]. In oil palm, to the best of our knowledge, the potential contribution of immature bunch photosynthesis is unknown.
4. Equation 1 only considers carbon so as to compute sucrose phloem flux. However, phloem sap also carries nitrogenous compounds such as amino acids, and it may be relevant to calculate phloem sap N flux too. To do so, an equation similar to Eq. 1 can be used:

$$N = \frac{\Upsilon \cdot x_n}{M_n \cdot T_{\mathrm{am}}}, \qquad (3)$$

where M_n = 14 g mol^{-1} (molecular weight of nitrogen) and other variables (Υ, T_{am}, x_n) have been defined above. Equation 2 yields the nitrogen flux in moles of N per surface area per time unit.

5. It can be useful to compute water movement in xylem too, in order to compare with phloem sap velocity and thereby check if the order of magnitude found for F_{phloem} is correct. To do so, we

adapt Eq. 2 to water, using an estimate of water consumption, E, here expressed in mol m^{-2} s^{-1}:

$$F_{\text{xylem}} = \frac{E}{10^3 \cdot C_{\text{water}} \cdot \sigma_{\text{xylem}}}, \tag{4}$$

where $C_{\text{water}} = 55.5$ mol L^{-1}. Xylem cross-sectional surface area, σ_{xylem}, should be obtained from anatomical data (adjusted to phloem surface area to account for grain number per spike in wheat; calculated using bundle number and metaxylem surface area per bundle in oil palm). E is either among the agronomic dataset (observed water consumption) or recalculated from Υ using observed water use efficiency. Similarly, F_{xylem} can be converted to cm h^{-1} by multiplying by 3.6×10^5.

Conflict of Interest

The authors have no conflicts of interest to declare that are relevant to the content of this chapter.

References

1. Reynolds M, Atkin OK, Bennett M, Cooper M, Dodd IC, Foulkes MJ et al. (2021) Addressing research bottlenecks to crop productivity. Trend Plant Sci 26(6):607–330
2. Braun DM, Wang L, Ruan Y-L (2014) Understanding and manipulating sucrose phloem loading, unloading, metabolism, and signalling to enhance crop yield and food security. J Exp Bot 65(7):1713–1735
3. Deng Y, Yu Y, Hu Y, Ma L, Lin Y, Wu Y et al. (2021) Auxin-mediated regulation of dorsal vascular cell development may be responsible for sucrose phloem unloading in large panicle rice. Front Plant Sci 12:630997
4. You C, Wang H, Huang Y, Xu P, Wu L, Yu F et al. (2021) Relationship between the vascular bundle structure of panicle branches and the filling of inferior spikelets in large-panicle *Japonica* rice. Front Plant Sci 12:774565
5. Zhai L, Yan A, Shao K, Wang S, Wang Y, Chen Z-H et al. (2023) Large vascular bundle phloem area 4 enhances grain yield and quality in rice via source–sink–flow. Plant Physiol 191(1):317–334
6. Lombardo F, Gramazio P, Ezura H (2021) Increase in phloem area in the tomato Hawaiian *skirt* mutant is associated with enhanced sugar transport. Genes 12(6):1464
7. Nam H, Gupta A, Nam H, Lee S, Cho HS, Park C et al. (2022) *JULGI*-mediated increment in phloem transport capacity relates to fruit yield in tomato. Plant Biotechnology Journal 20(8):1533–1545
8. Evans LT, Wardlaw IF (1996) Wheat. In: Zamski E, Schaffer A (eds) Photoassimilate distribution plants and crops source-sink relationships. Taylor and Francis, New York, pp 501–519
9. Evans LT, Dunstone RL, Rawson HM, Williams RF (1970) The phloem of the wheat stem in relation to requirements for assimilate by the ear. Aust J Biol Sci 23(4):743–752
10. Whingwiri EE, Kuo J, Stern WR (1981) The vascular system in the rachis of a wheat ear. Ann Bot 48(2):189–202
11. Adams WW, Cohu CM, Muller O, Demmig-Adams B (2013) Foliar phloem infrastructure in support of photosynthesis. Front Plant Sci 4:194
12. Adams WW, Watson AM, Mueh KE, Amiard V, Turgeon R, Ebbert V et al. (2007) Photosynthetic acclimation in the context of structural constraints to carbon export from leaves. Photosynth Res 94:455–456
13. Amiard V, Mueh KE, Demmig-Adams B, Ebbert V, Turgeon R, Adams WW (2005) Anatomical and photosynthetic acclimation to the light environment in species with differing mechanisms of phloem loading. Proc Natl Acad Sci 102(36):12968–12973
14. Peuke AD, Rokitta M, Zimmermann U, Schreiber L, Haase A (2001) Simultaneous measurement of water flow velocity and solute

transport in xylem and phloem of adult plants of *Ricinus communis* over a daily time course by nuclear magnetic resonance spectrometry. Plant Cell Environ 24(5):491–503

15. Windt CW, Vergeldt FJ, De Jager PA, Van As H (2006) MRI of long-distance water transport: a comparison of the phloem and xylem flow characteristics and dynamics in poplar, castor bean, tomato and tobacco. Plant Cell Environ 29(9):1715–1729
16. Ren Y, Zhang Z, Zhanakhmetova D, Li W, Chen S, Werner T et al. (2024) Fast and simple fluorometric measurement of phloem loading exposes auxin-dependent regulation of *Arabidopsis* sucrose transporter. Plant J 120(5):2305–2318.
17. Jeannette E, Rocher J-P, Prioul J-L (1995) Effect of an increased sink demand on the carbon metabolism and export of a maize source leaf. Physiol Plant 94(2):319–327
18. Rocher JP, Prioul J-L (1981) A three-compartment model for assimilate export from maize leaves. In: Metzner H (ed) Photosynthesis and plant productivity: joint meeting of the OECD and Studiezentrum Weikersheim. Wissenschaftliche Verlagsgesellschaft GmbH, Ettlingen pp 159–162
19. Morot-Gaudry JF, Farineau J, Rocher JP, Jolivet E (1981) Modifications of the first stages of photosynthetic carbon assimilation in maize as affected by genotype. Agronomie 1(9):739–743
20. Tcherkez G, Holloway-Phillips M, Lothier J, Limami A, Ball MC (2023) Revisiting yield in terms of phloem transport to grains suggests phloem sap movement might be homeostatic. Plant Cell Environ 46(9):2667–2679
21. Henderson W (2015) Oil palm (*Elaeis guineensis* Jacq.) bunch structure variations and limitations. Sci Res Jo 3(1):5–11
22. Moore HE, Uhl NW (1982) Major trends of evolution in palms. Bot Rev 48(1):1–69
23. Tomlinson PB (2006) The uniqueness of palms. Bot J Linn Soc 151(1):5–14
24. Nátrová Z (1991) Anatomical characteristics of the uppermost internode of winter wheat genoypes differing in stem length. Biol Plant 33(6):491–494
25. Nátrová Z, Nátr L (1993) Limitation of kernel yield by the size of conducting tissue in winter wheat varieties. Field Crops Res 31(1):121–130
26. Gifford RM (2003) Plant respiration in productivity models: conceptualisation, representation and issues for global terrestrial carbon-cycle research. Funct Plant Biol 30:171
27. Choudhury BJ (2000) A sensitivity analysis of the radiation use efficiency for gross photosynthesis and net carbon accumulation by wheat. Agric For Meteorol 101(2–3):217–234
28. Fisher DB, Gifford RM (1987) Accumulation and conversion of sugars by developing wheat grains. VII. Effect of changes in sieve tube and endosperm cavity sap concentrations on the grain filling rate. Plant Physiol 84(2):341–347
29. Fisher DB (1990) Measurement of phloem transport rates by an indicator-dilution technique. Plant Physiol 94(2):455–462
30. Hayashi H, Chino M (1986) Collection of pure phloem sap from wheat and its chemical composition. Plant Cell Physiol 27(7):1387–1393
31. Purusamata B, Kim J, Prasetia D, Savero A, Wistara N, Kim N (2024) Anatomical characteristics of unproductive *Elaeis guineensis* stems and their correlation with density. Bioresources 19(4):939
32. Parthasarathy M (1968) Observations on metaphloem in the vegetative parts of palms. Am J Bot 55(10):1140–1168
33. Zimmermann MH, Sperry JS (1983) Anatomy of the palm *Rhapis excelsa*. IX. Xylem structure of the leaf insertion. J Arnold Arboretum 64(4):599–609
34. Lamade E, Tcherkez G, Darlan NH, Rodrigues RL, Fresneau C, Mauve C et al. (2016) Natural ^{13}C distribution in oil palm (*Elaeis guineensis* Jacq.) and consequences for allocation pattern. Plant Cell Environ 39(1):199–212
35. Van Die J, Tammes P (1975) Phloem exudation from monocotyledonous axes. In: Zimmermann MH, Milburn J (eds) Transport in plants I: phloem transport. Encyclopedia of plant physiology. Springer Verlag, Berlin, pp 196–222
36. Taillez B, Coulibaly M, Bonny C, Jacquemard J (1993) Oil palm bunch ripening and harvesting criteria at Palmindustrie (Ivory Coast). Oléagineux 48(4):169–179
37. Tomlinson, PB, Horn, JW, Fisher, JB, 2011. The anatomy of palms: Arecaceae-Palmae. Oxford University Press, Oxford
38. Uhl N, Moore H (1971) The palm gynoecium. Am J Bot 58(10):945–992
39. Mat Rasat MS, Wahab R, Abdus Salam M, Razak MH, Ramle NH, Sulaiman MS et al (2022) Processing and properties of oil palm fronds composite boards from *Elaeis guineensis*. In: Kamyab H (ed) *Elaeis guineensis*. IntechOpen, Rijeka, p Chapter no. 13.
40. Waite P-A, Schuldt B, Mathias Link R, Breidenbach N, Triadiati T, Hennings N et al. (2019) Soil moisture regime and palm height influence embolism resistance in oil palm. Tree Physiol 39(10):1696–1712
41. Eyaa Ngombo C, Nsimi M, Ndele P, Bille N, Molo T, Mbo LF et al. (2023) A review of main factors involved in the maturation of oil palm (*Elaeis guineensis* Jacq.) fruit bunches. Am J Plant Sci 14:727–740

42. David O, Osonubi O, Olaiya C, Agbolade J, Ajiboye A, Komolafe R et al. (2017) Anatomical response of wheat cultivars to drought stress. Ife J Sci 19(2):323–332
43. Moore CE, Meacham-Hensold K, Lemonnier P, Slattery RA, Benjamin C, Bernacchi CJ et al. (2021) The effect of increasing temperature on crop photosynthesis: from enzymes to ecosystems. J Exp Bot 72(8):2822–44.
44. El-Afry M, El-Nady M, Abdelmonteleb E, Metwaly M (2012) Anatomical studies on drought-stressed wheat plants (*Triticum aestivum* L.) treated with some bacterial strains. Acta Biol Szeged 56(2):165–174
45. Stewart JJ, Polutchko SK, Demmig-Adams B, Adams WW (2018) *Arabidopsis thaliana* Ei-5: minor vein architecture adjustment compensates for low vein density in support of photosynthesis. Frontiers in Plant Science 9:693
46. Muller O, Cohu CM, Stewart JJ, Protheroe JA, Demmig-Adams B, Adams III WW (2014) Association between photosynthesis and contrasting features of minor veins in leaves of summer annuals loading phloem via symplastic versus apoplastic routes. Physiol Plant 152(1):174–183
47. Sevanto S (2014) Phloem transport and drought. J Exp Bot 65(7):1751–1759
48. Samsudin AI, Vetaryan S, Ahmad Mohktar MA, Abdul Rahim MF, Ting N-C, Zolkafli SH et al. (2025) Genome-wide association study for compactness traits in oil palm (*Elaeis guineensis* Jacq.). Euphytica 221(5):57
49. Serdari NM, Sanusi NSNM, Suzana M, Jalani BS, Mackay I, Singh R et al. (2024) Association mapping of selected oil palm germplasm reveals novel and known genomic regions influencing vegetative and bunch component traits. Gen Res Crop Evol 71(5):1709–1735
50. de Almeida Rios S, da Cunha RNV, Lopes R, Barcelos E, da Rocha RNC, de Lima WAA (2018) Correlation and path analysis for yield components in *Dura* oil palm germplasm. Ind Crops Prod 112:724–733
51. Lim S (1997) A note on the structure of oil palm trunk by scanning electron microscopy. J Trop Forest Prod 3(1):105–109
52. Jeje A, Odetola A, Zimmermann MH (1978) Transpiration and oil accumulation rates for developing oil palm fruits *Elaeis guineensis* Jacq. Acta Bot Neerland 27(3/4):213–228
53. Gebbing T, Schnyder H (2001) ^{13}C labeling kinetics of sucrose in glumes indicates significant refixation of respiratory CO_2 in the wheat ear. Funct Plant Biol 28(10):1047–1053
54. Gebbing T, Schnyder H, Kühbauch W (1998) Carbon mobilization in shoot parts and roots of wheat during grain filling: assessment by $^{13}C/^{12}C$ steady-state labelling, growth analysis and balance sheets of reserves. Plant Cell Environ 21(3):301–313